D0295894

Elementary
Technical Mathematics

Prentice-Hall Series in Technical Mathematics

Frank L. Juszli, *Editor*

THIRD EDITION

Elementary Technical Mathematics

Frank L. Juszli
New Hampshire Vocational-Technical College

Charles A. Rodgers
Hartford State Technical College

PRENTICE-HALL, INC., Englewood Cliffs, New Jersey 07632

QA
39.2
.J87
1980

Library of Congress Cataloging in Publication Data

Juszli, Frank L.
Elementary technical mathematics.

(Prentice-Hall series in technical mathematics)
Includes index.
1. Mathematics—1961- I. Rodgers, Charles A., joint author. II. Title.
QA39.2.J87 1980 512'.13 79-28260
ISBN 0-13-260869-3

32173

©1980 by Prentice-Hall, Inc., Englewood Cliffs, N.J. 07632

All rights reserved. No part of this book
may be reproduced in any form or
by any means without permission in writing
from the publisher.

Printed in the United States of America

10 9 8 7 6 5 4 3 2 1

Editorial/production supervision and interior
design by Zita de Schauensee
Cover design by Miriam Recio
Manufacturing buyer: Gordon Osbourne

Prentice-Hall International, Inc., *London*
Prentice-Hall of Australia Pty. Limited, *Sydney*
Prentice-Hall of Canada, Ltd., *Toronto*
Prentice-Hall of India Private Limited, *New Delhi*
Prentice-Hall of Japan, Inc., *Tokyo*
Prentice-Hall of Southeast Asia Pte. Ltd., *Singapore*
Whitehall Books Limited, *Wellington, New Zealand*

Contents

7 Quadratic equations 166

8 Identification and approximation of roots 213

9 Exponentials and logarithms 234

10 Miscellaneous topics in exponentials and logarithms 262

Preface

The third edition of *Elementary Technical Mathematics* is an integrated presentation of algebra, geometry, and trigonometry designed for use in colleges and industry.

Emphasis is on a student-oriented treatment of topics appropriate to a variety of technical disciplines. A careful, graphical approach offers abundant applications. Formulation and problem-solving methods involving elementary functions are stressed. Nearly 3000 exercises are provided, with answers to about half of them. Some 270 illustrations and line drawings combine with 300 solved examples to assist the student in use of the text.

The third edition, when compared with the second edition, reveals diminished precision in computations by reducing from 5-place to 4-place tables in both logarithms and trigonometry. It eliminates reference to the slide rule and introduces electronic calculator methods. Chapters 1 and 2 are revised to relate the laws of algebra to the laws and "rules" for handling real numbers in arithmetic. Chapter 3 provides increased attention to Standard International Units. Chapter 5 has been reoriented to the functional approach. Changes to 4-place tables have brought about substantial modification of Chapters 9, 10, 12, and 13.

New applications and illustrations have been added to complement those present in earlier editions.

New applications and illustrations have been added to complement those present in the first edition.

The authors are grateful to their colleagues and students in their respective institutions for valuable suggestions made toward this revised edition.

Claremont, New Hampshire Frank L. Juszli
Newington, Connecticut Charles A. Rodgers

1

Properties of real numbers

Algebra can be described as generalized arithmetic. Under this description the real numbers of arithmetic are given literal (alphabetic) designations and are manipulated through the laws and properties of real numbers. In this chapter such laws are discussed, together with various means of representing real numbers.

1-1 Real numbers

In mathematics courses preceding algebra we studied arithmetic that involved three different forms of real numbers—namely,

1. *Integers* (or counting numbers) like 1, 2, 3, 4, ...
2. *Fractions* (which are indicated divisions of integers) like $\frac{2}{3}, \frac{5}{9}, \frac{23}{10}, \ldots$.
3. *Decimals* (which are the results of having performed the divisions indicated in fractions), such as 3.14, 0.006, 9.125, ...

A main goal of arithmetic courses was to perform the *fundamental operations* (addition, subtraction, multiplication, and division) involving the three aforementioned number forms.

Clearly integers, fractions, and decimals are expressible by use of integers only. Consider three examples.

Example 1 Express the fraction three-fourths by using only integers.

Solution By definition, three-fourths is written $\frac{3}{4}$.

Example 2 Express 0.75 by using integers only.

Solution We read 0.75 as 75-hundredths, which is expressible as $\frac{75}{100}$. This answer is adequate in view of the way in which the question was asked. However, we usually choose lowest terms and reduce $\frac{75}{100}$ to $\frac{3}{4}$ by dividing numerator and denominator by 25 and offer the answer $\frac{3}{4}$.

Example 3 Given that $\frac{1}{11} = 0.090909 \cdots$, express $0.272727 \cdots$ as a fraction.

Solution Noting that $0.272727 \cdots = (3)(0.090909 \cdots)$, then

$$0.272727 \cdots = (3)\left(\frac{1}{11}\right) = \frac{3}{11}.$$

Rational and irrational numbers Because fractions and decimals are expressible using integers only, we call them *rational*—that is, they "make sense." By contrast, a group of real numbers exists that cannot be expressed as the quotient of two integers; they do not "make sense" and are called *irrational*. Examples of irrational numbers are π, $\sqrt{2}$, and $\sqrt[3]{7.5}$.

Signed numbers Any of the aforementioned number forms can take on a positive or a negative sense, using the symbols $+$ and $-$ to denote same. A number without a quality sign preceding it is considered positive, per long-standing usage, and may have the $+$ sign entered without error. Examples of signed numbers are -1, 6.4, $+\frac{3}{8}$.

Literal representation of real numbers Any real number, whether integral, fractional, decimal, rational, irrational, positive, or negative, can be represented by a literal (alphabetic) symbol. Consider a statement "Given the real numbers a and b." This statement simply requires that a and b are real but does not restrict their sign or rationality or basic number form. Thus a could conceivably have a negative integral value like -7, whereas b might be irrational like $\sqrt{23.6}$.

Hierarchy or order Given two real numbers a and b, one of three conditions exists regarding their relative size or magnitude:

$a > b$	"a is greater than b"	$a - b > 0$.
$a < b$	"a is less than b"	$a - b < 0$.
$a = b$	"a equals b"	$a - b = 0$.

The "lazy vee" is called an *inequality sign* and has the larger number at the larger (or open) end of the "vee." Combining symbols, $a \geqq b$ reads "a is equal to or greater than b."

Sometimes it is difficult to establish which of two real numbers is larger. The task of comparison can be accomplished by converting both numbers to the same number form and then comparing by subtraction. Consider this example.

Example 4 If $a = \sqrt{7}$ and $b = 2\frac{3}{5}$, which is the larger?

Solution Because a and b are in different number forms, we convert both to decimal form by use of a hand calculator. Now

$$a = \sqrt{7} = 2.6457513\cdots \quad \text{and} \quad b = 2\tfrac{3}{5} = 2.6000000\cdots$$

and, by subtraction, we see that $a - b > 0$, or $a > b$.

Locating numbers on a *number line* (see Sec. 1-2) is also a method of determining hierarchy or order.

1-2 The number line

The *number line* is introduced here to provide some basic elements of graphing plus a visual comparison of the magnitude of numbers. Figure 1-1 shows the conventional number line, which has the following characteristics.

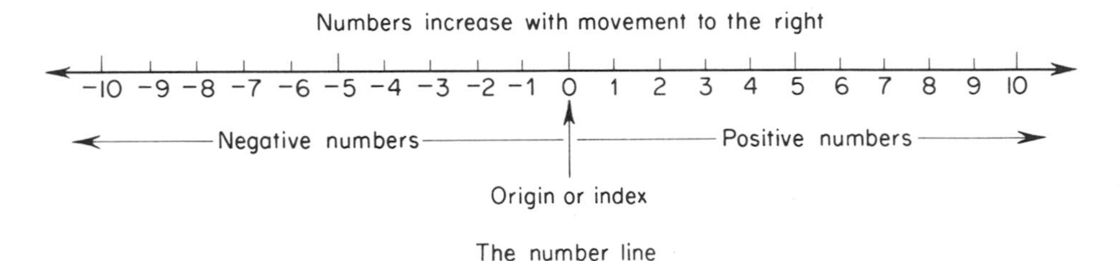

Figure 1-1

1. An origin (zero) is marked on a straight line to separate positive numbers from negative numbers.
2. Uniformly spaced marks are provided on both sides of zero to indicate number spacing or distances of equal magnitude. In Fig. 1-1 we chose to label these marks as consecutive integers.
3. Positive numbers are on the right and negative numbers on the left of zero.
4. Any movement to the *right* brings us to a *larger* number.

Note: In Fig. 1-1 we have shown only integers, but we recognize that decimals, fractions, and even irrational numbers may be located on our number line. See Example 5.

Example 5 Locate the numbers $A = 1.5$, $B = -2\frac{1}{3}$, $C = \sqrt{6}$ on a number line.

Solution Refer to Fig. 1-2, which locates A, B, and C.

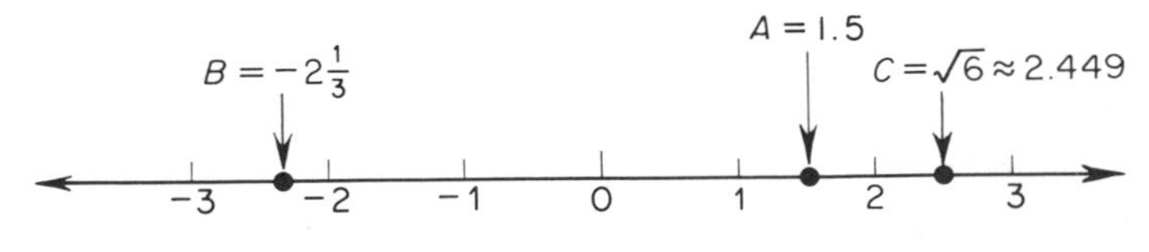

Figure 1-2

Points on a number line A statement like "$x = 1.5$" is displayed on a number line as a single point, since only one value of x (namely, 1.5) is indicated. Thus $x = \sqrt{6}$ graphs as the single point C in Fig. 1-2.

A continuity of points A statement like "graph *all* the numbers greater than -2 and less than $+1$" requires an infinite number of points in a continuity. The points would include all the integral, fractional, decimal, and irrational values between $x = -2$ and $x = +1$ and would logically be represented by a line segment as in Fig. 1-3.

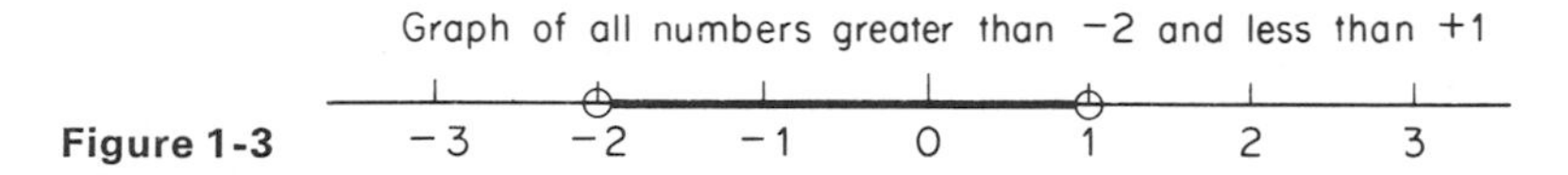

Figure 1-3

Observations on a continuity The graph in Fig. 1-3 may be said to represent x, where $-2 < x < +1$. We note that the solid line does not include the numbers -2 and $+1$ (here we used open dots to denote omission of these endpoints).

Example 6 Graph the statements (a) $x > 0$ and (b) $2 \leqq x \leqq -1$.

Solution

(a) If $x > 0$, we graph a continuity of all numbers greater than (but omitting) the number zero as in Fig. 1-4(a), where the arrow indicates an indefinite continuation to the right.

(b) If $2 \leqq x \leqq -1$, we graph *two* continuities, one for all x equal to or greater than 2 and the other for all x equal to or less than -1. See Fig. 1-4(b).

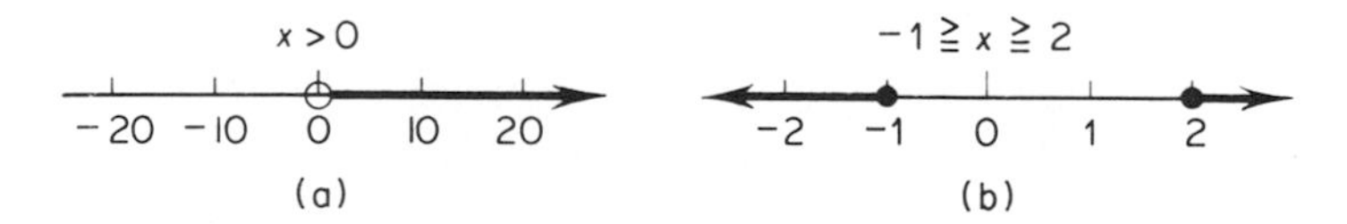

Figure 1-4

Absolute value The number of units between two points on a number line describes the distance between the points. This number is called the *absolute value* when it ignores direction. Absolute value is indicated by parallel vertical bars $|\ |$. Thus the directed distance from 0 to 5 is $+5$ and the directed distance from 5 to 0 is -5; if, however, we ignore direction, we see that

$$|+5| = |-5| = 5,$$

which reads "the absolute value of $+5$ equals the absolute value of -5."

Example 7 Graph the statements (a) $|x| = 3$ and (b) $|x| > 5$.

Solution

(a) The statement $|x| = 3$ graphs as *two points* $x = 3$ and $x = -3$ because absolute value can be considered as the distance from the origin and there are two points that are exactly three units from the origin. See Fig. 1-5(a).

(b) The statement $|x| > 5$ describes *all points* more than five units removed from the origin, going to both the left and the right as in Fig. 1-5(b).

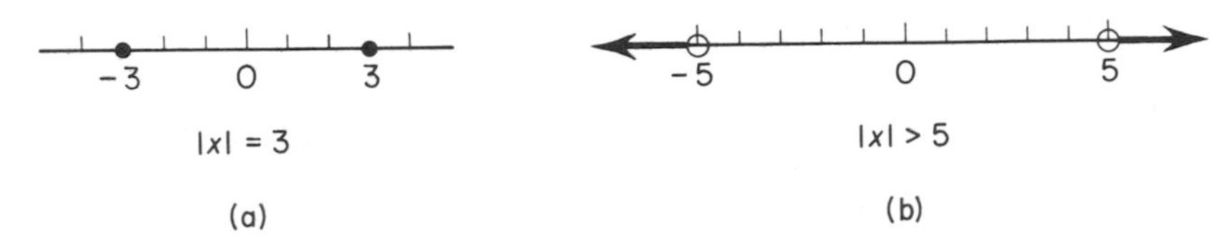

Figure 1-5

Example 8 Consider all the points less than three units removed from the point $P = 2$. Show this continuity on a number line and give an algebraic statement for the continuity.

Solution Because we are considering all the points that are less than three units from $P = 2$, we can consider $P = 2$ as the origin of our set and range less than three units in both directions. Referring to Fig. 1-6,

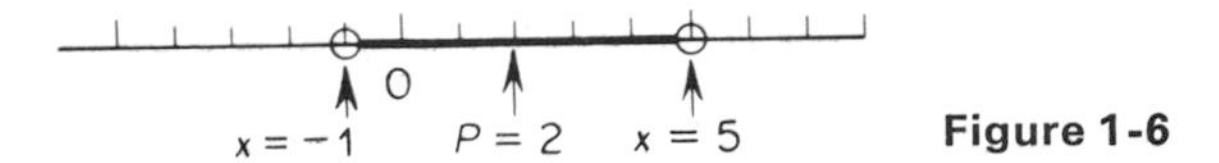

Figure 1-6

we see that all numbers greater than -1 and less than 5 are included in our continuity. This fact can be stated algebraically as $-1 < x < 5$ or, using absolute value notation, since all numbers are less than three units from $x = 2$, then $|x - 2| < 3$.

Exercises 1-2

1. Convert the given fractions to decimals.

   (a) $\frac{1}{2}$ (b) $-\frac{3}{4}$ (c) $2\frac{1}{8}$ (d) $\frac{11}{4}$ (e) $-\frac{9}{2}$

2. Convert the given decimals to fractions.

   (a) 0.5 (b) 3.2 (c) $-1.33333\cdots$ (d) -6.25 (e) 0.125

3. Arrange the given numbers in hierarchy from smallest to largest numbers.

(a) $2.65; \sqrt{7}; 2\frac{2}{3}; \frac{23}{9}; (6 - 3.31)$

(b) $-0.4; -\frac{1}{3}; -\frac{3}{4}; -2; (4 - 5.7)$

4. Plot the given statements on number lines.

(a) $x = 2$ (b) $x = -2$ (c) $x > 3.5$
(d) $x < -3$ (e) $x \geqq 1.3$ (f) $x \leqq 4.2$
(g) $|x| = 1.5$ (h) $-2 < x < 3$ (i) $-6 < x \leqq -1$
(j) $|x - 3| = 6$ (k) $|x - 3| < 6$ (l) $x + 4 > -3$

5. Plot the given statements on number lines and give algebraic statements for them.

(a) All numbers greater than 2.
(b) All numbers between -3 and 5.
(c) All numbers exactly three units from the origin.
(d) All numbers between zero and 1.
(e) All numbers less than -1.
(f) All numbers less than three units from the point where x equals 2.
(g) All negative numbers.
(h) All numbers.
(i) All numbers between 0.01 and 0.03.

1-3 Axioms of real numbers

An *axiom* is a statement that requires no proof because its truth is regarded as obvious. Certain properties of real numbers are in common usage without any challenge to their truth; one is involved in the method of checking the sum of a column of numbers where we are asked to add from top to bottom and then are advised to check our sum by adding from bottom to top. Here we are accepting without proof the notion that a group of numbers may be added in different orders without affecting their sum.

Certain axioms of real numbers are given with explanatory statements in Table 1-1. We have chosen to list those that are useful in future material in this book. Some comments on the axioms and properties in Table 1-1 are in order. Most of these axioms are useful in solving equations or manipulating algebraic quantities. We must remember that they are applicable to *real* numbers, regardless of their form, whether integral, fractional, decimal rational, irrational, positive, or negative. We emphasize, too, that throughout this book we are dealing with real numbers except where otherwise noted. Consider some examples.

Table 1-1 Axioms of Real Numbers

If a, b, c, and d are real numbers, the following are regarded as true.

1. SYMMETRIC PROPERTY *If* $a = b$, *then* $b = a$. An effect of this statement is that an equation may be read from left to right or from right to left.

2. TRANSITIVE PROPERTY *If* $a = b$ *and* $b = c$, *then* $a = c$. Here b is the "middleman" or transitional member involved in passing the property of equality from a to c. This property may be read as "Two quantities (a and c) that are equal to the same quantity (b) are equal to each other."

3. SUBSTITUTION PROPERTY

   A. ADDITION *If* $a + c = d$ *and* $a = b$, *then* $b + c = d$.

   B. MULTIPLICATION *If* $ac = d$ *and* $a = b$, *then* $bc = d$.

   The effect of this property is that a number b may be substituted for its equal a in either the addition or multiplication process.

4. COMMUTATIVE PROPERTY

   A. ADDITION $a + b = b + a$.

   B. MULTIPLICATION $ab = ba$.

   These assert that the sum (or product) of two numbers is not affected by the order in which the numbers are taken. This property can be extended to more than two elements:

   $$a + b + c = a + c + b = b + c + a, \text{ etc.}$$
   $$abc = acb = bca, \text{ etc.}$$

   This property is true for all real numbers (including negative and reciprocal numbers in particular), but is not true for subtraction and division.

5. ASSOCIATIVE PROPERTY

   A. ADDITION $a + (b + c) = (a + b) + c$.

   B. MULTIPLICATION $a(bc) = (ab)c$.

   These state that "Three (or more) numbers can be added (or multiplied) by any subgrouping we choose." In these examples b is associated in a subgroup with c in the left member of the equation and a in the right member.

6. IDENTITY PROPERTY

   A. ADDITION $a + 0 = a$.

   B. MULTIPLICATION $a \cdot 1 = a$.

   These assert that addition of zero to a number or multiplying a number by one leaves the number unchanged.

7. INVERSE PROPERTY

   A. ADDITIVE INVERSE $a + (-a) = 0$.

   B. MULTIPLICATIVE INVERSE $a \cdot \frac{1}{a} = a \cdot (a^{-1}) = 1$.

   These extremely useful properties assert that (1) addition of a number and its

Table 1-1 Axioms of Real Numbrs (*cont.*)

negative results in 0 (zero) and (2) multiplication of a number by its reciprocal (inverse) results in 1 (one).

8. DISTRIBUTIVE AXIOM OF MULTIPLICATION AND ADDITION

$$a(b + c) = ab + ac \quad \text{or} \quad (b + c)a = ab + ac$$

This property is called "distributive" because it "distributes" a multiplier (a) across a sum ($b + c$). It asserts: To multiply a number (a) by the sum of two other numbers (b and c), multiply each member of the sum by a and add the resulting products. This property is extended as

$$a(b + c + d + \cdots) = ab + ac + ad + \cdots.$$

9. PROPERTY OF EQUIVALENT FRACTIONS

$$\frac{a}{b} = \frac{a(c)}{b(c)} \quad (c \neq 0).$$

This property states that the numerator and denominator of a fraction may be multiplied by the same nonzero number without changing the value of the fraction.

Example 9 Identify the axiom or property that is used in each of the following statements.

(a) 3 times 4 is the same as 4 times 3.

(b) If $x + y = 5$ and $y = 2$, then $x + 2 = 5$.

(c) If $x - 2 = 5$, then $x - 2 + 2 = 5 + 2$.

(d) If $x - 2 + 2 = 7$, then $x = 7$.

(e) $2(3x - 2y) = 6x - 4y$.

(f) $\frac{3}{4} = \frac{9}{12}$.

(g) If $2x = 8$, then $x = 4$.

Solution

(a) Commutative property of multiplication.

(b) Substitution property in addition.

(c) If equals are added to equals, the sums are equal (here 2 is added to both sides of the equation).

(d) Additive inverse (note that $x + 0 = 7$ is the same as $x = 7$).

(e) Distributive law.

(f) Equivalent fraction property (numerator and denominator of the right member are exactly 3 times the corresponding items in the left member).

(g) If equals are multiplied by equals, the products are equal. (Here both members are multiplied by $\frac{1}{2}$, which is the multiplicative inverse of 2.)

Exercises 1-3

In Exercises 1–14, state which axiom or property of real numbers is involved in the given statement.

1. $6 + 4 + 3 + 7 = 10 + 10 = 13 + 7 = 6 + 14 = 20.$
2. If $F = \frac{9}{5}C + 32$ and $C = 20$, then $F = \frac{9}{5}(20) + 32.$
3. If $x + y = 10$ and $y = 2$, then $x + 2 = 10.$
4. If $x + y = 16$ and $x = y$, then $y + y = 16.$
5. If $5x = 15$, then $\left(\frac{1}{5}\right)(5x) = \left(\frac{1}{5}\right)(15)$ or $x = 3.$
6. $\frac{5}{10} = \frac{10}{20} = \frac{15}{30} = \frac{1}{2}.$
7. If $\frac{x}{3} = \frac{4}{6}$, then $x = 2.$
8. If $\frac{12}{x} = \frac{4}{20}$, then $x = 60.$
9. $5(x + y) = 5x + 5y$
10. $2(3 \cdot 4) = (2 \cdot 3) \cdot 4 = (2 \cdot 4) \cdot 3$
11. $x^2yz = yx^2z = yzx^2 = zyx^2 = zx^2y$
12. If $xy = z$, then $z = xy.$
13. If $3z = 27$, then $\left(\frac{1}{3}\right)(3z) = \left(\frac{1}{3}\right)(27)$ or $z = 9.$
14. If $\frac{m}{12} = 5$, then $12\left(\frac{m}{12}\right) = 12(5)$ or $m = 60.$

In Exercises 15–16, an equation is solved step by step by isolating the literal quantity. Each step involves one or more of the axioms or properties listed in Table 1-1. State which is used in moving from one step to the next.

15. Given $3x - 6 = 9$,
then $3x = 9 + 6$ or $3x = 15$
and $x = 5$.

16. Given $2(y + 1) = y + 6$,
then $2y + 2 = y + 6$
and $2y - y = 6 - 2$ or $y = 4$.

1-4 Addition and multiplication versus subtraction and division

A review of the axioms and properties of real numbers in Table 1-1 reveals that addition and multiplication are mentioned and that subtraction and division are not. This omission occurs because certain properties (notably the commutative property) are not true for subtraction and division. We note that

$$5 - 3 \neq 3 - 5 \text{ or } a - b \neq b - a \text{ except when } b = a,$$

and $$6 \div 2 \neq 2 \div 6 \text{ or } a \div b \neq b \div a \text{ except when } a = b \neq 0.$$

Yet addition and subtraction are often substituted for each other, as are division and multiplication. Let us examine this occurrence further by illustration in Example 10.

Example 10 Given $y + 4 = 6$, show that $y = 2$.

Solution 1 If $y + 4 = 6$, then $y + 4 + (-4) = 6 + (-4)$ by adding the additive inverse of $+4$ to each side, and $y = 2$.

Solution 2 If $y + 4 = 6$, then $y + 4 - 4 = 6 - 4$ by subtracting 4 from both sides, and $y = 2$.

Solution 3 If $y + 4 = 6$, then $y = 6 - 4$ by "transposing" $+4$ from the left member of the equation and calling it -4 when it arrives in the right member, and $y = 2$.

We see from Example 10 that the same net effect is achieved by (a) adding the additive inverse of 4 to both sides of the equation, or (b) subtracting 4 from both sides of the equation, or (c) "transposing" 4 from the left side to the right side with a sign change. We must agree with the net effect statement and also agree that subtraction and transposing were used for years in algebra books. Our statement on this subject is that Table 1-1 is confined to addition and multiplication for brevity, although we do not deny the existence and popularity of subtraction and transposing.

Notice also from Example 11 that the use of division can be successful as a substitute for multiplication.

Example 11 Given $2x = 6$, show that $x = 3$.

Solution 1 If $2x = 6$, then $(\frac{1}{2})(2x) = (\frac{1}{2})(6)$ by applying the multiplicative inverse and $x = 3$.

Solution 2 If $2x = 6$, then $(2x) \div 2 = (6) \div 2$ and $x = 3$.

We see in Example 11 that multiplication by $\frac{1}{2}$ and division by 2 have the same net effect and that we were successful because we did not apply an operation that was not true under division.

1-5 Ratio and proportion

When the magnitudes of two quantities are compared, two methods of comparison are commonly used.

1. *Subtract* the numbers to determine which is the larger or smaller. If the numbers are a and b and

$$a - b = 0, \text{ the numbers are equal.}$$

$$a - b > 0, \text{ then } a > b.$$

$$a - b < 0, \text{ then } a < b.$$

2. *Divide* the numbers to determine which is larger or smaller. If the numbers are a and b and *both are positive* and

$$\frac{a}{b} = 1, \text{ then the numbers are equal.}$$

$$\frac{a}{b} > 1, \text{ then } a > b.$$

$$\frac{a}{b} < 1, \text{ then } a < b.$$

The ratio of two numbers is comparison by division

The statement "The ratio of a to b" implies the division of a by b and is written $a \div b$ or $\frac{a}{b}$ or $a: b$.

A pure ratio is unitless; that is, the two measured items under comparison are the same kind of measure. So we can compare two lengths if they are both in meters or both in feet. Refer to Example 12.

Example 12 What is the ratio of (a) 2 ft to 3 ft (b) 7 ft to 2 yd (c) 2.5 meters (m) to 125 cm?

Solution

(a) $\frac{2\text{ ft}}{3\text{ ft}} = \frac{2}{3}$ (unitless)

(b) $\frac{7\text{ ft}}{2\text{ yd}} = \frac{7\text{ ft}}{6\text{ ft}} = \frac{7}{6}$ (unitless)

(c) $\frac{2.5\text{ cm}}{125\text{ cm}} = \frac{250\text{ cm}}{125\text{ cm}} = \frac{250}{125} = \frac{2}{1} = 2$ (unitless)

Frequently we compare by division two quantities that are of different units; in such cases, we carry the units along with the quotient as an impure ratio and commonly call the quotient a *rate*. Consider Example 13.

Example 13 If a car travels 100 miles in 2 hours, what is the ratio of distance to time?

Solution $\dfrac{100 \text{ miles}}{2 \text{ hr}} = \dfrac{50 \text{ miles}}{1 \text{ hr}} = 50$ miles per hour (called "rate").

Ratio is commonly used in geometry. The most popular geometrical figure to which the concept of ratio applies is the triangle. Two triangles that are similar in shape have sides that are in a fixed ratio. Thus in Fig. 1-7, if the ratio of d to a in the similar triangles is 3 to 2, then $e: b = 3: 2$, and $f: c = 3: 2$. This notion of equal ratios introduces proportions.

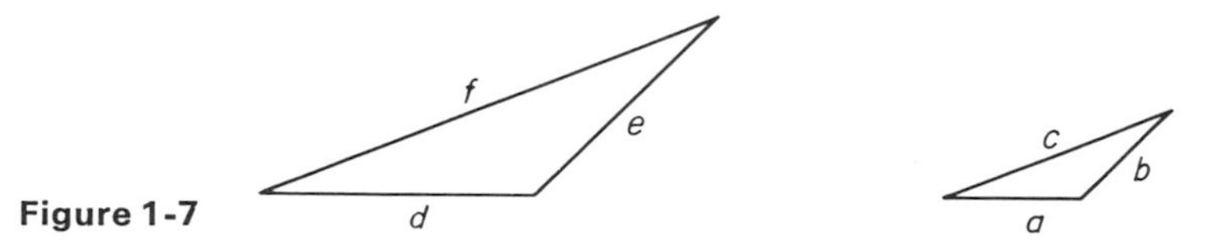

Figure 1-7

A proportion is the equality of two ratios

If two ratios $a: d$ and $c: f$ are equal, we have the proportion

$$\frac{a}{d} = \frac{c}{f}. \tag{1}$$

Some nomenclature and properties of (1) are as follows.

1. If (1) is written in a single line, it becomes $a: d = c: f$, where d and c are the middle terms (or *means*) and a and f are the end terms (or *extremes*). Also, a, c, d, and f are respectively called the first, second, third, and fourth proportions.
2. Since two fractions are equated, they are *equivalent fractions.* Thus in (1) if c is some multiple (k) of a, then f is the same multiple (k) of d.
3. Because of property (2) above, (1) can be written $a/d = ka/kd$, where the *extremes* are a and kd, the *means* are d and ka, and *the product of the means equals the product of the extremes*, or, in (1), $af = cd$.

Example 14 In Fig. 1-7, if the triangles are similar and $a = 6$, $b = 7$, $c = 9$, find e and f if $d = 9$.

Solution $a/d = b/e = c/f$ and $6/9 = 7/e = 9/f$. This multiple proportion can be separated into parts with each part solved separately. From the proportion $6/9 = 7/e$ we have (by the fact that the product of the means equals the product of the extremes) $(6)(e) = (9)(7)$ from which $e = 63/6 = 10.5$. Similarly, $6/9 = 9/f$ from which $(6)(f) = (9)(9)$ and $f = 81/6 = 13.5$.

Example 15 If 12 oranges cost \$1.50, what should 18 oranges cost?

Solution Setting up equal ratios into a proportion, we have

$$\frac{12 \text{ oranges}}{18 \text{ oranges}} = \frac{1.50 \text{ dollars}}{x \text{ dollars}}. \tag{2}$$

Now (2) reduces by equivalent fractions to $2/3 = 1.5/x$. Applying the fact that the product of the means equals the product of the extremes, we have $2x = (3)(1.5)$ from which $x = 2.25$.

A mean proportion is a proportion in which the means are equal

Frequently the second and third proportions (the means) are equal. In these cases, solution by cross product (product of means equals product of extremes) creates a square and requires a square root operation for solution. See Example 14.

Example 16 Find the mean proportion between 4 and 9.

Solution In a mean proportion, the means are equal; therefore the given proportion is set up as $4/x = x/9$. By cross products, we have $x^2 = 36$ from which $x = \pm\sqrt{36} = \pm 6$.

Percent is a ratio of parts to 100

If we have a number in fractional, decimal, or integral form and wish to express it as percent, we set it equal to the ratio $x/100$, since percent suggests "per centum" or "per 100." See Example 17.

Example 17 Express as a percent (a) 3/4 (b) 1.2 (c) 0.015.

Solution

(a) $3/4 = x/100$ from which $4x = 300$ and $x = 75\%$.

(b) Set $1.2 = x/100$ from which $x = 120\%$.

(c) Set $0.015 = x/100$ from which $x = 1.5\%$.

Exercises 1-4

In Exercises 1–10, which kind of comparison (subtraction or division) is suggested by the given statement?

1. The temperature went up 10°.
2. The cost of living increased by 8 cents.
3. He's twice her age.
4. Joe is nearly 10 years older than John.
5. The frequency was tripled.
6. This alloy is 20% tin.
7. The mixture is two parts water and three parts alcohol.
8. I'm getting 25 miles per gallon.
9. The average voltage dropped by 3.2 volts.
10. A cc of mercury weighs 13 times as much as a cc of water.

Given the proportion $x/10 = 12/5$ in Exercises 11–13, solve for x by using the fact that the

11. Product of the means equals the product of the extremes.
12. Bottom left (10) is twice the bottom right (5); therefore top left (x) is twice the top right (12).
13. Third proportion (12) is 2.4 times fourth proportion (5); therefore first proportion (x) is 2.4 times second proportion (10).

Express 14–21 as unitless ratios in lowest terms.

14. 2 ft to 3 ft
15. 600 km to 450 km
16. 8 ft to 2 yd
17. 2 miles to 8000 ft
18. 7 yd to 35 ft
19. 3 liters to 1500 cc
20. 3 ft to 1 meter
21. 1 qt to 1 liter

Express Exercises 22–30 as percent.

22. $\frac{4}{5}$
23. $\frac{5}{4}$
24. $\frac{3}{20}$
25. $\frac{1}{20}$
26. $\frac{21}{20}$
27. 2.5
28. 0.25
29. One part in 400
30. Three parts in 1000.
31. We have the instruction "Increase the power by a factor of 3." Is this an order to add 3 or to triple?
32. Given the two ratios $a : b$ and $c : d$, what is a simple method of determining if they make a true proportion?
33. A 6-ft man casts a 9-ft shadow. How tall is a tree that casts a 40-ft shadow at the same time?

34. Divide the number 960 into two parts that are in a 7: 5 ratio.

35. A gaseous mixture is three parts oxygen and five parts nitrogen by weight. How many grams of oxygen are present in 120 grams of mixture?

1-6 Scientific notation and laws of exponents

Numbers that are very large or very close to zero are frequently expressed in a shortened form called *scientific notation.* Examples of such numbers are Avogadro's constant, which is a 24-digit number starting with 6025 and followed by 20 zeros, and the mass of an electron, which is 90 billionths of one-billionth of one-billionth of a gram.

Use of scientific notation depends in part on knowledge and use of the definition of a *power*, particularly powers of 10. We define the nth power of 10 by the expression

$$10^n = 10 \times 10 \times 10 \times \cdots \times 10 \qquad (n \text{ a positive number}) \tag{3}$$

where the number 10 occurs n times as a factor in the right member of (3). Thus

$$10^2 = 10 \times 10 = 100 \tag{4}$$

$$10^5 = 10 \times 10 \times 10 \times 10 \times 10 = 100{,}000. \tag{5}$$

In (4), 100 is called the *second power* of 10; the superscript, 2, is called the *exponent*; and 10 is called the *base*. Thus in (5) the base is 10, the exponent is 5, and 100,000 is the fifth power of 10. Inspection of (4) and (5) shows that the number of zeros following the lead digit agrees with the exponent.

A number greater than 10 can be expressed as the product of a number between 1 and 10, multiplied by a positive integral power of 10. We have, as examples,

$$3856 = 3.856 \times 1000 = 3.856 \times 10^3 \tag{6}$$

$$680{,}000 = 6.8 \times 100{,}000 = 6.8 \times 10^5. \tag{7}$$

Expressions (6) and (7) are examples of scientific notation.

We define *scientific notation* as the method of expressing a number as the product of a number between 1 and 10, multiplied by the proper power of 10. *Note*: We use the qualifying term "proper power of 10" in the definition of scientific notation because of the possibility of using either positive or negative powers of 10.

A negative power of 10 implies a division or reciprocal. Thus

$$10^{-n} = \frac{1}{10^n} = \frac{1}{10 \times 10 \times 10 \times \cdots \times 10} \qquad (n \text{ a positive integer}) \tag{8}$$

where the number 10 occurs n times as a factor in the denominator of the right member of (8). Examples of negative powers of 10 are.

$$10^{-1} = \frac{1}{10} = 0.1$$

$$10^{-4} = \frac{1}{10 \times 10 \times 10 \times 10} = \frac{1}{10^4} = 0.0001.$$

Examples of scientific notation of numbers between 0 and 1 are

$$0.0056 = 5.6 \times 0.001 = 5.6 \times 10^{-3} \quad (9)$$

$$0.0000082 = 8.2 \times 0.000001 = 8.2 \times 10^{-6}. \quad (10)$$

Use of the *compensating factor* notion enables us to understand how decimal point relocation is related to the exponent over the number 10 as we move in either direction between ordinary and scientific notation. Consider numbered statement (6); in ordinary notation we had the number 3856 (where the decimal point follows the digit 6) and we moved the decimal point three places to the left in going to scientific notation. This movement of the decimal point three places to the left amounts to division by 1000 or multiplication by 10^{-3} and *must be compensated for* by inserting the factor 10^{+3} so that the value of the number remains unchanged. Similarly, in (10) as we proceed from ordinary to scientific notation, we move the decimal point six places to the right, which amounts to multiplication by 10^6 and must be compensated for by inserting the factor 10^{-6}. Use of the *compensating factor* is an application of the *multiplicative inverse.*

Changing scientific notation to ordinary notation is relatively simple if we take our cue from the exponent in the power of 10. We simply move the decimal point to the left for negative exponents and to the right for positive exponents; the numbers of places that we move the decimal point conforms with the absolute value of the exponent. Thus

$3.25 \times 10^3 = 3250.$ (move decimal point three places to right)

$6.85 \times 10^{-2} = 0.0685$ (move decimal point two places to left).

Laws of exponents Three laws of exponents, as they involve powers of 10, are given here.

LAW 1 $10^m \times 10^n = 10^{m+n}.$

This law states that multiplication of powers of 10 can be accomplished by adding exponents. Thus

$$10^3 \times 10^2 = 10^{3+2} = 10^5.$$

LAW 2 $10^m \div 10^n = 10^{m-n}$.

This law states that division of powers of 10 can be accomplished by subtracting exponents (subtracting the exponent of the divisor from the exponent of the dividend). Thus

$$10^6 \div 10^4 = 10^{6-4} = 10^2.$$

LAW 3 $(10^m)^n = 10^{mn}$.

This law states that raising a power to a power can be accomplished by multiplying exponents. Thus

$$(10^2)^3 = 10^{2 \cdot 3} = 10^6.$$

The preceding laws of exponents and the associative law are used when multiplying or dividing numbers that are expressed in scientific notation. Consider Examples 18 and 19.

Example 18 Multiply 8×10^5 by 2.5×10^2 and express the product in scientific notation.

Solution We can regroup the factors by using the associative law of multiplication:

$$8 \times 10^5 \times 2.5 \times 10^2 = (8 \times 2.5) \times (10^5 \times 10^2) = 20 \times 10^7.$$

Now $20 = 2 \times 10^1$ and so $20 \times 10^7 = 2 \times 10^1 \times 10^7 = 2 \times 10^8$.

Example 19 Perform the operation $(8 \times 10^5) \div (16 \times 10^3)$.

Solution

$$(8 \times 10^5) \div (16 \times 10^3) = \frac{8 \times 10^5}{16 \times 10^3} = \frac{8}{16} \times 10^{5-3} = 0.5 \times 10^2.$$

But $0.5 = 5 \times 10^{-1}$; so $0.5 \times 10^2 = 5 \times 10^{-1} \times 10^2 = 5 \times 10^{-1+2}$ $= 5 \times 10 = 50$.

Exercises 1-5

In Exercises 1–8, write the given number in scientific notation.

1. 93,000,000 **2.** 186,000

3. 0.0000315 **4.** 0.00000048

5. 0.00000001 **6.** 1,000,000

7. 3 billions **8.** 1.5 billionths

In Exercises 9–16, write the given number in ordinary notation.

9. 10^3 **10.** 10^{-4}

11. 6.25×10^{-5} **12.** 4.12×10^4

13. 5 millionths

14. 1.6 billions

15. $(10^3)^4$

16. $(10^{-3})^2$

Perform the indicated operations in Exercises 17–25. Express the results in scientific notation.

17. 3×200

18. 3000×2000

19. 0.002×0.0003

20. 2000×0.003

21. $1.5 \times 200 \times 30000$

22. $40 \times 500 \times 0.00001$

23. $(10^6 \div 10^{-2}) \div 0.0005$

24. $2 \times 10^{-1} + 2 \times 10^{-2} + 2 \times 10^{-3}$

25. $4 \times 10^3 + 8 \times 10^2 + 7$

2

Algebraic fundamentals

In this chapter the fundamental operations of literal (algebraic) quantities are discussed. These operations include addition, subtraction, multiplication, and division of various literal forms. The literal forms discussed here include the monomial and polynomial and fractional forms. Use is made of laws of operation borrowed from arithmetic and applied to literal quantities that represent real numbers.

2-1 Operations with signed numbers

The *fundamental operations* of arithmetic and algebra (addition, subtraction, multiplication, and division) are commonly denoted by the symbols

$+$ means "add"

$-$ means "subtract"

$\times$ means "multiply"

$\div$ means "divide."

These symbols, as noted, are *operational orders* or instructions.

Two of these symbols have alternate meanings—namely, $+$ and $-$—and are used to designate the *quality* (positive or negative) of a number. However, we frequently avoid use of the $+$ quality sign and are obliged to understand that a number without a quality sign is a positive number. Thus 4 means $+4$ and is located to the right of zero on the conventional number line. A number of negative quality *must* be preceded by a minus sign.

It is sometimes difficult to distinguish between an *operational* and a *quality* sign; this difficulty is eased when we accept the convention that a signless number is of positive quality. It is also eased when we are able to write the operational order as a word instead of a symbol; this practice is common when we write our numbers in a column,

$$\text{Add: } \begin{array}{r} 5 \\ \underline{6} \end{array} \qquad \text{means} \qquad \text{Add: } \begin{array}{r} +5 \\ \underline{+6} \end{array}$$

and is an order to add 5 and 6 where the operational order is the word "add" and the $+$ signs that appear are easily identified as quality signs. When written horizontally, the expression ordering the addition of 5 and 6 is

$$5 + 6 \text{ (simple version)}$$

or

$$+5 + (+6) \text{ (cluttered version).}$$

In the simple version, it is understood that 5 and 6 are both of positive quality and that the $+$ sign is an order to add. In the cluttered version, the first and third plus signs are quality signs and the center plus sign is the operational order to add. Note that parentheses were used in the cluttered version to clarify the picture.

Similarly, the instruction

$$\text{Subtract: } \begin{array}{r} -4 \\ \underline{-7} \end{array} \qquad \text{means} \qquad -4 - (-7).$$

Rules for adding signed numbers

1. To *add* two numbers of the *same sign*, *add* their *absolute values* and prefix the sum with the *common* sign.

2. To *add* two numbers of *different signs*, *subtract* their *absolute values* and prefix the difference with the *sign of the number with the greater absolute value*.

Example 1 Add.

(a) $\begin{array}{r} 4 \\ \underline{3} \end{array}$ (b) $\begin{array}{r} 7 \\ \underline{-2} \end{array}$ (c) $\begin{array}{r} -6 \\ \underline{9} \end{array}$ (d) $\begin{array}{r} -1.8 \\ \underline{-3.7} \end{array}$

Solution In all cases, the operational order is the word "add."

(a) Quality signs are the *same* (+4) and (+3). *Sum* of the absolute values is 7. Result is 7 or +7.

(b) Quality signs are *different* (+7) and (−2). *Difference* of the absolute values is 5. Because $|+7| > |-2|$, we choose the sign of the +7. Result is 5 or +5.

(c) Quality signs are *different* (−6) and (+9). *Difference* of the absolute values is 3. Because $|+9| > |-6|$, we choose the sign of +9. Result is +3.

(d) Quality signs are the *same* (−1.8) and (−3.7). *Sum* of the absolute values is 5.5. Result is −5.5.

Example 2 Simplify $4 + (-7) + 6 + 8 + (-5)$.

Solution All operational orders are "add." The 4, 6, and 8 are all of positive quality. The −7 and −5 are both of negative quality and were enclosed in parentheses for clarity. We may choose to apply the associative law of addition by rewriting the problem as

$$(4 + 6 + 8) + [(-7) + (-5)]$$

where our intent is to add the positive numbers as one subgroup and the negative numbers as another. Our intermediate result is $18 + (-12)$, and our final result is 6.

Alternate solution We might choose to apply the associative law of addition by working in pairs from left to right in the given expression. Now $4 + (-7) = -3$ and $6 + 8 = 14$, and our new version of the problem is $-3 + 14 + (-5)$. Subgrouping again, $-3 + 14 = 11$; our problem becomes $11 + (-5) = 6$, agreeing with the first solution.

We note in Example 2 that many different methods of subgrouping the five elements in the given problem are available under the associative law, meaning that there are many different routes to the solution of the problem.

Rule of subtracting signed numbers

To *subtract* two signed numbers, *change the sign* of the *subtrahend* and *proceed* as in *addition.*

We note that the subtrahend in subtraction is the number that is being subtracted from the other.

Example 3 Subtract.

(a) $\begin{array}{r} 4 \\ \underline{3} \end{array}$ (b) $\begin{array}{r} 7 \\ \underline{-2} \end{array}$ (c) $\begin{array}{r} -6 \\ \underline{9} \end{array}$ (d) $\begin{array}{r} -1.8 \\ \underline{-3.7} \end{array}$

Solution The operational order in all cases is "subtract"; therefore we change the sign of the subtrahend in each case and change the operational order to "add." Accordingly, the problems are restated (with results) as

(a) $\begin{array}{r} 4 \\ \underline{-3} \\ 1 \end{array}$ (b) $\begin{array}{r} 7 \\ \underline{2} \\ 9 \end{array}$ (c) $\begin{array}{r} -6 \\ \underline{-9} \\ -15 \end{array}$ (d) $\begin{array}{r} -1.8 \\ \underline{3.7} \\ 1.9 \end{array}$

Example 4 Perform the indicated operation.

$$6 - 5 + 4 - (-7) + (-8)$$

Solution Two subtractions are indicated here: subtracting 5 and subtracting -7. We can change these operational orders to addition (+), change the succeeding quality sign, and proceed as in addition:

$$6 + (-5) + 4 + 7 + (-8).$$

Here we have $6 + 4 + 7 = 17$ positive units and $-5 + (-8) = -13$ negative units or a final total of $17 + (-13) = 4$.

Rules for multiplication and division of signed numbers

Rules for multiplication (and division) of signed numbers are the same.

1. If two numbers are of the *same quality sign,* their *product* (or quotient) is *positive.*
2. If two numbers are of *different quality sign,* their *product* (or quotient) is *negative.*

Multiplication of real numbers is symbolized variously:

$$4 \times 3 = 4 \cdot 3 = (4) \cdot (3) = (4)(3) = (4) \times (3) = 12.$$

When these real numbers are represented by letters, we have

$$a \times b = a \cdot b = (a) \cdot (b) = (a)(b) = (a) \times (b) = ab.$$

Division of real numbers is symbolized

$$8 \div 2 = \frac{8}{2} = 2\overline{)8} = 4$$

or when using letters,

$$a \div b = \frac{a}{b} = b\overline{)a}.$$

Example 5 Perform the indicated operations.

(a) $(-6)(2)$ (b) $(-5) \div (-2)$ (c) $\frac{-5}{10}$ (d) $(-2)(-2)$

Solution All four examples involve either multiplication or division of signed numbers. In (b) and (d) the quality signs are the same and so the result is positive. In (a) and (c) the quality signs differ and so the result is negative. With these signs in mind, the results are

(a) -12 (b) 2.5 (c) -0.5 (d) 4.

Multiplication involving many factors of both positive and negative quality

A multiplication problem may involve many signed numbers as multipliers or factors. In such cases, it is recommended that the associative law of multiplication be applied in two stages, the first to obtain the sign of the product and the second to obtain the numerical product. See Example 6.

Example 6 Evaluate $(2)(-3)(-\frac{1}{2})(6)(-\frac{1}{4})$.

Solution First, obtain the *sign* of the product. We have five factors, three of which are negative. We can associate these five in any way that we wish; one way is to consider *pairs* of factors because the rule for multiplication of signed numbers involves pairs. Each pair of positive factors has a positive product; each pair of negative factors also has a positive product. It is readily seen that an *odd* number of negative factors has a negative product and an *even* number of negative factors has a positive product. We have *three* negative factors in our problem; so the sign of the product is negative. Meantime the five numbers multiply to a product of 4.5 (absolute value) and the product is -4.5.

Combinations of fundamental operations An arithmetic statement may require more than one of the fundamental operations, such as $4 + 3 - 2$, which requires an *addition* and a *subtraction* (answer: 5). Also, $(2)(3) + (4)(5)$ requires that the *product* 4 times 5 be *added* to the *product* 2 times 3 (answer:

26). A fractional statement like $(8 - 2)/(2)(3)$ requires that the *difference* 8 minus 2 be *divided by* the *product* 2 times 3 (answer: 1). The statement $3(5 - 2)$ may be handled by first *subtracting* 2 from 5 and then *multiplying* this difference by 3 (answer: 9); we note that the quantities (5 and 2) inside the parentheses are combinable under subtraction; had they not been combinable, a different approach (the distributive law) would have been applied. When combining numbers, it is useful to give first priority to evaluating fractional components and/or quantities inside of parentheses. Next, carry out multiplication and division and then, finally, addition or subtraction of the separate terms. Thus

$$\frac{3+6}{2-(-3)} + 3\left(\frac{6-2}{2}\right) = \frac{9}{5} + 3(2) = 1.8 + 6 = 7.8. \tag{1}$$

Calculator users must be aware that the operations in (1) cannot be handled without use of parentheses or memory, where the memory may be electronic storage or writing down an intermediate result that can be recalled for use later.

Exercises 2-1

Perform the indicated operations in Exercises 1–20.

1. $6 + 5$

2. $6 + (-5)$

3. $6 - (+5)$

4. $6 - (-5)$

5. $-7 + 4$

6. $-7 + (-4)$

7. $-7 - (+4)$

8. $-7 - (-4)$

9. $13 \times (-2)$

10. $(-12)(-5)$

11. $(-144) \div 12$

12. $(-72) \div (-8)$

13. $\dfrac{6 - (-5)}{4 + (-3)}$

14. $5 - \dfrac{6 + 3}{6 - (-3)}$

15. $\dfrac{-3 - (-12)}{4 + (-7)}$

16. $\dfrac{3 + (-12)}{-4 - (-7)}$

17. $(-8)(-3 + 5)$

18. $-10 + \dfrac{4 + (-8)}{-2}$

19. $\dfrac{-8 - (-4)}{-2}$

20. $\dfrac{-3 + (-7)}{2} - 5$

A ball rolling up an inclined plane slows at the rate of 5 ft/sec each second. It started (initial velocity) at 60 ft/sec. In Exercises 21–24,

21. What is its rate at the end of 5 sec? 20 sec?

22. Is the ball moving up or down the plane at the end of 15 sec? How fast?

23. How many seconds have elapsed when the ball is momentarily at rest?

24. At the end of 3 sec the ball is traveling up the plane at 45 ft/sec; at the end of how many seconds is the ball traveling down the plane at 45 ft/sec?

25. In a tug-of-war game, forces of 80, 60, and 75 lb are exerted to the right and forces of 65, 85, and 70 lb are exerted to the left. What is the net (resultant) force and in which direction does it act?

26. A water tank is fed by two pipes at the rate of 3.5 ft^3/min and 4.2 ft^3/min. Water is being removed by three users at 0.6 ft^3/min, 4.2 ft^3/min, and 3.2 ft^3/min. Is the tank filling or emptying? At what rate? If the 0.6 ft^3/min user were eliminated, would the tank be filling or emptying? At what rate?

27. A student obtains marks of 68, 73, 95, and 70 on quizzes. Without actually computing his average mark, determine whether his average is over or under 74; over or under 80; over or under 78.

28. A man has assets of $1000. He takes on a debt of $200 and "loses" a debt of $400. He then divides his assets equally among his three sons. How much does each son get?

29. Assume that time zones on the face of the earth are each 1000 miles wide and that a clock must be set back 1 hour for each time zone traversed by a westbound traveler. A pilot starts flying westward from point A at 12:00 noon. He flies westward for 6 hours to point B, rests there for 1 hour, and then flies eastward for 4 hours. If he travels at a uniform rate of 500 mph while he is in flight, what time did his clock read when he arrived at point B? What time should his clock have read to be correct for that time zone when he is at point B? What time should his clock read at the end of his journey?

2-2 Literal quantities; addition and subtraction

Use of alphabetic characters (letters) to represent numbers is a common practice in algebra. Letters are used in order to remain general rather than specific. A popular example of the use of letters is in the relationship "The circumference of a circle is π (pi) times the diameter," which we express as $C = \pi d$ in letter (or literal) form. This relationship is true for *any* circle and so d is *any* diameter; hence the use of a letter to remain general rather than to use a specific number.

In Table 1-1 of Chapter 1 the letters a, b, c, and d were used to denote real numbers. Some combinations of literal quantities were given in that table. These and other combinations are given here to provide illustrations of literal quantities.

	Literal expression	*Meaning*
A.	$a + b$	Sum of a and b
B.	$a - b$	Difference resulting from subtracting b from a
C.	$a \times b$ or $a \cdot b$ or ab	Product of a and b
D.	$\frac{a}{b}$ or $a \div b$	Quotient of a divided by b

E.	$5a$	Five times a or $a + a + a + a + a$
F.	$a \cdot a$ or a^2	a multiplied by itself or "a squared"
G.	$\sqrt{a}$	Square root of a
H.	$ab + cd$	Product of a and b added to the product of c and d
I.	$6abc^2$	Six times the product of a, b, and the square of c.

Clearly a large number of different types of algebraic expressions is possible. Nomenclature of these expressions is essential to facilitate communication.

A *term* consists exclusively of products of literal and/or ordinary numbers. Each multiplier is a *factor* of the term; thus $5x^2y$ has prime factors 5, x, x, and y, where a number is prime if it has no factors other than itself and the number 1.

A *monomial* is an expression that consists of only one term. In the list shown C, D, E, F, G, and I are examples of monomials.

A *binomial* is an expression that contains exactly *two* terms. These terms are joined by addition or subtraction. In the list A, B, and H are binomials.

A *trinomial* is an expression that consists of exactly three terms; a *polynomial* consists of two or more terms.

One factor of a term is called the *coefficient* of the remaining factors of that term. The *numerical coefficient* of a term is the *numerical factor*. Thus in the term $6abc^2$, 6 is the numerical coefficient. The numerical coefficient is customarily written first; thus $6x$ is preferred to $x6$.

Like terms differ only in their numerical coefficients. Thus $6abc^2$ and $-5abc^2$ are like terms, whereas $6abc^2$ and $-5ab^2c$ are unlike. In like terms, the literal portions must contain the same letters with the same exponents. *Only like terms may be added or subtracted.*

Symbols of grouping, such as parentheses ( ), brackets [ ], and braces { }, are used to convey the idea that the enclosed expression is to be regarded as a single quantity and to avoid confusion caused by the presence of both signs of quality and signs of operation. Thus "three times the sum of a and b" would be written $3(a + b)$, where the sum of a and b is regarded as a single quantity. Also, "4 diminished by -3" is written $4 - (-3)$, where parentheses are used around -3 to avoid confusion of signs.

Powers of literal quantities are written in a manner similar to that used in writing powers of 10 (see Sec. 1-6). They also obey the same laws regarding multiplication, division, and raising a power to a power. By definition,

$$a^n = a \times a \times a \times \cdots \times a \qquad (n \text{ factors}). \tag{2}$$

The *basic laws of exponents* include ($a \neq 0$)

$$\text{Multiplication:} \quad a^m \times a^n = a^{m+n} \tag{3}$$

$$\text{Division:} \quad a^m \div a^n = a^{m-n} \tag{4}$$

$$\text{Power to a power:} \quad (a^m)^n = a^{mn} \tag{5}$$

Conventions on choice of letters in expressions When we generalize by representing quantities with letters, several conventions are practiced.

1. In applications, use the first letter of the measurable quantity involved, choosing v for velocity, r for radius, m for mass, and so on. This convention is useful but must be modified when two or more quantities in the same problem might claim the same letter.
2. When the same quantity is given two or more times with different values, subscripts are often used. For instance, initial velocity and final velocity may be represented by v_i and v_f or v_1 and v_2.
3. When no particular applications are intended and a quantity is to vary, it is customary to represent that quantity with a letter chosen from the end of the alphabet, such as x or y.
4. When a quantity is to remain constant throughout a given problem, that constant is frequently represented by a letter chosen from the early end of the alphabet, like a or b.

These conventions, although useful, are not without exception and are not rigidly obeyed.

The process of devising a set of algebraic symbols to express a mathematical or physical relationship is called *formulation.* Technical studies rely heavily on this process. Consider Examples 7 and 8.

Example 7 Given that a rectangle is such that its length is twice its width, find an expression for its (a) perimeter, (b) area.

Solution Because the length is twice the width, we might represent the width as w and the length as $2w$ as shown in Fig. 2-1. The perimeter, being the sum of the four sides, would be

$$p = w + 2w + w + 2w = 6w.$$

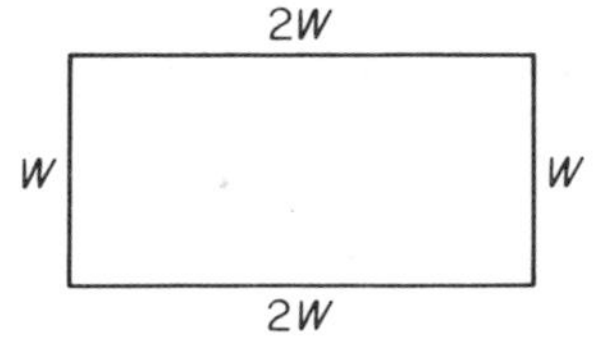

Figure 2-1

The area, being the product of the length and width, would be

$$A = (w)(2w) = 2w^2.$$

Example 8 If a dozen eggs cost d cents, how many cents will k eggs cost?

Solution Since a dozen (12) eggs cost d cents, the cost of one egg is $d/12$ cents and k eggs will cost $(k)(d/12) = kd/12$ cents.

Addition and subtraction of literal quantities

As noted, *only like terms can be combined under addition or subtraction.* Addition or subtraction of like terms is accomplished by combining coefficients according to the law of signed numbers. Thus $3x$ and $2y$ are *not* like terms; so their sum is simply the *indicated* sum, or $3x + 2y$. However, $3xy$ and $2xy$ are like terms and are combined under addition to $5xy$. Observe also that $3x^2yz$ and $-4zx^2y$ are like terms because multiplication is commutative.

If two polynomials are added, like portions are combined; thus $2x^2 + 3y$ and $5x^2 - 4y$ add to $7x^2 - y$. Additional samples are

1. $3x^2 - 4xy + y^2 - 2x^2 + 4xy + y^2 = x^2 + 2y^2.$
2. $4m - 2mn - 6m - 4mn = -2m - 6mn.$
3. Add: $\begin{array}{r} 4x^2 - 7xy \\ 2xy + y^2 \\ \hline 4x^2 - 5xy + y^2 \end{array}$ Subtract: $\begin{array}{r} 4x^2 - 7xy \\ 2xy + y^2 \\ \hline 4x^2 - 9xy - y^2 \end{array}$

Exercises 2-2

Given expressions A to D below, answer Exercises 1–5.

A. $3x^2 - 5y + z$ B. $6y^2z^3w$ C. $\frac{1}{2}mnp + 6$ D. $0.1x + 4.5y$

1. Which is a monomial? A trinomial?

2. Name the second term of D.

3. What is the numerical coefficient of the first term of C?

4. What are the prime factors of B?

5. What is the numerical coefficient of the last term of A?

Perform the operations given in Exercises 6-13.

6. Add $2x^2 - 2x + 3$ and $3x - 4 + x^2$.

7. Add $2xy + y^2 - 3x^2$ and $-2y^2 + 3xy$.

8. Add $a^2 - 2ab + b^2$ and $-a^2 + 2ab + 2b^2$.

9. Add $2(a + b)$ and $3(a + b)$ and $-4(a + b)$.

10. Subtract $m + n - p$ from $8m - 2n + p$.

11. Subtract $3pr - 6qr + 5$ from $4 - 2qr + pr$.

12. Subtract $0.3m^2 - 1.25mn$ from $1.2m^2 - 3.5mn$.

13. Subtract $\frac{1}{2}ab + \frac{1}{3}bc$ from $\frac{1}{4}bc - 2ab$.

Give literal expressions for the word statements in Exercises 14–33.

14. The product of velocity v and time t.

15. The sum of the two horizontal components F_{1x} and F_{2x}.

16. The sum of the squares of x and y.

17. The square root of the sum of the squares of a and b.

18. The product of r_1 and r_2 divided by their sum.

19. Four-thirds of the cube of the radius r, all multiplied by π.

20. The difference, x less y, divided by the sum of x and y.

21. The square of the sum of u and v.

22. The square root of the quotient of the power p divided by resistance r.

23. One-half the product of the graviational constant g multiplied by the square of time t.

24. The cube of the difference x less y.

25. The difference of the cube of x less the cube of y.

26. The product of π and the fourth power of diameter D, all divided by 32.

27. The fourth power of the product of $\pi/32$ and diameter D.

28. The sum of the diameter and one-half the circumference of a circle of radius r.

29. The surface area of a cube, each edge of which has the length x.

30. The surface area of a cube, each edge of which has the length $x + a$.

31. The square of one-half the sum of x and y.

32. One-half of the sum of the squares of x and y.

33. The square root of the product of x and y, all divided by the cube of their sum.

34. Give expressions for the perimeters and areas of the polygons shown in Fig. 2-2.

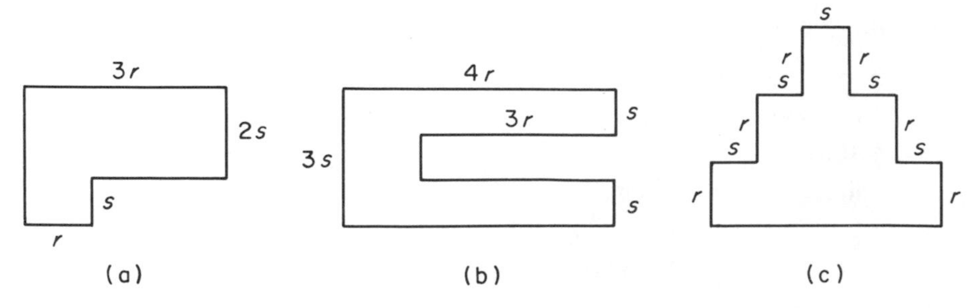

Figure 2-2

35. How much money is taken in at a theater if m adult tickets are sold at d dollars each and n children's tickets are sold at c cents each?

36. Three boxes of dimensions W, L, and H are stacked. Box A lies on the floor on its W by L side; box B stands on its L by H side atop box A; box C stands on its W by H side atop box B. How tall is the stack?

37. The turning moment of a lever is the product of the mass and the moment arm. A clockwise moment is called positive and a counterclockwise moment is called negative. If the mass is represented by m and the moment arm by d, give the expression for the total moment in Fig. 2-3.

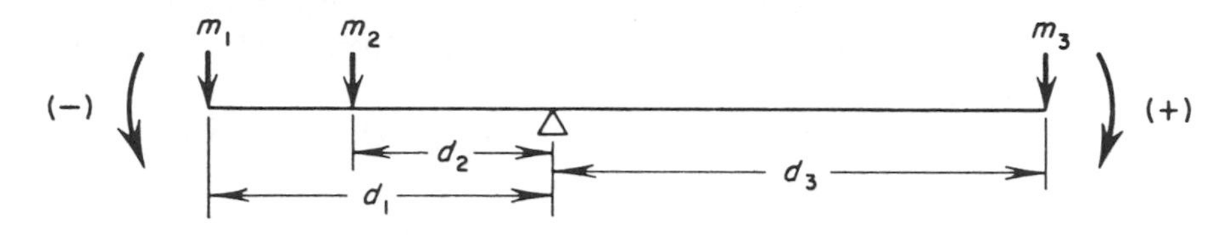

Figure 2-3

38. A metal bar of length L_0 when the temperature t is zero is heated. Heat expands the bar by an amount that is the product of L_0 by the change in temperature Δt by the coefficient of linear expansion c. Give a general expression for the length of the bar when t is not zero.

39. The area of a trapezoid is one-half the altitude h multiplied by the sum of bases b and b'. If b equals h and b' is twice h, give an expression for the area in terms of h alone.

40. A boat travels in still water at h mph. A river has a current of c mph. The boat is to make a one-way trip of m miles in the river. Give expressions for the length of time required for the trip if the trip is upstream; downstream.

41. The weight of an object may be described as the product of the volume and density of the object. Assuming a density d, give an expression for the weight of each of the solid figures shown in Fig. 2-4.

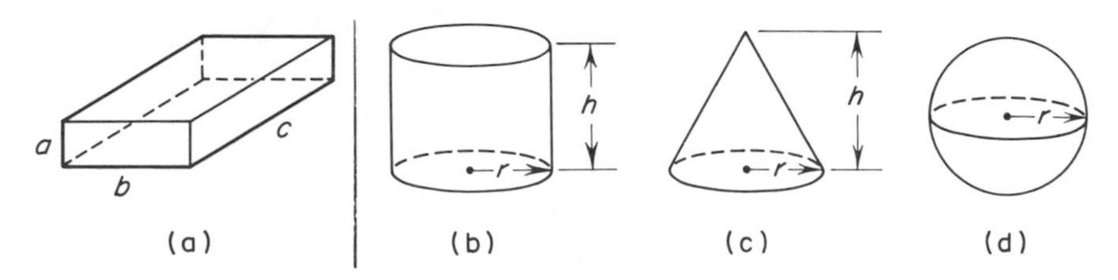

Figure 2-4

42. Give an expression for the surface area of the solid figures in Fig. 2-4(a) and (b).

43. A fancy box is built in the form of Fig. 2-4(a). The bottom costs r cents per area unit, the sides cost s cents per area unit, and the top costs t cents per area unit. Give an expression for the total cost of the box.

2-3 Grouping symbols

In Sec. 2-2 grouping symbols and reasons for their use were mentioned. Removal of grouping symbols is frequently required, particularly in order to simplify algebraic quantities. The process of removing grouping symbols is demonstrated in the *distributive law*

$$a(b + c + d + \cdots) = ab + ac + ad + \cdots \tag{6}$$

which states that "the product of a number a and the polynomial $b + c + d + \cdots$ is the sum of the several products resulting from multiplying a by each term of the polynomial." Note that the left member of (6) has grouping symbols, whereas the right member has none. Examples of removing grouping symbols follow.

Example 9 Clear the grouping symbols.

(a) $3(4 + 5) = 3 \cdot 4 + 3 \cdot 5 = 12 + 15 = 27$

(b) $2(x - y + 2z) = 2x - 2y + 4z$

(c) $-ab(a - 3b) = -a^2b + 3ab^2$

When one set of grouping symbols encloses another, the usual procedure for removal of the symbols is to remove the innermost set first.

Example 10 Remove grouping symbols and collect like terms.

$$2\{b - [3b + 4(b - 1)] + 2\}$$

Solution Inspection reveals that the parentheses enclosing $b - 1$ are innermost. They can be removed by multiplying $b - 1$ by 4, producing

$$2\{b - [3b + 4b - 4] + 2\}.$$

Collecting like terms inside the brackets, we have

$$2\{b - [7b - 4] + 2\}.$$

Now the brackets are innermost and are removed by multiplying $7b - 4$ by -1; the expression becomes

$$2\{b - 7b + 4 + 2\}.$$

Collecting like terms inside the braces gives

$$2\{-6b + 6\}.$$

The braces are removed by a third application of the distributive law, and we have the final solution

$$-12b + 12.$$

2-4 Operations with zeros

The fundamental operations (addition, subtraction, multiplication, and division) require special care for several reasons. Among them are (a) certain operations with zeros are impossible and (b) an answer of zero is not to be construed as "no answer."

Consider the following facts, where a is a real number.

Addition: $a + 0 = 0 + a = a$

Subtraction: $a - 0 = a$, but $0 - a = -a$

Multiplication: $a \cdot 0 = 0 \cdot a = 0$

Division: $\frac{0}{a} = 0$, but $\frac{a}{0}$ and $\frac{0}{0}$ are indeterminate

These addition, subtraction, and multiplication facts appear straightforward. Division by zero requires some explanation. If we start with the fact that $a \cdot 0 = 0$ where a is *any real number*, then (if division by zero were possible) by dividing both sides of this statement by zero, we would have $a = 0/0$, which asserts that $0 \div 0$ equals *any real number*. With this unusual condition, we conclude that $0 \div 0$ is indeterminate.

A solution of value zero is often improperly referred to as "no answer." Consider the question "What (x) do we add to 3 to get a sum of 3?" This is written $3 + x = 3$. Only one value of x satisfies this equation and answers the question posed; that value is $x = 0$, as opposed to the incorrect answer "no answer."

The zero exponent The law of division of powers of the same base is $a^m \div a^n = a^{m-n}$, where $a \neq 0$. If $m = n$, then $a^m \div a^m = a^{m-m} = a^0$; but we clearly have the case of a nonzero number being divided by itself, producing the answer 1, and so $a^0 = 1$.

Exercises 2-3 and 2-4

In Exercises 1–21, remove all the grouping symbols and collect like terms.

1. $3 + (4 - 5)$

2. $(8 - 2) + 2(5 - 7)$

3. $3 - (4 + 5)$

4. $(6 - 3) + 2(5 - 2)$

5. $7(9 - 5) + 3(6 - 9)$

6. $8 - 3 - 4(4 - 6) - 5$

7. $4(5 - 3) - 3(5 - 3)$

8. $3 + 2 - 3(4 + 2) - 1 + 2$

9. $9 - (-6 + 1)$

10. $a - b - (c - b)$

11. $3 - (-2 - 3)$

12. $a + b - (c + b)$

13. $-a - (b - a)$

14. $7x - [4x + 3(x - 2)]$

15. $a - (-b + a)$

16. $4x - 5\{x - 2[(4 - 2y) + 3(y - x)]\}$

17. $7x - 3\{x + 4[(x - y) - 2(3y - 2)]\}$

18. $3\{8 - [2(5 - x) - x]\}$

19. $4[4(4 - 1) - 1] - 1$

20. $6\{4 - [3 - (2 - x) + x] - 2x\}$

21. $-2\{-2[-2(-x + 1) + 1] + 1\} + 1$

In Exercises 22–25, we are given a triangle with sides a, b, and c, where c is the largest side and a is the smallest side. Formulate the expressions requested by using grouping symbols and then simplify where possible.

22. The perimeter less twice the sum of the largest and smallest sides.

23. The perimeter tripled, less the difference $c - 2b$.

24. The sum of the three quantities, where the three quantities are all the possible arrangements formed by subtracting twice the sum of two sides from the third side.

25. The sum of the squares of the three sides, diminished by $3d$, where d is the sum of the products of the sides taken two at a time. (*Note:* There are three such products.)

In Exercises 26–29, we are given the expression

$$y_{av} = \frac{y_1 + y_2 + y_3}{3}.$$

Find y_{av} for the given values of y_1, y_2, and y_3.

26. $y_1 = a, y_2 = 2a, y_3 = 3a$

27. $y_1 = a + b, y_2 = a - b, y_3 = a$

28. $y_1 = 2b - 3c, y_2 = 2(a - b), y_3 = c - b$

29. $y_1 = 2(a + b + c), y_2 = b - 3c, y_3 = a - (2b - c)$

Answer Exercises 30–34 as directed.

30. Joe has no money. He divides his money equally among four relatives. How much money does each relative get?

31. Hank doubles his money every day. At the start he has no money. How much money does he have after 5 days?

32. Which is the largest: 0×0 or 5×0 or -3×0?

33. Jane had $800. In two transactions she gained zero and then lost zero. What does she have left?

34. If $a \neq 0$ and $b \neq 0$, evaluate the following.

(a) a^0b^0 (b) $(ab)^0$ (c) $(a + b)^0$ (d) $a^0 + b^0$ (e) $(a^0 + b^0)^0$

2-5 Algebraic products

When algebraic products are studied, it is customary to progress from simple factors to more complicated factors involved in the multiplication process. Simple cases involve only monomial factors; the more complicated cases involve polynomials.

Product of two or more monomials Consider that a monomial may be a single number (like 2) or a single letter (like x) or a product of numbers and letters (like $24xy$) and may be negative (like $-2x^2y^3$). To multiply two monomials, we are involved with (a) the sign of the product, (b) the numerical coefficient of the product, and (c) the literal portion of the product. Consider two examples.

Example 11 Multiply $3x^2y$ by $-4xy^3$.

Solution Inspecting the factors to be multiplied, we make *three* observations.

1. $3x^2y$ is positive and $-4xy^3$ is negative; the product is therefore negative according to the laws of multiplication of signed numbers.
2. The numerical coefficients are 3 and 4 (ignoring signs) and their product is 12.
3. The literal portions are x^2y and xy^3; their product is x^3y^4 according to the law of multiplication of powers of literal quantities given in Sec. 2-2. Combining these three observations, we have

$$(3x^2y)(-4xy^3) = -12x^3y^4.$$

Example 12 Multiply $(-4a)(-3bc)(-2ac)(-2abc)$.

Solution Inspection shows that

1. since all four factors are negative, the product is positive;
2. the product of the unsigned numerical coefficients is

$$4 \cdot 3 \cdot 2 \cdot 2 = 48;$$

3. applying the law of multiplication of literal quantities, we have

$$(a)(bc)(ac)(abc) = a^3b^2c^3.$$

Combining these three observations gives

$$(-4a)(-3bc)(-2ac)(-2abc) = 48a^3b^2c^3.$$

The preceding discussion and Examples 11 and 12 suggest the following procedure.

To multiply two or more polynomials, consider three steps: (1) *determine the sign of the product by use of the laws of multiplication of signed numbers,* (2) *determine the numerical coefficient of the product by multiplying the numerical coefficients of the separate factors, and* (3) *multiply the literal portions according to the laws of multiplication of powers of like bases; then combine these three into a single product.*

Product of a monomial and a polynomial This product is an application of the distributive law

$$a(b + c + d + \cdots) = ab + ac + ad + \cdots$$

where a is the monomial and $b + c + d + \cdots$ is the polynomial. Close inspection of the distributive law shows that the monomial factor is multiplied by each term of the polynomial and that the separate products are joined by the proper signs.

Example 13 Multiply $2x$ by $3x - 5y + z$.

Solution This indicated product may be written $2x(3x - 5y + z)$. Multiplying each term of the polynomial by $2x$, we have the product

$$2x(3x - 5y + z) = 6x^2 - 10xy + 2xz.$$

Example 14 Multiply $-3ab^2c$ by $-6a^2b + 12b^3c^2$.

Solution Applying the distributive law,

$$-3ab^2c(-6a^2b + 12b^3c^2) = 18a^3b^3c - 36ab^5c^3.$$

A summary statement of procedure follows. *To multiply a monomial by a polynomial, apply the distributive law by multiplying each term of the polynomial by the monomial and join the separate products with the proper signs.*

Product of two polynomials This product, with some insight, may be seen as a special application of the distributive law. Consider the product

$$(r + s)(t + u + v),$$

where $r + s$ will become the multiplier of each term of the polynomial $t + u + v$ on application of the distributive law, and we have

$$\begin{aligned}(r + s)(t + u + v) &= (r + s)t + (r + s)u + (r + s)v \\ &= rt + st + ru + su + rv + sv. \qquad (7)\end{aligned}$$

Inspecting (7), we see *six* products and these six are the result of multiplying *each* term of the first factor $(r + s)$ by *each* term of the second factor $(t + u + v)$. Note that a polynomial with m terms multiplied by a polynomial with n terms will involve $m \cdot n$ separate products; *example*, a trinomial times a trinomial involves *nine* $(3 \cdot 3)$ separate products.

The polynomials in (7) are arranged horizontally. They may be arranged vertically as in "long" multiplication in arithmetic. Consider an arithmetic case.

Example 15 Multiply 247 by 32, expressing each factor as a polynomial.

Solution Since 247 is $200 + 40 + 7$ or $2(10)^2 + 4(10) + 7$ and 32 is $30 + 2$ or $3(10) + 2$, we will have the following polynomial expressions and six products.

$$\begin{array}{rrrrcr}
 & 2(10)^2 + & 4(10) + & 7 & & 247 \\
 & & 3(10) + & 2 & & 32 \\
\hline
 & 4(10)^2 + & 8(10) + & 14 & & 494 \\
6(10)^3 + & 12(10)^2 + & 21(10) & & & 741 \\
\hline
6(10)^3 + & 16(10)^2 + & 29(10) + & 14 & & 7904 \\
\text{or} \quad 6000 + & 1600 + & 290 + & 14 & = & 7904
\end{array}$$

Note here that the six separate products are written in rows and columns so as to facilitate future vertical addition. This orderliness, although not essential per the commutative law, is usually sought after.

Example 16 Multiply $(2x - 3y)(3x + 4y)$.

Solution 1 Setting down the two factors vertically and proceeding with the requirement that each term of the first factor be multiplied by each term of the second, we have

$$\begin{array}{rrr}
2x - & 3y & \\
3x + & 4y & \\
\hline
6x^2 - & 9xy & \\
 & 8xy - & 12y^2 \\
\hline
6x^2 - & xy - & 12y^2
\end{array}$$

Solution 2 Choosing to retain the horizontal arrangement, we have

$$(2x - 3y)(3x + 4y) = 6x^2 - 9xy + 8xy - 12y^2 \tag{8}$$

$$(2x - 3y)(3x + 4y) = 6x^2 - xy + 12y^2 \tag{9}$$

where inspection of (8) shows all four of the separate products of the two original binomials and (9) shows the final product with like terms collected.

Here is a summary statement. *To multiply two polynomials, multiply each term of one by each term of the other and collect like terms among the several products.*

Some special products Certain products involving monomials and polynomials occur repeatedly in albegra. Thorough knowledge of these products is extremely useful in a reverse process called *factoring* and discussed later in this chapter. Some of these products include

$$a(b + c + d + \cdots) = ab + ac + ad + \cdots. \tag{10}$$

$$(a + b)(a - b) = a^2 - b^2. \tag{11}$$

$$(a + b)^2 = a^2 + 2ab + b^2. \tag{12}$$

$$(a - b)^2 = a^2 - 2ab + b^2. \tag{13}$$

$$(ax + b)(cx + d) = acx^2 + (ad + bc)x + bd. \tag{14}$$

$$(x + y)(x^2 - xy + y^2) = x^3 + y^3. \tag{15}$$

$$(x - y)(x^2 + xy + y^2) = x^3 - y^3. \tag{16}$$

Exercises 2-5

Obtain the indicated products in Exercises 1–20. Simplify each of the products.

1. $3(2x - 5y)$

2. $3x(5ax - 7by)$

3. $(2x - 3)(x + 1)$

4. $(4t - 3s)(4t + 3s)$

5. $(4t - 3s)^2$

6. $(3t + 4s)^3$

7. $(a + 2)(a - 3)$

8. $(a + 2)(a + 6)$

9. $(3a + 2)(2a - 3)$

10. $5(x^2 - 3)(x + 1)$

11. $abc^2(a + bc - 2ac)$

12. $(x - y + 5)(x - y - 5)$

13. $(2ab)(-3a^2bc)(-4ab^2c^2)$

14. $(x^2 - 9y^2)^2$

15. $(4a - 3b + 2)(a - 2b)$

16. $\frac{x^2y^2}{ab}\left(ac + ab + \frac{a}{x}\right)$

17. $(1 - 9x + 4x^3)(3x + 2 - x^2)$

18. $[2a - (x - y)][2a + (x - y)]$

19. $(9s^2 - 6st^2 + 4t^4)(3s + 2t^2)$

20. $(x + y)(x - 2y)(x + 3y)$

In Exercises 21–26, show literal quantities for the given expressions. Simplify wherever possible.

21. The product of three successive integers if the least is x.

22. The product of three successive odd integers if the least is x.

23. The volume of a block whose length is twice its width and whose height exceeds its width by 3. Call the width w.

24. The volume of the pan formed from the sheet-metal rectangle shown in Fig. 2-5(a) if the four equal shaded squares are removed from the corners and the remaining sheet is folded along the dotted lines into the shape of a shallow pan.

25. The volume of the block shown in Fig. 2-5(b).

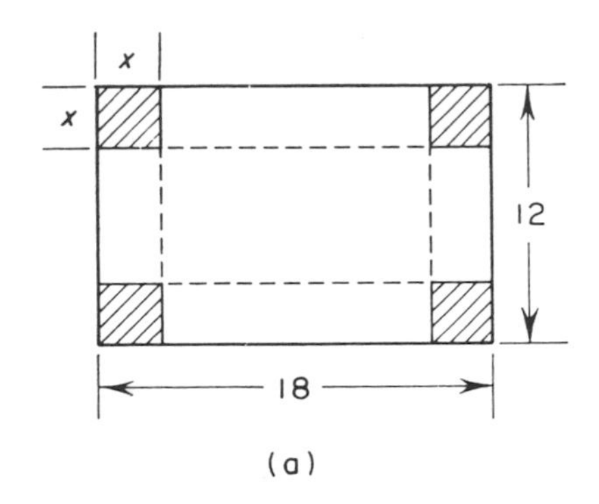

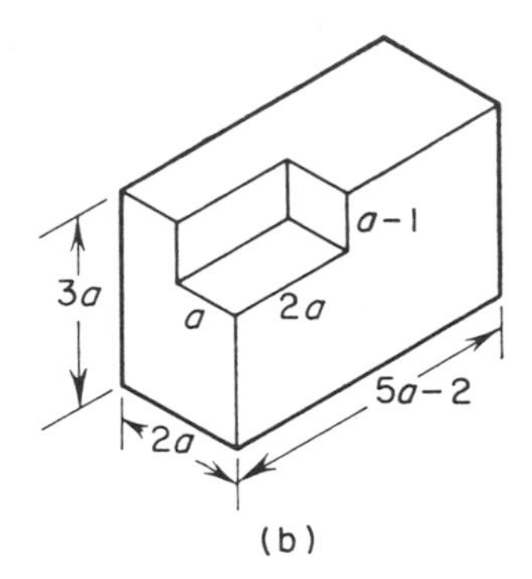

Figure 2-5

26. The total surface area (for the nine planes) of the block in Fig. 2-5(b).

27. The volume of the frustum of a cone is given as the product of two quantities. The first quantity is 1/3 of the product of π and the altitude; the second quantity is the sum of the squares of the radii of the bases, increased by the product of the radii of the bases. If the smaller radius is three units smaller than the larger radius R and the altitude is two units greater than R, give an expression for the volume in terms of R.

28. A group of people purchase tickets at a theater. The number of adult tickets purchased exceeds the number of children's tickets by 18. The adult price exceeds the children's price by 55 cents per ticket. Given an expression for the total cost of the tickets if m tickets are purchased and the cost of the adult tickets is c cents each.

29. Give an expression for the area of the hook shown in Fig. 2-6(a).

30. Give an expression for the area of the cross shown in Fig. 2-6(b).

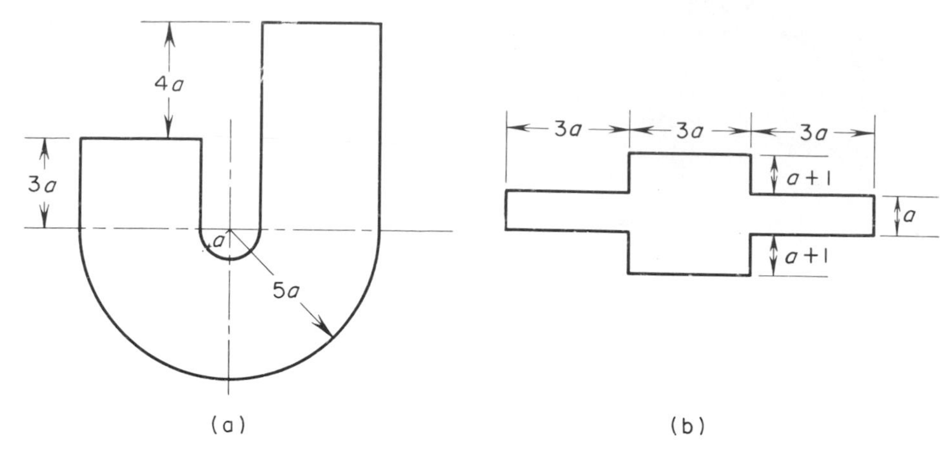

Figure 2-6

2-6 Algebraic division

Division of algebraic quantities is best discussed in two cases—that is, those involving a monomial divisor and those involving a polynomial divisor.

Division of a monomial by a monomial The process involved in division of a monomial by a monomial closely parallels the process of multiplying two monomials.

Example 17 Perform the division: $-35x^2y^4z \div -7xy^2z$.

Solution This problem may be written

$$\frac{-35x^2y^4z}{-7xy^2z} = 5x^{2-1}y^{4-2}z^{1-1} = 5xy^2. \tag{17}$$

Inspecting (17) closely, we see that the sign of the quotient is positive as the result of dividing a negative number by a negative number. The numerical coefficient of the quotient is 5, the result of dividing the coefficient 35 by the coefficient 7. The literal portion of the quotient is the result of the law of division of powers of literal quantities as described in Sec. 2-2.

The preceding example suggests the following procedure. *To divide a monomial by a monomial, consider three steps:* (1) *determine the sign of the quotient by the laws of division of signed numbers,* (2) *determine the numerical coefficient of the quotient by dividing the numerical coefficients as in ordinary arithmetic, and* (3) *divide the literal portions by using the division law of exponents and then combine these into a single quotient.*

Division of a polynomial by a monomial To perform the division of a polynomial by a monomial, we recall a procedure from arithmetic—*invert the divisor and proceed as in multiplication.* Thus

$$b \div a = b \cdot \frac{1}{a}$$

and, for a polynomial dividend, we have the distributive law:

$$(b + c + d + \cdots) \div a = \frac{1}{a}(b + c + d + \cdots).$$

Example 18 Perform the division: $(6x^2y - 9x^3y^3 + 12z) \div 3xy$.

Solution Inverting the divisor and applying the distributive law, we have

$$\begin{aligned} \frac{1}{3xy}(6x^2y - 9x^3y^3 + 12z) &= \frac{1}{3xy}(6x^2y) - \frac{1}{3xy}(9x^3y^3) + \frac{1}{3xy}(12z) \\ &= 2x - 3x^2y^2 + \frac{4z}{xy}. \end{aligned}$$

A summary statement follows. *To divide a polynomial by a monomial, invert the divisor and proceed as in multiplication, using the distributive law.*

Division of a polynomial by a polynomial Division of one polynomial by another is similar in many respects to "long division" in arithmetic. Example 19 gives a sample division problem; following the sample is a set of simple rules of procedure to follow in division.

Example 19 Divide $26x - 31x^2 + 6x^4 - 6$ by $1 + 2x^2 - 4x$.

Solution We rearrange the terms of the dividend and divisor in descending order of exponents and provide "placeholders" for "missing terms."

$$\begin{array}{lr|rrrrrl}
 & & & & 3x^2 & +\ 6x & -\ 5 & \text{(quotient)} \\ \cline{3-7}
\text{(divisor)} & 2x^2 - 4x + 1 & 6x^4 & +\ 0x^3 & -\ 31x^2 & +\ 26x & -\ 6 & \text{(dividend)} \\
 & & 6x^4 & -\ 12x^3 & +\ 3x^2 & & & \text{line A} \\ \cline{3-5}
 & & & 12x^3 & -\ 34x^2 & +\ 26x & & \text{line B} \\
 & & & 12x^3 & -\ 24x^2 & +\ 6x & & \text{line C} \\ \cline{4-6}
 & & & & -\ 10x^2 & +\ 20x & -\ 6 & \text{line D} \\
 & & & & -\ 10x^2 & +\ 20x & -\ 5 & \text{line E} \\ \cline{5-7}
 & & \text{(remainder)} & & & & -\ 1 & \text{line F}
\end{array}$$

Here are certain observations on this procedure, taken in the order of their appearance.

1. The first term ($3x^2$) of the quotient is the result of dividing the first term ($2x^2$) of the divisor into the first term ($6x^4$) of the dividend.
2. Line A is the product of the *first* term of the quotient ($3x^2$) and the entire divisor. This product, for convenience, is placed term by term under similar terms in the dividend above.
3. Line B is the result of subtracting line A from like terms in the dividend and "bringing down" the next term ($26x$) of the dividend.
4. The second term ($6x$) of the quotient is the result of dividing the first term ($2x^2$) of the divisor into the first term ($12x^3$) of line B.
5. Line C is the product of the *second* term of the quotient ($6x$) and the entire divisor.
6. Line D is the result of subtracting line C from line B and "bringing down" the next term (-6) of the dividend.
7. The third term (-5) of the quotient is the result of dividing the first term ($2x^2$) of the divisor into the first term ($-10x^2$) of line D.
8. Line E is the product of the *third* term (-5) of the quotient and the entire divisor.
9. Line F is the result of subtracting line E from line D. This terminates the operation, since there are no other terms in the dividend to "bring down" and the highest degree of x in line F is less than the highest degree of x in the divisor. The result in line F is called the *remainder*.

Here is a summary statement on how *to divide one polynomial by another*.

STEP 1 *Arrange the terms in the dividend and divisor in the same (descending or ascending) order of exponents. Provide placeholders with zero coefficients for missing terms.*

STEP 2 *Divide the first term of the divisor into the first term of the dividend; this result will be the first term of the quotient.*

STEP 3 *Multiply the entire divisor by the first term of the quotient, placing the product under similar terms of the dividend.*

STEP 4 *Subtract this latest product from the dividend and bring down the next term of the dividend; then repeat steps 2 and 3, using the difference obtained as the new dividend.*

STEP 5 *Continue the process until a zero remainder is obtained or until there are no remaining terms of the dividend to bring down.*

Note: Step 5 suggests termination of the division process when there is no term of the dividend to be brought down. Observation 9 following

Example 19 suggests termination when the highest degree of x in the remainder is less than the highest degree of x in the divisor. Both comments are based on the idea that the highest degree of x in the dividend is greater than the highest degree of x in the divisor; if the reverse is true, divisibility still exists, but negative powers of x will occur. An example is the division $1 \div (1 - x)$, which is quite possible through the steps suggested and will yield a never-ending quotient comparable to the decimal value of $1 \div 3$ in arithmetic.

Exercises 2-6

Perform the divisions in Exercises 1–24.

1. $12x^3y^2 \div 3x^2y$

2. $(16x - 12y) \div 4$

3. $(ax + ay - az) \div a$

4. $(x^2 - 4y^2) \div (x - 2y)$

5. $(x^2 - 4y^2) \div (x + 2y)$

6. $(x^2 + 6xy + 9y^2) \div (x + 3y)$

7. $(x^2 - 4xy + 4y^2) \div (x - 2y)$

8. $(x^5 - 1) \div (x - 1)$

9. $(x^3 - 1) \div (x - 1)$

10. $(x^3 + 1) \div (x + 1)$

11. $(x^5 + 1) \div (x + 1)$

12. $(\pi Rr + \pi r^2) \div \pi r$

13. $(a^2b^3c - 4ab^2c^2) \div abc$

14. $(x^2 + x - 2)(x - 3) \div (x + 2)$

15. $(x - 1)(x + 2)(x - 3) \div (x - 1)$

16. $(625s^4 - 81) \div (5s - 3)$

17. $(T_2^4 - T_1^4) \div (T_2 - T_1)$

18. $(16x^4 - 1) \div (2x + 1)$

19. $(625s^4 - 81) \div (5s + 3)$

20. $(a^2 + 5ab - ac + 6b^2 - 5bc - 6c^2) \div (a + 2b - 3c)$

21. $(x^2 - 4y^2 + 12yz - 9z^2) \div (x - 2y + 3z)$

22. $(4x^4 - 12x^3 + 15x^2 - 17x + 12) \div (2x - 3)$

23. $(x - 3x^2 + 4x^4 - 12) \div (2x + 3)$

24. $(3 + 4x - 5x^2 + 3x^3) \div (3 + x - x^2)$

25. Given are four consecutive integers with the least integer N. If the three smaller numbers are multiplied by each other and this product is divided by the largest integer, show that a remainder of -6 is obtained.

26. In Exercise 25, find the remainder if the least integer is $N - 1$.

Below are expressions in which a binomial variable is carried as a unit. A monomial could be substituted for the binomial to simplify the division indicated. Find these quotients.

27. $[(a + b)^2 - (c - d)^2] \div [(a + b) - (c - d)]$

28. $[(x + 2y)^2 + 3(x + 2y) - 4] \div (x + 2y + 4)$

29. $[x^2(a - 2b)^2 + 4x(a - 2b) + 4] \div x(a - 2b) + 2$

30. $[(a + b)^3 - (c + d)^3] \div (a + b - c - d)$

2-7 Factoring algebraic products

A product is known to be the result of multiplying two or more numbers together; these numbers are regarded as *factors* of the product. The process called *factoring* assumes that an expression is a product of two or more quantities and is the process of determining these quantities or factors. Clearly multiplication and factoring are inverse processes. The *prime factors* of a number include all the factors of the number except for the number 1 and the number itself. Thus the prime factors of 6 are 2 and 3, since $2 \times 3 = 6$ and we have avoided the choice $1 \times 6 = 6$.

Some numbers have no prime factors; they are called *prime numbers.* The numbers 2, 5, 7, 13 are examples of prime numbers. Other numbers have several prime factors: We note that 24 factors into 6 and 4, but 6 is factorable into 3 and 2, and 4 is factorable into 2 and 2, suggesting that the *prime factors* of 24 are 3, 2, 2, 2.

Factors of a monomial An algebraic monomial is readily factorable by factoring the numeric coefficient and then factoring the literal portion, using the definition of powers or exponents. Note that

$$18x^3y^2 = 3 \cdot 3 \cdot 2 \cdot x \cdot x \cdot x \cdot y \cdot y.$$

Polynomials with common factors The distributive law states that

$$a(b + c + d) = ab + ac + ad. \tag{18}$$

We can regard the right side of (18) as a product and the left side as the factors of that product. The key to the factorability of the right side of (18) is the presence of the factor a in each term, making a the *factor common to each term* or a *common factor*. We can remove the common factor by dividing each term by this common factor; the resulting quotient $(b + c + d)$ is the other factor.

Example 20 Factor $3x^2y - 6xy^2 - 3xy$.

Solution By inspection, we see that all terms of the given expression contain the quantity $3xy$ as a factor, making $3xy$ the *common* factor of the terms. This fact may be seen more clearly if each term is prime factored as

$$\begin{aligned} 3x^2y - 6xy^2 - 3xy = {} & \underline{3} \cdot \underline{x} \cdot x \cdot \underline{y} - 2 \cdot \underline{3} \cdot \underline{x} \cdot \underline{y} \cdot y \\ & - \underline{3} \cdot \underline{x} \cdot \underline{y}, \end{aligned} \tag{19}$$

where the common factor $3xy$ is underlined in each term to display commonality. To determine the other factor, we divide each term of (19) by $3xy$, obtaining the quotient $(x - 2y - 1)$, and we conclude that

$$3x^2y - 6xy^2 - 3xy = 3xy(x - 2y - 1).$$

Note that the factoring process is, in effect, a *division* process; this justifies the -1 as the third term of the second factor.

Expressions that are the product of two binomials At the end of Sec. 2-5 we listed several special products. Four of them, written "backward," are

$$a^2 - b^2 = (a + b)(a - b). \tag{20}$$

$$a^2 + 2ab + b^2 = (a + b)(a + b). \tag{21}$$

$$a^2 - 2ab + b^2 = (a - b)(a - b). \tag{22}$$

$$acx^2 + (ad + bc)x + bd = (ax + b)(cx + d). \tag{23}$$

In (20) we have a product $a^2 - b^2$ that is commonly classified as the *difference of two squares*, and we note that its factors are the *sum* of the *square roots* multiplied by the *difference* of the *square roots*.

Example 21 Factor $25m^2 - 16p^4$.

Solution Inspection shows that the first term is a square—that is, $25m^2$ is the square of $5m$—and the second term is also a square where $16p^4$ is the square of $4p^2$; so the given expression is the difference of two squares and by applying (20),

$$25m^2 - 16p^4 = (5m)^2 - (4p^2)^2 = (5m + 4p^2)(5m - 4p^2).$$

In (21) we have a product $a^2 + 2ab + b^2$ that is classified as a trinomial with first and last terms being the *sum of two squares* and the *middle term*, which is *twice the product of the square roots of the two end terms*. The factors of (21) are identical $(a + b)$ and $(a + b)$, meaning that $a^2 + 2ab + b^2$ is the *perfect square* of the sum of a and b.

Example 22 Factor $9x^2 + 12xyz^2 + 4y^2z^4$.

Solution Inspection shows that the two end terms are perfect squares—that is, $9x^2 = (3x)^2$ and $4y^2z^4 = (2yz^2)^2$. Also, the center term is exactly twice the product of the square roots of the end terms—that is, $12xyz^2 = (2)(3x)(2yz^2)$—and so by comparison to (21),

$$9x^2 + 12xyz^2 + 4y^2z^4 = (3x + 2yz^2)(3x + 2yz^2) = (3x + 2yz^2)^2.$$

Note that (21) and (22) are identical forms except that (21) is the square of a *sum*, whereas (22) is the square of a *difference*.

Expression (23) is a frequently occurring product that requires special attention. We refer to the product in Example 16 (Sec. 2-5), where $6x^2 - xy - 12y^2$ is shown to be the product of factors $(2x - 3y)$ and $(3x + 4y)$ or

$$6x^2 - xy - 12y^2 = (2x - 3)(3xy + 4y). \qquad (24)$$

Comparing (23) and (24), we have

$$acx^2 + (ad + bc)x + bd = (ax + b)(cx + d)$$
$$6x^2 - 1xy - 12y^2 = (2x - 3y)(3x + 4y).$$

Looking at the left members, we have $ac = 6$, $ad + bc = -1y$, and $bd = -12y^2$. We are left with a tortuous demand: "Give me a pair of numbers (a and c) such that $ac = 6$ and *another* pair (b and d) such that $bd = -12y^2$ and the *two pairs* satisfy the condition $ad + bc = -1y$." The only numbers that satisfy these conditions are $a = 2$, $b = -3y$, $c = 3$, and $d = +4y$, giving the right side of (24) as the factors of the left side. See Example 23 for an orderly approach to this type of problem.

Example 23 Factor $3x^2 + 5x - 2$. (25)

Solution Expression (25) is comparable to (23), suggesting that

$$3x^2 + 5x - 2 = (ax + b)(cx + d),$$

where now $ac = 3$ and $bd = -2$. These values of ac and bd identify the possible values of a, b, c, and d and the factors resulting from these possibilities. A complete listing follows.

	a	b	c	d	$ad + bc$
$(3x + 2)(x - 1)$	3	2	1	-1	-1
$(3x - 2)(x + 1)$	3	-2	1	1	$+1$
$(3x + 1)(x - 2)$	3	1	1	-2	-5
$(3x - 1)(x + 2)$	3	-1	1	2	$+5$

Since all four choices satisfy conditions $ac = 3$ and $bd = -2$ but *only* the *last* choice satisfies the third condition that $ad + bc = 5$, then the factors of (24) are

$$3x^2 + 5x - 2 = (3x - 1)(x + 2).$$

Example 24 Factor $x^2 - 2x - 3$. (26)

Solution Expression (26) is a simplified version of (23) with $ac = 1$, $bd = -3$, and $ad + bc = -2$. The choices are reduced to either

$(x - 1)(x + 3)$ or $(x + 1)(x - 3)$, and we choose $(x + 1)(x - 3)$ because the middle term must be -2 and $(1)(1) + (-3)(1) = -2$.

Comment: The process of factoring a trinomial like (23) involves some guesswork. Clearly three conditions must be satisfied in making our choices. Special techniques, such as one that bears the acronym FOIL, provide alternate explanations. Also, some trinomials like $x^2 + 2x + 7$ are not factorable with real or rational factors. A test for factorability of the form under discussion involves the discriminant of a quadratic that is treated in a later chapter. Furthermore, the method in Examples 22 and 23 is a process that can be used for all the forms (20) to (23).

The sum and differences of two cubes Two special products given at the end of Sec. 2-5 involve cubes. As products classified for factoring, we have

$$x^3 + y^3 = (x + y)(x^2 - xy + y^2) \tag{27}$$

$$x^3 - y^3 = (x - y)(x^2 + xy + y^2). \tag{28}$$

Example 25 Factor $8m^3 - 27p^6$. (29)

Solution We observe that (29) is the difference of two cubes with $8m^3 = (2m)^3$ and $27p^6 = (3p^2)^3$; this is comparable with (28) with $x = 2m$ and $y = 3p^2$. By substitution, we have

$$\begin{aligned} 8m^3 - 27p^6 &= (2m - 3p^2)[(2m)^2 + (2m)(-3p^2) + (-3p^2)^2] \\ &= (2m - 3p^2)(4m^2 - 6mp^2 + 9p^4). \end{aligned}$$

Expressions with several factors Frequently the factoring process yields factors that are not prime—that is, they are further factorable. In these cases, we naturally continue the factoring process until prime factors are reached. In all cases of factoring, it is best to remove the common factors first.

Example 26 Factor $a^2x^4 - 16a^2y^4$.

Solution The factor a^2 is common to both terms and

$$a^2x^4 - 16a^2y^4 = a^2(x^4 - 16y^4).$$

Now we observe that the binomial factor above is the difference of two squares and

$$a^2(x^4 - 16y^4) = a^2(x^2 - 4y^2)(x^2 + 4y^2).$$

Again, one of the factors above—$x^2 - 4y^2$—is the difference of two squares and

$$a^2(x^2 - 4y^2)(x^2 + 4y^2) = a^2(x - 2y)(x + 2y)(x^2 + 4y^2).$$

Exercises 2-7

Factor the given expressions in Exercises 1–50. Obtain the simplest factors in each case.

1. $\pi r^2 + 2\pi r$

2. $a^2b^2 - 2abc$

3. $-a^2x^2 + a^2$

4. $Bh + 4Mh + bh$

5. $nx + x - fnx$

6. $12x^2 + 6xy - 18x$

7. $(a + b)x - (a + b)y$

8. $a^2 - b^2$

9. $a^2 - 16b^2$

10. $64b^2 - 81y^2$

11. $(x + y)^2 - x^2$

12. $16z^8 - 1$

13. $x^2 + 5x + 6$

14. $x^2 + 7x + 6$

15. $x^2 - x - 6$

16. $x^2 + x - 6$

17. $x^2 - 5x - 6$

18. $x^2 + 5x - 6$

19. $m^2 - 10m + 25$

20. $s^2 - 12s + 36$

21. $a^3 - b^3$

22. $8x^3 + 27y^3$

23. $(a + b)^3 - (a - b)^3$

24. $0.001m^6 - 64n^{12}$

25. $m^6n^3 - m^3n^6$

26. $(a + b)^2 - 5(a + b) - 24$

27. $12x^2 + x - 20$

28. $12x^2 - 239x - 20$

29. $12x^2 + 14x - 20$

30. $81m^2 - 90mn + 25n^2$

31. $5s^2 - 20s + 20$

32. $64x^4 - 16$

33. $x^3 - x$

34. $x^3 + 2x^2 + x$

35. $4m^2 - 100$

36. $(30)^2 - (29)^2$

37. $(\frac{5}{6})^2 - (\frac{1}{6})^2$

38. $121y^2 - 88y + 16$

39. $12a^2b^2 - 24a^2b - 36a^2$

40. $(a + b)^2 - (a - b)^2$

41. $(r + s)(a + b) - t(a + b)$

42. $4a^2 + a + 2a^3$

43. $n + n(n - 1) + n(n - 1)(n - 2)$

44. $(r + s)^2 - 2a(r + s) - 3a^2$

45. $s^2z^2 - 6sz(s + z) + 9(s + z)^2$

46. $3r^4 - 2r^2 - 40$

47. $3m^4 - 10m^2n^2 + 3n^4$

48. $4a^2 - 8 + 6a - 3a^2 - 8a$

49. $-20a - 4 - 16a^2$

50. $m - mn$

It is recognized in squaring a binomial that the result is a trinomial whose first and last terms are squares of the first and last terms, respectively, of the binomial and whose middle term is twice the product of the two terms of the binomial. Using this information, provide the missing terms of the following expressions, which are given as perfect squares.

51. $x^2 - 2x + (\ \ )$

52. $m^2 - (\ \ ) + 16$

53. $s^2 + (\ \ ) + 49t^2$

54. $a^2 + 16ab + (\ \ )$

55. $4k^2 + (\ \ ) + 9h^2$

56. $25m^2n^2 - (\ \ ) + 36b^4$

57. $(s + t)^2 - (\quad) + 4y^2$

58. $x^2 - 5x + (\quad)$

59. $4x^2 - 7x + (\quad)$

60. $r^2 - (\quad) + \dfrac{t^2}{4}$

2-8 Simplification, multiplication, and division of fractions

Algebraic fractions are simplified (reduced to lowest terms) by procedures similar to those used in arithmetic. In all cases, the Fundamental Principle of Fractions is employed.

FUNDAMENTAL PRINCIPLE OF FRACTIONS *If both the numerator and denominator of a fraction are multiplied or divided by the same nonzero quantity, the value of the fraction is unchanged.*

Simplification of fractions Fractions may be simplified (reduced to lowest terms) by *dividing* the numerator and denominator (application of the fundamental principle of fractions) by a quantity that will divide without remainder into each. This divisor must obviously be a *factor* of both the numerator and denominator, thus suggesting the following procedure.

To simplify a fraction, factor the numerator and denominator into their prime factors and then apply the Fundamental Principle of Fractions by dividing the numerator and denominator by any factors common to both.

Example 27 Simplify.

(a) $\dfrac{6}{10}$ (b) $\dfrac{24x^2y}{36xy^2}$ (c) $\dfrac{x^2 - 25}{x^2 - 3x - 10}$

Solution In each case, factor the numerator and denominator into prime factors and then divide both the numerator and denominator by factors common to both.

(a) $\dfrac{6}{10} = \dfrac{3 \cdot \cancel{2}}{5 \cdot \cancel{2}} = \dfrac{3}{5}$ (common factor was 2).

(b) $\dfrac{24x^2y}{36xy^2} = \dfrac{\cancel{2} \cdot \cancel{2} \cdot 2 \cdot \cancel{3} \cdot \cancel{x} \cdot x \cdot \cancel{y}}{\cancel{2} \cdot \cancel{2} \cdot 3 \cdot \cancel{3} \cdot \cancel{x} \cdot \cancel{y} \cdot y} = \dfrac{2x}{3y}$ (common factor was $2 \cdot 2 \cdot 3 \cdot x \cdot y$ or $12xy$).

(c) $\dfrac{x^2 - 25}{x^2 - 3x - 10} = \dfrac{\cancel{(x - 5)}(x + 5)}{\cancel{(x - 5)}(x + 2)} = \dfrac{x + 5}{x + 2}$ (common factor was $x - 5$).

Comment: In Example 27 slash marks are used to denote common factors of the numerator and denominator that were "canceled." The term *cancel* is commonly used, but one meaning of the term suggests that usage in this context leads to misconceptions. Cancel means "void" or "nullify," suggesting a zero result, which is *not* the case here. Actually, our "cancellation" was an application of the multiplicative inverse (Sec. 1-3), bringing about results of 1 (one) rather than 0 (zero).

Multiplication of fractions The product of two or more fractions is a fraction; the numerator of the product is the product of the numerators of the separate fractions, and the denominator of the product is the product of the denominators of the separate fractions, or

$$\frac{a}{b} \cdot \frac{c}{d} \cdot \frac{e}{f} = \frac{a \cdot c \cdot e}{b \cdot d \cdot f}.$$

This property of fractions suggests the following procedure for the multiplication of fractions:

To multiply two or more fractions, express the numerator of the product as the product of the separate numerators, express the denominator of the product as the product of the separate denominators, and then simplify the result if possible.

Example 28 Multiply $\dfrac{x^2 - 9}{6x^2}$ by $\dfrac{3x}{x^2 - x - 6}$.

Solution First we show the product as a single fraction:

$$\frac{x^2 - 9}{6x^2} \cdot \frac{3x}{x^2 - x - 6} = \frac{(x^2 - 9)(3x)}{(6x^2)(x^2 - x - 6)}.$$

Factoring the resulting product and simplifying, we have

$$\frac{\cancel{(x - 3)}(x + 3)\cancel{(3)}\cancel{(x)}}{\cancel{(3)}(x)(2)\cancel{(x)}\cancel{(x - 3)}(x + 2)} = \frac{x + 3}{2x(x + 2)}.$$

Division of fractions Following the procedure of division of fractions as used in arithmetic, we have

To divide one fraction by another, invert the divisor and proceed as in multiplication.

Example 29 Perform $\dfrac{x^2 - 4}{x^2 - 2x - 6} \div \dfrac{x^2 + 5x + 6}{x^2 - 9}$.

Solution Inverting the divisor, factoring, and simplifying, we have

$$\frac{(x-2)(x+2)}{(x-3)(x+2)} \cdot \frac{(x-3)(x+3)}{(x+2)(x+3)} = \frac{x-2}{x+2}.$$

Exercises 2-8

In Exercises 1–12, reduce the given expressions to lowest terms.

1. $\dfrac{12xy}{16x^2y^3}$

2. $\dfrac{2x+2y}{4x^2-4y^2}$

3. $\dfrac{2x+2y}{4x^2-16y^2}$

4. $\dfrac{a^2-4ab+4b^2}{a^2-4b^2}$

5. $\dfrac{m^2-3m-4}{m^2-2m-8}$

6. $\dfrac{3+2r-r^2}{r^2+r-12}$

7. $\dfrac{r^3-s^3}{r^2-s^2}$

8. $\dfrac{m^3+m^2n-6mn^2}{m^2n-2n^2m}$

9. $\dfrac{(m^2+mn+n^2)(x+y)}{m^3-n^3}$

10. $\dfrac{(a-2b)^3}{a^2b-4ab^2+4b^3}$

11. $\dfrac{a-a^2}{a-1}$

12. $\dfrac{(m+n)^2-5(m+n)+6}{(m+n)^2+3(m+n)-10}$

Perform the indicated operations, reducing the solutions to the simplest form.

13. $\dfrac{4a}{x^3} \cdot \dfrac{12bx^2}{16a^2}$

14. $\dfrac{x^3}{4ay^2} \cdot \dfrac{20a^2y}{5xy^2}$

15. $\dfrac{m-n}{mn} \cdot \dfrac{m^2n}{m-n}$

16. $\dfrac{m^2-n^2}{m} \cdot \dfrac{m^2}{m^2-mn}$

17. $\dfrac{6R-18}{5} \cdot \dfrac{10R}{2R-6}$

18. $\dfrac{2x+y}{4x^2-y^2} \cdot (8x^3-y^3)$

19. $\dfrac{a^2-b^2}{a+b} \cdot \dfrac{a^2+ab+b^2}{a^3-b^3}$

20. $\dfrac{a^2-2a-3}{a^2-9} \cdot \dfrac{a^2+6a+9}{a^2-a}$

21. $\dfrac{a^3-ax^2}{a^2+ax} \cdot \dfrac{ax^2-x^3}{x^2}$

22. $\dfrac{R^2+3R-10}{R^2-6R-7} \cdot \dfrac{R-7}{R+5}$

23. $\dfrac{x^2-5x-6}{x^2-6x} \cdot \dfrac{x^2+3x-10}{x^2-x-2}$

24. $\dfrac{2x^2-x-1}{3x^2+7x+2} \cdot \dfrac{x^2+7x+10}{2x^2+5x+2} \cdot \dfrac{3x^2+x}{x^2+4x-5}$

25. $\dfrac{x^2+7x+12}{x^2+2x-15} \cdot \dfrac{x+5}{x+4}$

26. $\dfrac{8-2x-x^2}{x^2+5x+4} \cdot \dfrac{2x^3+2x^2}{2x-x^2}$

27. $\dfrac{3+2x-x^2}{x^2+x-2} \cdot \dfrac{1-x^2}{x^2-3x}$

28. $\left(\dfrac{a^2+ab}{b}\right)\left(\dfrac{a}{b^2+ab}\right)$

29. $\dfrac{h^2-k^2}{h^2k} \cdot \dfrac{hk^2}{h+k}$

30. $\dfrac{ax+bx^2}{(c+x)^2} \cdot \dfrac{c+x}{x^3} \cdot \dfrac{cx+x^2}{a+bx}$

31. $\left(\dfrac{m-1}{m^2-4mn+4n^2}\right)\left(\dfrac{2n-m}{n}\right)\left(\dfrac{m-2n}{n}\right)$

32. $\dfrac{3abc}{5d} \div \dfrac{2d}{15a^2b^2}$

33. $\dfrac{8xy^2}{6z} \div \dfrac{2x^3y^3}{3x^2z^2}$

34. $\dfrac{x^2-y^2}{2x+3y} \div \dfrac{x-y}{4x+6y}$

35. $\left(\dfrac{4Rt^2}{9mp} \cdot \dfrac{18m^2}{7R^2}\right) \div \left(\dfrac{Rt^2}{14m^2p^2}\right)$

36. $\left(\dfrac{a-b}{a}\right)\left(\dfrac{b}{b-a}\right)\left(\dfrac{a^2}{b}\right)$

37. $\dfrac{15-2h-h^2}{6h^2-h-1} \div \dfrac{3h^2+16h+5}{4h^2-1}$

38. $\dfrac{k^4-kh^3}{4h^2-k^2} \div \dfrac{k^4-k^2h^2}{k+2h}$

39. $\dfrac{2a^2-6a}{10a^2-a-2} \div \left(\dfrac{-a^2+6a-9}{5a^2+2a} \cdot \dfrac{2a^3+a^2}{5a^4-2a^3}\right)$

40. $\left[(c^4-16d^4) \div \dfrac{c^2-d^2}{cd}\right] \div c^2(c^2+d^2)$

2-9 Addition and subtraction of fractions

In Sec. 2-2 we noted that "only like terms can be added or subtracted." The "likeness" of fractions is determined by their denominators, for denominators "give name" to fractions. Numerators "enumerate" or "tell how many" of the items named by the denominators are present. If the denominators are alike, we can add the numerators over the common denominators. If the denominators are different, we can make them alike by inserting well-chosen factors into the denominators where needed; according to the Fundamental Principle of Fractions, however, it is necessary to insert the same factors in the corresponding numerators so that the values of the fractions remain unchanged.

Example 30 Add $\dfrac{3}{8}+\dfrac{7}{12}+\dfrac{2}{3}$. (30)

Solution The separate fractions in (30) have different denominators (or names), 8, 12, and 3, and cannot be added as they stand. If the denominators were the same, the numerators could be added over the common denominator. Factoring the denominators, (30) becomes

$$\frac{3}{2\cdot 2\cdot 2}+\frac{7}{2\cdot 2\cdot 3}+\frac{2}{3}. \tag{31}$$

Careful inspection of the denominators of (31) shows that we can *insert* well-chosen factors into each denominator and achieve the goal of making the three denominators alike, observing, however, that we must insert the same factors in the respective numerators so that, under the Fundamental Principle of Fractions, the values of the fractions remain unchanged. It is easy to see that the common denominator $2 \cdot 2 \cdot 2 \cdot 3$ is our goal; so we rewrite (31) as (32), where the inserted factors are displayed in parentheses:

$$\frac{3 \cdot (3)}{2 \cdot 2 \cdot 2 \cdot (3)} + \frac{7 \cdot (2)}{2 \cdot 2 \cdot 3 \cdot (2)} + \frac{2 \cdot (2) \cdot (2) \cdot (2)}{3 \cdot (2) \cdot (2) \cdot (2)}. \tag{32}$$

Next, (32) can be rewritten and added as

$$\frac{9}{24} + \frac{14}{24} + \frac{16}{24} = \frac{9 + 14 + 16}{24} = \frac{39}{24} = 1\frac{15}{24},$$

where the numerators are added over the common denominator. The denominator in (32) is called the *least common denominator* (LCD) and is the *least common multiple* (LCM) or smallest number into which the original denominators 8, 12, and 3 will divide without remainder.

The preceding commentary and Example 30 can be summarized in the following procedure statement. *To combine fractions under addition or subtraction, insert factors into the separate denominators with the goal of making the denominators alike; then insert these same factors into the corresponding numerators under the Fundamental Principle of Fractions and combine the numerators over the common denominator.*

Example 31 Perform the indicated operations and simplify:

$$\frac{x}{x + 1} - \frac{3}{x} + \frac{2x}{x - 1}. \tag{33}$$

Solution Inspection of (33) shows three fractions to be combined under subtraction and addition and also that the three denominators are different. By inserting factors for the purpose of making the denominators alike, (33) becomes

$$\frac{(x)[x][x - 1]}{(x + 1)[x][x - 1]} - \frac{3[x + 1][x - 1]}{(x)[x + 1][x - 1]} + \frac{(2x)[x][x + 1]}{(x - 1)[x][x + 1]}. \tag{34}$$

Note in (34) that the inserted factors are shown in brackets; these are also inserted into the respective numerators per the Fundamental Principle of Fractions. Next, the numerators of (34) can be combined over the common denominator

$$\frac{x(x)(x-1) - 3(x+1)(x-1) + (2x)(x)(x+1)}{x(x+1)(x-1)} \tag{35}$$

and (35) simplifies to

$$\frac{x^3 - x^2 - 3x^2 + 3 + 2x^3 + 2x^2}{x^3 - x} = \frac{3x^3 - 2x^2 + 3}{x^3 - x}.$$

2-10 Complex fractions

A *complex fraction* is defined here as one that contains a fraction in the numerator, denominator, or both. Complex fractions frequently originate in algebraic manipulations and must usually be simplified for clarity. Two methods of simplifying such fractions are discussed through examples.

Example 32 Simplify the fraction

$$\frac{1 + \frac{1}{y}}{1 - \frac{1}{x}}. \tag{36}$$

Solution 1 By the preceding definition, (36) is a complex fraction. We regard $1 + \frac{1}{y}$ as the *primary numerator* and $1 - \frac{1}{x}$ as the *primary denominator*. The *secondary denominators* present are x and y. It is apparent that by multiplying the primary numerator and denominator by $x \cdot y$, the secondary denominators vanish and the resulting fraction is no longer complex. Thus

$$\frac{\left(1 + \frac{1}{y}\right)(xy)}{\left(1 - \frac{1}{x}\right)(xy)} = \frac{xy + x}{xy - y}.$$

Solution 2 We can regard (36) as an exercise in division, rewriting it as

$$\left(1 + \frac{1}{y}\right) \div \left(1 - \frac{1}{x}\right). \tag{37}$$

Writing the dividend and divisor as single fractions, (37) becomes

$$\frac{y+1}{y} \div \frac{x-1}{x}.$$

Inverting the divisor and proceeding as in multiplication, we have

$$\frac{y+1}{y} \cdot \frac{x}{x-1} = \frac{x(y+1)}{y(x-1)} = \frac{xy + x}{xy - y}.$$

Neither method of solving Example 32 is championed here. Both are quite acceptable and both apply the Fundamental Principle of Fractions.

Exercises 2-9 and 2-10

Find the least quantity into which the given quantities in Exercises 1–5 will divide evenly; that is, find the least common multiple of the given numbers.

1. a^2b, ab^2, a^3b
2. $2, 4, 8, 18$
3. $x + y, x^2 - y^2, x^2 + 2xy + y^2$
4. $x^4 - x^2, x^3 - x, x^2 - 2x + 1, x^3$
5. $a^2 - b^2, b - a, 2a + 2b, ab$

In Exercises 6–27, simplify the given expressions to single fractions.

6. $\frac{a}{4} - \frac{3a}{8} + \frac{5a}{2}$
7. $\frac{3}{2a} - \frac{4}{5a}$
8. $\frac{x}{2b} - \frac{3x}{4b}$
9. $\frac{\pi r^2}{4} - \frac{4\pi r^3}{3}$
10. $\frac{2x^2 + 1}{3x^2} + \frac{x - 4}{6x} - \frac{5}{12}$
11. $\frac{x - 4y}{3} - \frac{2x + 3y}{5}$
12. $\frac{1}{r_1 + r_2} - \frac{1}{r_1 - r_2}$
13. $3 - \frac{4}{a^2}$
14. $x + y - \frac{x}{4y}$
15. $\frac{x + y}{x - y} - \frac{x - y}{x + y}$
16. $\frac{5}{a^2 - b^2} - \frac{4}{a - b}$
17. $\frac{a + b}{x - y} - \frac{a - b}{x + y}$
18. $\frac{x + 4y}{x^2 - 16y^2} - \frac{5}{4y - x}$
19. $r - 2 + \frac{1}{r}$
20. $2m^2 + 6m + 6 + \frac{2}{m - 1}$
21. $\frac{x + 1}{x^2 - 2x + 1} + \frac{x - 2}{x^2 - x} - \frac{2 - x}{x^2 - 5x + 6}$
22. $\frac{m + p}{(m - n)(m - p)} - \frac{p}{(n - m)(m - p)} + \frac{m}{(p - m)(m - n)}$
23. $\frac{1}{a + b} - \frac{2a}{(a + b)^2} - \frac{a - b}{(a + b)^2}$
24. $\frac{x + 4}{x + 3} + \frac{x - 3}{x - 4} - \frac{4}{3}$
25. $\frac{c(t_1 - t_2)}{t_3 - t_2} - 1$
26. $\frac{x - b}{b} - \frac{1 + x - a}{a}$
27. $\frac{m + n}{m - n} - \frac{n - m}{m + n} + \frac{3n}{n - m}$

Simplify the given complex fractions.

28. $\dfrac{\frac{a}{b}}{a}$
29. $\dfrac{a}{\frac{b}{a}}$
30. $\dfrac{a}{\frac{1}{b}}$
31. $\dfrac{\frac{a}{b}}{\frac{1}{b}}$

32. $\dfrac{-\dfrac{1}{a}}{\dfrac{1}{ab}}$ **33.** $\dfrac{\dfrac{1}{x}}{\dfrac{1}{x-y}}$ **34.** $\dfrac{\dfrac{1}{x}-\dfrac{1}{y}}{\dfrac{1}{xy}}$ **35.** $\dfrac{\dfrac{1}{x}-\dfrac{1}{y}}{\dfrac{1}{x^2}-\dfrac{1}{y^2}}$

36. $\dfrac{\dfrac{1}{x}-\dfrac{1}{y}}{\dfrac{1}{x^2y^2}}$ **37.** $\dfrac{\dfrac{m}{m-n}}{\dfrac{m+n}{n}}$ **38.** $\dfrac{\dfrac{m}{m+n}}{\dfrac{n}{m-n}}$

39. $\dfrac{\dfrac{a^2}{b^2}-2\dfrac{a}{b}-3}{\dfrac{a-3b}{ab}}$ **40.** $\dfrac{\dfrac{x^2}{y^2}-3\dfrac{x}{y}-4}{\dfrac{x+y}{y}}$ **41.** $\dfrac{3(x^2+2x-15)}{\dfrac{x-3}{2}}$

42. $\dfrac{r_1}{\dfrac{r_1}{r_2+1}}$ **43.** $1-\dfrac{1}{1-\dfrac{1}{1-\dfrac{1}{4}}}$

44. $\dfrac{1+\dfrac{3}{x}-\dfrac{1}{x^2}-\dfrac{3}{x^3}}{x^2+2x-3}$ **45.** $\dfrac{1-\dfrac{3}{2}}{-1-\dfrac{1}{2-\dfrac{5}{2}}}$

46. $2-\dfrac{2}{2-\dfrac{2}{2-\dfrac{1}{2}}}$ **47.** $\dfrac{1+\dfrac{3}{x}+\dfrac{3}{x^2}+\dfrac{1}{x^3}}{1+\dfrac{2}{x}+\dfrac{1}{x^2}}$

3

Physical units

An important part of the solution of a physical problem is the selection of compatible dimensional units for the various physical quantities represented in the equation to be solved.

The technician and engineer should be well acquainted with dimensional units, their conversion, and the procedure of inspecting equations for compatibility of units.

With the United States in the process of joining other nations in the adoption of SI (Système International) units, it is important that textbooks dealing with dimensional analysis help pave the way during the transition period. This chapter includes introductory material on SI units.

During the transition period, however, there will be much need for technicians and engineers to convert various data from the old English and metric units to the new SI units.

3-1 Introduction

The application of mathematics to a physical problem requires more than the correct manipulation of numbers and equations. Physical problems involve quantities expressed in terms of numbers and units, such as 10 ft, 6 kw-hr, and 10^{-6} N•s/m^2. A correct numerical procedure is valueless if an incorrect combination of units is used in an equation relating physical quantities.

The last quantity given above is expressed in SI unit symbols and represents newton-seconds per square meter. SI symbols are not ordinary abbreviations and therefore do not include periods. A centered dot, as in N•s, is used to indicate a product of two units.

Although the number of basic dimensions is not large, consisting only of length, mass, time, and a few others, several different units or combinations of units can be used for the same type of dimension. It is usually necessary for the two sides of an equation to have a dimensional equality, such as 5 sec + 2 sec = 7 sec. However, it is obviously incorrect to state that 5 sec + 2 min = 7 sec, even though both sides have the dimension of time. An equality, or balance, of units is necessary. Consequently, some of the known quantities in a physical relation must often be converted to other units.

With the adoption of SI (*Système International*) units in the United States, we are faced not only with the changeover to SI units but also with the adoption of SI rules for notation of units. These rules are necessary for international consistency. In this chapter some of the more commonly used SI units and the associated rules of notation are presented.

Our purpose is to describe units and procedures in order to assist the reader in the following steps:

1. Converting units.
2. Checking equations for balance of units.
3. Creating unit equations to help solve problems.
4. Becoming acquainted with SI units and rules for notation.

3-2 Converting simple quantities

The ability to convert a quantity of given units to an equivalent value in terms of other related units is often required, especially in technical and scientific work. Although various books contain tables of conversion factors, an enormous set of tables would be needed to include all possible conversions. Numerous conversions can be made quite conveniently, however, with the aid of only a relatively few basic unit relations.

In time the eventual exclusive use of SI units will greatly reduce the need for converting units. During the transition period, however, engineers and technicians will be concerned with the problem of converting various data and specifications from old unit systems to the new one.

The material in this section is directed toward problems involving some largely familiar units of both the English and old metric systems. In Sec. 3-3 the SI units and related rules of notation are discussed.

Simple quantities expressed in terms of one unit are easily converted to an equivalent value in terms of another unit if the relation between the two units is known. A length of 3 ft is readily converted to the equivalent value of 36 in. if it is known that there are 12 inches per foot. However, many quantities are expressed in terms of complex combinations of units, and errors are more likely to occur when converting the units of such a quantity.

In order to avoid errors, a methodical procedure for converting units is not only desirable but almost necessary, for many units and several unit systems have been in common technical use in the United States. Some of the more common units of these systems are presented later in this chapter. In addition, within a given system of units, various combinations of units can often be used for the same concept or type of quantity. Energy units commonly used in the old English system, for instance, include the foot-pound, British thermal unit, and horsepower-minute.

The examples that follow demonstrate a methodical procedure for converting units. Although quite elementary, they serve as steppingstones to more complex problems in which the same simple procedure may be used. Familiarity with the methodical procedure is emphasized.

In each example the problem is to convert a quantity of given units to its equivalent value in terms of some other related units. The process involves multiplying the original number and units by various ratios of related units until the desired value is obtained. The units are treated as algebraic quantities and are canceled out until the desired units are obtained.

Example 1 Convert a length of 3 ft to its equivalent in centimeters. There are 2.54 centimeters per inch.

Solution

$$(3\text{ ft})\left(12\,\frac{\text{in.}}{\text{ft}}\right)\left(2.54\,\frac{\text{cm}}{\text{in.}}\right) = 91.4\text{ cm}$$

If the units in the equation are treated as algebraic quantities, the foot and inch units can be canceled.

$$(\cancel{\text{ft}})\left(\frac{\cancel{\text{in.}}}{\cancel{\text{ft}}}\right)\left(\frac{\text{cm}}{\cancel{\text{in.}}}\right) = \text{cm}$$

A balanced equation of units is obtained, cm = cm. To distinguish the word *in* from the abbreviation for inch (in.), a period is used.

In each example a unit in the second enclosure serves to cancel a unit in the first. Some conversions require a sequence of several enclosures, but the units of each are selected so that at least one undesired preceding unit is canceled. It is suggested that a line be drawn through each unit that may be canceled. These lines do not remove or annihilate the units but only serve to aid in the search for a balance of units. The ratio of units used in an enclosure will depend on which ratios are conveniently remembered or available. We can often use different sequences and ratios to achieve the same conversion.

Example 2 Convert a force of 0.25 ton to its equivalent in newtons. There are 4.45 newtons of force per pound.

Solution

$$(0.25\ \text{ton})\left(2000\,\frac{\text{lb}}{\text{ton}}\right)\left(4.45\,\frac{\text{newtons}}{\text{lb}}\right) = 2225\ \text{newtons}$$

$$(\cancel{\text{ton}})\left(\frac{\cancel{\text{lb}}}{\cancel{\text{ton}}}\right)\left(\frac{\text{newtons}}{\cancel{\text{lb}}}\right) = \text{newtons}$$

The reciprocal of a known ratio of units can be used. For instance, the ratio of 60 sec/min can be inverted to read 1 min/60 sec.

Example 3 Convert a speed of 100 ft/min to its equivalent in centimeters per second.

Solution

$$\left(100\,\frac{\text{ft}}{\text{min}}\right)\left(12\,\frac{\text{in.}}{\text{ft}}\right)\left(2.54\,\frac{\text{cm}}{\text{in.}}\right)\left(\frac{1}{60}\,\frac{\text{min}}{\text{sec}}\right) = 50.8\,\frac{\text{cm}}{\text{sec}}$$

$$\left(\frac{\cancel{\text{ft}}}{\cancel{\text{min}}}\right)\left(\frac{\cancel{\text{in.}}}{\cancel{\text{ft}}}\right)\left(\frac{\text{cm}}{\cancel{\text{in.}}}\right)\left(\frac{\cancel{\text{min}}}{\text{sec}}\right) = \frac{\text{cm}}{\text{sec}}$$

Example 4 An engine consumes fuel at the rate of 0.5 gal/min. Convert this rate to tons per day. The fuel weighs 6.5 lb/gal and the engine runs 8 hr/day.

Solution

$$\left(\frac{\text{gal}}{\text{min}}\right)\left(\frac{\text{lb}}{\text{gal}}\right)\left(\frac{\text{tons}}{\text{lb}}\right)\left(\frac{\text{min}}{\text{hr}}\right)\left(\frac{\text{hr}}{\text{day}}\right) = \frac{\text{tons}}{\text{day}}$$

$$(0.5)(6.5)\left(\frac{1}{2000}\right)(60)\,(8) = 0.78$$

Exercises 3-2

For each of the conversions indicated, write a sequence of units and numbers as in the preceding examples.

1. A machine processes a material at the rate of 5 oz/sec. Convert to tons per hour.
2. A section of pipe weighs 12 oz per inch of length. Convert to newtons per meter. There are 4.45 newtons/lb.
3. An electric current of 3.2 amperes (A) can be expressed as 3.2 coulombs per second. Convert to electrons per second. There is $1.6(10)^{-19}$ coulomb of electric charge per electron.
4. Find the equivalent of 10,000 Btu/hr in horsepower. Use the relations 778 ft-lb/Btu and 33,000 ft-lb/(hp-min).
5. A pumping station consumes 200 kw-hr of energy each day, working 8 hr/day. Find the rate of energy consumption (power) in foot-pounds per minute. There are 778 ft-lb/Btu and 3413 Btu/kw-hr.
6. Avogadro's number, $6.02(10)^{23}$, gives the number of atoms per gram-atomic weight of any element. Convert this information to atoms per cubic centimeter for aluminum, which has a density of 2.7 g/cm^3 and a value of 27 g/g-atomic wt.
7. An airplane traveling at 700 kilometers per hour (km/hr) consumes fuel at the rate of 120 liters/min. Find the consumption of fuel in kilograms per kilometer (kg/km). There is 0.8 kg of fuel per liter.
8. There is one curie of radioactivity in an object when its atoms are disintegrating at the rate of $3.7(10)^{10}$ disintegrations per second. The probability that an atom of I^{131} (iodine isotope) will disintegrate at any time is about $9.9(10)^{-7}$ disintegrations/(sec-atom). How many atoms of I^{131} are present in a one-curie specimen? Write disintegrations per second per curie as dis./(sec-curie).
9. A cable drum is rotating at 90 rpm. The drum surface turns through a distance of 1.22 meters/rev. Using inches per second, how fast is the cable being wound onto the drum?
10. A furnace consumes 500 lb coal/hr. Assume that there is 0.9 lb carbon/lb coal and that the carbon requires 3 lb oxygen/lb carbon. There is about 0.2 lb oxygen/lb air and 0.076 lb air/ft^3 of air. Arrange this information to find the combustion air required in cubic feet per minute.

3-3 SI units

The system of units known as SI (*Système International d'Unités*) was derived primarily from certain metric units. Established by international agreement, it has become the standard international system of units. In the United States the Metric Conversion Act was passed in 1975.

Two of the basic SI units are the *meter* and the *kilogram*. The meter is familiar as being a little more than 3 feet. The kilogram (a unit of mass, not weight or force) may be regarded, roughly, as the mass of a quart of water. The SI unit of force is the *newton*. It is somewhat small, compared to the pound. About $4\frac{1}{2}$ newtons are equivalent to one pound of force. However, the old English and non-SI metric units will probably fade into oblivion after the period of conversion to SI units.

SI is based on seven fundamental units and two supplementary units for angular measure. The rest of the system is derived from them. Table 3-1 gives the names and SI symbols for these nine units. As indicated in the table, periods are not used with SI symbols.

Table 3-1 SI Base and Supplementary Units

Quantity	*SI name*	*Symbol*
Length	meter	m
Mass	kilogram	kg
Time	second	s
Temperature	kelvin	K
Electric current	ampere	A
Amount of substance	mole	mol
Luminous intensity	candela	cd
	Supplementary units	
Plane angle	radian	rad
Solid angle	steradian	sr

The symbols for kelvin (K) and ampere (A) are capitalized because they are named after persons. Seventeen SI derived units have been given special names. All but two are named after persons, and the corresponding symbols are therefore capitalized. Table 3-2 presents only the more commonly used derived units that have been given special names, instead of all seventeen.

As indicated in Table 3-2, the units of SI are related to each other

Table 3-2 Some SI Derived Units Having Special Names

Quantity	*Name*	*Symbol*
Force	newton	N (= kg·m/s²)
Energy	joule	J (= newton-meter, N·m)
Power	watt	W (= joule per second, J/s)
Pressure, stress	pascal	Pa (= newton per square meter, N/m²)
Frequency	hertz	Hz (= events per second, 1/s)
Electric charge	coulomb	C (= ampere-second, A·s)
Electric potential	volt	V (= watt per ampere, W/A)
Electric resistance	ohm	Ω (= volt per ampere, V/A)

by unity. This fact simplifies many calculations. For instance, a force of one newton acting through a distance of one meter involves energy of one joule. If the action takes place during one second, the power involved is one watt. In other words, a newton meter equals one joule, and a joule per second is one watt of power. Only the watt (W) is used for power in SI, regardless of whether it is electrical, mechanical, or thermal power.

All other SI units are derived from the 26 base, supplementary, and special-name units. The unit for specific heat, for instance, is joule per kilogram kelvin, or J/(kg·K). In this example note that products of symbols in a denominator are enclosed in parentheses.

Some non-SI units are so convenient and so widely used that it is not practical to abandon them at the moment. Such units are the minute (min), hour (h), and day (d). Also, the degree of angle will probably be used, but as decimal degrees rather than being subdivided into minutes and seconds of angle. The liter (L), another non-SI unit, will probably be used by the general public as a unit of volume of fluids. One liter (a little more than the old quart of volume) equals 1000 cubic centimeters. The symbol L is capitalized because the lowercase letter l can be confused with the number 1.

Also, degree Celsius (°C) will probably find continued popular usage along with the SI unit of temperature, the kelvin (K). Kelvin temperature is obtained, closely, by adding 273 to Celsius temperature. Conveniently, any change in temperature on one scale is numerically equal to a change on the other.

An appropriate prefix should be used before a unit so that the number to be written will be between 0.1 and 1000. Table 3-3 presents some of the more useful prefixes for SI units.

Table 3-3 Frequently Used Prefixes for SI Units

Value	*Prefix*	*Symbol*
10^9	giga	G
10^6	mega	M
10^3	kilo	k
10^{-3}	milli	m
10^{-6}	micro	μ
10^{-9}	nano	n

The prefix symbols G and M are in capital letters to avoid confusion because g is used for gram and m is used for both meter and the prefix milli.

Example 5 Convert the following values to numbers between 0.1 and 1000 by using appropriate prefixes: 6400 joules, 0.072 meter, 0.025 millisecond.

Solution

$$6400 \text{ J} = 6.4(10)^3 \text{ J} = 6.4 \text{ kJ}$$
$$0.072 \text{ m} = 72(10)^{-3} \text{ m} = 72 \text{ mm}$$
$$0.025 \text{ ms} = 25(10)^{-3} \text{ ms} = 25 \ \mu\text{s}$$

As shown in Example 5, between a prefix and a unit symbol there is no space or raised dot.

ms = millisecond m·s = meter-second

Also, double prefixes tend to be awkward. For instance,

use ns, not μms (micro-millisecond),
use GJ, not kMJ (kilo-megajoule).

A prefix in the denominator of a compound unit is undesirable. Restricting prefixes to the numerator helps clarify analysis and reduces the number of compound-unit arrangements encountered. Write MJ/s (megajoules per second), not J/μs (joules per microsecond). An exception involves kilogram (kg), which is a base SI unit. Write J/kg (joules per kilogram), not mJ/g (millijoules per gram).

The following rules are presented for the consistent use of SI units and symbols.

1. The words for SI units are written in lowercase letters. The symbols are capitalized if they represent units named after persons.
2. Symbols are never written in plural form. Write 10 kg, not 10 kgs.
3. A space must be left between a numerical value and its unit symbol. 5 m means 5 meters.
4. In writing the product of a number and an algebraic quantity, no space is left between the two parts. The algebraic quantity is in *italics* (sloping letters). $5m$ means 5 times the quantity m, where m might represent the mass of an object, for example.

   $N = 300$ N means the force N equals 300 newtons.

5. Numbers having several digits, on either side of the decimal point, are written with a space instead of a comma after every third digit away from the decimal point. (In some countries a comma stands for a decimal point.) A correctly written number would be 64 520.123 49, for instance. A four-digit number can be written without the space, as in 7258.

When making calculations, prefixes should be converted to powers of 10 and the solution should be given with a single prefix, if needed.

Example 6 Power multiplied by time equals energy. Convert the product of 6 kilowatts and 20 microseconds to its equivalent energy with a single prefix.

Solution

$$\begin{aligned}(6\text{kW})(20\mu\text{s}) &= (6(10)^3\text{W})(20(10)^{-6}\text{s}) \\ &= 120(10)^{-3}\ \text{W}\cdot\text{s} \\ &= 120\ \text{mW}\cdot\text{s (milliwatt-seconds)}\end{aligned}$$

When working with such quantities as area and volume, where units are raised to a power, care should be taken if prefixes are used with the units. For instance, does 5 km^2 mean an area of 5000 square meters or 5 square kilometers? The difference is great. If we write 5 k(m)^2, it clearly means 5000 square meters. If instead we write 5 $(\text{km})^2$, it is clear that we mean 5 square kilometers. Generally, however, it is assumed that when we write 5 km^2, it means 5 $(\text{km})^2$. Or if we write 8 cm^2, it means 8 $(\text{cm})^2$ and so on.

Example 7 Convert an area of 5 square kilometers to its equivalent in square meters by using prefix values. Make sure that all meanings are clear.

Solution

$$5\ \text{km}^2 = 5\ (\text{km})^2 = 5\ \text{k}^2\text{m}^2 = 5(10^3)^2\ \text{m}^2 = 5(10)^6\ \text{m}^2 = 5\ \text{M(m)}^2$$

Exercises 3-3

Answer the following questions, using Table 3-2 and SI notation and rules.

1. A force of 800 N moves through a distance of 20 m. How many joules of energy are involved?
2. A force of 200 kN moves through a distance of 2 mm. How many joules of energy are involved?
3. An engine does 900 kJ of work every minute. How much power is it producing?
4. An engine does 300 MJ of work each minute. How much power is it producing?
5. The horizontal flat bottom of a water tank has an area of 50 m^2. The total force of the water acting on the bottom of the tank is 250 kN. What is the pressure, in pascals, at the bottom of the tank?
6. A force of 40 kN is exerted against an area of $8(10)^{-4}$ m^2. What is the amount of pressure involved?
7. If 100 V (volts) is required to cause a current of 50 mA to flow through a resistor, what is its resistance?

8. If 140 V is required to cause a current of 70 μA to flow through a resistor, what is its resistance?
9. What is the equivalent, in cubic meters, of $5(10)^6$ $(mm)^3$? Solve by using prefix values. Make sure all meanings are clear.
10. What is the equivalent, in square meters, of $8(10)^5$ $(mm)^2$? Solve by using prefix values. Make sure all meanings are clear.

3-4 Older unit systems

Before the creation of SI units by an international conference in 1960, three systems of units were commonly used in technical work: English, MKS, and CGS. English units were used by some English-speaking nations and include the familiar foot and pound units. The units enjoying the widest worldwide usage were metric units, which make up the MKS and CGS systems. MKS is the abreviation for meter-kilogram-second. CGS represents centimeter-gram-second. The English system is often called the FPS system, for foot-pound-second.

Table 3-4 gives the units for some of the various physical concepts in English, SI, MKS, and CGS systems. Because several of the base units and specially named units of the MKS system appear in SI, Table 3-4 does not include a separate heading for MKS units. For the brief list of units presented, the MKS and SI unit names are the same, indicating the close relationship between the two systems. However, SI is different from the old MKS system. In creating SI, some new unit names appeared, some units were abandoned, and internationally acceptable rules for symbols and notation were adopted.

Table 3-4 A Brief Comparison of Unit Systems

Quantity	*English*	SI *and* MKS	CGS
Length	foot	meter	centimeter
Mass	slug*	kilogram	gram
Force	pound	newton	dyne
Energy	ft-lb	joule	erg (= dyne-cm)
Power	$\frac{\text{ft-lb}}{\text{sec}}$	watt	$\frac{\text{erg}}{\text{sec}}$
Density	$\frac{\text{slug}}{\text{ft}^3}$	$\frac{\text{kg}}{\text{m}^3}$	$\frac{\text{gram}}{\text{cm}^3}$
Charge		coulomb	statcoulomb
Current		ampere	statampere

*One slug is the mass of an object that would weight about 32.2 lb at sea level.

When numerical values are to be inserted for the various quantities of a scientific law or relation, it is usually necessary to avoid mixing units of the several different systems. However, some laws are expressed in such a way that mixed units are allowed, an example being the gas law $P_1V_1/T_1 = P_2V_2/T_2$. In this case, the gas pressure P might be in pounds per square inch, whereas the volume V might be in cubic meters or any other unit of volume. The units on the left side, though, must be the same as those on the right.

Exercises 3-4

In answering the following questions, it may be helpful to refer to Tables 3-1, 3-2, and 3-4.

1. The English units for moment of inertia of a rotating object are often given as slug-ft^2. What would be the corresponding SI and CGS units?

2. Coulomb's electric force law contains a special constant k if MKS units are to be used. If Q, R, and F have units of charge, length, and force, respectively, what are the units of k to make a balanced equation of units?

$$F = \frac{kQ_1Q_2}{R^2}$$

3. Newton's gravitational constant, in SI, may have the units N·m^2/kg^2. Give the units for the CGS and English systems.

4. The rate of energy flow through a square meter of area may be given in J/(s·m^2). What units could be used in the English and CGS systems?

5. In the equation for horsepower, F, S, and T have units of force, length, and time, respectively. The constant K must have units that cause the right side of the equation to be in net units of horsepower. What are the units of K in the English system?

$$\text{hp} = \frac{FS}{TK}$$

6. In the equation for power P in watts, F, S, and T have units of force, length, and time, respectively. Fill in the equation with SI unit symbols. The unit equation may be said to balance. Why?

$$P = \frac{FS}{T}$$

7. If viscosity can be measured in terms of lb-sec/ft^2, what are the corresponding SI and CGS units? For SI, use symbols.

3-5 Relations for converting units

Table 3-5 contains unit relations that may be useful in the remaining exercises of this chapter. Many more might be listed, but a few relations committed to memory can greatly reduce dependence on tables.

Table 3-5 Some Useful Unit Relations

Length	*Force*
1 in. = 2.54 cm	1 newton = 10^5 dynes
1 mile = 5280 ft	1 lb = 4.45 newtons
	1 ton = 2000 lb
Mass	
1 slug = 14.6 kg	*Volume*
	1 m^3 = 35.3 ft^3
Energy	1 gal = 231 $in.^3$
1 joule = 10^7 ergs	1 liter = 1000 cm^3
1 ft-lb = 1.36 joules	
1 Btu = 778 ft-lb	*Power*
1 calorie = 4.18 joules	1 hp = 550 ft-lb/sec
1 kilowatt-hour = 3413 Btu	1 hp = 0.75 kilowatt
1 Btu = 252 calories	1 kilowatt = 3413 Btu/hr

Example 8 Convert $2(10)^7$ ft-lb of energy to kilowatt-hours.

Solution 1

$$(\text{ft-lb})\left(\frac{\text{hp-sec}}{\text{ft-lb}}\right)\left(\frac{\text{kW}}{\text{hp}}\right)\left(\frac{\text{hr}}{\text{sec}}\right) = \text{kW-hr}$$

$$(2(10)^7)\left(\frac{1}{550}\right)(0.75)\left(\frac{1}{3600}\right) = 7.5$$

Solution 2

$$(\text{ft-lb})\left(\frac{\text{Btu}}{\text{ft-lb}}\right)\left(\frac{\text{kW-hr}}{\text{Btu}}\right) = \text{kW-hr}$$

$$(2(10)^7)\left(\frac{1}{778}\right)\left(\frac{1}{3413}\right) = 7.5$$

In Solution 1 we note that there is 1 hp-sec of energy per each 550 ft-lb of energy. A hp-sec would be the amount of work done by a 1-hp engine during one second and would equal 550 ft-lb.

A quantity known as the *gmole* is often encountered in chemistry and physics. A gmole of a substance (element or compound) is the mass that, if measured in grams, is equal to the molecular weight of the substance. Oxygen, for instance, has a molecular weight of 32 and requires 32 grams to constitute a gmole. A kgmole of oxygen is 32 kg.

The gmole is important because, for all substances, a gmole always contains the same number of molecules. This number, $6.02(10)^{23}$, is the familiar Avogadro's number. The number of atoms in a gram-atomic weight is the same as the number of molecules in a gmole (gram-molecular weight), being Avogadro's number in both cases. A gram-atomic weight of oxygen is 16 grams, corresponding to the atomic weight of oxygen.

and Kelvin, are associated with metric
at the freezing point of water, whereas
o, the lowest temperature theoretically
nge on the Celsius scale equals a degree

emperature scales are associated with
eads 32° at the freezing point of water,
ibsolute zero. A degree of temperature
ils a degree of change on the Rankine

used for converting temperatures. The
temperature on the Celsius scale.

) $T_k = T_c + 273$

$T_r = T_f + 460$

should be distinguished from that of
temperature, the degree mark is placed
C. A temperature change is often indicat-
the scale symbol, appearing, for example,
change are related as follows:
°/R°, 5 C°/9 F°, 5 K°/9 R°.

of an object is increased from 28°C to
ure change in K°, F°, and R°.

$28°C = 50C° = 50K°$

$\left(\frac{9F°}{5C°}\right) = 90F° = 90R°$

ber that C° represents a greater change of

ieat of aluminum is about 0.21 cal/(g-C°),
ie of heat is required to raise the tempera-
1 C°. What is the equivalent value express-

$\left(\frac{\text{oules}}{\text{cal}}\right)\left(\frac{\text{g}}{\text{kg}}\right)\left(\frac{\text{C°}}{\text{K°}}\right) = \frac{\text{joules}}{\text{kg-K°}}$

$1)(4.18)(1000)(1) = 878$

rough all units

to the equiva-

R°

and units as in

C°). Convert to

ohm/(ohm-C°)
ce of 0.4% per

ded in order to
rt this informa-
V-hr).

s of statampere-

per gmole of a
molecules per
re, a condition
rs per gmole of

momentum of

ert a force of
s per inch.

°). What is the
problem at the

To verify that the unit equation balances, draw a line t
that can be canceled on the left side of the equation.

Example 11 Convert the value obtained in Example 1
lent in Btu/(slug-R°).

Solution

$$\left(\frac{\text{joules}}{\text{kg-K}^\circ}\right)\left(\frac{\text{ft-lb}}{\text{joule}}\right)\left(\frac{\text{Btu}}{\text{ft-lb}}\right)\left(\frac{\text{kg}}{\text{slug}}\right)\left(\frac{\text{K}^\circ}{\text{R}^\circ}\right) = \frac{\text{Bt}}{\text{slug-}}$$

$$(878)\left(\frac{1}{1.36}\right)\left(\frac{1}{778}\right)(14.6)\left(\frac{5}{9}\right) = 6.7$$

Exercises 3-5

For each of the conversions indicated, write a sequence of number
the preceding examples.

1. How many Btu/hr are equivalent to 5 MW of power?
2. The ideal gas constant has the value $8.32(10)^7$ ergs/(gmole
joules/(kgmole K°).
3. The temperature coefficient of electrical resistance is about 0.00
for pure metals. It may be regarded as a change in resista
Celsius degree. Convert to %/F° and %/K°.
4. Find the equivalent of one horsepower in Btus per minute.
5. An estimated $4(10)^7$ ft-lb of energy, supplied to a motor, is ne
pump a certain quantity of water to a higher elevation. Conve
tion to cost in cents. The electric company charges 4 cents/(k
6. A battery discharge of 5 ampere-hours is to be expressed in term
seconds. One ampere equals $3(10)^9$ statamperes.
7. Avogadro's number, $6.02(10)^{23}$, is the number of molecules
substance. Convert Avogadro's number to find the number c
cubic centimeter, for a gas at standard temperature and press
under which it is known that gases occupy a volume of 22.4 lit
gas. There are 1000 cm³/liter.
8. Momentum is the product of mass times velocity. Convert a
50 kg·m/s to its equivalent in slug-foot per minute.
9. A magnetic force is distributed along a conductor so as to e
80 N/m of length. Convert to dynes per centimeter and poun
10. The thermal conductivity of steel is about 320 Btu-in./(ft²-hr-
equivalent value in cal-cm/(cm²-sec-C°)? (It may simplify this
present time to know that there are about 930 cm²/ft².)

11. Each foot of length along a new road requires 20 ft^3 of concrete, which a machine pours at the rate of 24 tons/hr. Find the length of road poured in an 8-hr day. Assume that the concrete weighs 150 lb/ft^3.

12. A certain refrigeration machine requires 220 ft-lb of compressor work per Btu of heat removal. Convert this information to kilowatts of power required per ton of refrigeration. Heat removal occurs at the rate of 200 Btu/min per ton of refrigeration. This fact may be expressed as 200 Btu/(min-ton).

3-6 Units and exponents

Many quantities are expressed in terms of several units, some of which may be squared, cubed, or raised to higher powers. The procedure of the previous examples can be used to convert the units of such a quantity. As noted, the net units of products or quotients of physical quantities are obtained by treating the units of the various quantities algebraically. For instance, the expression involving pressure, volume, and temperature from the ideal gas law may have units obtained as follows:

$$\frac{PV}{T} = \frac{(\text{lb/ft}^2)(\text{ft}^3)}{^\circ\text{R}} = \frac{\text{lb ft}}{^\circ\text{R}}$$

When a ratio of units is raised to some power, the units are also raised to the power.

$$\left(12\,\frac{\text{in.}}{\text{ft}}\right)^2 = (12)^2\left(\frac{\text{in.}}{\text{ft}}\right)^2 = 144\,\frac{\text{in.}^2}{\text{ft}^2}$$

The preceding equation indicates a useful procedure. To illustrate, the number of cubic centimeters per cubic inch is obtained by cubing the number of centimeters per inch.

$$\left(2.54\,\frac{\text{cm}}{\text{in.}}\right)^3 = (2.54)^3\left(\frac{\text{cm}}{\text{in.}}\right)^3 = 16.4\,\frac{\text{cm}^3}{\text{in.}^3}$$

This is the same as the product of the three edges of a one-inch cube, with each edge taken in centimeters.

Example 12 Convert a specific weight of 0.50 lb/in.^{3} to its equivalent in tons/ft^3.

Solution

$$\left(\frac{\text{lb}}{\text{in.}^3}\right)\left(\frac{\text{in.}^3}{\text{ft}^3}\right)\left(\frac{\text{tons}}{\text{lb}}\right) = \frac{\text{tons}}{\text{ft}^3}$$

$$(0.50)(12)^3\left(\frac{1}{2000}\right) = 0.43$$

Example 13 A length of pipe has an outside surface area of 4 m^2. The outside diameter is 3 in. What is the length in inches?

Solution

$$\text{Area} = (\text{circumference}, \pi D)(\text{Length}, L)$$

$$(\text{m}^2)\left(\frac{\text{cm}^2}{\text{m}^2}\right)\left(\frac{\text{in.}^2}{\text{cm}^2}\right) = (\text{in.})(\text{in.})$$

$$(4)(100)^2\left(\frac{1}{2.54}\right)^2 = (\pi 3)(L)$$

$$L = 658 \text{ in.}$$

Example 14 The Stefan-Boltzmann constant, used in the equation for rate of heat transfer by radiation, has the value $5.7(10)^{-12}$ watts/(cm^2-°K^4). Convert the constant to its value in Btu/(hr-ft^2-°R^4).

Solution

$$\left(\frac{\text{watts}}{\text{cm}^2\text{-°K}^4}\right)\left(\frac{\text{kW}}{\text{watt}}\right)\left(\frac{\text{Btu}}{\text{kW-hr}}\right)\left(\frac{\text{cm}^2}{\text{in.}^2}\right)\left(\frac{\text{in.}^2}{\text{ft}^2}\right)\left(\frac{\text{°K}^4}{\text{°R}^4}\right) = \frac{\text{Btu}}{\text{hr-ft}^2\text{-°R}^4}$$

$$(5.7(10)^{-12})\left(\frac{1}{1000}\right)(3413)(2.54)^2(12)^2\left(\frac{5}{9}\right)^4 = 1.73(10)^{-9}$$

In this example we have used the fact that the ratio of any Kelvin temperature to its corresponding Rankine temperature is five to nine, or 5/9.

Exercises 3-6

Write a sequence of numbers and units for each conversion, crossing out units that cancel until the desired units are obtained. Equations for areas, volumes, and similar quantities can be found in Table A at the back of the book.

1. The speed of a rocket increases by 300 ft/sec each second. We can say that it has an acceleration of 300 ft/(sec-sec) or 300 ft/sec^2. Convert to miles/(hr-min).
2. How many lb/ft^2 are equivalent to a pressure of 20,000 dynes/cm^2?
3. When the sun is directly overhead, a square centimeter of the earth's surface receives energy at the rate of about 2 calories/min. Convert this information to hp/mile^2.
4. If T_2 is 310 K and T_1 is 300 K, what is the numerical difference between $(T_2 - T_1)^4$ and $(T_2^4 - T_1^4)$? If the Kelvin temperatures were converted to Celsius, which expression would be unchanged in value?
5. A painter in a housing development finds that he uses 5 gallons of paint per

house. The painted area is 1200 ft^2 per house. Find the cubic centimeters of paint used per square foot of painted area. There are 231 in.^{3}/gal.

6. The viscosity of a certain fluid is given as 0.50 dyne-sec/cm^2. Find the viscosity in English units, lb-sec/ft^2.

7. If a substance contains $(10)^{18}$ atoms/in.^{3}, what is the number of atoms/cm^3?

8. The velocity of a freely falling object, dropped from a height h, can be obtained from $V = (2gh)^{1/2}$. Using English and SI units, show that this expression has units of velocity (speed). The letter g represents acceleration due to gravity, such as ft/sec^2.

9. When an electric current of I amperes flows through a resistance R for t seconds, the amount of heat produced can be calculated from I^2Rt. If R has the units volts per ampere and an ampere is a coulomb of electric charge per second, show that I^2Rt may be in net units of joules. (A joule equals a volt-coulomb.)

10. For a uniformly loaded simple beam, the maximum deflection is $D = 5WL^4/384EI$. If L is the length in inches, E the modulus of elasticity for the beam material in lb/in.^{2}, and I the moment of inertia of the beam cross section in in.^{4}, what units are required for the load W in order that D may be in inches?

11. The maximum shearing stress due to torsion in a drive shaft is often calculated from $S = TR/J$. If S, T, and R have, respectively, the units lb/in.^{2}, in.-lb, and in., what are the units of the polar moment of inertia J?

12. *Reynolds number* is the name for a dimensionless quantity used in the study of fluid flow, as in pipes. If the Reynolds number is $\rho VD/\mu$, show that it is dimensionless by using CGS units. V is the fluid velocity in centimeters per second, ρ (rho) is its density, μ (mu) is its viscosity in g/(cm-sec), and D is the pipe diameter.

13. Show that Reynolds number, described in Exercise 12, is dimensionless by using SI units.

14. A cylindrical can has a volume of 1000 cm^3 and a height of 6 in. What is its diameter in inches?

15. A parking lot in the shape of a half-circle has an area of 10 k(m)^2. What is the total perimeter in feet?

16. A solid object in the shape of a half-sphere has a volume of 1000 cm^3. What is the total surface area, including the flat surface, in square inches?

17. A water pump is pumping 2 liters per second. Convert this amount to gallons per minute.

18. An automobile engine has a piston displacement of 3 liters and is rated at 120 kW of power. How many watts are produced per cubic centimeter of displacement?

3-7 Mass, weight, and force

It should be noted that *mass* and *weight* are different concepts. Weight is force, the force of gravity exerted by two objects on each other. If an object is said to weigh 16 lb, it means that 16 lb of gravity force is acting on both the object and the earth, which tends to bring them together. The gravity force, or weight, is not a property of the object alone. It depends on the mass of both the object and the earth and the distance between them.

Mass can be considered as the amount of matter in an object and is a property of the object. If the mass of an object is 10 kg, the gravity force (weight) acting on the object at the surface of the earth is 98 newtons. If the object were on the moon, its mass would still be 10 kg, but its weight would be reduced considerably.

Units of force have been defined in terms of mass and acceleration units in Newton's second law of motion, $F = Ma$. One newton, for instance, is defined as the amount of unbalanced force that will cause a mass of 1 kg to have an acceleration of 1 m/s^2. How many newtons should be required to give the kilogram an acceleration of 9.8 m/s^2? Since the force is proportional to acceleration ($F = Ma$), the required force is 9.8 newtons. It happens that the acceleration of a falling object, caused by gravity force at the surface of the earth, is 9.8 m/s^2. Therefore the gravity force acting on 1 kg at the surface of the earth is 9.8 newtons. Table 3-6 gives the gravity forces acting on various masses at the surface of the earth.

Table 3-6 Gravity Forces Acting on Units of Mass at the Earth's Surface

System	*Mass*	*Gravity force*	*Acceleration* (g)
MKS and SI	1 kg	9.8 newtons	9.8 m/s^2
CGS	1 g	980 dynes	980 cm/sec^2
English	1 slug	32.2 lb	32.2 ft/sec^2
	1 kg	2.2 lb	
	454 g	1 lb	

If we use the letter g to represent the acceleration due to gravity and refer to the gravity force as weight W, Newton's law becomes $W = Mg$. The units of weight, mass, and g must all be from the same system of units. This relation ($W = Mg$) will give the weight of a mass if the value of g is known for the location of the mass.

Example 15 What is the weight of a 5-kg mass (a) at the surface of the earth and (b) on a planet where g is 12 m/s^2?

Solution

(a) $$W = Mg = (5 \text{ kg})\left(9.8 \frac{\text{m}}{\text{s}^2}\right) = 49 \text{ newtons}$$

(b) $$W = Mg = (5 \text{ kg})\left(12 \frac{\text{m}}{\text{s}^2}\right) = 60 \text{ newtons}$$

At first the units of Example 15 do not appear to form an equality or balance of units. We have noted, however, that force units are defined in terms of mass and acceleration units in such a way ($F = Ma$) that

$$1 \text{ newton} = 1 \text{ kg}\cdot\text{m/s}^2$$
$$1 \text{ dyne} = 1 \text{ gm-cm/sec}^2$$
$$1 \text{ lb} = 1 \text{ slug-ft/sec.}^2$$

These equivalencies can be used when converting units and checking equations for balance of units.

Example 16 The gravity force on a satellite is 200 lb at an altitude where g is 25 ft/sec.^{2} (*a*) What is its mass? (*b*) What would it weigh at the earth's surface?

Solution

(a)
$$W = Mg$$
$$\text{lb} = (\text{slug})(\text{ft/sec}^2)$$
$$200 = (M)(25)$$
$$M = 8 \text{ slugs}$$

(b)
$$W = Mg = (8)(32.2)$$
$$W = 258 \text{ lb } (= 258 \text{ slug-ft/sec}^2)$$

Many balances used for weighing objects give readings in grams. The gram is not a unit of weight or force but a mass unit. Although it is common practice to speak of an object as "weighing" so many grams, it is actually the mass that has been determined.

3-8 Equations and equality of units

The physical quantities represented on one side of an equation usually must have units arranged so as to be identical or equivalent to the arrangement of units on the other side. Only quantities having identical or equivalent units may be added, subtracted, or set equal. Consequently, the units of an equation must form an equality, such as Btu + Btu = Btu. For convenience, we shall refer to this condition of dimensional equality as a *balance of units.*

As a routine but important step in the solution of a physical problem, the equations used should always be checked for balance of units at the time of inserting known values. If two or more equations are to be solved simultaneously, the same units must be used for like quantities in all equations. For instance, if time and displacement are the unknowns in two equations, the units of time and displacement involved in known quantities must be the same for both equations.

When both mass and force units appear in an equation, it should be remembered that they are related by the expression $F = Ma$, or $W = Mg$, as indicated in Sec. 3-7.

Example 17 Show a balance of units for the kinetic energy equation, $E = \frac{1}{2}mv^2$, using the English system. E, m, and v represent energy, mass, and velocity, respectively.

Solution

$$E = \frac{1}{2}(m)(v)^2$$

$$\text{ft-lb} = (\text{slug})\left(\frac{\text{ft}^2}{\text{sec}^2}\right)$$

Using the relation $F = Ma$ from Sec. 3-7, we can state that 1 lb = 1 slug-ft/sec^2 or, by transposing, 1 slug = 1 lb-sec^2/ft. Either the lb or slug may be replaced in order to form an obvious balance of units. Replacing the slug with its equivalent gives

$$\text{ft-lb} = \left(\frac{\text{lb-sec}^2}{\text{ft}}\right)\left(\frac{\text{ft}^2}{\text{sec}^2}\right)$$

$$\text{ft-lb} = \text{ft-lb}$$

Exercises 3-7 and 3-8

1. Show a balance of SI units for the energy-mass relation $E = Mc^2$. The letter c represents the speed of light.
2. Do Exercise 1, using CGS and English units.
3. An expression for estimating the power of a nuclear reactor is $P = FEk$, where P is the power in watts and E is the energy released from each fission. If E has the units Mev/fission and k is the conversion factor joules/Mev, what are the units of F? The Mev (million electron volts) is an energy unit.
4. Show that the units of E/B, an electric field strength divided by a magnetic field strength, are those of velocity in the SI system. E may have the units volts per meter and B may be in newtons per ampere-meter. An ampere is a coulomb per second.

5. Impulse is equated to change of momentum in the equation $Ft = M(V_2 - V_1)$. F, t, and M are force, time, and mass, respectively. Show a balance of CGS units. The quantity $(V_2 - V_1)$ represents velocity change and has the units cm/sec. It is not necessary to write (cm/sec − cm/sec) for the units of $(V_2 - V_1)$.
6. Do Exercise 5, using SI and English units.
7. The useful output power of a pump can be expressed as $P = WQ$, where W is the useful work done on each pound of fluid, or ft-lb/lb. If power P is in ft-lb/sec, what are the units of Q?
8. Show a balance of SI units for the equation of motion $S = Vt + \frac{1}{2}at^2$. S, V, t, and a have units of length, velocity, time, and acceleration (m/s²), respectively.
9. In certain cases, the pressure drop due to fluid flow in a horizontal pipe is given by $(P_1 - P_2) = 32uLV/D^2$. V and u are the fluid velocity and viscosity, whereas L and D are the length and diameter of pipe. If $(P_1 - P_2)$ is in dynes/cm² (CGS units), what are the units of viscosity?
10. Do Exercise 8 in English and CGS units.
11. Convert the answer of Exercise 9 to an alternate form involving a mass unit. Use units from $F = Ma$.
12. From kinetic theory, the average gas molecule kinetic energy is a function of absolute temperature, or $\frac{1}{2}mV^2 = 3kT/2$. Show a balance of SI units. V, m, and T have units of velocity, mass, and absolute temperature. The Boltzmann constant k has the units j/K.
13. The expression $LI^2/2$ represents the magnetic field energy of an inductor. I is in amperes, and the inductance L may have the units volts per ampere per second (volt-sec/ampere). Show that the units of the expression are equivalent to joules. An ampere is a colomb per second.
14. In the equation $wh - Fs = wV^2/2g$ each term represents energy. Show a balance of SI units. F and w have units of force, whereas h and s have units of length. V has units of velocity and g is the acceleration of gravity (m/s²).
15. Do Exercise 14 in CGS and English units.
16. The speed of sound through a material can be calculated from $V = (E/d)^{1/2}$. Show that V will be in ft/sec if E is in lb/ft² and the material density d is in slugs/ft.³ Replace the lb unit with its equivalent from $F = Ma$.

3-9 The problem of the radian

In the study of technical subjects, particularly in physics and engineering mechanics, equations dealing with the angular unit of measure called the *radian* are often encountered. In checking these equations for a balance of units, it frequently seems that the units cannot be balanced. Our purpose here is to present some of the reasons for this apparent lack of balance in units and to illustrate briefly the manner in which the units can be shown to balance.

The radian is defined as the angle between radii of a circle that mark off on the circumference a length equal to the radius.

In Fig. 3-1 the angle A is one radian if the arclength r is numerically equal to the radius length R. However, it should be noted that r and R do not have the same units. The radius R has units of simple length, such as meters or feet. The arclength per radian, r, has units like meters per radian of angle or feet per radian of angle.

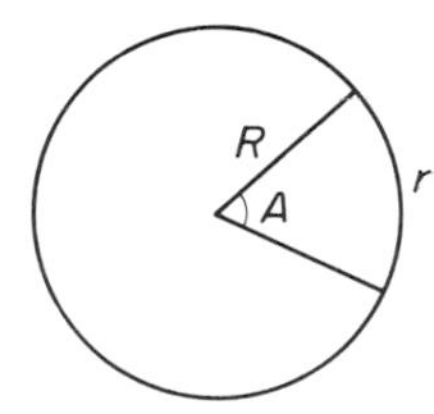

Figure 3-1

Some equations relating physical quantities contain various combinations of the radian unit of angular measure and the radius R or the arclength r per radian. It is in such equations that problems often occur in obtaining a balance of units. The usual situation results in a radian unit on one side of the equation and none on the other side. This situation often arises because some equations are treated as containing the radius R when it is actually r, the arclength per radian, that is involved.

A simple example is the equation for circular motion, $S = \theta r$, which is often treated as $S = \theta R$. The angle θ (theta) must be in radians.

Example 18 In Fig. 3-2 the radius R is 3 meters and the angle θ is 2 radians. What is the arclength S?

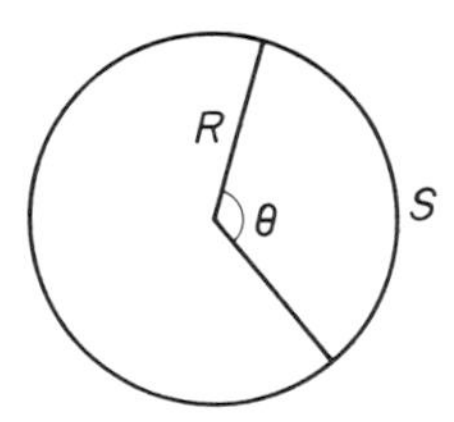

Figure 3-2

Solution

$$S = (\theta)(r)$$

$$\text{m} = (\text{rad})\left(\frac{\text{m}}{\text{rad}}\right)$$

$$S = (2)(3) = 6 \text{ m}$$

Here r is the arclength per radian, not the radius R. A balanced equation of units is obtained.

Example 19 Use the equation of Example 18 to find the circumference C of the same circle. There are 2π radians per revolution. Show a balanced equation of units.

Solution

$$S = (\theta)(r)$$
$$C = (2\pi)(r)$$
$$\frac{\text{m}}{\text{rev}} = \left(\frac{\text{rad}}{\text{rev}}\right)\left(\frac{\text{m}}{\text{rad}}\right)$$
$$C = (2\pi)(3)$$
$$C = 6\pi \text{ m/rev}$$

A second reason for the difficulty in balancing units when radians are involved occurs because in the derivation of some equations a ratio such as R/r will appear. Since R and r are numerically equal, they are commonly canceled out of the equations. But because they have different units, as noted, the units do not completely cancel. The result of canceling out the R/r ratio is an unbalance of radian units in the equation.

To illustrate, the angle of twist θ in a drive shaft is often calculated from $\theta = TL/EJ$, where T is the torque transmitted and L, E, and J are the length, shear modulus, and polar moment of inertia of the shaft. For our purposes, it is not necessary to define the shear modulus and polar moment of inertia. The units, which might appear as follows, do not balance.

$$\theta = \frac{TL}{EJ}$$
$$\text{rad} = \frac{(\text{lb-in.})(\text{in.})}{(\text{lb/in.}^2)(\text{in.}^4)}$$
$$\text{rad} = \frac{\text{lb-in.}^2}{\text{lb-in.}^2}$$

It can be shown in the derivation of this equation that $\theta = TLR/EJr$ and that the units involved do balance.

Example 20 Using information from the preceding paragraph, show a balance of units for $\theta = TLR/EJr$.

Solution

$$\theta = \frac{TLR}{EJr}$$

$$\text{rad} = \frac{(\text{lb-in.})(\text{in.})(\text{in.})}{(\text{lb-in.}^2)(\text{in.}^4)(\text{in./rad})}$$

$$\text{rad} = \frac{(\text{lb-in.}^3)(\text{rad})}{(\text{lb-in.}^3)}$$

$$\text{rad} = \text{rad}$$

A third situation in which the radian problem appears occurs in trying to balance units in an equation that has been modified for technical convenience. An example is $P = T\omega$, in which P is the power transmitted by a rotating drive shaft, T is the torque in the shaft, and ω(omega) is the rotational speed in radians per second. Written in the form above, the units do not balance.

The equation should read

$$P = (W)(\omega)$$

$$\frac{\text{ft-lb}}{\text{sec}} = \left(\frac{\text{ft-lb}}{\text{rad}}\right)\left(\frac{\text{rad}}{\text{sec}}\right),$$

where W is the work done per radian of rotation. It is known from engineering mechanics that the work done per radian of rotation is numerically equal to the torque in the shaft. Because engine dynamometers can measure torque directly, the equation is modified for convenience to $P = T\omega$, thus resulting in an unbalance of units.

In closing, we consider the problem of the units for the quantity known as the *moment of inertia* for rotating objects, usually indicated by the letter I. We may think of the moment of inertia I as a measure of rotational inertia, as for a flywheel or rotor. A flywheel of large radius R and mass M would have a large resistance to change in rotational speed and would have a large I value.

A typical expression used to calculate a moment of inertia is kMR^2, where k is a constant that depends on the shape of the object and its axis of rotation. Accordingly, the units commonly given for moment of inertia in the English system, for instance, are slug-ft^2. These units, however, generally cause an unbalance of units in equations that contain the quantity I.

Consider equation $T = I\alpha$, for example. T may be the lb-ft of torque causing a rotor with a moment of inertia I to have an angular acceleration of α (alpha) rad/sec^2. The units of the equation do not balance if slug-ft^2 is inserted for I. Nevertheless, in deriving equation $T = I\alpha$, it can be shown that the form of the expression used for I is actually $kMRr$ (instead of kMR^2). The corresponding units for I would then be (slug)(ft)(ft/rad), or slug-ft^2/rad.

Example 21 Using information from the preceding paragraph, show a balance of units for equation $T = I\alpha$.

Solution

$$T = I\alpha$$

$$\text{lb-ft} = \left(\frac{\text{slug-ft}^2}{\text{rad}}\right)\left(\frac{\text{rad}}{\text{sec}^2}\right)$$

$$\left(\text{from } F = Ma,\ 1 \text{ slug} = 1\,\frac{\text{lb-sec}^2}{\text{ft}}\right)$$

$$\text{lb-ft} = \left(\frac{\text{lb-sec}^2}{\text{ft}}\right)\left(\frac{\text{ft}^2}{\text{rad}}\right)\left(\frac{\text{rad}}{\text{sec}^2}\right)$$

$$\text{lb-ft} = \text{lb-ft}$$

Many students of technical subjects will encounter the equation for rotational kinetic energy, KE $= \frac{1}{2}I\omega^2$, where ω is the rotational speed of a rotor in radians per second and KE may be in foot-pound. If, as in Example 21, we accept the units of I as being slug-ft^2/rad (or kg·m^2/rad in SI), then it would seem that these I units should provide a balance of units for the kinetic energy equation. Yet they don't. It is left to the reader to show this unbalance.

What is thc reason for this predicament? In the derivation of the equation for rotational kinetic energy we find that the I term is inserted to replace a quantity of form kMr^2 instead of $kMRr$ or kMR^2. This insertion of the I term is done for convenience and provides a correct numerical result because R and r are numerically equal. But the units for kMr^2 are slug-ft^2/rad^2. If these units are used for I in equation KE $= \frac{1}{2}I\omega^2$, then a balance of units may be obtained.

Example 22 Show a balance of English units for KE $= \frac{1}{2}I\omega^2$, using the special units required for I in this particular equation.

Solution

$$\text{KE} = \frac{1}{2}I\omega^2$$

$$\text{ft-lb} = \left(\frac{\text{slug-ft}^2}{\text{rad}^2}\right)\left(\frac{\text{rad}^2}{\text{sec}^2}\right)$$

$$\left(\text{from } F = Ma,\ 1 \text{ slug} = 1\,\frac{\text{lb-sec}^2}{\text{ft}}\right)$$

$$\text{ft-lb} = \left(\frac{\text{lb-sec}^2}{\text{ft}}\right)\left(\frac{\text{ft}^2}{\text{rad}^2}\right)\left(\frac{\text{rad}^2}{\text{sec}^2}\right)$$

$$\text{ft-lb} = \text{ft-lb}$$

If problems are encountered in obtaining a balance of units in equations that contain the radian unit, the solution can be found by following through the derivations of these equations. If a derivation uses any of the relations $S = \theta r$, $C = 2\pi r$, $V = \omega r$, and $a = \alpha r$, it should be remembered that the r in each case has such units as ft/rad or m/rad.

3-10 Formulation of units

Often it is possible to construct an equation of units that may expedite the solution of a physical problem. This procedure is of considerable value in a problem in which no specific physical law or relation immediately appears to apply. The units of the known pieces of information are assembled, along with the units of the unknown, to form an equation of units. The known numerical values are then placed in position as indicated by the units.

It should be noted that a balanced equation of units may be formed that does not correspond to a correct relation of physical quantities. For instance, the lb-ft of torque in a drive shaft might erroneously be set equal to the ft-lb of energy transmitted by the shaft during some period of time. Torque and energy are not like quantities and should not be equated. Also, a process that takes place at a variable rate may incorrectly be treated as operating at a constant rate. Consequently, the physical quantities involved must be well understood when formulating units, and the proposed unit equation should be checked for physical sense.

Example 23 Pump A can fill a tank in 4.200 hr. Pump B can fill the tank in 9.700 hr. How many hours will it take to fill the tank if both A and B are pumping? The pumps have constant delivery rates.

Solution A pumps at the rate of one tank per 4.200 hr, or 0.2381 tanks/hr. B pumps at the rate of one tank per 9.700 hr, or 0.1031 tanks/hr. An equation of units may be set up based on the central idea that the amount pumped by A plus the amount pumped by B equals 1 tank volume.

$$\left(\frac{\text{tanks}}{\text{hr}}\right)(\text{hr}) + \left(\frac{\text{tanks}}{\text{hr}}\right)(\text{hr}) = \text{tanks}$$

$$(0.2381)(T) + (0.1031)(T) = 1$$

$$T = \frac{1}{0.3412} = 2.931 \text{ hr}$$

This procedure is correct only if the pumps have constant delivery rates.

Example 24 Find the amount contributed to the tank by each pump in Example 23 and check the solution.

Solution

Pump A:

$$\left(\frac{\text{tanks}}{\text{hr}}\right)(\text{hr}) = \text{tanks}$$

$$(0.2381)(2.931) = 0.698$$

Pump B: $$(0.1031)(2.931) = 0.302$$

Check: The sum of 0.698 and 0.302 tanks is 1 tank, the specified total.

Formulation of units can be used as an aid in remembering the laws and relations of science. Assume that you have temporarily forgotten the correct expression for centripetal acceleration and are considering $a = v^2R$ and $a = v^2/R$, where v represents velocity and R has units of length. The SI units for v^2R would be $(\text{m}^2/\text{s}^2)(\text{m})$, or m^3/s^2. The units for v^2/R would be $(\text{m}^2/\text{s}^2)(1/\text{m})$, or m/s^2, the correct units for acceleration.

Dimensional analysis has been used extensively to help in clarifying the relations of fluid flow, heat transfer, electricity, mechanics, and other fields of study.

Example 25 Using the observation that the nonrotational kinetic energy of an object is a function of its mass and velocity, develop the form of the expression for kinetic energy using SI units. Use the letter C for an unknown constant.

Solution

$$\text{KE} = f(m, v)$$

$$\text{joules} = \text{C}(m)^{\text{a}}(v)^{\text{b}}$$

$(1\ \text{J} = 1\ \text{N}\cdot\text{m})$ $$\text{N}\cdot\text{m} = (\text{kg})^{\text{a}}\left(\frac{\text{m}}{\text{s}}\right)^{\text{b}}$$

$\left(1\ \text{N} = 1\,\frac{\text{kg}\cdot\text{m}}{\text{s}^2}\right)$ $$\left(\frac{\text{kg}\cdot\text{m}}{\text{s}^2}\right)(\text{m}) = (\text{kg})^{\text{a}}\left(\frac{\text{m}}{\text{s}}\right)^{\text{b}}$$

$$\frac{\text{kg}\cdot\text{m}^2}{\text{s}^2} = (\text{kg})^{\text{a}}\left(\frac{\text{m}}{\text{s}}\right)^{\text{b}}$$

$$\text{a} = 1,\ \text{b} = 2$$

$$\text{KE} = Cmv^2 \qquad \left(\frac{1}{2}mv^2\right)$$

Exercises 3-10

1. A large batch of metal parts can be processed by machine A in 3 days. Machine B would take 4 days, and machine C would take 5 days. How many days will it take if all three machines operate?

2. If the batch of the preceding problem consists of 10,000 parts, how many are handled by each machine?

3. A beam of electrons has a speed of $2.0(10)^6$ m/s and is measured as 0.0010 A. For every ampere, there is a flow of $6.28(10)^{18}$ electrons/s. Find the number of electrons per centimeter length of the beam (electrons/cm).

4. The equation $H = CAT^4$ is used in calculating the rate at which energy (heat) is radiated from an object. If A is the surface area in ft^2, T the temperature in °R, and H the rate of radiation in Btu/hr, what are the units of the constant C?

5. If peanuts sell at 95 cents/lb and cashews sell at 375 cents/lb, how many lb (W) of cashews should be mixed with 25 lb of peanuts to sell at 210 cents/lb? (*Note:* Write an expression for the total cost in cents and divide it by the expression for total pounds. The total cents/total lb may be set equal to the selling cents/lb of the mix.)

6. Two drops of a certain fluid form a thin film that covers 2400 cm^2 of a water surface. If the volume of a drop is 0.001 cm^3, what is the film thickness?

7. A jet plane consumed 2400 gal of fuel at the rate of 120 lb/min while on a flight of 1200 miles. What was its speed in mph? Assume that the fuel weighs 6 lb/gal.

8. From the observation that the velocity of a freely falling object is a function of g (acceleration of gravity) and h (height from which it has dropped), develop the equation for its velocity. The proportionality constant k is $(2)^{1/2}$.

9. If there are 63.6 g/g-atomic weight and 8.89 g/cm^3 for copper, what is the number of atoms per cubic centimeter of copper? Using Avogadro's number, there are $6.02(10)^{23}$ atoms/g-atomic weight.

10. The following ratios are associated with events that occur in a nuclear reactor.
    N = neutrons produced by thermal fission/neutron absorbed in fuel.
    P = total neutrons produced/thermal neutron absorbed in anything.
    f = neutrons absorbed in fuel/thermal neutron absorbed in anything.
    E = total neutrons produced/neutron produced by thermal fission.
    Write an equation giving f in terms of N, E, and P.

11. Period T of a vibratory motion is the time required for a complete oscillation, such as that of a mass suspended from a spring. Which of the following equations is correct if T is in seconds, m is in kilograms, and K is a spring constant in newtons per meter? C is a constant without dimensions.
    (a) $T = C(mK)^{1/2}$ (b) $T = C(m/K)^{1/2}$ (c) $T = C(K/m)^{1/2}$

12. Let H represent the Btu per hour of heat energy conducted through a solid wall of thickness L and area A, due to a temperature difference ΔT across the wall thickness. The conductivity of the wall material is k, a quantity with the units Btu-in./ft^2-F°-hr. By formulating units, obtain the equation giving H in terms of L, A, k, and ΔT.

4

First-degree equations in one unknown

In Chapter 2 we dealt with the manipulation of algebraic quantities, usually in the form of addition, subtraction, multiplication, and division of those quantities. These operations involved primary collection and simplification of quantities. Little mention was made of equalities and the solutions of equations.

The simplest of the algebraic equations are first-degree equations in one unknown, and in this chapter we discuss such equations and their solutions. First-degree equations are important to students of technical studies because they occur frequently in such topics as heat, electricity, chemistry, mechanics, light, and finance, to mention a few.

4-1 Definitions

In algebra an *equation* is a statement that two quantities are equal. Equations may involve no literal quantities, as in $2 + 3 = 5$; they may involve one literal quantity, as in $x - 4 = 7$; and they may involve more than one literal quantity, as in $x + 2y - 3z = 4$.

A literal quantity in an equation can be identified as the unknown; it may be first degree (x), second degree (x^2), or any degree. The *degree of an equation* is the same as the degree of the term of highest degree. Thus with one unknown shown, $x + 2 = 5$ is a first-degree equation, and $x^2 + 3x = 9$ is a second-degree equation. In this chapter our discussion is restricted to first-degree equations involving one unknown.

An equation in one unknown is either an *identity* or a *conditional equation.* An *identity* is an equation that is true for all values of the variable, and the customary equal sign is a triple bar ($\equiv$); thus $2x + 2 \equiv 2(x + 1)$ is true for any value of x and is called an identity. In contrast to the identity is the *conditional equation*, which is true for a restricted number of values of the unknown; thus $x + 4 = 7$ is true *on the condition* that $x = 3$. The value of the unknown that satisfies the equation is called the *solution* or *root* of the equation. The number of roots agrees with the degree of the equation; so a first-degree conditional equation will have exactly one solution.

There may be more than one alphabetic letter in an equation; then the unknown must be pointed out by the writer or a convention for identifying the unknown can be used (see Sec. 2-2). In most cases, the unknown is a letter chosen from the end of the alphabet nearest z; a solution may involve a letter rather than an ordinary number. Thus the equation $x + a = b$ has the solution $x = b - a$.

4-2 Solution of first-degree equations in one unknown

As noted, a *solution* of an equation is a value of the unknown for which the equation is true. The process of finding a solution is called *solving the equation.* Simple equations can be solved by inspection; for $x + 2 = 5$, we ask the question "To what number do we add 2 to get 5?" Inspection reveals an answer of $x = 3$ without algebraic manipulation. The inspection process fails in more complicated equations like $3x + 0.5 = 6x - 12.4$, particularly when the solution is not an integer.

In *solving* an equation, the *goal* is to *isolate* the unknown on one side of the equal sign. The *tools* that are used are the axioms of real numbers given in Chapter 1 and the following *principle of manipulating equations.*

If the same operation is performed on both sides of an equation, the equation is still true.

Typical operations referred to in the foregoing statement of principle are addition, subtraction, multiplication, division (except by zero), squaring, taking the square root, taking the logarithm, and so on. Two of the most popular operations involve the *additive inverse* $a + (-a) = 0$ and the multiplicative inverse $a \cdot 1/a = 1$. As noted in Sec. 1-4, they achieve the same effect as subtraction and division.

Strategy Because the goal of solving an equation is to isolate the unknown on one side of the equal sign, the strategy of solution involves a step-by-step removal of unwanted addends and multipliers. To perform this removal, inverse operations are used; that is, add away subtractors, subtract away adders, multiply away dividers, and divide away multipliers. Consider some examples.

Example 1A Solve equation $2x - 7 = 17$.

Solution Given $2x - 7 = 17$, then

$2x = 24$ (by adding 7 to both sides)

and $x = 12$ (by dividing both sides by 2).

Example 1B Solve equation $3x + 4x - 9 = 2x + 21$.

Solution Given $3x + 4x - 9 = 2x + 21$, then

$3x + 4x - 9 - 2x = 21$ (by subtracting $2x$ from both sides)

and $5x - 9 = 21$ (by collecting like terms)

and $5x = 30$ (by adding 9 to both sides)

from which $x = 6$ (by dividing both sides by 5).

Example 1C Solve equation $3(x - 4) = 4x - 5$.

Solution Given $3(x - 4) = 4x - 5$, then

$3x - 12 = 4x - 5$ (distributive law removes parentheses)

and $3x = 4x + 7$ (by adding 12 to both sides)

and $-x = 7$ (by subtracting $4x$ from both sides)

and $x = -7$ (by dividing both sides by -1).

Minor complications, such as involvement of literal quantities and ending up with the unknown on the right-hand side of the equal sign, should not be troublesome when solving equations. Literal quantities are presumed to be real numbers through most of this text and should be treated according to the axioms of real numbers. The symmetric property (if $a = b$, then $b = a$) allows the unknown to end up on either side of the equal sign.

Example 2 Solve for n if $l = a + (n - 1)d$.

Solution Given $l = a + (n - 1)d$, then

$$l - a = (n - 1)d \quad \text{(by subtracting } a \text{ from both sides)}$$

and $$l - a = nd - d \quad \text{(by applying the distributive law)}$$

and $$l - a + d = nd \quad \text{(by adding } d \text{ to both sides)}$$

and $$\frac{l - a + d}{d} = n \quad \text{(by dividing both sides by } d\text{).}$$

Checking solutions To check the solution of an equation, substitute the solution into the *original* equation and determine the truth of the resulting statement. If the statement is still true, the solution is probably correct. Two reservations are noted here in the use of the word "probably." First, the arithmetic of the checking process might contain an error that nullifies an error in the solution, leading to the incorrect assumption that the solution is correct. Secondly, that same arithmetic might contain an error and cause the incorrect assumption that the solution was wrong.

Example 3 Check the solution of Example 1C above.

Solution Given equation $3(x - 4) = 4x - 5$ and solution $x = -7$, substitute -7 for x at every place that x occurs in the given statement, obtaining

$$3(-7 - 4) \stackrel{?}{=} 4(-7) - 5$$

where the question mark above the equal sign means "do they equal each other?" Consequently,

$$3(-11) \stackrel{?}{=} 4(-7) - 5$$

from which

$$-33 \stackrel{?}{=} -28 - 5$$

and $$-33 = -33.$$

Because the final statement is true, we are reasonably well assured that $x = -7$ is the solution of the given equation.

Exercises 4-2

In Exercises 1–24, find the solution of the given equation. Check the solution obtained by substitution into the given equation.

1. $5a - 7 = 13$

2. $2a + 7 = 13$

3. $3m + 5 = 14$

4. $4k - 2 = 26$

5. $7x - 8 = 3x$

6. $5y - 8 = 13 - 2y$

7. $4m + 2 = 2m - 9$

8. $3p + 2p = 4p + 12$

9. $4(m - 3) = 6m$

10. $9y = 3(8 + 2y)$

11. $6(k + 4) = 3k + 27$

12. $2(x - 5) = 32 - x$

13. $6(4 - m) + 8m = 36$

14. $9y = 36 - 3(4 - y)$

15. $4(x - 3) + 3(x + 8) = 5x$

16. $12(z - 9) - 14z = 100$

17. $5.5(k + 3) = 2.5(40 - k)$

18. $0.3(T - 40) = 5(0.4T) - 15$

19. $45.6(R + 12) - 109.2 = 16.5(R - 3)$

20. $0.05(M + 5) = 6(0.09M) - 4.1$

21. $6(1.05T - 4) - 0.2(3 - 7T) = 0$

22. $3(k - \frac{1}{2}) - 5(k - \frac{2}{3}) = 1$

23. $9(1.08y + 2) - 0.4(6y + 2) = 7$

24. $2(y + \frac{3}{2}) - 6(y - \frac{2}{3}) = 4$

25. The sum of what two consecutive odd integers is 76?

26. The sum of what three consecutive integers equals 120?

27. The dimensions of a circle are such that the circumference exceeds the radius by 8 in. Find the radius.

28. The length of a rectangle is 3 in. less than twice the width. Its perimeter is 24 in. Find the dimensions.

29. The length of a metallic bar increases by 0.008% of its original length for each centigrade degree increase in temperature. After a 132°C increase in temperature, the length of the bar is 24.7873 in. What was the original length?

30. The perimeter of the object sketched in Fig. 4-1 is 40 in. Find the height of the object.

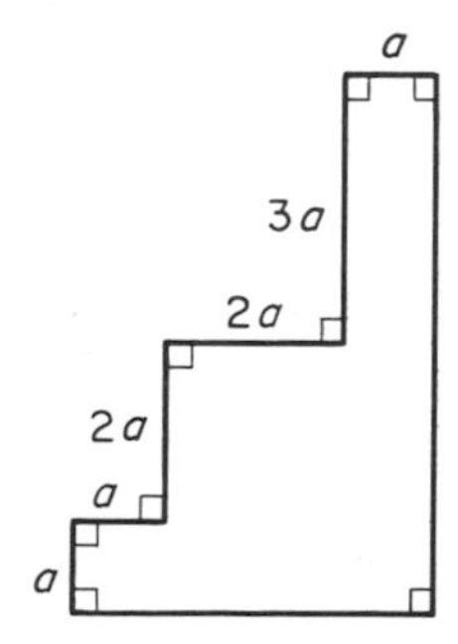

Figure 4-1

31. The perimeter of the Norman window shown in Fig. 4-2 is 22 ft. Find the radius.

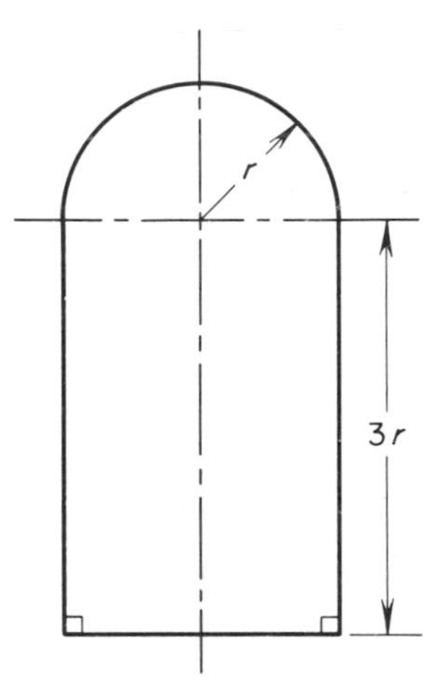

Figure 4-2

32. The perimeter of the racetrack shown in Fig. 4-3 is one mile. Find the radius of the circular ends.

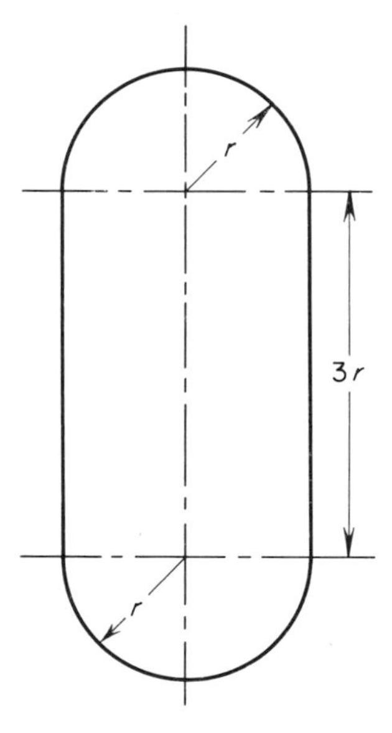

Figure 4-3

33. The Fahrenheit temperature reading always exceeds 9/5 of the centigrade reading by 32. Find the centigrade reading when the Fahrenheit reading is 197°.

34. In Fig. 4-4 the distance $AD = 36$ ft. Find BC and CD.

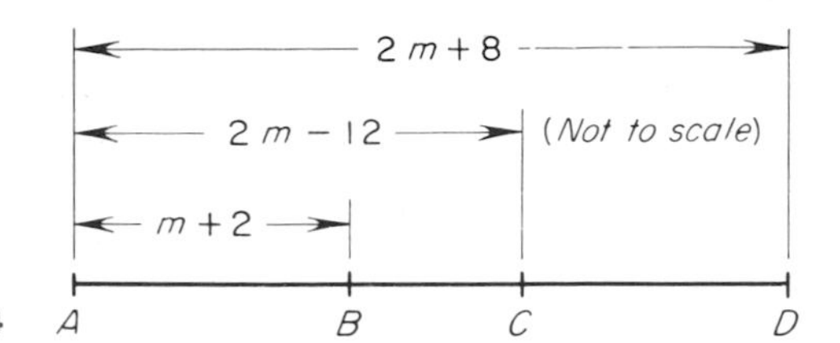

Figure 4-4

35. In simple interest computations, the amount equals the principal plus the interest. The interest accumulates at the rate of 6% of the original principal per year. The amount at the end of 8 years is $512. Find the principal.

36. A triangle is such that side b is 2/3 of side a and side c is 2/3 of side b. The perimeter is 50 in. Find side c.

37. In Fig. 4-5 the angle ABC is 2.5 times the size of angle CBD. If angle ABD is 56°, find angle ABC.

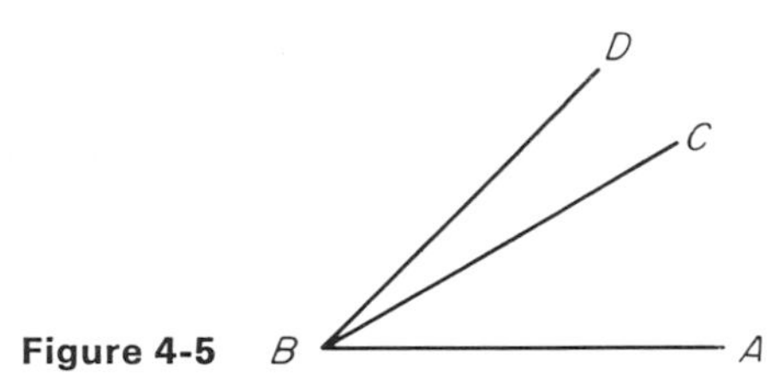

Figure 4-5

4-3 Equations involving fractions

Many fractional equations in one unknown can be reduced to the form

$$Ax + B = 0$$

by clearing fractions. Fractions can be cleared by multiplying each term of the given equation by the least common multiple of the denominators involved. We mentioned the least common multiple briefly in Sec. 2-10 and now we will elaborate on it. The LCM of a group of numbers is the smallest number that is divisible without remainder by each of the numbers. Thus the LCM of the numbers 2, 3, 12, 16, and 48 is 48 because 48 is the smallest number into which each of the given numbers will divide exactly.

Example 4 Find the LCM of 15, 20, 12, and 18.

Solution Let us list the prime factors of each of the given numbers.

$$15 = 3 \cdot 5$$

$$20 = 2 \cdot 2 \cdot 5$$

$$12 = 2 \cdot 2 \cdot 3$$

$$18 = 2 \cdot 3 \cdot 3$$

Note that the factor 2 occurs at most twice in any set of factors; this is in $20 = 2 \cdot 2 \cdot 5$ and in $12 = 2 \cdot 2 \cdot 3$. The factor 3 occurs at most twice; this is in $18 = 2 \cdot 3 \cdot 3$. The factor 5 occurs at most once. The smallest number into which 15, 20, 12, and 18 each will divide evenly contains as its factors the sets of factors cited in the preceding three sentences. So the desired LCM is

$$2 \cdot 2 \cdot 3 \cdot 3 \cdot 5 = 180.$$

If we multiply each term of a fractional equation by the LCM of the denominators of the separate terms of the equation, we will immediately clear all the denominators. Refer to Example 5.

Example 5 Find the solution of

$$\frac{x}{2} + \frac{3x + 1}{8} = \frac{2(x - 4)}{8} - \frac{x + 18}{20}. \tag{1}$$

Solution Using methods shown in Example 4, we find that the LCM of 2, 8, 5, and 20 is 40. Now we will multiply each term of (1) by 40, giving us

$$\frac{(40)(x)}{2} + \frac{(40)(3x + 1)}{8} = \frac{(40)(2)(x - 4)}{5} - \frac{(40)(x + 18)}{20}. \tag{2}$$

Next, each denominator of (2) divides exactly into 40, which is the numerator factor just inserted. The result is an equation without denominators:

$$20x + 5(3x + 1) = (8)(2)(x - 4) - 2(x + 18). \tag{3}$$

We solve (3) by using methods shown in the preceding section, obtaining the solution $x = -5$.

Check: It is well to mention the method of checking the solution $x = -5$. We properly check $x = -5$ against the *original* equation (1), as opposed to any of the other expressions that we have obtained. If we choose to check the solution against (2) or (3) or any other equation that is the result of our manipulation of (1), we may be misled by our conclusion. Thus if (3) were wrong (contained an error that we introduced) and all manipulations between (3) and the final answer were done properly, the final wrong answer would check against (3) but would not check against the original expression (1).

In Example 5 none of the denominators contained a literal quantity. In cases of reciprocals and inverse variation, an unknown may appear in the denominator. This occurrence does not alter the method of solution.

Example 6 Find the solution of

$$\frac{3}{2x} + \frac{2x - 3}{x^2 - 1} = \frac{7}{2(x - 1)}. \tag{4}$$

Solution First we find the LCM of the denominators of (4). Examining the factors, we have

$$\begin{aligned} 2x &= (2)(x), \\ x^2 - 1 &= (x - 1)(x + 1), \\ 2(x - 1) &= (2)(x - 1), \end{aligned} \tag{5}$$

where, from (5), the smallest number into which all of the denominators will divide without remainder is the quantity $2(x)(x - 1)(x + 1)$; this is the LCM by which we will multiply all terms of (4). After multiplying and canceling, we obtain

$$3(x^2 - 1) + 2x(2x - 3) = 7x(x + 1)$$

or

$$3x^2 - 3 + 4x^2 - 6x = 7x^2 + 7x. \tag{6}$$

The second-degree terms of (6) cancel additively (if they did not cancel, we would require different methods of solution), giving

$$13x = -3$$

and the desired solution

$$x = -\frac{3}{13}.$$

Exercises 4-3

In Exercises 1–20, find the value of the unknown that satisfies the given equation. Check all answers by substituting the solution back into the original equation.

1. $\frac{a}{10} = \frac{3}{5}$
2. $\frac{4}{b} = \frac{32}{12}$
3. $\frac{7}{91} = \frac{12}{w}$
4. $\frac{8}{15} = \frac{m + 2}{9}$
5. $\frac{11}{2y} = \frac{5}{18}$
6. $\frac{9}{2(x + 1)} = \frac{3}{8}$
7. $\frac{2k - 4}{6} = \frac{14 - k}{9}$
8. $\frac{4}{a + 1} = \frac{5}{a - 1}$
9. $\frac{m + 5}{4} + \frac{1}{2} = \frac{2m - 3}{6}$
10. $\frac{r - 3}{6} - \frac{2}{3} = \frac{2r - 9}{4}$
11. $\frac{b - 2}{6} - \frac{12}{21} = \frac{b - 4}{7}$
12. $\frac{m + 10}{3} - \frac{3}{9} = \frac{m + 5}{2}$
13. $\frac{3}{x} - \frac{3}{x^2 - x} = \frac{4}{3x}$
14. $(m - 1)^2 - 3(m^2 + 2m) - m(1 - 2m) = 0$

15. $\dfrac{6}{x-1} - \dfrac{4}{x+4} = \dfrac{2}{x}$

16. $\dfrac{b}{3} - \dfrac{b-3}{4} = \dfrac{b+4}{8}$

17. $\dfrac{6}{s+2} = \dfrac{8}{s-3} - \dfrac{2}{s-1}$

18. $\dfrac{2x+5}{x^2+3x+2} + \dfrac{3x-2}{x^2-1} = \dfrac{5x}{x^2+x-2} + \dfrac{2}{x^2+5x+6}$

19. $\dfrac{7-m}{m^2-1} - \dfrac{m+1}{m^2+5m+4} = \dfrac{-2m+1}{m^2+3m-4}$

20. $\dfrac{4}{a+2} + \dfrac{1}{a-2} = \dfrac{12}{4(a-2)}$

21. The total resistance R of a circuit with resistances r_1 and r_2 in parallel is given as

$$\frac{1}{R} = \frac{1}{r_1} + \frac{1}{r_2}.$$

Find the value of r_1 if $R = (r_1 + 3)/2$ and $r_2 = r_1 - 3$.

22. The angle ϕ between two intersecting lines may be found by the equation

$$\tan \phi = \frac{m_2 - m_1}{1 + m_1 m_2},$$

where m_1 and m_2 are the slopes of the lines. Find m_1 if $\tan \phi = 3.22$ and $m_2 = 1.04$.

23. The sum S of n terms of a geometric progression is given by the equation

$$S = \frac{a - rl}{1 - r},$$

where a is the value of the first term, r the common ratio, and l the value of the nth term. Find the ratio if $a = 27$, $l = 16/3$, and $S = 211/3$.

24. The length of a certain rectangle exceeds its width by 4 ft. If the width is increased by 4 ft and the length decreased by 3 ft, the area is increased by 5 sq ft. Find the dimensions of the original rectangle.

25. In Fig. 4-6, EC is parallel to AB and AB exceeds EC by 5 in. The area of triangle ABD is three times the area of triangle ECD. Find AB.

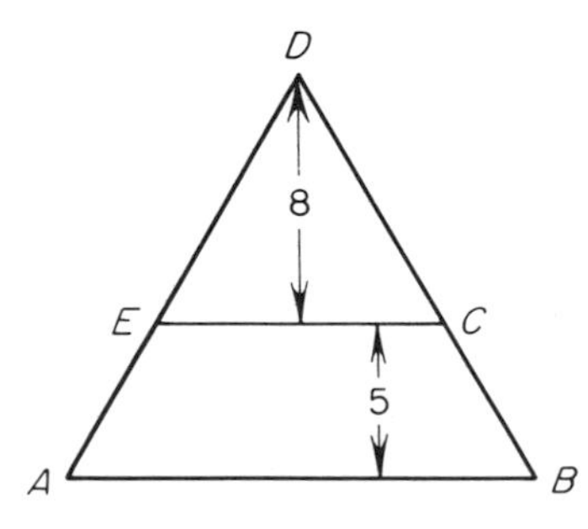

Figure 4-6

26. In Fig. 4-7, forces A and B are applied to a lever as shown. If A is 3 greater than B, find both forces, assuming the lever is in balance.

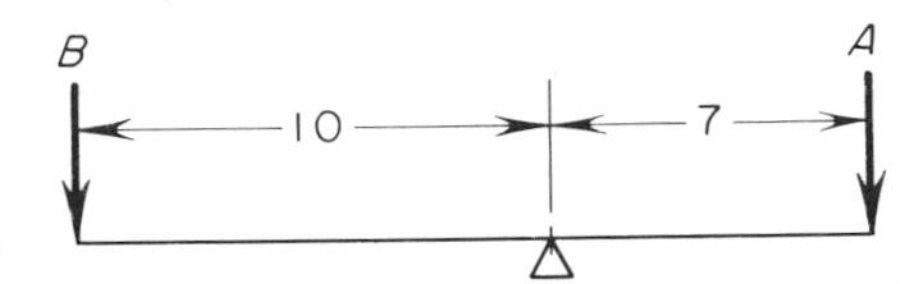

Figure 4-7

4-4 Literal equations and formulas

In the exercises and examples given earlier we confined ourselves to equations that involved one or more unknown terms, but each unknown term was described in terms of the same literal quantity. Most formulas taken from the sciences, however, involve more than one literal quantity. Our goal is to solve, for one of the literal quantities, an equation involving two or more literal quantities. It will still be an equation in one unknown, and the method of solution will remain unaltered.

We list here a suggested set of operations that are useful in obtaining the solution of an equation in one unknown. There are other ways of reaching a solution; the steps listed are intended only as a guide.

STEP 1 *Clear fractions by multiplying each term by the least common multiple of the denominators.*

STEP 2 *Clear parentheses.*

STEP 3 *Add quantities to each side (or subtract them) so that all terms involving the unknown will collect in one member of the equation and all others in the other member.*

STEP 4 *Factor out the unknown as a common factor of the one or more terms involving the unknown.*

STEP 5 *Divide both members by the coefficient of the unknown.*

Consider some examples.

Example 7 Solve for m_2 if

$$\tan \phi = \frac{m_2 - m_1}{1 + m_1 m_2}. \tag{7}$$

Solution Applying step 1 to (7), we multiply both members by $1 + m_1 m_2$ to obtain an expression with the fraction cleared:

$$(1 + m_1 m_2) \tan \phi = m_2 - m_1. \tag{8}$$

Applying step 2 to (8), we remove parentheses to obtain

$$\tan \phi + m_1 m_2 \tan \phi = m_2 - m_1.$$

Now we apply step 3. The goal here is to isolate the unknown m_1; that is, we wish to transpose all terms containing m_1 to one side of the equal sign and all others to the other side. We can do so by adding m_1 to each side and subtracting $\tan\phi$ from each side:

$$m_1 m_2 \tan\phi + m_1 = m_2 - \tan\phi. \tag{9}$$

Factoring the left side of (9) in compliance with step 4 gives

$$m_1(m_2 \tan\phi + 1) = m_2 - \tan\phi. \tag{10}$$

Following step 5, we divide each member of (10) by the coefficient of m_1—namely, $m_2 \tan\phi + 1$—obtaining the desired result

$$m_1 = \frac{m_2 - \tan\phi}{1 + m_2 \tan\phi}. \tag{11}$$

Check: To check our solution, we substitute (11) into (7) to see if the latter satisfies the former. This is similar to the method described in preceding sections, but the algebra here is much more difficult.

$$\tan\phi \stackrel{?}{=} \frac{m_2 - \dfrac{m_2 - \tan\phi}{1 + m_2 \tan\phi}}{1 + \left(\dfrac{m_2 - \tan\phi}{1 + m_2 \tan\phi}\right) m_2} \tag{12}$$

Multiplying the numerator and denominator of the right member of (12) by $1 + m_2 \tan\phi$, results in

$$\tan\phi \stackrel{?}{=} \frac{m_2 + m_2^2 \tan\phi - m_2 + \tan\phi}{1 + m_2 \tan\phi + m_2^2 - m_2 \tan\phi} \tag{13}$$

Collecting terms, factoring, and canceling in (13), we have

$$\tan\phi = \tan\phi.$$

In Example 7 we solved an example that required the use of all five of the steps mentioned earlier. Many equations require fewer steps; note Example 8.

Example 8 Solve for r if

$$F = \frac{Wv^2}{gr}.$$

Solution First we clear fractions, obtaining

$$Fgr = Wv^2.$$

There are no parentheses to clear; neither transposition nor factoring is necessary. We divide now by the coefficient of r to obtain the solution

$$r = \frac{Wv^2}{Fg}.$$

Example 9 A metallic bar is originally of length L_0. The length changes c times the original length for each degree centigrade change in the temperature. Find the new length L after a temperature change of Δt and solve the resulting equation for L_0.

Solution Formulating the given information, we have

$$L = L_0 + L_0 c\,\Delta t.$$

Factoring the right side results in

$$L = L_0(1 + c\,\Delta t).$$

Dividing by $1 + c\,\Delta t$ gives

$$L_0 = \frac{L}{1 + c\,\Delta t}.$$

Example 10 A uniformly accelerating body increases its velocity from v_0 to v_t over a given time period. Its average velocity $\bar{v}$ during that time period equals one-half the sum of the initial and final velocities. Obtain an expression for v_0 in terms of $\bar{v}$ and v_t.

Solution From the given information,

$$\bar{v} = \frac{v_0 + v_t}{2}. \tag{14}$$

Clearing fractions and transposing in (14), we have

$$v_0 = 2\bar{v} - v_t.$$

Exercises 4-4

In Exercises 1–30, solve the given equation for the literal quantity indicated in parentheses. Each equation may be encountered in either mathematics or science.

1. $E = IR$, (I)
2. $I = \dfrac{E}{R + r}$, (R)
3. $I = \dfrac{2E}{R + 2r}$, (r)
4. $I = \dfrac{E}{R + r/2}$, (r)
5. $w = 2\pi f$, (f)
6. $x = 2\pi fL - \dfrac{1}{2\pi fC}$, (C)

7. $\frac{P_1}{P_2} = \frac{V_2}{V_1}$, (V_1)

8. $P_t V_t = P_0 V_0 \left(1 + \frac{t}{273}\right)$, (t)

9. $W = RI^2 t$, (t)

10. $\frac{v_1^2}{2g} + \frac{p_1}{\gamma} + x_1 = \frac{v_2^2}{2g} + \frac{p_2}{\gamma} + x_2 + h_L$, (p_1)

11. $V_s = \pi[(r+h)^2 - r^2]$, (r)

12. $R = \frac{r_1 r_2}{r_1 + r_2}$, (r_2)

13. $k = 0.4\left(1 - \frac{A_1}{A_2}\right)$, (A_2)

14. $a^2 = b^2 + c^2 - 2bc \cos A$, $(\cos A)$

15. $\frac{1}{R} = \frac{1}{r_1} + \frac{1}{r_2} + \frac{1}{r_3}$, (R)

16. $s = \frac{gt^2}{2} + v_0 t + s_0$, (g)

17. $A = \frac{h}{2}(b_1 + b_2)$, (b_1)

18. $k = \frac{a-b}{a+b}$, (b)

19. $A = 2(xy + xz + yz)$, (z)

20. $A = 2\pi r^2 + 2\pi rh$, (h)

21. $V = \pi h^2\left(r - \frac{h}{3}\right)$, (r)

22. $R = \frac{P}{L}(L - b + a)$, (L)

23. $h = x - \frac{y'[1 + (y')^2]}{y''}$, (y'')

24. $r^2 = \frac{(s-a)(s-b)(s-c)}{s}$, (a)

25. $\delta = \frac{Px^2}{48EI}(3l - 4x)$, (l)

26. $E = k(T_1^4 - T_2^4)$, (T_1^4)

27. $\bar{e} = L \cdot \frac{i_2 - i_1}{t_2 - t_1}$, (t_2)

28. $S = \frac{a - rl}{1 - r}$, (r)

29. $l = a + (n-1)d$, (n)

30. $S = \frac{n}{2}(a + l)$, (a)

31. In Fig. 4-8 two concentric circles differ in radius by an amount h. Find r in terms of both h and area A of the ring.

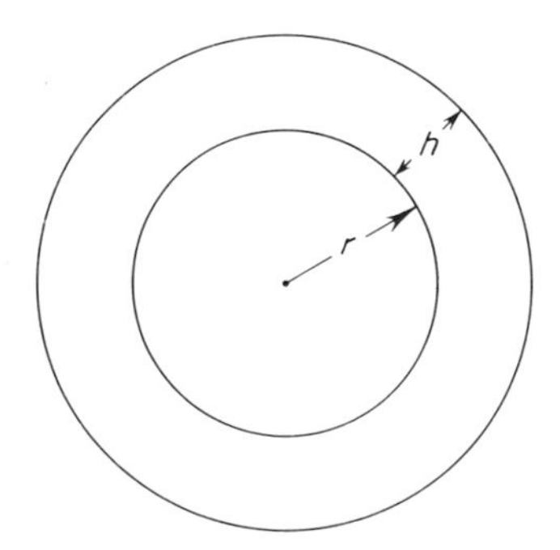

Figure 4-8

32. In Fig. 4-9 a tracer starts at A, traces once around the circle of radius r, proceeds to B, goes once around the square of side s, back to B, and then

returns to A. Give an expression for r in terms of s, h, and D, where D is the total distance traversed by the tracer.

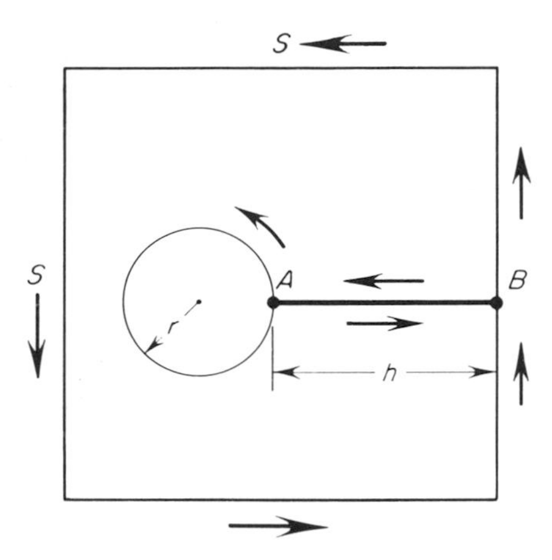

Figure 4-9

33. In Fig. 4-10 a trapezoid with bases a and b is mounted on a rectangle of sides b and c. Give an expression for c in terms of a, b, h, and A, where A is the area of the combined figures.

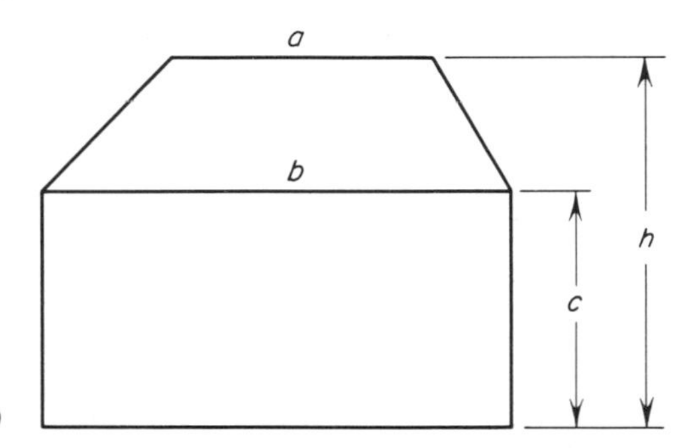

Figure 4-10

34. Total resistance R in a circuit is given by

$$\frac{1}{R} = \frac{1}{r_1} + \frac{1}{r_2},$$

where r_1 and r_2 are resistances in parallel. If $r_1 = k$ and r_2 exceeds r_1 by 3, find R in terms of k alone.

35. A boat cruises at a uniform rate of R mph in still water. It makes a round trip of m miles up and m miles down a river. The river's current flows at a uniform rate of r mph. The time required to make the round trip is T hours. Using this information, obtain the equation

$$m = \frac{T}{2R}(R^2 - r^2).$$

36. A square metallic plate of side s is heated, causing each side to increase in length by an amount Δs and changing the area of the plate by an amount ΔA. Show that

$$\frac{\Delta A}{\Delta s} = 2s + \Delta s.$$

37. A vehicle travels h miles in t_1 hours and then k miles in t_2 hours, resulting in an average speed $\bar{v}$ for the total trip. Formulate the equation

$$h = \bar{v}(t_1 + t_2) - k.$$

38. Figure 4-11 shows a seesaw arrangement. Forces A, B, and C are applied in the positions shown. The turning effect (moment) provided by a force equals the product of the force times the distance of that force from the fulcrum F. To have equilibrium, the sum of the clockwise moments equals the sum of the counterclockwise moments. Using this information, obtain an expression for x in terms of A, B, C, h, and k.

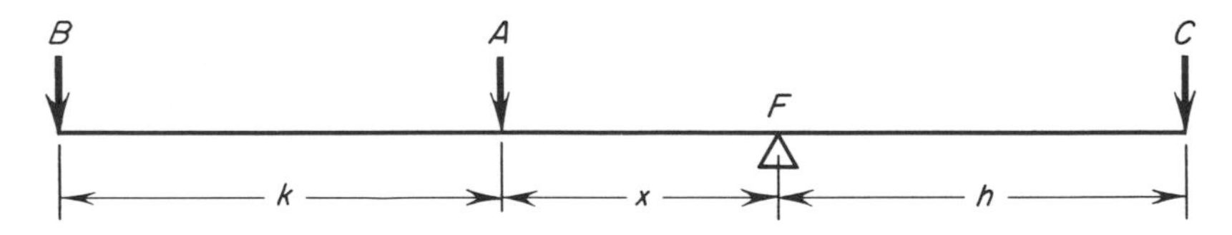

Figure 4-11

5

Linear equations

The more common types of variation that occur in technical applications are classified as *algebraic, logarithmic, exponential,* and *trigonometric.* Perhaps the simplest algebraic variation is the so-called *linear* type. A common instance is temperature measurement, in which two different scales are popularly used: the Fahrenheit and the centigrade. A change in temperature changes both the Fahrenheit and centigrade temperature readings. Therefore it is most useful to be able to (a) produce a relationship between Fahrenheit and centigrade for any temperature, (b) find a centigrade reading mathematically if we know the Fahrenheit reading and vice versa, (c) construct a graph or picture of the variation involved in order to make further interpretations easily, and (d) manipulate the Fahrenheit–centigrade relationship in broadening its use.

In this chapter we are concerned with deriving and plotting linear forms and with applying these forms to certain useful applications.

5-1 Functions and graphing

In the preceding chapter we dealt with equations that contained only one unknown. Most were *conditional* equations—that is, equations that were true under the condition that the unknown took on a specific value, which we called the *solution*. In this chapter we consider the simplest equations in two unknowns, which (because they graph as straight lines) are called *linear equations*. Consider a physical example of such an equation.

Functions If a car moves at a steady rate of 30 mph, the distance traveled in miles is always 30 times the elapsed time in hours. Using s for distance and t for time, the relationship between s and t is expressed by the simple equation $s = 30t$, which clearly contains both variables. From this equation we observe that an *indefinite* number of *number pairs* satisfies the equation; examples include (a) when $t = 2$, $s = 60$, (b) when $t = 4$, $s = 120$, and (c) when $t = 1.6$, $s = 48$. We note, too, the reverse condition; if we have a set of number pairs (2, 60), (4, 120), (1.6, 48), a relationship $s = 30t$ is suggested. This relation is called a *function*. We state here that s is a function of t, that s and t are the variables in the relationship, and that, as we assign an arbitrarily chosen value to the *independent variable* t, the *dependent variable* s is obliged to take on exactly one value if the equation is to remain true and if the equation represents a true function.

Designating a variable as independent or dependent is sometimes troublesome. In an equation like $x + y = 12$ we can solve for y and obtain $y = 12 - x$ or we can solve for x and obtain $x = 12 - y$. It is often useful to regard the *dependent variable* as the *variable that has been solved for*; thus x or y might qualify as dependent in our example. In the equation $y = 2x - 1$ we say that y is a function of x, that y is the dependent variable, and that x is the independent variable. Exercising the independence of x, we can assign arbitrarily chosen values to x and obtain the corresponding dependent values of y, thereby producing a table of values as follows.

x	10	5.1	2	1	0	-2	$-\frac{1}{4}$	-100	(independent)
y	19	9.2	3	1	-1	-5	$-1\frac{1}{2}$	-201	(dependent)

The general notation that expresses a dependent variable y in terms of an independent variable x is $y = f(x)$, which is read "y is a function of x." *Note*: $f(x)$ must *not* be considered *f times x*.

A function may be valid only for certain values of the independent variable. Consider a worker's earnings (E) as a function of time (t) if he is paid \$5 per hour. His total earnings will be expressed by the equation $E = 5t$, *assuming that t is 40 hours or less*. Clearly if t is greater than 40 hours in a week, then a different equation would express the worker's weekly earnings. Thus a function can be defined for only a restricted set of values of the independent variable; this set is called the *domain of definition* of the function. Concurrently, the dependent variable becomes confined along with the restrictions on the independent variable; this limitation imposed by the domain is called the *range*.

Cartesian coordinate plane The preceding discussion of functions clearly indicates that we are dealing with two variables in this chapter. In Sec. 2-1 we introduced the number line and were able to plot points along it; we described these points in terms of a single variable like x. Now we are faced with plotting number pairs or graphs of equations suggested by number pairs. In order to do so, we must abandon the number line and look to a graphing system that will take care of two variables at the same time. The system that we use here is called the *cartesian coordinate system* or *rectangular coordinate system*. A description of this system follows with reference to Fig. 5-1. Several characteristics should be noted.

Characteristics of the cartesian coordinate system (see Fig. 5-1)

1. Two axes, intersecting at right angles, are constructed on a plane surface. On these axes uniformly spaced calibrations are marked off (similar to the number line in Sec. 1-2).

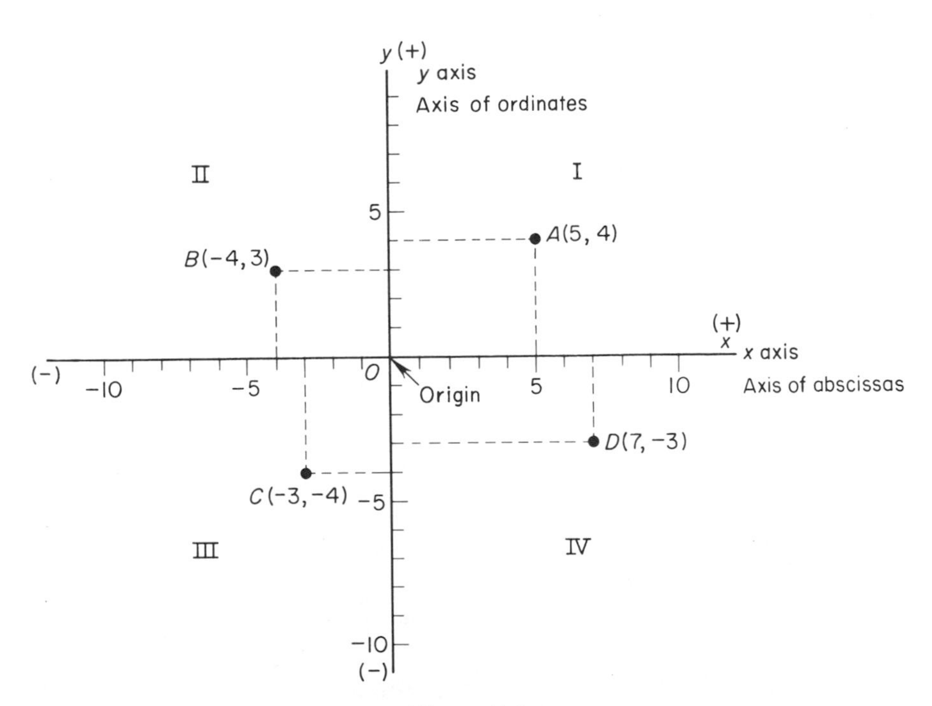

Figure 5-1

2. Positive directions are assigned as upward and to the right. Negative directions are down and left.
3. The axes intersect at the zeros of the two numbering systems introduced. This point of intersection is called the *origin.*
4. The *horizontal* axis is called the *axis of abscissas.* It is generally designated the x *axis* when a function of x is graphed. Because x is usually the independent variable in a function, the horizontal axis normally belongs to the independent variable.
5. The *vertical* axis is the *axis of ordinates* or y *axis* or axis for the dependent variable.
6. The axes divide the plane into four portions called *quadrants.* These quadrants are usually numbered with roman numerals, starting with the upper-right quadrant as number I and proceeding in a counterclockwise direction.

Conventions on plotting number pairs A statement that "x is 5 when y is 4" suggests a number pair that is symbolized as (x, y) generally and as $(5, 4)$ specifically. (Note that the x value is given *first* in the parentheses for the sake of order.) To plot $(5, 4)$ on the cartesian coordinate plane, we observe that the resulting *point* is located five units to the *right* of the origin and four units *above* the origin; this is shown as the point $A(5, 4)$ in Fig. 5-1. Similarly, the point $(-4, 3)$ is four units to the *left* and three units above the origin as point B. The pair $(-4, 3)$ describes the *coordinates* of point B.

Graphing a function Because a function ordinarily satisfies an indefinite number of number pairs, it may be graphed by plotting several of these pairs as points on a cartesian plane and then joining these points with a smooth line or curve. The number pairs or points can be determined by arbitrarily assigning values to the independent variable and then substituting these values into the function to obtain the corresponding values of the dependent variable. Consider Example 1.

Example 1 Graph equation $y = 2x - 1$.

Solution We give ourselves arbitrarily chosen values of x as shown in the upper half of the table.

x	-2	0	1	2	4	5	(independent)
y	-5	-1	1	3	7	9	(dependent)

The y values in the lower half are generated by substituting the x values into equation $y = 2x - 1$. Points $(-2, -5)$, $(0, -1)$, $(1, 1)$, and so on suggested by the table are plotted on the cartesian plane and joined by a smooth line as in Fig. 5-2. This smooth line is given equation

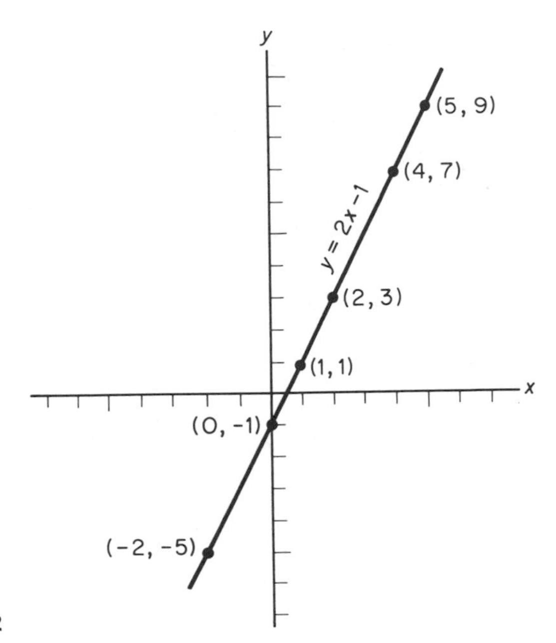

Figure 5-2

$y = 2x - 1$ because all the number pairs satisfying it will plot as points on this line.

Note: The values of x in Example 1, although chosen arbitrarily, are not chosen without some restraint. We customarily hover near the origin when graphing (there are notable exceptions); so we choose values of x near the origin. It might be regarded as absurd to choose fractional, decimal, very large, and very small values of x.

In this chapter we are concerned primarily with linear equations that graph as straight lines.

Exercises 5-1

In which quadrants are the points shown in Exercises 1–5?

1. $(3, -5)$ **2.** $(-2, -4)$ **3.** $(-3.5, 4.2)$

4. $(9.8, \pi)$ **5.** $\left(\frac{33}{5}, -\frac{19}{8}\right)$

6. If a point lies on the x axis, what is its ordinate?

7. If a point lies on the y axis, what is its abscissa?

8. If a point lies on both the x axis and the y axis, what are its coordinates?

9. What is the sign of the ordinate of a point if it lies in either Quadrant III or IV?

In Exercises 10–12, you are given point $P(3, k)$ and $Q(3, k + 1)$.

10. Are P and Q on the same vertical line?

11. Are P and Q on the same horizontal line?

12. Does P lie above or below Q?

13. Plot a triangle ABC with vertices $A(-3, -6)$, $B(-4, 7)$, and $C(5, 0)$.

14. Three vertices of a square are at $(-1, 3)$, $(4, 3)$, and $(4, -2)$. What are the coordinates of the fourth vertex?

15. The line segment AB terminates at $A(-3, -8)$ and $B(-3, 4)$. What are the coordinates of the midpoint of AB?

16. A particle starts at $A(8, -10)$ and moves in such a way that each move covers three units to the left and one unit upward. What are the coordinates of the point after six such moves?

17. The coordinates of the vertices of a square $ABCD$ are $A(3, 0)$, $B(0, 3)$, $C(3, 6)$, and $D(6, 3)$. What are the coordinates of the point of intersection of its diagonals?

18. A circle of radius five units is constructed on a cartesian coordinate plane. How many intersections does it have with the x axis if its center is at (a) $(3, 4)$, (b) $(-3, 5)$, (c) $(-9, 6)$, and (d) $(6, 0)$?

19. Graph the given equations.

(a) $y = x$	(b) $y = 2x$	(c) $y = x + 1$
(d) $y = 2x - 1$	(e) $y = -\frac{1}{2}x + 1$	(f) $y = -\frac{1}{2}x - 3$
(g) $y = \frac{3}{4}x + 2$	(h) $y = 3x - 6$	(i) $y = 0.1x + 0.5$
(j) $y = \frac{9}{5}x + 32$	(k) $y = -2x - 0.75$	

5-2 Distance between two points; slope

Given any two points P_1 and P_2 along a horizontal line, we can readily find the distance between them. In Fig. 5-3 the distance from P_1 to P_2 is $+7$ units of length; it can be determined by counting the units from P_1 to P_2. The

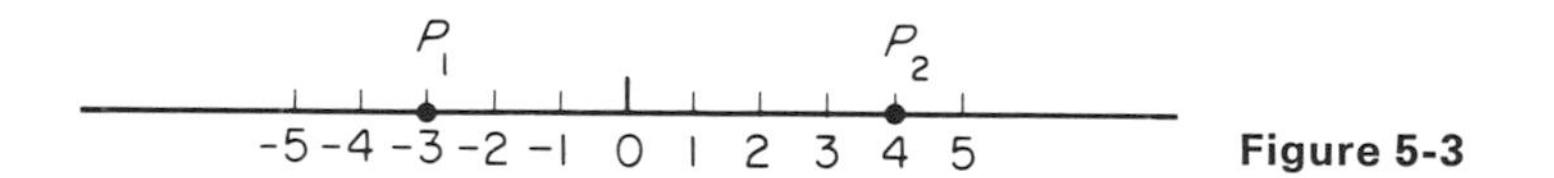

Figure 5-3

choice of $+7$ units, as opposed to the choice of -7 units, is made because we are proceeding to the right (a positive direction). This $+7$ can also be shown as

$$P_1P_2 = 4 - (-3) = +7,$$

where the value of the initial point (-3) is subtracted from the value of the terminal point $(+4)$. Reversing directions in Fig. 5-3, the distance P_2P_1 can be shown as

$$P_2P_1 = -3 - (+4) = -7.$$

Distances along horizontal and vertical lines in a cartesian coordinate plane are found as shown in Fig. 5-3.

Let us assume that the two points under consideration do not lie on a vertical or horizontal line but on a slanting line. Refer to Fig. 5-4. Here we are asked to find the distance from P_1 to P_2. The *horizontal* distance P_1M is the x value at $P_2(x_2 = 10)$ less the x value at $P_1(x_1 = 1)$ or

$$P_1M = 10 - (1) = 9.$$

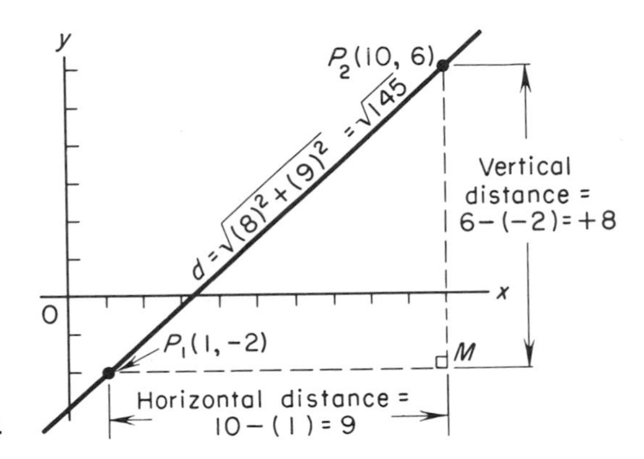

Figure 5-4

Similarly, the *vertical* distance from P_1 to P_2 is the y value at $P_2(y_2 = 6)$ less the y value at $P_1(y_1 = -2)$ or

$$MP_2 = 6 - (-2) = 8.$$

By the Pythagorean Theorem, the distance is

$$d = \sqrt{(\text{horizontal distance})^2 + (\text{vertical distance})^2} \qquad (1)$$

or $$d = \sqrt{(8)^2 + (9)^2} = \sqrt{145}.$$

In the generalized sense, the distance between two points is shown in Fig. 5-5. There the given points are $P_1(x_1, y_1)$ and $P_2(x_2, y_2)$. The *horizontal*

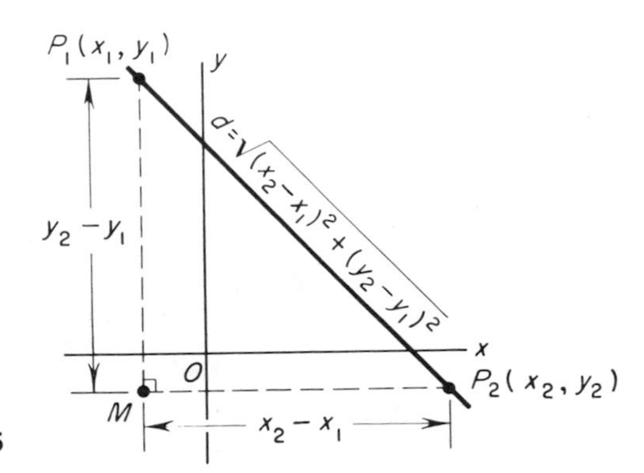

Figure 5-5

distance from P_1 to P_2 is

$$MP_2 = x_2 - x_1 \tag{2}$$

and the *vertical* distance between P_1 and P_2 is

$$P_1M = y_2 - y_1. \tag{3}$$

By the Pythagorean Theorem and Eqs. (1), (2), and (3),

$$d = \sqrt{(x_2 - x_1)^2 + (y_2 - y_1)^2}, \tag{4}$$

where (4), called the *distance equation*, is useful in finding the distance between any two points $P_1(x_1, y_1)$ and $P_2(x_2, y_2)$.

Example 2 Find the distance between $(-6, -2)$ and $(-1, 12)$.

Solution Call either point P_1. Let us choose $P_1(x_1, y_1) = (-6, -2)$ and $P_2(x_2, y_2) = (-1, 12)$. Now we have the assignments $x_1 = -6$, $y_1 = -2$, $x_2 = -1$, and $y_2 = 12$. Substituting into (4), we have

$$d = \sqrt{[-1 - (-6)]^2 + [12 - (-2)]^2} = \sqrt{(5)^2 + (14)^2} = \sqrt{221}.$$

We can find the distance mentally. Consider simply that the x value changes from -6 to -1, or 5 units, and the y value changes from -2 to $+12$, or 14 units. Then

$$d = \sqrt{5^2 + 14^2} = \sqrt{221}.$$

Example 3 Find the distance from $(-2, 9)$ to $(-2, -7)$.

Solution Note that the x value of both points is the same; therefore the line joining the two points is vertical. Now the change in y is the distance. This change in y is

$$-7 - (+9) = -16,$$

which is the distance requested.

The *slope* of a line is defined with reference to Fig. 5-6. There $P_1(x_1, y_1)$ and $P_2(x_2, y_2)$ are any two given points on a line. The slope m of the line is defined by the relationships taken from Fig. 5-6:

$$m = \frac{y_2 - y_1}{x_2 - x_1} = \frac{\text{change in } y}{\text{change in } x} = \frac{\Delta y}{\Delta x} = \frac{\text{rise}}{\text{run}}. \tag{5}$$

Example 4 Find the slope of the straight line passing through $P_1(4, -3)$ and $P_2(-5, 9)$.

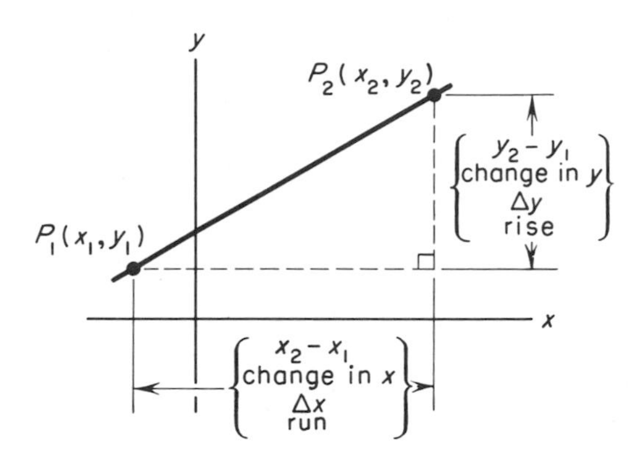

Figure 5-6

Solution Using P_1 as the initial point and P_2 as the terminal point,

$$y_2 - y_1 = \Delta y = \text{rise} = \text{change in } y = 9 - (-3) = +12,$$

$$x_2 - x_1 = \Delta x = \text{run} = \text{change in } x = -5 - (4) = -9.$$

Applying (5), we have

$$m = \frac{y_2 - y_1}{x_2 - x_1} = \frac{+12}{-9} = -\frac{4}{3}.$$

Discussion: If we had chosen P_2 as the initial point and P_1 as the terminal point, we would have had

$$m = \frac{y_1 - y_2}{x_1 - x_2} = \frac{-12}{+9} = -\frac{4}{3}.$$

This asserts that the slope between two points is unchanged if we reverse the assignments of the initial and terminal points.

Example 5 A vehicle is accelerating uniformly; that is, during equal time intervals its velocity changes by equal amounts. If velocity $v =$ 4 fps when time $t = 3$ sec and $v = 14$ fps when $t = 8$ secs, find the rate of change of velocity per unit of time.

Solution Figure 5-7 shows a graphical solution. In $8 - 3 = 5$ sec the velocity has changed $14 - 4 = 10$ fps. This means that a 5-sec change in t is accompanied by a 10-fps change in v, or

$$\frac{\Delta v}{\Delta t} = \frac{10 \text{ fps}}{5 \text{ sec}} = 2 \text{ ft/sec}^2 \tag{6}$$

Solution (6) is shown in Fig. 5-7, where we see a 2-fps change in the velocity for each 1-sec change in time. We also observe from the units in (6) that $\Delta v/\Delta t$ is acceleration.

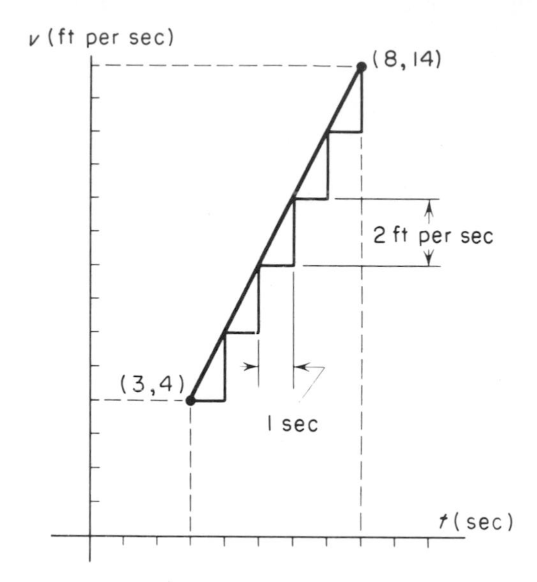

Figure 5-7

Discussion: Using the definition of slope,

$$m = \frac{v_2 - v_1}{t_2 - t_1} = \frac{14 - 4}{8 - 3} = 2,$$

where m can be interpreted as acceleration. Actually, the slope of a line can assume various interpretations, depending on the quantities assigned to the abscissa and ordinate axes.

Exercises 5-2

In Exercises 1–13, find the distance between the given points. Also find the slope of the straight line joining the points.

1. (0, 0), (3, 4) **2.** (0, 0), (4, 3) **3.** (−5, 12), (0, 0)

4. (−5, −12), (0, 0) **5.** (6, −14), (−3, 4) **6.** (4, 5), (6, −7)

7. (−6, 4), (3, −9) **8.** (7, −9), (−4, −6) **9.** (12, −3), (15, −3)

10. (5.2, 4), (4.2, −3.6) **11.** (6, −3), (6, 7) **12.** $(a, b), (b, a)$

13. $(a + b, b), (a, a + b)$

In Exercises 14-20, what are the units of the slope of a line if the abscissa and ordinate units are as given?

14. Abscissa, time; ordinate, distance.

15. Abscissa, number of tickets; ordinate, total cost.

16. Ordinate, force; abscissa, area.

17. Abscissa, time; ordinate, number of electrons passing a point.

18. Abscissa and ordinate units the same.

19. Ordinate, interest accumulated; abscissa, time.

20. Ordinate, cubic feet; abscissa, square feet.

21. Show by use of the definition of slope that points $(-2, -5)$, $(1, -4)$, $(4, -3)$ lie on the same straight line.

22. Using the distance formula, show that points $(-2, -5)$, $(1, -4)$, and $(0, -1)$ are the vertices of a right triangle. Without sketching the triangle, identify the right angle.

23. A vehicle travels at a uniform velocity. The distance d traveled is 24 ft when time t is 4 sec. Also, $d = 52$ ft when $t = 11$ sec. What is the velocity of the vehicle? Using the velocity determined, find d when $t = 10$ sec. What was d when t was zero?

24. Two vehicles A and B travel the same route with the same velocity. For A, $d = 120$ ft when $t = 2$ sec and $d = 300$ ft when $t = 6$ sec. For B, $d = 80$ ft when $t = 2$ sec. Find d for vehicle B when $t = 6$ sec. At any instant, by how many feet does A lead B?

25. A straight line passes through $(3, 0)$ with a slope of $-1/2$. A point P on the line has an abscissa of 9. What is the ordinate of P?

26. A straight line passes through $(8, 6)$ with a slope of -4. How many units from the origin is the point where the line crosses the y axis?

27. If a line passes through $(6, 4)$ and $(-3, -7)$, at what point does it intercept the y axis?

5-3 Division of a line segment

A line segment can be divided into any number of equal parts. Let us consider the division of a line segment by referring to the method of division used in plane geometry.

Example 6 Divide line segment AB_5 shown in Fig. 5-8 into five equal parts.

Solution Using a plane geometry method, we construct any line AC forming an angle with AB_5. Along AC mark off five equal units of length of any convenient size; these equal lengths are AA_1, A_1A_2, A_2A_3, etc. in Fig. 5-8. Draw the line A_5B_5. Now construct A_4B_4, A_3B_3, etc. parallel to A_5B_5. Now B_1, B_2, B_3, and B_4 divide AB_5 into five equal parts. This method uses the theorem that asserts that "if parallel lines intercept equal line segments on one transversal, they will intercept equal line segments on any other transversal."

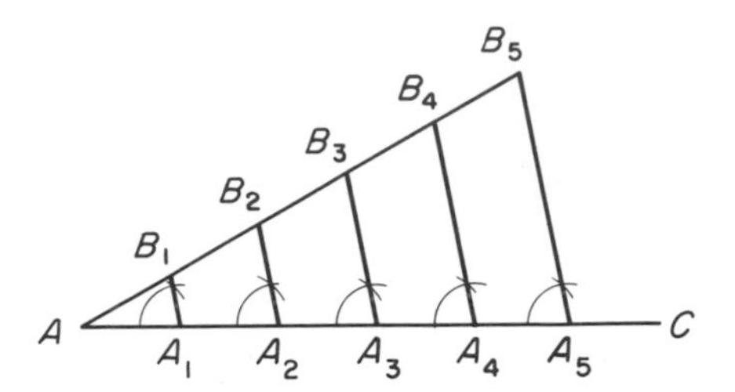

Figure 5-8

The method shown in Fig. 5-8 can be used, slightly modified, in determining the coordinates of the points of division of a line segment if the coordinates of the endpoints are known. Before presenting a general case, let's consider a specific one.

Example 7 Given the line segment AB with endpoints $A(-6, 2)$ and $B(4, 7)$, divide the line segment into five equal parts and give the coordinates of each of the points of division.

Solution Refer to Fig. 5-9, where we have plotted A and B. The horizontal distance from A to B is

$$4 - (-6) = +10$$

units. These 10 units can be divided into five equal parts of 2 units each. A horizontal line AA_5 is divided into five equal parts. Vertical lines A_1B_1, A_2B_2, etc. are drawn, dividing AB into five equal parts according to the geometric method shown in Example 6. The x coordinates of B_1, B_2, etc. are

$$x_a = -6 + \tfrac{1}{5}(10) = -4,$$
$$x_b = -6 + \tfrac{2}{5}(10) = -2,$$
$$x_c = -6 + \tfrac{3}{5}(10) = 0,$$
$$x_d = -6 + \tfrac{4}{5}(10) = 2.$$

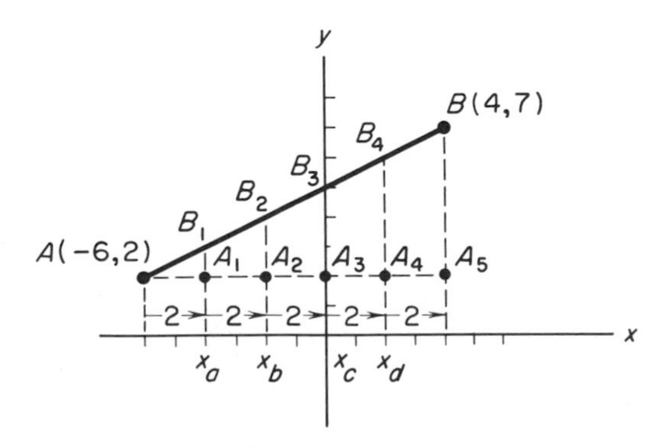

Figure 5-9

The y coordinates of the points of division can be found analogously. Coordinates of the points of division are $(-4, 3)$, $(-2, 4)$, $(0, 5)$ and $(2, 6)$.

Consider next a general expression for division of a line segment. Given segment P_1P_2, where $P_1(x_1, y_1)$ and $P_2(x_2, y_2)$ are the endpoints, we wish to find the coordinates of the rth point of division removed from P_1 if P_1P_2 is divided into k equal parts. To find x_r (the x coordinate of the rth point of division removed from P_1), we proceed as follows.

First, the horizontal distance from P_1 to P_2 is $x_2 - x_1$. Now we are dividing P_1P_2 into k equal parts so that each horizontal part is of width $(x_2 - x_1)/k$ and r of these parts accumulate to $r(x_2 - x_1)/k$ units. The rth point removed from P_1 has x coordinate

$$x_r = x_1 + \frac{r}{k}(x_2 - x_1). \tag{7}$$

By similar reasoning,

$$y_r = y_1 + \frac{r}{k}(y_2 - y_1), \tag{8}$$

where (7) and (8) give the coordinates of the rth point of division removed from P_1 if line segment P_1P_2 is divided into k equal parts.

The midpoint of a line segment If we wish to find the midpoint (x_m, y_m) of line segment P_1P_2, then using (7) and (8), $k = 2$ and $r = 1$, so that

$$x_m = x_1 + \frac{1}{2}(x_2 - x_1) = \frac{x_1 + x_2}{2}, \tag{9}$$

$$y_m = y_1 + \frac{1}{2}(y_2 - y_1) = \frac{y_1 + y_2}{2}, \tag{10}$$

where (9) and (10) give the coordinates requested. The conclusion in (9) and (10) is not surprising because a midcoordinate is an *average* coordinate. To find the average of two quantities, we add them together and divide them by 2, as shown in (9) and (10).

Applications of Eqs. (7) and (8) are called *interpolation* and are particularly useful when values are desired that lie between two values listed in a table of logarithms, trigonometric functions, and others. Interpolation is considered more fully in a later chapter.

Exercises 5-3

1. Sketch on graph paper the straight line connecting points $A(2, 1)$ and $B(11, 6)$. Find the midpoint of AB.
2. Sketch on graph paper the straight line connecting points $A(-3, 2)$ and $B(5, 5)$. Find the midpoint of AB.
3. For line segment AB of Exercise 1, find the coordinates of the second point of division removed from A if AB is divided into three equal parts.

4. For line segment AB of Exercise 2, find the coordinates of the second point of division removed from B if AB is divided into three equal parts.

5. Sketch on graph paper the straight line connecting points $A(6, -2)$ and $B(-4, 3)$. Find the coordinates of the third point of division removed from A if AB is divided into five equal parts.

6. Sketch on graph paper the straight line connecting points $A(7, -1)$ and $B(-3, -6)$. Find the coordinates of the third point of division removed from A if AB is divided into five equal parts.

7. Line segment AB has its midpoint at $(2, 4)$ and point A at $(-3, 7)$. Find the coordinates of point B.

8. Line segment AB has its midpoint at $(-2, -3)$ and point A at $(-8, -7)$. Find the coordinates of point B.

9. In Fig. 5-10, find the midpoints of AB, BC, and AC.

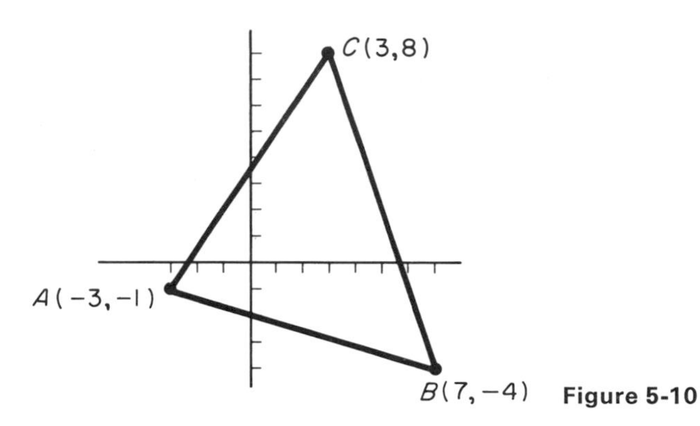

Figure 5-10

10. In Fig. 5-10, find the coordinates of the second point of division removed from A if AB is divided into six equal parts.

11. A certain investment P grows by the same amount each year. At the end of year 4 its value is \$1200; at the end of year 9 its value is \$1550. What is its value at the end of year 7; year 13; year 2?

12. Figure 5-11 shows a small portion of curve $y = \log x$. From it we read $\log 15.03 = 1.17696$ and $\log 15.04 = 1.17725$. Assuming for the brief interval shown

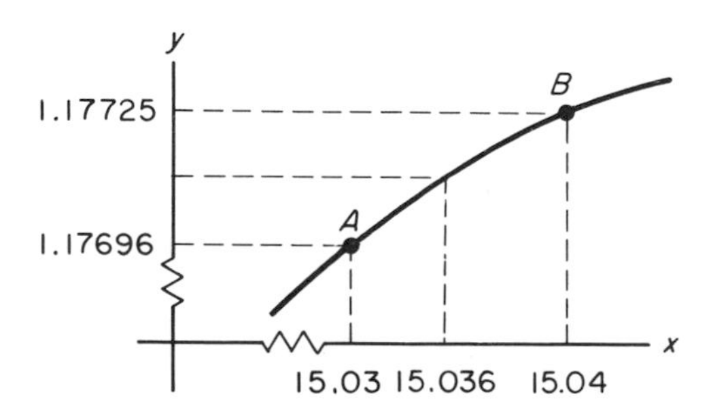

Figure 5-11

that AB is a straight line, use the division of a line segment to find the approximate value of log 15.036.

13. From trigonometry we are given ctn $16°45'00'' = 3.3226$ and ctn $16°45'60'' = 3.3191$. Using division of a line segment, find the approximate value of ctn $16°45'28''$, assuming that, for the brief interval under consideration, we have a straight line.

14. Modify Eqs. (7) and (8) to locate the $(k - r)$th point of division from P_2. Show that the resulting expressions are equivalent algebraically to (7) and (8).

5-4 Determining linear equations from given conditions

Suppose that we have two quantities x and y that are related by the fact that y is always three units greater than x. This means that when $x = 2$, $y = 5$; when $x = 7.2$, $y = 10.2$; when $x = -3$, $y = 0$; and so on. Actually, any (x, y) pair that satisfies equation

$$y = x + 3 \tag{11}$$

is a suitable number pair. There are indefinitely many such pairs. If all were plotted on a cartesian coordinate plane, they would lie in a straight line, as shown in Fig. 5-12, where only certain of the number pairs appear. Any point along the line can be designated as (x, y), where the x and y of (x, y) are the same x and y used in (11).

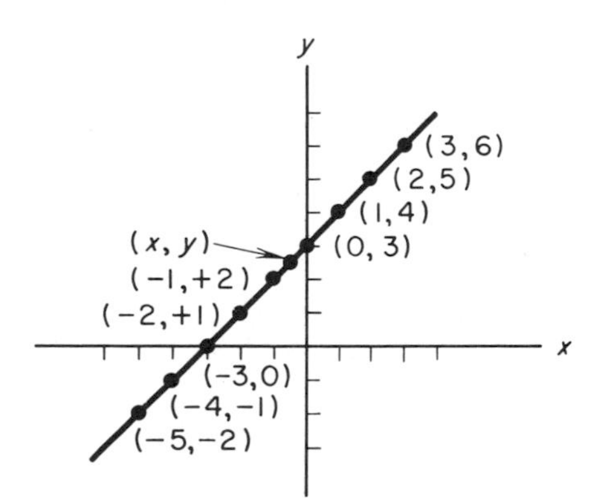

Figure 5-12

We note, too, from Fig. 5-12 that the slope of the line is unity, as can be determined by selecting any two points and applying (5). Let us create a special description of the slope, using the general point (x, y) and any point shown in Fig. 5-12, say (3, 6). We have

$$\frac{y - 6}{x - 3} = \text{slope.}$$

But the slope is $m = 1$ so that

$$\frac{y-6}{x-3} = 1. \tag{12}$$

Simplifying (12) gives

$$y = x + 3,$$

which agrees with (11). Here we demonstrated that, given the slope of a line ($m = 1$) and a point (3, 6) on the line, we can obtain the equation of the line. The slope, incidentally, was obtained from knowing the coordinates of any two points on the line.

We can obtain the equation of a straight line knowing *one* point on the line and the slope of the line. The equation of the line can be written various ways, as shown below.

Two-point form

Given any two points $P_1(x_1, y_1)$ and $P_2(x_2, y_2)$ on a straight line, let us derive the equation of the line. We do so by equating two expressions for the slope. (See Fig. 5-13.) Using the given points,

$$m = \frac{y_2 - y_1}{x_2 - x_1}. \tag{13}$$

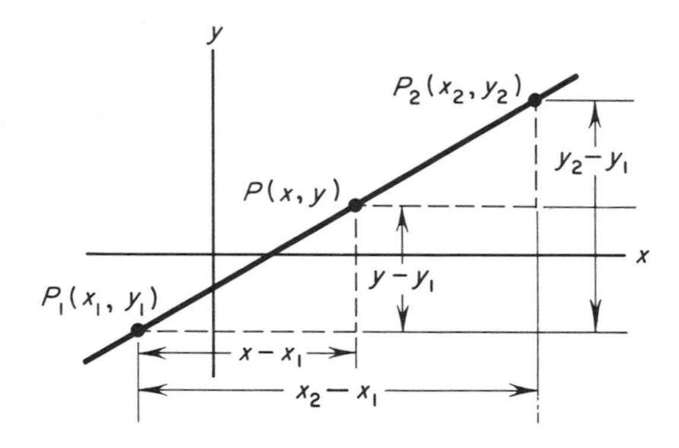

Figure 5-13

Using the general point $P(x, y)$ and either of the given points (we chose P_1 in Fig. 5-13).

$$m = \frac{y - y_1}{x - x_1}. \tag{14}$$

Equating (13) and (14), we obtain

$$\frac{y - y_1}{x - x_1} = \frac{y_2 - y_1}{x_2 - x_1}, \tag{15}$$

where (15) is the *two-point form* of a straight line.

Example 8 Find the equation of the line passing through $(-4, 3)$ and $(2, -5)$.

Solution We choose to designate $(-4, 3)$ as P_1 so that $x_1 = -4$ and $y_1 = 3$ and $(2, -5)$ as P_2 so that $x_2 = 2$ and $y_2 = -5$. Substituting into (15), we have

$$\frac{y - (3)}{x - (-4)} = \frac{-5 - (3)}{2 - (-4)}$$

or

$$\frac{y - 3}{x + 4} = -\frac{8}{6}$$

from which

$$3y + 4x = -7$$

is the equation of the line in question.

Point-slope form

Let us suppose that we know the slope m of a straight line and a point $P_1(x_1, y_1)$ through which it passes. We can obtain the equation of the line readily by referring to (15). There the right member is an expression for the slope, but we are given the slope as m. Replacing the right member of (15) by m, we have

$$\frac{y - y_1}{x - x_1} = m, \tag{16}$$

where (16) is the equation of a line known to pass through $P_1(x_1, y_1)$ with slope m and is called the *point-slope form.*

Example 9 Find the equation of the line passing through $(-4, 3)$ with a slope of -1.

Solution Call $(-4, 3)$ the point $P_1(x_1, y_1)$ so that $x_1 = -4$ and $y_1 = 3$. The slope is $m = -1$. Substituting into (16) gives

$$\frac{y - (+3)}{x - (-4)} = -1.$$

Simplifying, we have

$$y = -x - 1.$$

Slope-intercept form

Suppose again that we are given the slope of a line and a point on the line, as in the point-slope case. This time, however, the given point will be the *y intercept*—that is, the distance from the origin to the point where the line crosses the y axis. Call the coordinates of the y intercept $(0, b)$. The y intercept is therefore called b and the slope is once again m. Referring to (16), we have $x_1 = 0$ and $y_1 = b$. Then substituting into (16) results in

$$\frac{y - b}{x - 0} = m$$

from which

$$y = mx + b, \tag{17}$$

where (17) is the *slope-intercept form* of a straight line.

Example 10 Find the equation of a straight line with slope -3 and y intercept $+5$.

Solution Referring to (17), we have $m = -3$ and $b = +5$. Then substituting, we have

$$y = -3x + 5.$$

Intercept form

Suppose that we are given the intercepts of a straight line. The line crosses the x axis at $x = a$ and the y axis at $y = b$, where a is the x intercept and b is the y intercept. The coordinates of the x intercept are $(a, 0)$ and the coordinates of the y intercept are $(0, b)$. Now we have the two-point form with $x_1 = a$, $y_1 = 0$, $x_2 = 0$, and $y_2 = b$. Substituting into the two-point form (15), we obtain

$$\frac{y - 0}{x - a} = \frac{b - 0}{0 - a},$$

which can be simplified to read

$$\frac{x}{a} + \frac{y}{b} = 1, \tag{18}$$

where (18) is the *intercept form* of a straight line.

Example 11 Find the equation of the straight line with x intercept 12 and y intercept -6.

Solution Referring to (18), we have $a = 12$ and $b = -6$. Substituting gives

$$\frac{x}{12} + \frac{y}{-6} = 1$$

from which

$$x - 2y = 12.$$

Example 12 Find the slope of the line

$$3x - 4y = 12. \tag{19}$$

Solution If we solve (19) for y, we will have the slope-intercept form. Solving,

$$y = \frac{3}{4}x - 3. \tag{20}$$

Comparing (20) to (17), the coefficient of x is 3/4 and is also the requested slope.

Example 13 What is the y intercept of the line passing through $(-3, -6)$ and $(-2, -3)$?

Solution By using the two-point form to obtain the equation of the line and then manipulating the equation into slope-intercept form, we will be able to recognize the y intercept. Substituting into (15),

$$\frac{y + 3}{x + 2} = \frac{-3 - (-6)}{-2 - (-3)}$$

from which

$$\frac{y + 3}{x + 2} = 3$$

and

$$y = 3x + 3. \tag{21}$$

Comparing (21) to (17), we recognize the y intercept as $b = +3$.

Exercises 5-4

In Exercises 1–12, find the equation of the straight line satisfying the given conditions.

1. Passes through (0, 0) and (3, 2).
2. Passes through (0, 0) and $(-3, 2)$.
3. Passes through $(3, -5)$ and $(-4, 6)$.
4. Passes through $(2, -9)$ with slope 1/2.
5. Passes through (5, 0) with slope 1/5.
6. Passes through $(-7, 2)$ with slope 3/4.
7. Has slope $-1/2$ and y intercept 4.
8. Has slope $m = -3/5$ and y intercept -5.
9. Has slope $-5/3$ and x intercept 10/3.
10. Has x intercept 5 and y intercept -2.
11. Has x intercept 1/2 and y intercept -4.
12. The x intercept is twice the y intercept and the line passes through $(-9, 6)$.

In Exercises 13–18, given the equation

$$Ax + By = C, \tag{22}$$

provide answers for the questions asked.

13. What is the slope of (22)?

14. What is the y intercept of (22)?

15. What is the x intercept of (22)?

16. What is the relationship between A and B if the slope of (22) is $-3/2$?

17. What is the relationship between A and B if the slope of (22) is $+1$?

18. Which constant must be zero if (22)
(a) passes through the origin?
(b) is vertical?
(c) is horizontal?

19. Show that all lines passing through $P(3, 2)$ in Fig. 5-14 can be represented by the equation

$$y = m(x - 3) + 2.$$

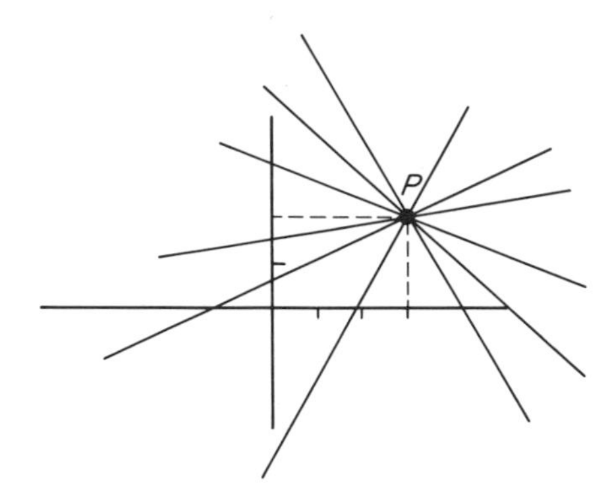

Figure 5-14

20. Show that the equations of all parallel lines shown in Fig. 5-15 can be represented by the equation

$$y = -\tfrac{3}{4}x + b,$$

where b is a y intercept.

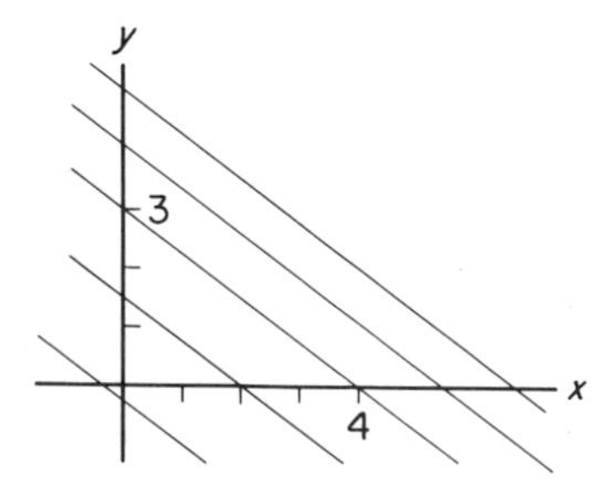

Figure 5-15

21. Jones and Smith design the temperature scales shown in Fig. 5-16. Draw up an equation relating degrees Smith (S) to degrees Jones (J). What equation relates Smith to Fahrenheit? Jones to centigrade?

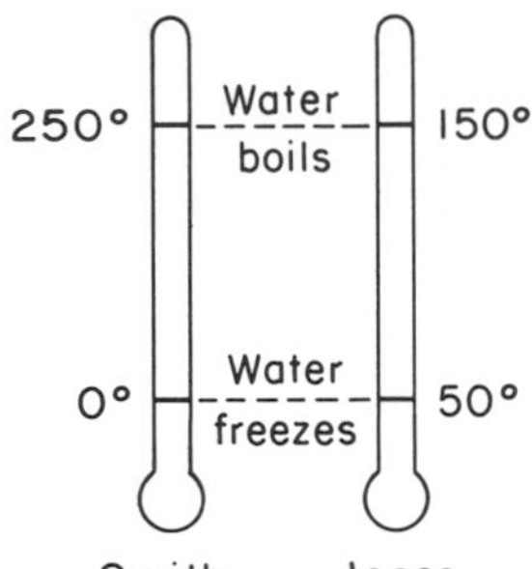

Figure 5-16

22. Figure 5-17 shows a belt system. Wheels of radii R and r rotate, giving motion to a nonslipping belt. Show that the distances traveled by P_R and P_r in a given amount of time are equal; show also that the number of circumferences completed by P_R and P_r has the ratio $r : R$.

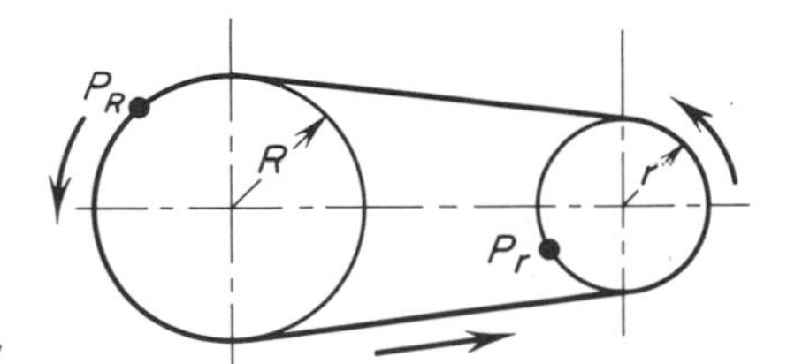

Figure 5-17

5-5 Graphing linear equations

Each of the forms of the straight line derived in Sec. 5-4 can be reduced to the form

$$Ax + By = C. \tag{23}$$

Certain observations of (23) are in order. First, A, B, and C are constants. Second, A and B cannot both be zero. Finally, x and y are variables and they are present in the first degree at the highest.

Graphing (23) can be a simple task. Since two points determine a straight line, the x intercept and the y intercept can be readily determined, plotted, and a straight line passed through them. To find the x intercept, let $y = 0$, from which $x = C/A$; to find the y intercept, let $x = 0$, from which $y = C/B$.

The intercepts of (23) may not be distinct or they may be unfortunately close together. If such is the case, we can solve for y so that

$$y = -\frac{A}{B}x + \frac{C}{B} \tag{24}$$

where (24) is in slope-intercept form, $y = mx + b$, with slope $m = -A/B$ and y intercept C/B. To plot (24), we can plot the y intercept and proceed from that point in the direction required by m.

Example 14 Graph the equation

$$3x + 2y = 6. \tag{25}$$

Solution To graph (25), determine the intercepts. If $x = 0$, $y = 3$; and if $y = 0$, $x = 2$. We plot the intercepts and pass a line through them as in Fig. 5-18. Note that we plotted a third point $(4, -3)$ that satisfies (25) and was arbitrarily chosen. This third point falls on the given line, giving us confidence that the line was properly graphed. If the three points are not collinear (that is, do not fall in a straight line), all three bear checking.

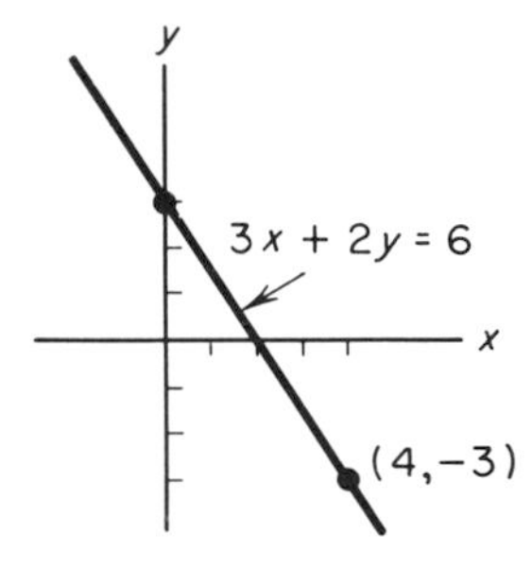

Figure 5-18

Example 15 Graph the equation

$$2y - x = 0. \tag{26}$$

Solution Solving (26) for y, we have

$$y = \tfrac{1}{2}x.$$

Next, we observe that the line passes through the origin with a slope of 1/2. We can start our graphing at the origin and move off at a slope of 1/2—that is, run 2 and rise 1 or run -2 and rise -1 as shown in Fig. 5-19.

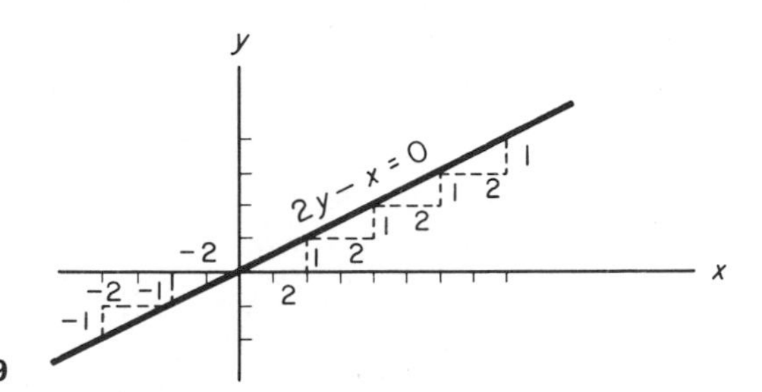

Figure 5-19

Examining (23), we observe more properties. First, if $A = 0$, then $By = C$ is a horizontal line C/B units removed from the x axis. Second, if $B = 0$, then $Ax = C$ is a vertical line C/A units removed from the y axis. Finally, if $C = 0$, the intercepts are both zero and the line passes through the origin.

Exercises 5-5

Graph the given pair of equations on one set of axes.

1. $y = x$
$y = x + 1$

2. $y = x$
$y = -x$

3. $y = 2x + 1$
$y = 2x + 3$

4. $x + y = 5$
$x + y = 3$

5. $2x + 3y = 6$
$2x + 3y = 12$

6. $2x + 3y = 6$
$4x + 6y = 12$

7. $y = mx + 3 \quad (m = 1)$
$y = mx + 3 \quad (m = 2)$

8. $y = \frac{1}{2}x + b \quad (b = 1)$
$y = \frac{1}{2}x + b \quad (b = 2)$

5-6 Parallel and perpendicular lines

If two lines are parallel, their slopes are equal; thus for equations

$$Ax + By = C \tag{27}$$

and

$$Dx + Ey = F, \tag{28}$$

we have parallelism if certain conditions are met. If we solve (27) and (28) for y, we have

$$y = -\frac{A}{B}x + \frac{C}{B}, \tag{29}$$

$$y = -\frac{D}{E}x + \frac{F}{E}. \tag{30}$$

The slopes of (29) and (30) are, respectively, $-A/B$ and $-D/E$; if the lines are parallel, the slopes are equal. So

$$-\frac{A}{B} = -\frac{D}{E}, \tag{31}$$

where (31) provides a test for parallelism of two straight lines. Note that (31) involves neither C nor F. Note further that (31) asserts that two lines are parallel if the ratio of the x coefficient to the y coefficient is the same for both lines. Thus

$$3x - 2y = 5$$
$$-6x + 4y = 25$$

are parallel, since

$$\frac{3}{-2} = \frac{-6}{4}.$$

Example 16 Give the equation of the line parallel to $2x + y = 9$ and passing through (5, 4).

Solution Since the desired line parallels $2x + y = 9$, it has the form

$$2x + y = k, \tag{32}$$

where the ratio of the coefficients of the x term and the y term is undisturbed. Next, (32) passes through the point where $x = 5$ and $y = 4$. Substituting,

$$2(5) + 4 = k$$

from which $k = 14$. Substituting $k = 14$ into (32), the desired solution is

$$2x + y = 14.$$

Two lines are perpendicular if their slopes are negative reciprocals of each other. This situation is shown below.

Given line l_1 (Fig. 5-20) of the form $y = mx + b$. A point $P(x_0, y_0)$ is on l_1. A line perpendicular to l_1 is drawn through P; call this line l_2. Our

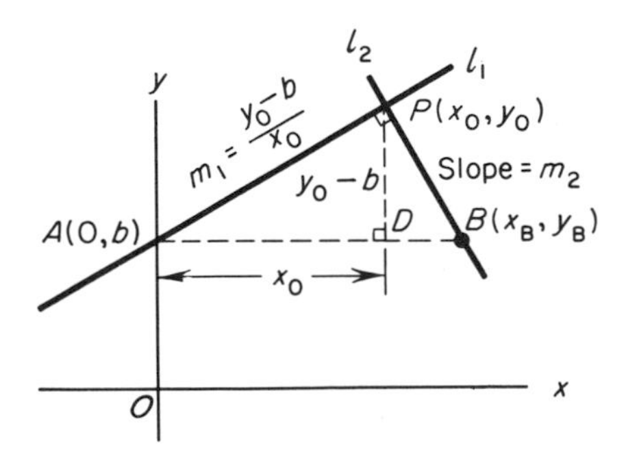

Figure 5-20

goal is to show that the slope of l_2 is $m_2 = -1/m_1$, where m_1 is the slope of l_1 and m_2 is the slope of l_2. Let us proceed by first finding the coordinates of B.

From plane geometry PD is the mean proportional between AD and DB. That is,

$$\overline{PD^2} = (AD)(DB)$$

from which

$$(y_0 - b)^2 = (x_0)DB \quad \text{and} \quad DB = \frac{(y_0 - b)^2}{x_0}.$$

Next, the coordinates of B are

$$x_B = x_0 + DB = x_0 + \frac{(y_0 - b)^2}{x_0}, \qquad y_B = b.$$

We find the slope of PB by using the coordinates of P and B:

$$m_2 = \frac{y_0 - y_B}{x_0 - x_B} = \frac{y_0 - b}{x_0 - \left[x_0 + \frac{(y_0 - b)^2}{x_0}\right]} = \frac{y_0 - b}{-\frac{(y_0 - b)^2}{x_0}} = -\frac{x_0}{y_0 - b}.$$

Comparing m_1 to m_2, we obtain

$$m_2 = -\frac{1}{m_1} \quad \text{or} \quad m_1 m_2 = -1. \tag{33}$$

The conclusion reached in (33) is: *if two lines are perpendicular, their slopes are negative reciprocals.*

Example 17 Give the equation of the lines perpendicular to

$$y = 2x + 5.$$

Solution The slope of the given line is $m = +2$; therefore the slope of the perpendicular, by (33), is $-1/2$. Since we require *any* line perpendicular to $y = 2x + 5$, let the y intercept be b, producing the solution

$$y = -\frac{1x}{2} + b$$

In Fig. 5-21 we see four of the family of lines perpendicular to $y = 2x + 5$.

Example 18 Give the equation of the line perpendicular to $y = 2x + 5$ and passing through (2, 9).

Solution From Example 17 the equation of any line perpendicular to $y = 2x + 5$ is $y = -1x/2 + b$. The point (2, 9) is on $y = -1x/2 + b$ so that

$$9 = -\tfrac{1}{2}(2) + b$$

from which $b = 10$, and the equation of the desired line is

$$y = -\tfrac{1}{2}x + 10,$$

which is shown in Fig. 5-21.

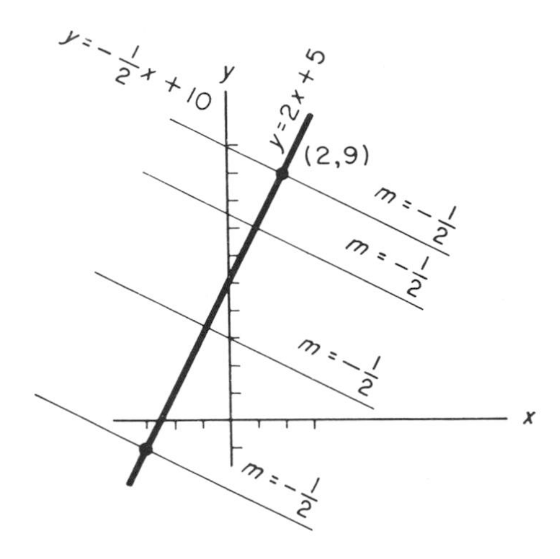

Figure 5-21

Exercises 5-6

1. Given equation $ax + by = c$, show that multiplying each term by a constant k does not affect either the slope or the y intercept.

Given equations $ax + by = c$ and $dx + ey = f$ in Exercises 2–4, answer these questions.

2. What relationship involving the constants must exist if lines are parallel?
3. What relationship involving the constants must exist if lines are perpendicular?
4. What relationship involving the constants must exist if lines have the same y intercept?
5. Find the equation of the line perpendicular to $2x - 3y = 6$ and passing through the origin.
6. Given l_1 as $Ax + By = C$, show that line l_2, obtained by interchanging A and B and changing the sign of either A or B, will be perpendicular to l_1.

Given line l_1 as $2x - y = 6$, find the equation of the line in Exercises 7–10.

7. Parallel to l_1 and passing through (2, 2).
8. Perpendicular to l_1 and passing through (2, 2).
9. Parallel to l_1, with y intercept -8.
10. Perpendicular to l_1 at (6, 6).
11. Find the equation of the perpendicular bisector of the line segment terminating at $P(-1, 3)$ and $R(5, 5)$.

12. Given a triangle with vertices $A(-3, -4)$, $B(0, 4)$, and $C(4, -2)$, find the equation of the altitude drawn from B to AC.

13. It is known that line $x + ky = 7$ is perpendicular to line $3x - 5y = 6$. Find k.

14. Given l_1 as $4x + y = 9$ and l_2 as $8x + 2y = k$, what value of k will cause l_1 and l_2 to be graphed as the same line?

15. Given l_1 as $4y - 3x = 12$ and l_2 as $ax + by = c$, if l_1 is perpendicular to l_2, what is the ratio $a:b$?

16. Given line l_1 as $2y = x + 2$, lines l_2 and l_3 are drawn perpendicular to l_1 at points (2, 2) and (4, 3), respectively. What is the distance between the y intercepts of l_2 and l_3?

5-7 Simultaneous solutions

If two straight lines are graphed on a cartesian coordinate plane, they will intersect in one point, provided that they are not parallel. This point of intersection possesses a set of coordinates that simultaneously satisfies both equations and is called the *simultaneous solution* of the two lines. The simultaneous solution can be found both graphically and algebraically. Both methods are discussed here.

Example 19 Using graphical methods, find the point of intersection of

$$x + y = 8, \tag{34}$$

$$x - y = 4. \tag{35}$$

Solution Refer to Fig. 5-22, where the graphs of (34) and (35) are shown. The lines intersect at $P(6, 2)$. Note that the coordinates (6, 2) satisfy both (34) and (35), since $6 + 2 = 8$ and $6 - 2 = 4$. No other point on (34) is also on (35).

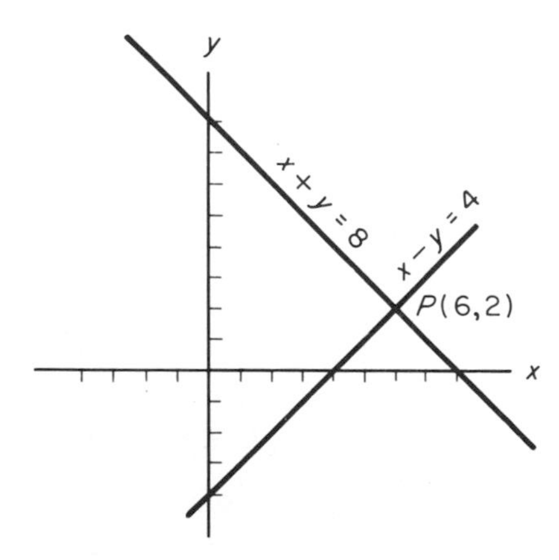

Figure 5-22

We can make an observation about the graphical method of solution shown in Example 19. The method may not be exact, due to inaccuracies that develop in the graphing process. This situation would be more apparent if the points of intersection were represented by fractional or irrational coordinates.

Next we discuss three algebraic methods of obtaining simultaneous solutions of linear equations. Our approach is a general one, followed by example problems.

Solution by substitution

Given two nonparallel lines

$$Ax + By = C, \tag{36}$$

$$Dx + Ey = F, \tag{37}$$

we can solve either equation for either unknown and substitute the result into the other equation. Proceeding, we solve (36) for y so that

$$y = -\frac{Ax}{B} + \frac{C}{B}. \tag{38}$$

Substituting (38) into (37) results in

$$Dx + E\left(-\frac{Ax}{B} + \frac{C}{B}\right) = F$$

from which, after several algebraic steps,

$$x = \frac{EC - FB}{AE - BD}. \tag{39}$$

If we substitute (39) into either (36) or (37), we obtain a value for y. Choosing (36), we have

$$A\left(\frac{EC - FB}{AE - BD}\right) + By = C$$

from which we obtain the value

$$y = \frac{AF - CD}{AE - BD}. \tag{40}$$

Expressions (39) and (40) are the x coordinate and y coordinate, respectively, of the point of intersection of (36) and (37). Figure 5-23 shows all the important expressions on a cartesian coordinate plane. Note that the intersection point P is actually the intersection of lines (39) and (40). At one point in our solution we solved the vertical line (39) against $Ax + By = C$ to obtain point P.

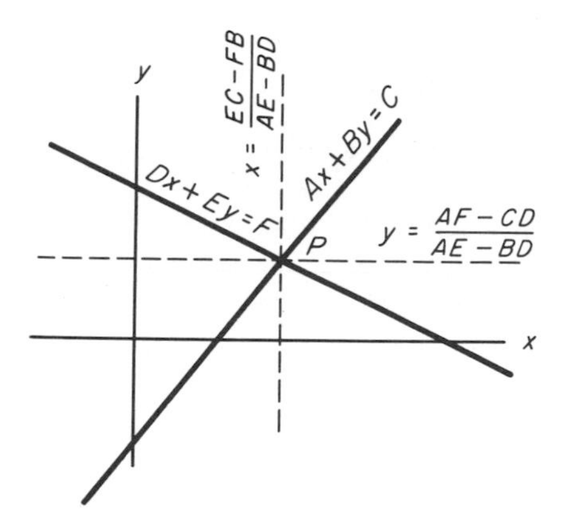

Figure 5-23

Solution by comparison

We can find the point of intersection of (36) and (37) by solving both equations for the same unknown and then comparing results. Choosing to solve both for y, we have

$$y = -\frac{Ax}{B} + \frac{C}{B} \tag{41}$$

and

$$y = -\frac{Dx}{E} + \frac{F}{E}. \tag{42}$$

Comparing (41) and (42) gives

$$-\frac{Ax}{B} + \frac{C}{B} = -\frac{Dx}{E} + \frac{F}{E} \tag{43}$$

from which we get

$$x = \frac{EC - FB}{AE - BD},$$

which is identical with (39), obtained by the substitution method. From this point on we complete the solution as in the substitution method.

Solution by addition or subtraction

We can choose to eliminate one of the unknowns by subtracting one equation from the other. Elimination of an unknown by subtraction occurs only if that unknown has the same coefficient in both equations. If the coefficients are the same but opposite in sign, addition will eliminate the unknown. Let us proceed, choosing to eliminate the y term by multiplying (36) by E and (37) by B; thus

$$AEx + BEy = CE, \tag{44}$$

$$BDx + BEy = BF. \tag{45}$$

Subtracting (45) from (44), we have

$$AEx - BDx = CE - BF, \tag{46}$$

and solving (46) for x, we have (39). To obtain y, we proceed as in previous methods.

Which method should be used in solving a system? The answer depends largely on the nature of the coefficients or the way in which the equations are given. Thus if both equations are already solved for one of the unknowns—that is, for the same unknown—comparison may be most useful. If one equation is readily solved for an unknown and the other cannot easily be solved for that unknown, then substitution may be the easiest. If both equations are given in the same algebraic form, addition or subtraction may be the easiest.

Example 20 Find the simultaneous solution of

$$3x - 5y = 23, \tag{47}$$

$$x + 4y = 2. \tag{48}$$

Solution Let us multiply (48) by 3 and subtract.

$$\begin{array}{r} 3x - 5y = 23 \\ 3x + 12y = 6 \\ \hline -\ 17y = 17 \end{array}$$

from which $y = -1$. Substituting $y = -1$ into (48) gives

$$x + 4(-1) = 2,$$

from which $x = 6$, and the complete solution is the point $(6, -1)$.

Discussion: We chose solution by subtraction, which required multiplying all terms of (48) by 3. There may be some question about the validity of such a multiplication. Equations

$$Ax + By = C \quad \text{and} \quad kAx + kBy = kC$$

graph as the same equation and are called *equivalent equations;* this fact can be demonstrated by obtaining the slope-intercept form of both equations and comparing the slopes and y intercepts.

Example 21 Two vehicles A and B proceed along the same route. Vehicle A travles at a constant speed of 30 mph; vehicle B travels at a constant speed of 40 mph. How long does it take B to catch A if A had a 50-mile head start? Where is the point at which B catches A?

Solution Upon formulating the distance-time relationships, we have for A

$$s = 30t + 50 \tag{49}$$

and for B

$$s = 40t, \tag{50}$$

where t is time in hours and s is distance traveled in miles. Figure 5-24 shows the graphical solution of (49) and (50). We can solve the system algebraically by comparison, where

$$30t + 50 = 40t$$

from which $t = 5$. Substituting $t = 5$ into (50), we obtain $s = 200$. The solution ($t = 5$ hr, $s = 200$ miles) means that B and A are simultaneously at the same point exactly 5 hours and 200 miles from the starting position, which was taken as $t = 0$ and $s = 0$ for B and $t = 0$, $s = 50$ for A.

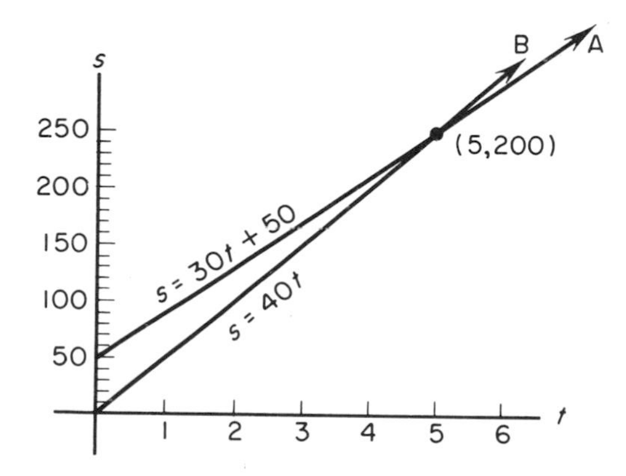

Figure 5-24

Exercises 5-7

In Exercises 1–20, solve the given systems algebraically. Choose any of the three methods discussed. It is suggested that graphical solutions also be shown to check for gross errors and assist in interpretation.

1. $x + y = 7$
$x - y = 3$

2. $y - x = 9$
$x + y = 15$

3. $y = 8 - x$
$y - 2 = x$

4. $11 - y = x$
$7 + x = y$

5. $2x + y = 7$
$x - 2y = 6$

6. $x + 4y = 7$
$3x - y = 8$

7. $2x + 3y - 14 = 0$
$3x - 4y - 4 = 0$

8. $3x + 4y = 4$
$4x + 3y = 10$

9. $3x + 5y = 9$

$4x - 6y = -7$

10. $5m - 2n = \frac{7}{6}$

$3m + 7n = -\frac{3}{4}$

11. $8r - \frac{1}{2}P = 60$

$3r + 2P = 40$

12. $2t - \frac{1}{3}x = 9$

$7t - 2x = 19$

13. $\mathrm{F} = \frac{9}{5}\mathrm{C} + 32$

$\mathrm{F} = \mathrm{C}$

14. $5(k - 1) + \frac{3}{2}(p + 2) = 9$

$\frac{k}{2} + \frac{2}{3}(p + 10) = \frac{17}{2}$

15. $3(x - 10) - \frac{1}{3}(y + 2) = 17$

$x + 2y = -1$

16. $\mathrm{C} = \frac{5}{9}(\mathrm{F} - 32)$

$\mathrm{F} = 2\mathrm{C}$

17. $r = \frac{2}{3}t + 21$

$t = 4r - 3$

18. $14.2x - 1.3y = 14$

$3.7x + 11.5y = 23$

19. $1.4x - 1.7y = 22$

$2.5x + 1.2y = 20$

20. $2(x - 3y) + y = 19$

$\frac{1}{2}(x + 3y) = -9$

21. The sum of two numbers is 24. Their difference is 12. What are the numbers?

22. The perimeter of a rectangle is 30 ft. The length is 4 ft less than twice the width. What are the dimensions of the rectangle?

Two racers A and B leave a point P traveling the same route. Racer A travels at a constant rate of 3.5 mph and departs from P at 12:00 noon. Racer B travels at 4 mph and departs from P at 2:00 P.M. on the same day. In Exercises 23–28,

23. Draw up the distance-time equation for A.

24. Draw up the distance-time equation for B.

25. The 2-hour head start by A is how large in terms of distance?

26. At what time will B catch A?

27. How far from P will B catch A?

28. Show the solution to Exercises 26 and 27 graphically.

Two racers A and B travel along the same route according to equations

$$s = v_a t + s_a \quad \text{and} \quad s = v_b t + s_b$$

where s is distance, v is average speed (called constant here), and t is time. For $t \geqq 0$, what conditions involving v_a, v_b, s_a, and s_b must exist if

29. A remains a constant distance from B?

30. A is initially ahead of B, but B eventually passes A?

31. A is always ahead of B?

32. A starts ahead of B and continues to move further ahead of B?

33. A and B travel in exactly opposite directions?

34. A fund grows by simple interest according to equation

$$A = P + Pni,$$

where i is the yearly interest rate, n is the number of years of growth, P is the original principal, and A is the amount to which the principal grows. Two accounts are established. The first account, with principal \$1000, grows at the rate of 3% annually. The second account, with principal \$850, grows at 5% annually. What is the amount when the two accounts are equal, and how many years pass before the accounts become equal?

35. Three radios and four clocks cost a total of \$128. Two of the same radios and seven of the same clocks cost a total of \$133. What are the individual costs of the clocks and the radios?

36. Two forces A and B are applied to a seesaw system as shown in Fig. 5-25. Force A is 6 lb applied at an unknown distance from the fulcrum F; it is in equilibrium with an unknown force y at B that is 4 ft from F. If A is increased by 3 lb, force y must be increased by 2.25 lb to maintain equilibrium. Find x and y.

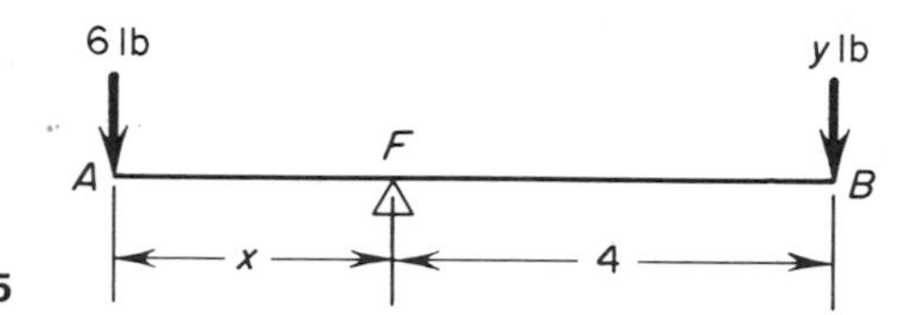

Figure 5-25

37. A boat travels at a constant rate in still water. It travels downstream 18 miles in two hours; the return trip requires six hours. Find the rate of the boat in still water and the rate of the current.

38. A certain alloy is 10% zinc and 6% tin; 25 lb of the alloy has some pure zinc and pure tin added, producing another alloy that is 25% zinc and 20% tin. How much pure zinc and pure tin are added?

39. A man invests one-fifth of his money at 5% per year and the remainder at 4%. His yearly income from the investments is \$240. How much did he invest at each rate?

40. Eleven coins consisting of quarters and dimes have a value of \$2.15. How many dimes and quarters are involved?

5-8 Systems involving more than two unknowns

Systems involving three or more linear equations can be solved (if the solutions exist) by reducing the number of unknowns through methods shown in Sec. 5-7. The procedure is somewhat involved but orderly. The following example illustrates the process.

Example 22 Find the solution of the system

$$x + y - z = 4, \tag{51}$$

$$2x - 3y + z = 1, \tag{52}$$

$$x - 4y - 2z = -7. \tag{53}$$

Solution Here we are instructed to find an (x, y, z) set that simultaneously satisfies Eqs. (51), (52), and (53). We can proceed by successively eliminating unknowns until we are left with one equation in one unknown. Here we will choose to eliminate x from the system first. If we subtract (53) from (51), we have

$$\begin{array}{r} x + y - z = 4 \\ \underline{x - 4y - 2z = -7} \\ 5y + z = 11 \end{array} \tag{54}$$

where (54) is an equation in y and z. If we choose to eliminate x from another pair of the given equations, we will obtain a second equation in y and z that can be solved against (54). Proceeding, let us double all terms in (53) and subtract the result from (52), obtaining

$$\begin{array}{r} 2x - 3y + z = 1 \\ \underline{2x - 8y - 4z = -14} \\ 5y + 5z = 15. \end{array} \tag{55}$$

Next, we observe that (54) and (55) constitute two equations in two unknowns and can be solved by methods discussed in the preceding section. Subtracting (55) from (54), we have

$$\begin{array}{r} 5y + z = 11 \\ \underline{5y + 5z = 15} \\ -4z = -4 \end{array}$$

from which $z = 1$.

At this point we have found a value of z that, with the proper x and y companions, constitutes the solution. We can find the proper value of y by substituting $z = 1$ into either (54) or (55). Choosing (54), the result is

$$5y + (1) = 11$$

from which $y = 2$.

Then we substitute $y = 2, z = 1$ into any of the original equations to obtain x. Choosing (51),

$$x + 2 - 1 = 4$$

from which

$$x = 3.$$

The desired (x, y, z) set is $(3, 2, 1)$. To check this set, we must show that it satisfies *all* the given equations. We already know that it satisfies (51) because (51) was used in finding x from the derived y and z. Substituting into (52),

$$2(3) - 3(2) + 1 \stackrel{?}{=} 1,$$

which checks as

$$1 = 1.$$

Substituting into (53),

$$3 - 4(2) - 2(1) \stackrel{?}{=} -7,$$

which checks as

$$-7 = -7.$$

The method used in Example 22 proves very orderly. It is useful for n equations in n unknowns, where, by subtracting equation pairs, we can produce $n - 1$ equations in $n - 1$ unknowns. These $n - 1$ equations can also be subtracted in pairs to produce $n - 2$ equations in $n - 2$ unknowns. The process is repeated until we are left with one equation in one unknown.

Prudent selection of the original pairs to be subtracted can often shorten the labor involved. Consider Example 23.

Example 23 Solve the system

$$a + b - 3c = 7, \tag{56}$$

$$a + b + 4c = 0, \tag{57}$$

$$5a \qquad + 2c = 8. \tag{58}$$

Solution Inspecting the system carefully, we note that subtracting (57) from (56) eliminates both a and b, producing the result $-7c = 7$, from which $c = -1$. Substituting $c = -1$ into (58) gives $a = 2$; substituting $a = 2$, $c = -1$ into either (56) or (57) produces $b = 2$, and the solution is $(a, b, c) = (2, 2, -1)$.

Simultaneous systems are frequently solved by use of determinants, which are discussed in Secs. 5-9 and 5-10.

Exercises 5-8

In Exercises 1–8, determine the solution of the given system.

1. $x + y = 9$
$x - y + 3z = 2$
$4y - 3z = 5$

2. $2x - 4y + 3z = 1$
$5x - 2y + 6z = 4$
$y - z = \frac{1}{12}$

3. $$\begin{aligned} m - n \quad\quad &= -6 \\ m \quad\quad + p &= -9 \\ 3n + p &= 1 \end{aligned}$$

4. $$\begin{aligned} 3r + 2s + 3t &= -7 \\ 5r - 3s + 2t &= -4 \\ 7r + 4s + 5t &= 2 \end{aligned}$$

5. $$\begin{aligned} r + s + t &= 2.7 \\ 6r + 7s - 5t &= -8.8 \\ 10r + 16s - 3t &= -6.4 \end{aligned}$$

6. $$\begin{aligned} 2x + 3y - 7z &= 13 \\ -5x + 2y + 2z &= -5 \\ x - 4y - 3z &= 16 \end{aligned}$$

7. $$\begin{aligned} 2r - 3s + 3t &= 7 \\ 3r + s - 2t &= -11 \\ 5r - 2s + 4t &= 11 \end{aligned}$$

8. $$\begin{aligned} 4x - 3y - 4z &= -5 \\ 3x + 2y + 2z &= 14.5 \\ 2x + 5y - 3z &= -13.5 \end{aligned}$$

9. Let C, F, and W represent, respectively, the prices in cents per pound of cashews, filberts, and walnuts. The cost for a mixture containing 1 lb of each is \$2.40. The cost for a mixture containing 2 lb of cashews, 2 lb of filberts, and 1 lb of walnuts is \$4. A mixture containing 3 lb of cashews, 1 lb of filberts, and 2 lb of walnuts cost \$4.60. Find C, F, and W.

10. Let S, C, and W represent, respectively, the densities, in pounds per cubic foot of steel, concrete, and wood. The total weight of 1 ft^3 of each is 650 lb. The weight of 2 ft^3 of steel, 3 ft^3 of concrete, and 4 ft^3 of wood is 1550 lb. The weight of 3 ft^3 of steel, 4 of concrete, and 5 of wood is 2200 lb. Find S, C, and W. This exercise demonstrates a system of *inconsistent* equations that defies solution by ordinary means.

11. The standard equation of a circle is given as

$$x^2 + y^2 + ax + by + c = 0.$$

A certain circle passes through points $P(-2, 3)$, $Q(5, 4)$, and $R(6, -3)$. If we substitute the coordinates of P, Q, and R into the standard equation, we will obtain three equations in a, b, and c from which numerical values of a, b, and c can be obtained. By substituting these derived values into the standard equation, we will obtain the equation of the circle passing through P, Q, and R. The method described here is the *method of undetermined coefficients* and is useful in determining equations from experimental data. Find the equation in question.

12. The standard form of a parabola with a vertical axis is given as

$$y = ax^2 + bx + c.$$

A certain parabola with a vertical axis passes through points $(3, -3)$, $(1, -2)$ and $(5, -2)$. Find the equation of the parabola in question by using the method described in Exercise 6.

13. In Fig. 5-26 a schematic is shown to illustrate Kirchhoff's law. From the schematic we can obtain equations

$$\begin{aligned} I_1R_1 + I_3R_3 &= E_2, \\ I_2R_2 - I_3R_3 &= E_1, \\ I_1 - I_2 - I_3 &= 0. \end{aligned}$$

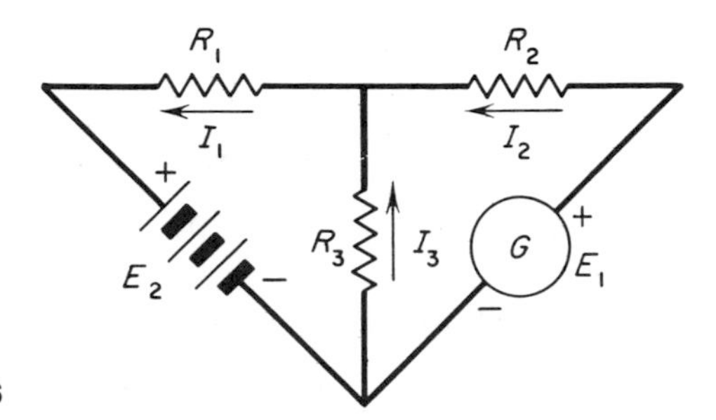

Figure 5-26

Find numerical values of I_1, I_2, and I_3 if we are given the values $R_1 = 1.1$, $R_2 = 0.8$, $R_3 = 1.0$, $E_1 = 10.5$, and $E_2 = 7.4$.

14. A man has coins consisting entirely of quarters, dimes, and nickels. He has 12 coins of value \$1.45. If the number of quarters and nickels were interchanged, the value would be \$1.65. How many coins of each denomination did he originally have?

15. An equation of form $ax + by + cz = d$ graphs as a plane surface. Three planes can intersect in one point; this point is the solution of the system. Find the point of intersection of

$$2x - 3y + z = 0,$$
$$5x + 4y - 2z = 7,$$
$$x + 8y - 5z = 4.$$

5-9 Solutions of systems by use of determinants

In Sec. 5-7 we showed that the solution of the system

$$Ax + By = C$$
$$Dx + Ey = F$$

has the (x, y) values

$$x = \frac{EC - FB}{AE - BD} \qquad y = \frac{AF - CD}{AE - BD}. \tag{59}$$

In this discussion it is important to note that each of the numerators and denominators of (59) is the difference of two products; such expressions can be rearranged into arrays called *determinants.*

Switching our given equations into notation more commonly used in discussions involving determinants, solution of the system

$$\begin{aligned} a_1x + b_1y &= c_1 \\ a_2x + b_2y &= c_2 \end{aligned} \tag{60}$$

is readily found (by elimination methods) to be

$$x = \frac{c_1b_2 - c_2b_1}{a_1b_2 - a_2b_1} \qquad y = \frac{a_1c_2 - a_2c_1}{a_1b_2 - a_2b_1}. \tag{61}$$

We introduce the symbol

$$\begin{vmatrix} a_1 & b_1 \\ a_2 & b_2 \end{vmatrix} \tag{62}$$

which is called a *determinant* or, more specifically, a *determinant of second order*.

By definition, determinant (62) is of value equal to the difference of the diagonal products or

$$\begin{vmatrix} a_1 & b_1 \\ a_2 & b_2 \end{vmatrix} = a_1b_2 - a_2b_1. \tag{63}$$

Note that the denominators of (61) have precisely the value cited in (63) and can therefore be expressed as determinants exactly like (62). The numerators of (61), being the difference of products, are also readily expressed as determinants.

Referring to (62), each of the four numbers a_1, a_2, b_1, b_2 is called an *element;* a_1 and a_2 constitute the *first column;* b_1 and b_2 constitute the *second column.* Elements a_1 and b_1 are the *first row;* a_2 and b_2 are the *second row.* Elements a_1 and b_2 are along the *principal diagonal;* a_2 and b_1 are along the *secondary diagonal.*

We note particularly that the value of the determinant in (63) is the product along the principal diagonal less the product along the secondary diagonal.

Example 24 Using (63), evaluate the determinant that has $a_1 = 1$, $b_1 = -1$, $a_2 = -2$, and $b_2 = 3$.

Solution Substituting the known elements into (63), gives

$$\begin{vmatrix} a_1 & b_1 \\ a_2 & b_2 \end{vmatrix} = \begin{vmatrix} 1 & -1 \\ -2 & 3 \end{vmatrix} = (1)(3) - (-2)(-1) = 3 - (2) = 1.$$

Example 25 Evaluate the determinant

$$\begin{vmatrix} 3 & -4 \\ 1 & 0 \end{vmatrix}.$$

Solution Using (63), we have

$$\begin{vmatrix} 3 & -4 \\ 1 & 0 \end{vmatrix} = (3)(0) - (1)(-4) = 0 + 4 = 4.$$

Next, we apply determinants to express the solution of system (60) by structuring (61) in determinant form. Examination shows that (61) can be written

$$x = \frac{\begin{vmatrix} c_1 & b_1 \\ c_2 & b_2 \end{vmatrix}}{\begin{vmatrix} a_1 & b_1 \\ a_2 & b_2 \end{vmatrix}} \qquad y = \frac{\begin{vmatrix} a_1 & c_1 \\ a_2 & c_2 \end{vmatrix}}{\begin{vmatrix} a_1 & b_1 \\ a_2 & b_2 \end{vmatrix}}. \tag{64}$$

Moreover, we see how (64) can be expressed from (60) by inspection. First, the denominators in (64) are arrayed precisely as the coefficients of x and y are arrayed in (60). Second, the numerator of x in (64) is similar to the denominator with c's replacing a's. Finally, the numerator of y in (64) is similar to the denominator with c's replacing b's.

Example 26 Using determinants, find the solution of

$$y = \frac{3x}{2} - 8$$

$$y = \frac{x}{3} - \frac{10}{3}.$$

Solution Revising the given equations to the form of (60), we have

$$3x - 2y = 16$$

$$-x + 3y = -10$$

and the solutions are

$$x = \frac{\begin{vmatrix} 16 & -2 \\ -10 & 3 \end{vmatrix}}{\begin{vmatrix} 3 & -2 \\ -1 & 3 \end{vmatrix}} = \frac{(16)(3) - (-10)(-2)}{(3)(3) - (-1)(-2)} = \frac{48 - 20}{9 - 2} = \frac{28}{7} = 4,$$

$$y = \frac{\begin{vmatrix} 3 & 16 \\ -1 & -10 \end{vmatrix}}{\begin{vmatrix} 3 & -2 \\ -1 & 3 \end{vmatrix}} = \frac{(3)(-10) - (-1)(16)}{7} = \frac{-30 + 16}{7} = \frac{-14}{7} = -2.$$

Special conditions are required of (60) to ensure solutions. We recall that a solution exists if the graphs of the two straight lines intersect in one point. We recall also from Sec. 5-6 that the lines are parallel if the ratio of the x coefficient to the y coefficient is the same for both lines. In other words, if $a_1/b_1 = a_2/b_2$, then Eqs. (60) are parallel (or coincident). Using this result, we obtain

$$a_1b_2 - a_2b_1 = 0. \tag{65}$$

Comparing (65) with the denominator in (61), we see that a zero denominator yields a no-solution condition.

Exercises 5-9

In Exercises 1–10, find the coordinate pairs satisfying the given equations. Use determinants for the solutions. If no solution exists, explain why.

1. $x + y = 5$
$x + 2y = 7$

2. $a + b = 12$
$2a + b = 19$

3. $x + y = 5$
$x - y = 1$

4. $a - b = 2$
$a + b = 12$

5. $x - 3y = 7$
$2x + 5y = -8$

6. $m - 3n = -12$
$2m + 3n = 21$

7. $3a = 2b + 6$
$b = a - 1$

8. $5F = 9C + 160$
$F = C$

9. $0.3m = 1.2n + 0.5$
$n - 1.5m = 0.4$

10. $ax + by = c$
$kax + kby = d$

11. Using determinants to solve systems

$$\begin{aligned} x - y &= 5 \\ 2x - 2y &= 10 \end{aligned} \quad \text{and} \quad \begin{aligned} x - y &= 5 \\ 2x - 2y &= 15 \end{aligned}$$

results in two different confusions. Explain why, using their graphs as a basis of explanation.

12. For additional exercises, refer to Exercises 1–20 following. Sec. 5-7. Use determinants for their solutions.

5-10 Determinants of higher order

Given three equations in three unknowns

$$\begin{aligned} a_1x + b_1y + c_1z &= d_1 \\ a_2x + b_2y + c_2z &= d_2 \\ a_3x + b_3y + c_3z &= d_3, \end{aligned} \tag{66}$$

a third-order determinant of coefficients of x, y, and z would appear as

$$\begin{vmatrix} a_1 & b_1 & c_1 \\ a_2 & b_2 & c_2 \\ a_3 & b_3 & c_3 \end{vmatrix} = \Delta. \tag{67}$$

Solution of (66) by elimination methods would yield similar denominators (Δ) for x, y, and z such that

$$\Delta = a_1b_2c_3 + b_1c_2a_3 + c_1a_2b_3 - a_3b_2c_1 - b_3c_2a_1 - c_3a_2b_1, \tag{68}$$

which may be seen as taking the three triple products along the solid diagonal lines in (69) as positive and the three along the dotted diagonals as negative.

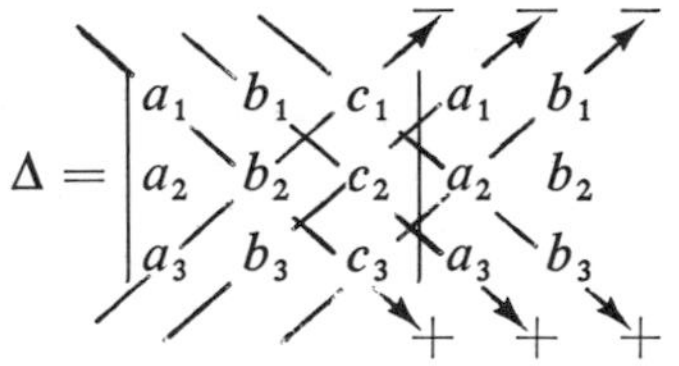

$$\Delta = \begin{vmatrix} a_1 & b_1 & c_1 \\ a_2 & b_2 & c_2 \\ a_3 & b_3 & c_3 \end{vmatrix} \begin{matrix} a_1 & b_1 \\ a_2 & b_2 \\ a_3 & b_3 \end{matrix} \tag{69}$$

$- \quad - \quad -$

$+ \quad + \quad +$

The actual solution of (66) can be shown to be

$$x = \frac{\begin{vmatrix} d_1 & b_1 & c_1 \\ d_2 & b_2 & c_2 \\ d_3 & b_3 & c_3 \end{vmatrix}}{\Delta}, \quad y = \frac{\begin{vmatrix} a_1 & d_1 & c_1 \\ a_2 & d_2 & c_2 \\ a_3 & d_3 & c_3 \end{vmatrix}}{\Delta}, \quad z = \frac{\begin{vmatrix} a_1 & b_1 & d_1 \\ a_2 & b_2 & d_2 \\ a_3 & b_3 & d_3 \end{vmatrix}}{\Delta}, \tag{70}$$

Note from (70) that the x numerator is the same as Δ with d's replacing a's; the y numerator is the same as Δ with d's replacing b's; the z numerator is the same as Δ with d's replacing c's.

Example 27 Evaluate the third-order determinant

$$\begin{vmatrix} 1 & 3 & 2 \\ 0 & 2 & -2 \\ 2 & -1 & 0 \end{vmatrix}.$$

Solution Referring to (69), repeating the first and second columns for clarity, taking the "down-right" diagonal products as positive and the "up-right" diagonal products as negative, we have

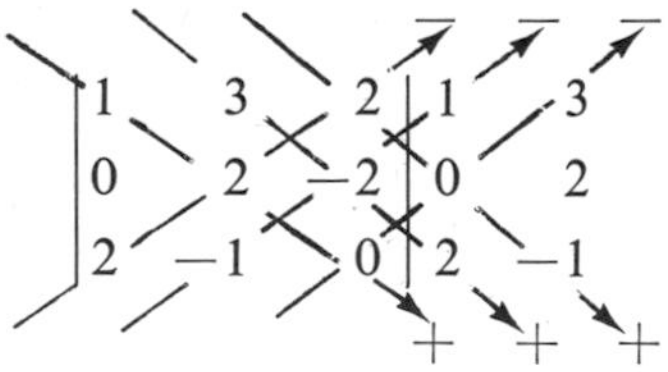

$$\begin{vmatrix} 1 & 3 & 2 \\ 0 & 2 & -2 \\ 2 & -1 & 0 \end{vmatrix} \begin{matrix} 1 & 3 \\ 0 & 2 \\ 2 & -1 \end{matrix}$$

$- \quad - \quad -$

$+ \quad + \quad +$

and the value is

$$(1)(2)(0) + (3)(-2)(2) + (2)(0)(-1) - (2)(2)(2) - (-1)(-2)(1) - (0)(0)(3)$$

or $$0 - 12 + 0 - 8 - 2 - 0 = -22.$$

Example 28 Using determinants, solve the system

$$\begin{aligned} x + y - 2z &= 1 \\ 2x - y \qquad &= 4 \\ x \qquad + 3z &= -2. \end{aligned}$$

Solution Using (70) and (67), we have solutions

$$x = \frac{\begin{vmatrix} 1 & 1 & -2 \\ 4 & -1 & 0 \\ -2 & 0 & 3 \end{vmatrix}}{\begin{vmatrix} 1 & 1 & -2 \\ 2 & -1 & 0 \\ 1 & 0 & 3 \end{vmatrix}} = \frac{(-3) + (0) + (0) - (-4) - (0) - (12)}{(-3) + (0) + (0) - (2) - (0) - (6)}$$

$$= \frac{-11}{-11} = 1,$$

$$y = \frac{\begin{vmatrix} 1 & 1 & -2 \\ 2 & 4 & 0 \\ 1 & -2 & 3 \end{vmatrix}}{\Delta} = \frac{(12) + (0) + (8) - (-8) - (0) - (6)}{-11}$$

$$= \frac{22}{-11} = -2,$$

$$z = \frac{\begin{vmatrix} 1 & 1 & 1 \\ 2 & -1 & 4 \\ 1 & 0 & -2 \end{vmatrix}}{\Delta} = \frac{(2) + (4) + (0) - (-1) - (0) - (-4)}{-11}$$

$$= \frac{11}{-11} = -1.$$

Substituting the triple $(1, -2, -1)$ into all three original equations will serve as a check of the solution.

Minors

Inspection of (67) and (68) reveals an interesting reduction of a third-order determinant to the algebraic sum of three second-order determinants. Referring to (68), we factor a_1 from two terms, $-a_2$ from two terms, and a_3 from two terms, obtaining

$$\begin{vmatrix} a_1 & b_1 & c_1 \\ a_2 & b_2 & c_2 \\ a_3 & b_3 & c_3 \end{vmatrix} = a_1(b_2c_3 - b_3c_2) - a_2(b_1c_3 - b_3c_1) + a_3(b_1c_2 - b_2c_1). \tag{71}$$

Now the expanded portion of (71) can be seen as the algebraic sum of three second-order determinants with a's as coefficients, or

$$\begin{vmatrix} a_1 & b_1 & c_1 \\ a_2 & b_2 & c_2 \\ a_3 & b_3 & c_3 \end{vmatrix} = a_1\begin{vmatrix} b_2 & c_2 \\ b_3 & c_2 \end{vmatrix} - a_2\begin{vmatrix} b_1 & c_1 \\ b_3 & c_3 \end{vmatrix} + a_3\begin{vmatrix} b_1 & c_1 \\ b_2 & c_2 \end{vmatrix}, \tag{72}$$

which is the *expansion by minors* of the third-order determinant. Detailed inspection of (72) shows that the second-order determinant with a_1 as coefficient can be found by crossing out the row and column containing a_1 in the third-order determinant; the second-order determinant with $-a_2$ as coefficient can be found by crossing out the row and column containing a_2; and the second-order determinant with a_3 as coefficient can be found by crossing out the row and column containing a_3. Expression (72) shows the third-order determinant expanded into minors along a column. Such expansions are possible along rows and diagonals. Selective manipulation of (67) into the minor form (72) will bear this out.

Example 29 Using minors, evaluate the determinant given in Example 27.

Solution Using (72), we have

$$\begin{vmatrix} 1 & 3 & 2 \\ 0 & 2 & -2 \\ 2 & -1 & 0 \end{vmatrix} = 1\begin{vmatrix} 2 & -2 \\ -1 & 0 \end{vmatrix} - 0\begin{vmatrix} 3 & 2 \\ -1 & 0 \end{vmatrix} + 2\begin{vmatrix} 3 & 2 \\ 2 & -2 \end{vmatrix}$$
$$= 1(0-2) - 0(0+2) + 2(-6-4)$$
$$= -2 - 20 = -22.$$

Note: Any higher-order determinant can be reduced to determinants of lower order by use of the method of minors. Any third-order determinant (or second) can be evaluated by the diagonal product method.

Caution: The diagonal product method is restricted to determinants of third or second order. Determinants of fourth order or higher must be reduced by the method of minors.

Example 30 Evaluate the determinant

$$\Delta = \begin{vmatrix} 1 & 2 & 3 & 0 \\ 2 & 1 & 0 & 3 \\ 3 & 0 & 1 & 2 \\ 4 & 3 & 2 & 1 \end{vmatrix}.$$

Solution Expanding to third-order minors, we have

$$\Delta = \begin{vmatrix} 1 & 0 & 3 \\ 0 & 1 & 2 \\ 3 & 2 & 1 \end{vmatrix} - 2\begin{vmatrix} 2 & 3 & 0 \\ 0 & 1 & 2 \\ 3 & 2 & 1 \end{vmatrix} + 3\begin{vmatrix} 2 & 3 & 0 \\ 1 & 0 & 3 \\ 3 & 2 & 1 \end{vmatrix} - 4\begin{vmatrix} 2 & 3 & 0 \\ 1 & 0 & 3 \\ 0 & 1 & 2 \end{vmatrix}$$
$$= 1(-12) - 2(12) + 3(12) - 4(-12) = 48.$$

A host of additional properties of determinants is available; however, the reader is referred to other texts for them. The discussion in this section is sufficient to provide the tools needed to solve determinants derived from certain physical situations, particularly electrical circuits.

Exercises 5-10

In Exercises 1–9, evaluate the given determinant by use of the method of minors.

1. $\begin{vmatrix} 2 & 3 & 1 \\ 1 & 3 & 2 \\ 3 & 0 & 0 \end{vmatrix}$ **2.** $\begin{vmatrix} 1 & 2 & 1 \\ 2 & 0 & 3 \\ 4 & 0 & 5 \end{vmatrix}$ **3.** $\begin{vmatrix} 2 & 3 & 0 \\ -2 & 5 & 0 \\ -1 & 1 & 6 \end{vmatrix}$

4. $\begin{vmatrix} 0 & -1 & -2 \\ 0 & 2 & 3 \\ -1 & 2 & 2 \end{vmatrix}$ **5.** $\begin{vmatrix} 0 & 4 & 5 \\ 1 & 0 & 6 \\ 2 & 3 & 0 \end{vmatrix}$ **6.** $\begin{vmatrix} 1 & 2 & 3 \\ 1 & 2 & 3 \\ 1 & 2 & 3 \end{vmatrix}$

7. $\begin{vmatrix} 1 & 2 & 5 & 6 \\ 0 & 3 & 2 & 1 \\ 0 & -1 & 8 & 0 \\ 0 & 2 & -3 & 0 \end{vmatrix}$ **8.** $\begin{vmatrix} 1 & 2 & 0 & 4 \\ 1 & 0 & 1 & 0 \\ 1 & 0 & -1 & -1 \\ 1 & 3 & 0 & 1 \end{vmatrix}$ **9.** $\begin{vmatrix} 1 & -2 & 0 & 3 \\ -1 & 2 & 0 & 1 \\ 1 & -2 & 3 & 2 \\ -1 & 2 & 0 & 2 \end{vmatrix}$

In Exercises 10–15, use determinants to solve the given systems.

10. $\begin{aligned} x + y - z &= 1 \\ y + 4z &= 0 \\ 3x + y &= 0 \end{aligned}$

11. $\begin{aligned} m + n + 3p &= 3 \\ 2m - n &= 0 \\ 4n - p &= 8 \end{aligned}$

12. $\begin{aligned} a + 2b + c &= 2 \\ 3a + 3b &= 0 \\ 2a - c &= -1 \end{aligned}$

13. $\begin{aligned} w + x - y + z &= 2 \\ 2w - x - 3y + 4z &= -2 \\ x - 2y + 2z &= 0 \\ w - 3y + 6z &= 1 \end{aligned}$

14. $\begin{aligned} 3r + s + 4t &= 1 \\ 10r + s - t &= -1 \\ r - 3s - 15t &= 1 \end{aligned}$

15. $\begin{aligned} 2a - b + 2d &= 2 \\ 4a - 2b + 3c &= -7 \\ b + c + d &= 6 \\ 6a - 4c + 5d &= 23 \end{aligned}$

In Exercises 16–20, you are given the determinant in expression (67).

16. Show that $\Delta = 0$ if $a_1 = a_2 = a_3 = 0$.

17. Show that $\Delta = 0$ if $b_1 = ka_1$, $b_2 = ka_2$, and $b_3 = ka_3$.

18. Show that $\Delta = a_1(b_2c_3 - b_3c_2)$ if $a_2 = a_3 = 0$.

19. Find Δ if $a_1 = b_2 = c_3 = 0$.

20. Interchange the a column and the b column and determine the effect on Δ.

21. Given the two determinants

$$\begin{vmatrix} a & b & c \\ d & e & f \\ g & h & i \end{vmatrix} \quad \text{and} \quad \begin{vmatrix} a + kb & b & c \\ d + ke & e & f \\ g + kh & h & i \end{vmatrix}$$

where the second determinant is created from the first by multiplying each element of the second column by a factor k and adding the results to the corresponding elements of the first column. Show that the two determinants are of equal value.

22. As a challenge to the assertion that the diagonal method does not work for determinants of order greater than three, try to evaluate Exercise 9 by the diagonal method and compare the result with that obtained by the method of minors.

6

Exponents and radicals

In Chapter 1 we discussed positive and negative powers of base 10 in standard notation. In this chapter we introduce the general laws of exponents, using any suitable base, and operations involving radicals. These operations will be the fundamental ones, including addition, subtraction, multiplication, and division. First, we describe certain appropriate laws and definitions.

6-1 Definitions and the laws of exponents

Many problems occur in mathematics when a number is multiplied by the same factor several times. An example is the case of starting today with one penny, doubling the amount to two pennies tomorrow, doubling again on the following day to four pennies, and so forth. After ten such doubling operations, the amount can be described as

$$1 \cdot 2 \cdot 2 \cdot 2 \cdot 2 \cdot 2 \cdot 2 \cdot 2 \cdot 2 \cdot 2 \cdot 2 \qquad (1)$$

pennies. Expression (1) is seen to contain the factor 2 a total of ten times. If we called for 100 doubling operations, it would be tedious to write an expression with 100 factors in the form of (1). For this and other reasons, exponential notation is introduced. Expression (1) is compactly written

$$1 \cdot (2)^{10},$$

where the number 2 is called the *base*, the number 10 the *exponent*, and the total expression $(2)^{10}$ "the tenth power of 2."

If we wish to indicate the product of n like factors where each factor is of value a, we have the definition

$$\underbrace{a \cdot a \cdot a \cdot \cdots \cdot a}_{n \text{ factors}} = a^n. \qquad (2)$$

Example 1 Evaluate (a) $(3)^4$ and (b) $(\frac{1}{2})^3$.

Solution

(a) $(3)^4 = 3 \cdot 3 \cdot 3 \cdot 3 = 81$ by definition (2).

(b) $(\frac{1}{2})^3 = \frac{1}{2} \cdot \frac{1}{2} \cdot \frac{1}{2} = \frac{1}{8}$ by definition (2).

Several laws of exponents are listed in Table 6-1; they are tabulated there for ready reference. Discussion and examples of these laws follow.

Table 6-1 Laws of Exponents (a and b are any nonzero number; m and n are integers)

Law 1:	$a^m \cdot a^n = a^{m+n}$
Law 2a:	$\frac{a^m}{a^n} = a^{m-n}$
Law 2b:	$a^0 = 1$
Law 2c:	$\frac{1}{a^n} = a^{-n}$
Law 3:	$(a^m)^n = a^{mn}$
Law 4:	$(ab)^n = a^n b^n$
Law 5:	$\left(\frac{a}{b}\right)^n = \frac{a^n}{b^n}$

Referring to Table 6-1, we have the following verbal statements for the laws given.

LAW 1 *In multiplying powers of like bases, add the exponents.*

LAW 2a *In dividing powers of like bases, subtract the exponent of the divisor from the exponent of the dividend.*

LAW 2b *Any nonzero base raised to the zero power has the value 1.*

LAW 2c *A power in the denominator of a fraction can become a factor of the numerator by changing the sign of the exponent.*

LAW 3 *To raise a power to a power, multiply exponents.*

LAW 4 *The power of the product of two numbers is the product of the powers of the two numbers.*

LAW 5 *The power of the quotient of two numbers is the quotient of the powers of the two numbers.*

Law 1 is readily apparent from definition (2). We have

$$a^m \cdot a^n = \underbrace{a \cdot a \cdot a \cdot \cdots}_{n \text{ factors}} \cdot \underbrace{a \cdot a \cdot a \cdot a \cdot \cdots \cdot a}_{m \text{ factors}} = a^{m+n}, \tag{3}$$

where the number of factors in (3) is apparently $m + n$.

Law 2 is also apparent from definition (2). We have

$$\frac{a^m}{a^n} = \frac{a \cdot a \cdot a \cdot \cdots \cdot a \quad (m \text{ factors})}{a \cdot a \cdot a \cdot \cdots \cdot a \quad (n \text{ factors})} = a^{m-n}. \tag{4}$$

In (4) several factors of the center equality are common to the top and bottom; the exact number is either m or n, depending on which is smaller. If $m > n$, then $m - n > 0$. If $m < n$, then $m - n < 0$. If $m = n$, then $m - n = 0$ and Law 2b holds.

Accepting Laws 2a and 2b, Law 2c follows:

$$\frac{1}{a^n} = \frac{a^0}{a^n} = a^{0-n} = a^{-n}.$$

Let's consider some example problems involving Laws 1 through 2c from Table 6-1.

Example 2 In parts (a) through (f) we show solved examples that require application of the laws of exponents drawn from Table 6-1. In each case, the object is simplification.

(a) $x^2 \cdot x^4 = x^{2+4} = x^6$ (Law 1)

(b) $x^{n+1} \cdot x^{n-1} = x^{(n+1)+(n-1)} = x^{2n}$ (Law 1)

(c) $\dfrac{y^3}{y^{-5}} = y^{3-(-5)} = y^{3+5} = y^8$ (Law 2a)

(d) $\frac{1}{x^{-2}} = x^{-(-2)} = x^2$ (Law 2c)

(e) $(a + b)^0 = 1$ (Law 2b)

(f) $p^{-m+2} \cdot p^{m-2} = p^{-m+2+m-2} = p^0 = 1$ (Laws 1 and 2b)

Continuing to demonstrate the laws in Table 6-1, we refer to Law 3:

$$(a^m)^n = \underbrace{a^m \cdot a^m \cdot a^m \cdot \cdots \cdot a^m}_{n \text{ factors}} = a^{mn}. \tag{5}$$

From (5) we see that there are n factors each containing m identical a factors, or a total of mn identical a factors.

From Law 4 we see that

$$\begin{aligned}(ab)^n &= (ab \cdot ab \cdot ab \cdot \cdots \cdot ab) \qquad (n \text{ factors}) \\ &= (\underbrace{a \cdot a \cdot a \cdot \cdots \cdot a}_{n \text{ factors}}) \cdot (\underbrace{b \cdot b \cdot b \cdot \cdots \cdot b}_{n \text{ factors}}) = a^n b^n,\end{aligned} \tag{6}$$

where $(ab)^n$ is the product of n factors with each factor of value ab; in other words, both a and b are present n times as factors, or the last term of (6) holds. Law 5 can be discussed in terms similar to those used for Law 4.

Additional examples illustrating the use of the laws of exponents follow.

Example 3 Simiplify the given expressions, using the laws of exponents in Table 6-1.

(a) $(k^2)^4 = k^{2 \cdot 4} = k^8$ (Law 3)

(b) $(6x)^2 = 6^2x^2 = 36x^2$ (Law 4)

(c) $\left(\frac{3k}{2p}\right)^3 = \frac{27k^3}{8p^3}$ (Law 5)

Many simplification operations require use of more than one of the laws in Table 6-1. Some are shown in Example 4.

Example 4 Simplify the expression

$$\frac{(3ab)^2}{6a^2b^3} \cdot \frac{9(a^2b)^3}{(3ab^2)^3}.$$

Solution Removing parentheses by Laws 3 and 4, we have

$$\frac{9 \cdot a^2 \cdot b^2 \cdot 9 \cdot a^6 \cdot b^3}{6 \cdot a^2 \cdot b^3 \cdot 27 \cdot a^3 \cdot b^6}. \tag{7}$$

Next, expression (7) reduces to expression (8) by Law 1,

$$\frac{9 \cdot 9 \cdot a^8 \cdot b^5}{6 \cdot 27 \cdot a^5 \cdot b^9}, \tag{8}$$

and expression (8) simplifies, by Law 2a, to the solution

$$\frac{a^3}{2b^4}.$$

Occasionally products and quotients of *unlike* bases can be simplified by changing to similar bases and applying Laws 1, 2a, and 3. Observe Example 5.

Example 5 Simplify the expression

$$\frac{(2)^{3x-3} \cdot (4)^{4x+2}}{(8)^{2x+5}}. \tag{9}$$

Solution The laws of multiplication and division in Table 6-1 are not appropriate here because they pertain only to powers of like bases. We note that the bases in (9) are 2, 4, and 8, which are not alike. However, 4 and 8 can be expressed as powers of the base 2, Now

$$4 = (2)^2 \quad \text{and} \quad (4)^{4x+2} = (2^2)^{4x+2} = (2)^{8x+4}$$

and

$$8 = (2)^3 \quad \text{and} \quad (8)^{2x+5} = (2^3)^{2x+5} = (2)^{6x+15}$$

and (9) becomes

$$\frac{(2)^{3x-3} \cdot (4)^{4x+2}}{(8)^{2x+5}} = \frac{(2)^{3x-3} \cdot (2)^{8x+4}}{(2)^{6x+15}} = \frac{(2)^{11x+1}}{(2)^{6x+15}} = (2)^{5x-14}.$$

The method used in Example 5 is important in understanding logarithms in a later chapter. In Example 5 we converted all bases to the same base 2. We chose base 2 simply because it seemed to be the most convenient base; that is, all other bases were powers of the number 2. Tables of logarithms are set up in such a way that, through the table, we can rapidly convert all bases to the base used in the table of logarithms. Two bases are popular in table of logarithms—base 10 or the *common* base and base *e* or the *natural* base. They are discussed extensively in the chapter on logarithms.

Exercises 6-1

Perform the indicated operations by the laws of exponents. Reduce answers to the simplest terms.

1. $m^2 \cdot m^4$

2. $p^3 \cdot p^2 \cdot p$

3. $(3m^2)^2$

4. $-(\frac{1}{2}x^2)^3$

5. $3(m^2)^2$

6. $-\frac{1}{2}(x^2)^3$

7. $3(m^2 + 2)^0$

8. $\dfrac{5(b + 2)^0}{10}$

9. $[4(p^2 + 2p)]^0$

10. $m^2n^3 \div mn$

11. $(z^2 \cdot z^3)^2$

12. $a^5 \div a^{-3}$

13. $\left(\dfrac{z^3}{z^{-2}}\right)^2$

14. $(r_1 r_2^{\frac{2}{3}})^2$

15. $\left(\dfrac{-4F}{F^2}\right)^3$

16. $\left(\dfrac{m^{-1}n^{-2}}{2n}\right)^3$

17. $(b^3 \cdot b^2 \cdot b^{-5})^5$

18. $[(m^2 - n^2)(p + 2)]^0$

19. $\dfrac{b^2}{ac} \cdot \dfrac{a^2c}{b} \cdot \dfrac{c}{2b}$

20. $(-b)(-b^2)(-b)^3$

21. $b^{-x} \cdot b^{1+x}$

22. $b^{3x} \cdot b^{x-2}$

23. $R^{2x+5} \cdot R^{x-5}$

24. $R^{2x} \cdot R^{x-3} \cdot R^{-x+2}$

25. $(ab)^{2k-3} \div (ab)^{3k-4}$

26. $(y^{c-6})(y^5) \div y^{c+2}$

27. $(x^2y^n)^{-5} \div 5x^3y^{3n}$

28. $[(x^2)^2]^3$

29. $(m^m)^m$

30. $(5)^{k+2} \cdot (25)^k$

31. $(3)^{2c+4} \cdot (9)^{1-c}$

32. $(9)^2 \cdot (3)^{k-4}$

33. $(125)^{m-n} \cdot (25)^{m+n} \cdot (5)^2$

34. $(p^{2+k}) \div (p^{3+k})$

35. $\dfrac{(36)^{q-3} \cdot (6)^{q+7}}{(216)^q}$

36. $\left(\dfrac{k^{m+2}}{k^{2m-1}}\right)^2$

37. $\left(\dfrac{b^{3k+2}}{b^{2k-2}}\right)^3$

38. $\left(\dfrac{x^{k-3}}{x^{k+3}}\right) \div (x)^3$

39. $\left(\dfrac{m^{2n-p}}{m^{p+n}}\right) \div \left(\dfrac{m^{p+3n}}{m^{2n-3p}}\right)$

40. $(p^2 - p^{-2})(p^2 + p^{-2})$

6-2 Radicals and fractional exponents

In Sec. 6-1 we treated only integral exponents, both positive and negative; we made no reference to fractional exponents. In this section we introduce fractional exponents and we also elaborate on the relationship between radical quantities and quantities raised to fractional powers.

If given the expression

$$y = b^n \qquad (b > 0), \tag{10}$$

where n is a positive integer, then

$$y = \underbrace{b \cdot b \cdot b \cdot \cdots \cdot b}_{n \text{ factors}} \tag{11}$$

according to definition (2). Now from (10) the number y (more specifically, y to the first power) can be expressed as the product of n factors

$$\underbrace{y^{1/n} \cdot y^{1/n} \cdot y^{1/n} \cdot \cdots \cdot y^{1/n}}_{n \text{ factors}} = (y^{1/n})^n = y^1, \tag{12}$$

and (10) can be written, using (11) and (12),

$$\underbrace{y^{1/n} \cdot y^{1/n} \cdot y^{1/n} \cdot \cdots \cdot y^{1/n}}_{n \text{ factors}} = \underbrace{b \cdot b \cdot b \cdot \cdots \cdot b}_{n \text{ factors}}. \tag{13}$$

From (13) there is a one-to-one correspondence between the positive factors $y^{1/n}$ and b such that

$$y^{1/n} = b. \tag{14}$$

Next, from (10) and (14), if y is the product of n factors, each of value b, then b is of value $y^{1/n}$. We can state that if y is the nth power of b, then b is the nth root of y.

The nth root of y is symbolized as $y^{1/n}$ as in (14) and is also symbolized in radical form as $\sqrt[n]{y}$, where y is called the *radicand*, the symbol $\sqrt{\ }$ is a *radical*, and n is the *index*. We have the equivalency

$$y^{1/n} = \sqrt[n]{y}. \tag{15}$$

Example 6 Express one of the five equal positive factors of r in both the radical and exponent form.

Solution In the exponent form, we have $r^{1/5}$ as one of the five equal factors of r because the product involving five such factors has the value r or

$$(r^{1/5})(r^{1/5})(r^{1/5})(r^{1/5})(r^{1/5}) = (r^{1/5})^5 = r.$$

In the radical notation, from (15) we have

$$r^{1/5} = \sqrt[5]{r}.$$

Referring back to (15), we note that the denominator of the fractional exponent is the index (or indicates the order) of the radical. It is possible that the exponent does not have a unity numerator, as in $y^{2/3}$. By Law 3 in Table 6-1, where

$$(a^m)^n = a^{mn},$$

we have

$$y^{2/3} = (y^2)^{1/3} = (y^{1/3})^2. \tag{16}$$

The center portion of (16) asserts that y to the 2/3 power is the same as the cube root of y^2, and the right portion of (16) asserts that y to the 2/3 power is the same as the square of the cube root of y. Writing (16) in radical notation, we have

$$y^{2/3} = \sqrt[3]{y^2} = (\sqrt[3]{y})^2.$$

In general, it can be seen that the following equalities are true:

$$a^{m/n} = \sqrt[n]{a^m} = (\sqrt[n]{a})^m. \tag{17}$$

Example 7 Evaluate $(16)^{3/2}$.

Solution From (17) we see that

$$(16)^{3/2} = \sqrt[2]{(16)^3}$$

[that is, $(16)^{3/2}$ equals the square root of the cube of 16] or

$$(16)^{3/2} = (\sqrt[2]{16})^3$$

[that is, $(16)^{3/2}$ equals the cube of the square root of 16]. Of the two preceding choices, we select the latter purely for its simplicity because the square root of 16 is 4 and the cube of 4 is 64, or

$$(16)^{3/2} = (\sqrt[2]{16})^3 = (4)^3 = 64.$$

Two more laws regarding radicals require some inspection. Let us consider the nth root of the product of two factors. Here

$$\sqrt[n]{ab} = (ab)^{1/n} = a^{1/n} \cdot b^{1/n} = \sqrt[n]{a} \cdot \sqrt[n]{b}, \tag{18}$$

where (18) asserts that the positive nth root of the product of two factors is the same as the product of the positive nth roots of the two factors. Also,

$$\sqrt[n]{\frac{a}{b}} = \left(\frac{a}{b}\right)^{1/n} = \frac{a^{1/n}}{b^{1/n}} = \frac{\sqrt[n]{a}}{\sqrt[n]{b}}, \tag{19}$$

where (19) asserts that the nth root of the quotient of two factors is the quotient of the nth roots of the two factors.

Definitions and Laws of Radicals (12), (15), (17), (18) and (19) are complied into Table 6-2. *Observation*: When the index is 2, the radical is often written without the index, so $\sqrt{x}$ and $\sqrt[2]{x}$ are the same.

Table 6-2 Definitions and Laws of Radicals (a and b are positive numbers; m and n are integers)

Definition 1:	$a^{1/n} = \sqrt[n]{a}$
Definition 2:	$(a^{1/n})^n = (a^n)^{1/n} = a$
Law 1:	$a^{m/n} = \sqrt[n]{a^m} = (\sqrt[n]{a})^m$
Law 2:	$\sqrt[n]{ab} = \sqrt[n]{a} \cdot \sqrt[n]{b}$
Law 3:	$\sqrt[n]{\frac{a}{b}} = \frac{\sqrt[n]{a}}{\sqrt[n]{b}}$

Example 8 Using the laws of exponents and radicals, we show some simplification operations.

(a) $\sqrt[3]{27a^6b^3} = (3^3a^6b^3)^{1/3} = 3^{3/3}a^{6/3}b^{3/3} = 3a^2b$

(b) $(64)^{2/3} = (4^3)^{2/3} = 4^{6/3} = 4^2 = 16$

(c) $\sqrt[2]{\dfrac{a^4c^2}{b^6}} = \dfrac{\sqrt[2]{a^4c^2}}{\sqrt[2]{b^6}} = \dfrac{(a^4c^2)^{1/2}}{(b^6)^{1/2}} = \dfrac{a^2c}{b^3}$

In all the parts of Example 8, the entire radicand was removable from under the radical sign; this constituted a step in simplification. In many cases, only portions of the radicand can be removed; once again this is simplification, but it is not as complete in that some radical quantities will still remain in the final simplified form.

Example 9 Simplify (a) $\sqrt{48}$, (b) $\sqrt[3]{a^4b^5}$, and (c) $\sqrt[4]{32a^7/c^4}$.

Solution In (a) and (b) we use Law 2 of Table 6-2 and Definition 2 of the same table.

$$\sqrt{48} = \sqrt{16 \cdot 3} = \sqrt{16} \cdot \sqrt{3} = 4\sqrt{3}$$

Note that the number 48 is expressed as the product of the two factors 16 and 3, where 16 is the largest perfect square that is a factor of 48. The reason for choosing the factor 16 lies in Definition 2 of Table 6-2, since

$$\sqrt{16} = (16)^{1/2} = (4^2)^{1/2} = 4.$$

In part (b), we must select the largest perfect *cube* that is contained as a factor of the radicand, since we are taking a cube root. So

$$\sqrt[3]{a^4b^5} = \sqrt[3]{a^3b^3 \cdot ab^2} = \sqrt[3]{a^3b^3} \cdot \sqrt[3]{ab^2} = ab\sqrt[3]{ab^2}.$$

In part (c), we must select the largest fourth-power factor of the radicand or

$$\sqrt[4]{\frac{32a^7}{c^4}} = \frac{\sqrt[4]{32a^7}}{\sqrt[4]{c^4}} = \frac{\sqrt[4]{16a^4 \cdot 2a^3}}{\sqrt[4]{c^4}} = \frac{2a\sqrt[4]{2a^3}}{c}.$$

In some examples where simplification is requested by removing a factor from the radicand, the index involved is literal. The procedures involved are similar to those in Example 9.

Example 10 Here are three examples of simplification of radicals with literal indices.

(a) $\sqrt[n]{c^{n+3}} = \sqrt[n]{c^n \cdot c^3} = \sqrt[n]{c^n} \cdot \sqrt[n]{c^3} = c\sqrt[n]{c^3}.$

(b) $\sqrt[n+1]{b^{2n+7}} = \sqrt[n+1]{b^{2n+2} \cdot b^5} = \sqrt[n+1]{b^{2n+2}} \cdot \sqrt[n+1]{b^5} = b^2(\sqrt[n+1]{b^5}).$

(c) $\sqrt[2n-1]{\dfrac{a^{2n+1}}{b^{4n-2}}} = \sqrt[2n-1]{\dfrac{a^{2n-1} \cdot a^2}{b^{4n-2}}} = \sqrt[2n-1]{\dfrac{a^{2n-1}}{b^{4n-2}}} \cdot \sqrt[2n-1]{a^2} = \dfrac{a}{b^2}(\sqrt[2n-1]{a^2}).$

In all the parts of Example 10 one intermediate goal is evident. In each case, the radicand is factored into two factors. One of these factors has an exponent that is an integral multiple of the index, and the exponent of the other factor is such that the sum of the exponents of the two factors equals the original exponent. To illustrate, we see in part (b) that the radicand is b^{2n+7} and the index is $n + 1$. The radicand is expressed as $b^{2n+2} \cdot b^5$, where $2n + 2$ is exactly twice $n + 1$ and $2n + 2 + 5$ adds to the original exponent $2n + 7$.

In Examples 9 and 10 simplification of the radical quantities was performed by removing a factor of the radicand from the radicand and making it a multiplier of the remaining radical quantity. Before introducing a set of exercises, we consider another simplification procedure, one involving fractional radicands or radical quantities in the denominators of fractions. From Law 3 of Table 6-2 we have

$$\sqrt[n]{\frac{a}{b}} = \frac{\sqrt[n]{a}}{\sqrt[n]{b}},$$

where the left side shows that the root of a quotient of two numbers is the quotient of the roots of the two numbers. In the simplified form, it is desirable to eliminate fractions from radicands and to eliminate radical quantities from denominators. Refer to Examples 11 and 12 for illustrations.

Example 11 Simplify the quantity $\sqrt[4]{ab^2/cd^2}$.

Solution Using Law 3 of Table 6-2, we have

$$\sqrt[4]{\frac{ab^2}{cd^2}} = \frac{\sqrt[4]{ab^2}}{\sqrt[4]{cd^2}},$$

where the denominator of the right side is a radical quantity. The process of eliminating the radical in the denominator is called *rationalizing the denominator* and can be accomplished by multiplying the denominator by a conveniently chosen quantity. If the exponents of c and d in the denominator were each 4 (or integral multiples of the index 4,) then the fourth root of the radicand would not involve a radical. If we multiply the denominator by $\sqrt[4]{c^3d^2}$, the new denominator will be

$$\sqrt[4]{cd^2} \cdot \sqrt[4]{c^3d^2} = \sqrt[4]{c^4d^4} = cd,$$

where cd is rational. We must remember, however, that if we multiply the denominator of a fraction by a factor, we must provide the numerator with the same factor so that the value of the original fraction will be unchanged. Proceeding,

$$\sqrt[4]{\frac{ab^2}{cd^2}} = \frac{\sqrt[4]{ab^2} \cdot \sqrt[4]{c^3d^2}}{\sqrt[4]{cd^2} \cdot \sqrt[4]{c^3d^2}} = \frac{\sqrt[4]{ab^2c^3d^2}}{cd}.$$

Example 12 Here are three more examples involving the process of rationalizing denominators.

(a) $\dfrac{3}{\sqrt{2}} = \dfrac{3}{\sqrt{2}} \cdot \dfrac{\sqrt{2}}{\sqrt{2}} = \dfrac{3\sqrt{2}}{2}$

(b) $\dfrac{3}{\sqrt[3]{2}} = \dfrac{3}{\sqrt[3]{2}} \cdot \dfrac{\sqrt[3]{4}}{\sqrt[3]{4}} = \dfrac{3\sqrt[3]{4}}{2}$

(c) $\dfrac{3}{\sqrt[5]{2}} = \dfrac{3}{\sqrt[5]{2}} \cdot \dfrac{\sqrt[5]{16}}{\sqrt[5]{16}} = \dfrac{3\sqrt[5]{16}}{2}$

Note in Example 12 that the rationalizing factor varies according to the nature of the root in the original denominator. In (a) the rationalizing factor involves a square root, in (b) a cube root, and in (c) a fifth root.

Exercises 6-2

The quantities given in Exercises 1–5 are in exponent form. Express them in radical form.

1. $b^{3/2}$ **2.** $(m + n)^{2/3}$ **3.** $(c^2d)^{2/5}$

4. $m^{1/4}n^{3/4}p^{-1/4}$ **5.** $\left(\dfrac{1 + x^2}{x}\right)^{2/3}$

The quantities given in Exercises 6–10 are in radical form. Express them in exponent form.

6. $\sqrt[3]{b^2c}$ **7.** $\sqrt[4]{a^3b}$ **8.** $\sqrt[3]{(m + n)^2}$

9. $\sqrt[4]{\dfrac{b}{(c + d)^3}}$ **10.** $\sqrt[5]{p^2q^{-3}m}$

Simplify the expressions in Exercises 11–40 by removing factors from the radicands.

11. $\sqrt{12}$ **12.** $\sqrt{18}$ **13.** $\sqrt{24}$

14. $\sqrt{72}$ **15.** $\sqrt{75}$ **16.** $\sqrt{54}$

17. $\sqrt{27}$ **18.** $\sqrt{288}$ **19.** $\sqrt{125}$

20. $\sqrt{108}$ **21.** $\sqrt[3]{54}$ **22.** $\sqrt[3]{81}$

23. $\sqrt[3]{24}$ **24.** $\sqrt[3]{128}$ **25.** $\sqrt[4]{32}$

26. $\sqrt[3]{320}$ **27.** $\sqrt[4]{162}$ **28.** $\sqrt[4]{48}$

29. $\sqrt[6]{576}$ **30.** $\sqrt[4]{a^5b}$ **31.** $\sqrt[3]{a^5b^4c^3}$

32. $\sqrt{ab^2c^3}$ **33.** $\sqrt{12m^3p^5}$ **34.** $\sqrt[3]{12m^3p^5}$

35. $\sqrt[4]{243a^{13}b^{14}}$ **36.** $\sqrt[n]{x^{3n}y^{n+1}}$ **37.** $\sqrt[n]{2^{n+1}x^{4n+5}}$

38. $\sqrt[n+1]{a^{2+n}b^{2n+3}}$ **39.** $\sqrt[2n-1]{a^{2n}b^{4n-1}}$ **40.** $\sqrt[2n+1]{a^{2n+2}b^{3n+5}}$

Simplify Exercises 42–70 by rationalizing denominators.

41. $\frac{1}{\sqrt{3}}$ **42.** $\sqrt{\frac{4}{9}}$ **43.** $\sqrt{\frac{1}{2}}$

44. $\sqrt{\frac{3}{4}}$ **45.** $\sqrt{\frac{3}{2}}$ **46.** $\sqrt{\frac{4.5}{2}}$

47. $\sqrt{\frac{5}{4}}$ **48.** $\sqrt{\frac{24}{1.5}}$ **49.** $\sqrt{\frac{5}{32}}$

50. $\sqrt{\frac{40}{18}}$ **51.** $\sqrt{\frac{7}{8}}$ **52.** $\sqrt{\frac{5}{8}}$

53. $\sqrt{\frac{a^2}{bc}}$ **54.** $\sqrt{\frac{4b^3}{9c^3}}$ **55.** $\sqrt{\frac{9ab}{98c^3}}$

56. $\sqrt[3]{\frac{2x^2}{9y^4}}$ **57.** $\sqrt[4]{\frac{b^{4n}}{c^{3n}}}$ **58.** $\sqrt[n]{\frac{1}{x^{n-1}}}$

59. $\sqrt[n+1]{\frac{1}{x^n}}$ **60.** $\sqrt[2n+1]{\frac{x^2}{y^{4n+4}}}$ **61.** $\frac{1}{\sqrt[3]{3}}$

62. $\frac{\sqrt{3}}{\sqrt{8}}$ **63.** $\frac{\sqrt[4]{3}}{\sqrt[4]{4}}$ **64.** $\sqrt{\frac{m+n}{m-n}}$

65. $\sqrt{\frac{m-n}{m+n}}$ **66.** $\sqrt[3]{\frac{m}{3n+2}}$ **67.** $\sqrt{\frac{5}{72}}$

68. $\sqrt{\frac{x^2+4x^2y}{8}}$ **69.** $\sqrt{\frac{(a+b)^3}{a}}$ **70.** $\sqrt[3]{\frac{1}{4x^2y}}$

6-3 Multiplication and division of radical quantities

The laws that are most useful in the multiplication and division of radical quantities are

$$\sqrt[n]{a} \cdot \sqrt[n]{b} = \sqrt[n]{ab}, \tag{20}$$

$$\sqrt[n]{a} \div \sqrt[n]{b} = \sqrt[n]{\frac{a}{b}}. \tag{21}$$

Note that (20) and (21) have similar indices and different radicands. This observation is valuable in identifying the proper procedure to use.

Appropriate also are the laws of exponents where

$$a^m \cdot a^n = a^{m+n}, \tag{22}$$

$$a^m \div a^n = a^{m-n}. \tag{23}$$

Expressions (22) and (23) are applied to radical quantities that, after they are converted to exponent form, have unlike exponents and similar bases.

We can organize our material to involve four cases for multiplication and division of radical quantities.

CASE 1 If the radicands are different and the indices are the same, apply (20) or (21).

CASE 2 If the radicands are the same and the indices different, convert to the exponential form and apply (22) or (23).

CASE 3 If the radicands are the same and the indices the same, apply (20), (21), (22), or (23).

CASE 4 If the radicands are different and the indices are different, similar indices can be obtained by converting to the exponential form and establishing an LCD of the exponents and then converting back to radicals and applying (20) or (21). If the bases are numerical and the indices are different, conversion to base 10 can be made through logarithm tables and (22) or (23) can be applied.

Example 13 In each example here the indices are the same and the radicands are different. This condition suggests that Case 1 is appropriate and so we will use either (20) or (21) in the solution.

(a) $\sqrt{2} \cdot \sqrt{3} = \sqrt{6}$

(b) $\sqrt[3]{5a^2b} \cdot \sqrt[3]{50a^2b^2} = \sqrt[3]{250a^4b^3} = \sqrt[3]{125a^3b^3} \cdot \sqrt[3]{2a} = 5ab\sqrt[3]{2a}$

(c) $\sqrt[n]{6a^{2n-3}} \div \sqrt[n]{3a^{n-6}} = \sqrt[n]{\dfrac{6a^{2n-3}}{3a^{n-6}}} = \sqrt[n]{2a^{n+3}} = a\sqrt[n]{2a^3}.$

Notice in Example 13 that, with the indices the same and the radicands different, we multiply or divide the radicands without modification of the index.

Example 14 Here we give two examples in which the radicands are similar and the indices different. Case 2 is appropriate.

(a) $\sqrt[2]{2} \cdot \sqrt[3]{2} = 2^{1/2} \cdot 2^{1/3} = 2^{(1/2)+(1/3)} = 2^{5/6} = \sqrt[6]{2^5} = \sqrt[6]{32}$

(b) $\sqrt[3]{18a} \div \sqrt[4]{18a} = (18a)^{1/3} \div (18a)^{1/4} = (18a)^{(1/3)-(1/4)}$
$= (18a)^{1/12} = \sqrt[12]{18a}$

The bases are the same and the indices are different. On conversion to the exponent form we added or subtracted the exponents by conversion to an LCD. After collecting the exponents, we converted back to the radical form.

Example 15 This example illustrates the multiplication of two quantities in which the radicands are similar and the indices are similar. This is Case 3. By (20),

$$\sqrt[5]{3a^2b} \cdot \sqrt[5]{3a^2b} = \sqrt[5]{9a^4b^2}$$

or by (22)

$$\sqrt[5]{3a^2b} \cdot \sqrt[5]{3a^2b} = (3a^2b)^{1/5} \cdot (3a^2b)^{1/5} = (3a^2b)^{2/5} = \sqrt[5]{9a^4b^2}.$$

Note that this is perhaps the simplest of the operations. It amounts merely to a squaring operation.

Example 16 Here we consider two examples in which the radicands and indices are different. This is Case 4. It can be solved by converting to exponent form, combining exponents, and then converting back to the radical form.

(a) $$\sqrt[3]{4} \cdot \sqrt[2]{2} = (4)^{1/3} \cdot (2)^{1/2} = (4)^{2/6} \cdot (2)^{3/6} = (4^2)^{1/6} \cdot (2^3)^{1/6}$$
$$= \sqrt[6]{4^2 \cdot 2^3} = 2\sqrt[6]{2}$$

(b) $$\sqrt[5]{6} \div \sqrt[3]{3} = (6)^{1/5} \div (3)^{1/3} = (6)^{3/15} \div (3)^{5/15}$$
$$= (6^3)^{1/15} \div (3^5)^{1/15} = \sqrt[15]{\frac{6^3}{3^5}} = \sqrt[15]{\frac{8}{9}}$$
$$= \frac{1}{3}\sqrt[15]{8 \cdot 3^{13}}$$

Exercises 6-3

Perform the indicated operations. Simplify the results wherever possible.

1. $\sqrt{3} \cdot \sqrt{15}$
2. $\sqrt{6} \cdot \sqrt{15}$
3. $\sqrt[3]{4} \cdot \sqrt[3]{12}$
4. $\sqrt[4]{12} \cdot \sqrt[4]{20}$
5. $\sqrt{3x^3y} \cdot \sqrt{6xy^5}$
6. $\sqrt[4]{x^5y^5} \cdot \sqrt[4]{x^3y^7}$
7. $\sqrt{8x^3y^3} \div \sqrt{2xy^3}$
8. $\sqrt{50} \div \sqrt{2}$
9. $\sqrt{48ab} \div \sqrt{12ab}$
10. $\sqrt[3]{a^2b} \div \sqrt[3]{ab}$
11. $\sqrt[2]{6} \cdot \sqrt[3]{12}$
12. $\sqrt[3]{12} \div \sqrt[2]{6}$
13. $\sqrt[4]{54} \div \sqrt[2]{27}$
14. $\sqrt{32} \cdot \sqrt{32}$
15. $\sqrt[3]{35} \div \sqrt[3]{35}$
16. $(2\sqrt[3]{5})^3$
17. $(\sqrt[3]{9x^2})^2$
18. $(\sqrt{5} - \sqrt{3})^2$
19. $(\sqrt{2} + \sqrt{3})^2$
20. $(3 - \sqrt{2})^2$
21. $(3 + \sqrt{2})(3 - \sqrt{2})$
22. $(2\sqrt{3} + 4)(2\sqrt{3} - 4)$
23. $(\sqrt{3} - 3)(\sqrt{12} + 2)$
24. $(\sqrt{x + y} + z)(\sqrt{x + y})$

25. $\sqrt{3-\sqrt{3}}\cdot\sqrt{3+\sqrt{3}}$

26. $\sqrt{\frac{A}{\pi}}\cdot\sqrt{\frac{\pi}{A}}$

27. $(\sqrt{x^2-y^2}-\sqrt{x^2+y^2})^2$

28. $\sqrt[3]{a^2b}\div\sqrt[2]{ab^2}$

29. $\sqrt[2]{a^3b}\div\sqrt[3]{ab}$

30. $\sqrt[4]{m^3n}\div\sqrt[3]{mn}$

31. $\sqrt[n]{x^{n+1}y^{n-1}}\cdot\sqrt[n]{x^ny^{n+3}}$

32. $\sqrt[n]{a^2b}\div\sqrt[n]{ab}$

33. $\sqrt[n+1]{a^{n-1}b^n}\div\sqrt[n]{a^{n+1}b^{n-1}}$

34. $\sqrt[n]{a^{n+1}b^{n-1}}\div\sqrt[n-1]{a^nb^n}$

6-4 Addition and subtraction of radical quantities

Two radical quantities are alike only if their radicands are the same and their indices are the same. If the quantities are expressed in exponential form, they are alike only if the bases are the same and the exponents are the same. *Radical quantities can be added or subtracted only if they are alike or can be made alike.*

Example 17 Perform the indicated operations.

(a) $\sqrt{3}+2\sqrt{3}$

(b) $\sqrt{2}+\sqrt{50}-\sqrt{18}$

(c) $2\sqrt{a^3}-\sqrt{a^5}-3\sqrt{a}$

Solution

(a) Since the radicands are the same and both quantities involve the same (square) root, the radical quantities are of the same kind and can be added by adding the coefficients.

$$\sqrt{3}+2\sqrt{3}=3\sqrt{3}$$

(b) The indices are the same but the radicands appear different. If we remove factors from the last two radicands, we will have similar radicands.

$$\sqrt{2}+\sqrt{50}-\sqrt{18}=\sqrt{2}+5\sqrt{2}-3\sqrt{2}=3\sqrt{2}$$

(c) The method of solution here is similar to that used in part (b); that is, factors are removed from the radicands.

$$\begin{aligned}2\sqrt{a^3}-\sqrt{a^5}-3\sqrt{a}&=2a\sqrt{a}-a^2\sqrt{a}-3\sqrt{a}\\&=-\sqrt{a}\,(a^2-2a+3)\end{aligned}$$

Example 18 Perform the indicated operation.

$$5\sqrt{6}-4\sqrt{\tfrac{3}{2}}+2\sqrt[4]{36}$$

Solution Similar indices and similar radicands can be obtained by standard simplification procedures.

$$5\sqrt{6} - 4\sqrt{\tfrac{3}{2}} + 2\sqrt[4]{36} = 5\sqrt{6} - 4\sqrt{\tfrac{6}{4}} + 2(36)^{1/4}$$
$$= 5\sqrt{6} - \tfrac{4}{2}\sqrt{6} + 2(6^2)^{1/4}$$
$$= 5\sqrt{6} - 2\sqrt{6} + 2\sqrt{6} = 5\sqrt{6}$$

Note that $\sqrt[4]{36}$ was modified to $\sqrt[2]{6}$. Changing from the higher index 4 to the lower index 2 is called *reduction in order*.

Exercises 6-4

Perform the indicated addition and subtraction operations wherever possible.

1. $\sqrt{3} - 3\sqrt{3} + 4\sqrt{3}$
2. $3\sqrt[3]{5} + 2\sqrt[3]{5} - \sqrt[3]{5}$
3. $a\sqrt[n]{x} - b\sqrt[n]{x} + c\sqrt[n]{x}$
4. $\sqrt{4x} - \sqrt{x}$
5. $\sqrt{2} - \sqrt{8}$
6. $\sqrt{3} - \sqrt{27}$
7. $\sqrt{12} - \sqrt{3}$
8. $\sqrt{32} - \sqrt{18}$
9. $\sqrt{27} + \sqrt{\dfrac{1}{3}}$
10. $3\sqrt{32} - 2\sqrt{50}$
11. $3\sqrt{50} - 2\sqrt{8}$
12. $\sqrt{8} + \sqrt{\dfrac{1}{2}}$
13. $2\sqrt{98} + 3\sqrt{72} - 5\sqrt{50}$
14. $\sqrt{80} - \sqrt{20} - \sqrt{\dfrac{5}{4}}$
15. $2\sqrt{48} + \sqrt{75} - 2\sqrt{27}$
16. $\sqrt{12} + \sqrt{27} - \sqrt{\dfrac{1}{3}}$
17. $\sqrt{28} - \sqrt{\dfrac{7}{9}} = \sqrt{\dfrac{1}{7}}$
18. $\sqrt[3]{16} + \sqrt[3]{54} - \sqrt[3]{128}$
19. $\sqrt[3]{32} + \sqrt[3]{500}$
20. $2\sqrt[3]{\dfrac{1}{2}} + 3\sqrt[3]{\dfrac{4}{27}}$
21. $3\sqrt[3]{\dfrac{1}{4}} - 2\sqrt[3]{\dfrac{2}{27}} + \sqrt[3]{\dfrac{4}{27}}$
22. $\sqrt{a^2b} - \sqrt{a^4b^3} + 2\sqrt{b^5}$
23. $\sqrt{x^3y} - \sqrt{xy^3} + \sqrt{xy}$
24. $\sqrt{x^5y^4} + \sqrt{x^3y^6}$
25. $\sqrt[3]{a^4b^5} - 3\sqrt[3]{a^7b^5} + 2\sqrt[3]{a^4b^2}$
26. $\sqrt[3]{(m+n)^2} - 2\sqrt[3]{\dfrac{1}{m+n}}$
27. $\sqrt{a+b} - 2\sqrt{\dfrac{1}{a+b}}$
28. $3\sqrt[3]{40x} + \sqrt[3]{135y^3x}$
29. $\sqrt[4]{(a^2-b^2)^2} + \dfrac{1}{3}\sqrt{a^2-b^2}$
30. $\sqrt{mn} - \sqrt{m^3n} + 4\sqrt{\dfrac{n}{m}}$
31. $\sqrt{75a^2x - 50a^4} + \sqrt{48x^5 - 32a^2x^4}$
32. $4\sqrt[3]{24c^4d^2} + 3\sqrt[3]{81cd^2}$
33. $\sqrt{\dfrac{1}{m}} - 3\sqrt{\dfrac{1}{m^3}} + 2\sqrt{\dfrac{1}{m^5}}$

34. $\sqrt{\frac{a-b}{a+b}} - \sqrt{\frac{a+b}{a-b}}$

35. $\sqrt[9]{(x+y)^{12}} + \sqrt[3]{64xy^3 + 64y^4}$

36. $\sqrt{(a+b)^3} - \sqrt{a^3b^2 + a^2b^3}$

37. $\sqrt{m^2 - m^2n^2} - \sqrt{\frac{1+n}{1-n}} + \sqrt{n^2 - n^4}$

38. $\sqrt[3]{1 - \frac{1}{x}} - \frac{1}{x}\sqrt[3]{x^3 - x^2}$

39. $\frac{x^2}{2}\sqrt{\frac{2y^3}{x^5}} - \sqrt{\frac{y^3}{2x}} + x\sqrt{\frac{y^3}{2x^3}}$

40. $3\sqrt[4]{\frac{a^2}{4b^2}} + \frac{b^2}{2}\sqrt{\frac{2a^3}{b^5}}$

6-5 More on rationalizing denominators: A summary

When radical expressions occur, a question regarding their simplification also occurs. A primary reason for certain standards in simplification is to leave results in the same form. Often two students will arrive at equivalent answers to a problem but fail to recognize the equivalency of their answers because the answers are left in two different algebraic forms. Other reasons for certain simplification standards are also given. If a student is without computational aids like calculators or tables, numerical radicands are best expressed in small numbers; that is, given

$$\sqrt{12} = \sqrt{4} \cdot \sqrt{3} = 2\sqrt{3},$$

it is less likely that the student knows the approximate decimal value of $\sqrt{12}$ than the decimal value of $\sqrt{3}$.

There are also cases involving division of radicals. Once again, if we are without computational aids, given

$$\frac{1}{\sqrt{3}} = \frac{1}{\sqrt{3}} \cdot \frac{\sqrt{3}}{\sqrt{3}} = \frac{\sqrt{3}}{3} = \frac{1.732}{3} = 0.577$$

it is easier to divide 1.732 by 3 than to divide unity by 1.732. So we attempt to eliminate radicals from denominators.

Four simplification operations are summarized here. We will accept them as operations that are standard but not absolutely essential.

1. *Reduce radicands to the simplest form.* When factors can be removed from radicands, remove them. Such simplifications depend on the law:

$$\sqrt[n]{ab} = \sqrt[n]{a} \cdot \sqrt[n]{b}.$$

   Many exercises were provided in Sec. 6-2.

2. *Do not leave a fraction under a radical sign.* Given the nth root of a fraction, express it as the quotient of the nth roots of the numerator

and denominator; then rationalize the denominator. That is,

$$\sqrt[n]{\frac{a}{b}} = \frac{\sqrt[n]{a}}{\sqrt[n]{b}} = \frac{\sqrt[n]{a}}{\sqrt[n]{b}} \cdot \frac{\sqrt[n]{b^{n-1}}}{\sqrt[n]{b^{n-1}}} = \frac{\sqrt[n]{ab^{n-1}}}{b}. \tag{24}$$

The preceding operation involved the division of a radical expression. Many exercises were provided in Secs. 6-2 and 6-3.

3. *Rationalize denominators whenever it is done simply.* Two common operations are involved here. In (24) we saw a denominator containing a monomial quantity $\sqrt[n]{b}$. The procedure used in simplification there was discussed previously in examples in Secs. 6-2 and 6-3. It is emphasized that those examples embraced only monomial denominators.

   The second operation regarding rationalization of denominators involves a binomial denominator in which either or both terms in the denominator involve a square root radical. Before showing some examples, let us reflect on a certain algebraic operation: if the sum of two numbers is multiplied by the difference of the same two numbers, the product is the difference of the squares of the two numbers. If either or both of the numbers is a square root, the operation eliminates the square root radical. That is,

$$(\sqrt{a} + b)(\sqrt{a} - b) = a - b^2,$$
$$(a + \sqrt{b})(a - \sqrt{b}) = a^2 - b,$$
$$(\sqrt{a} + \sqrt{b})(\sqrt{a} - \sqrt{b}) = a - b.$$

   Any two numbers of forms $x + \sqrt{y}$ and $x - \sqrt{y}$ are *conjugate* numbers. That is, $x + \sqrt{y}$ is the conjugate of $x - \sqrt{y}$ and $x - \sqrt{y}$ is the conjugate of $x + \sqrt{y}$.

Example 19 Rationalize the denominators of the expressions

(a) $\dfrac{3}{3 - \sqrt{5}}$, (b) $\dfrac{\sqrt{2}}{\sqrt{2} + 1}$, (c) $\dfrac{\sqrt{2} - \sqrt{3}}{\sqrt{2} + \sqrt{3}}$.

Solution In each case, rationalization is accomplished by multiplying the denominator by the conjugate of the denominator. In order to preserve the value of the fraction, the numerator is multiplied by the same quantity.

(a)
$$\frac{3}{3 - \sqrt{5}} \cdot \frac{3 + \sqrt{5}}{3 + \sqrt{5}} = \frac{3(3 + \sqrt{5})}{3^2 - (\sqrt{5})^2} = \frac{9 + 3\sqrt{5}}{9 - 5}$$
$$= \frac{9 + 3\sqrt{5}}{4}$$

(b) $$\frac{\sqrt{2}}{\sqrt{2}+1}\cdot\frac{\sqrt{2}-1}{\sqrt{2}-1}=\frac{\sqrt{2}(\sqrt{2}-1)}{(\sqrt{2})^2-1^2}=\frac{2-\sqrt{2}}{2-1}$$
$$=2-\sqrt{2}$$

(c) $$\frac{\sqrt{2}-\sqrt{3}}{\sqrt{2}+\sqrt{3}}\cdot\frac{\sqrt{2}-\sqrt{3}}{\sqrt{2}-\sqrt{3}}=\frac{(\sqrt{2}-\sqrt{3})^2}{(\sqrt{2})^2-(\sqrt{3})^2}$$
$$=\frac{2-2\sqrt{6}+3}{2-3}=-5+2\sqrt{6}$$

4. *Reduction of order.* The last standard operation discussed is called reduction of order. A cube root is a radical of third order, a square root is of second order, and so on. Often a higher-order expression can be reduced to a lower order by methods shown in Example 20.

Example 20 Reduce to simpler order expressions (a) $\sqrt[4]{25}$, (b) $\sqrt[6]{81}$, (c) $\sqrt[8]{a^4b^{12}}$.

Solution

(a) $\sqrt[4]{25}=(25)^{1/4}=(5^2)^{1/4}=5^{2/4}=5^{1/2}=\sqrt{5}$

(b) $\sqrt[6]{81}=(81)^{1/6}=(3^4)^{1/6}=(3)^{4/6}=(3)^{2/3}=(3^2)^{1/3}=\sqrt[3]{9}$

(c) $\sqrt[8]{a^4b^{12}}=b\sqrt[8]{a^4b^4}=b(ab)^{4/8}=b(ab)^{1/2}=b\sqrt{ab}$

In general, we have the problem of reducing in order the expression $\sqrt[n]{N}$. If N can be written in the form $N=(M)^a$, then

$$\sqrt[n]{N}=\sqrt[n]{M^a}=M^{a/n}.$$

Now if a/n is reducible to a simpler form, then $\sqrt[n]{N}$ can be reduced in order.

Exercises 6-5

In Exercises 1–27, simplify the given expressions.

1. $\sqrt{\frac{2}{5}}$ **2.** $\sqrt{\frac{3}{5}}$ **3.** $\sqrt{\frac{3}{8}}$

4. $\sqrt{\frac{4}{5}}$ **5.** $\sqrt{\frac{5}{8}}$ **6.** $\sqrt{\frac{3}{16}}$

7. $\sqrt{\frac{5}{12}}$ **8.** $\sqrt{\frac{5}{9}}$ **9.** $\sqrt[3]{\frac{5}{8}}$

10. $\sqrt{\frac{3}{32}}$ **11.** $\frac{1}{\sqrt{2}}$ **12.** $\frac{2}{\sqrt{3}}$

13. $\frac{2}{\sqrt{8}}$ **14.** $\frac{3}{\sqrt{5}}$ **15.** $\frac{3}{\sqrt[3]{2}}$

16. $\dfrac{6}{\sqrt[3]{4}}$ 17. $\dfrac{4}{\sqrt[4]{8}}$ 18. $\dfrac{5}{\sqrt[5]{16}}$

19. $\dfrac{1}{2+\sqrt{5}}$ 20. $\dfrac{8}{6+\sqrt{6}}$ 21. $\dfrac{4}{4-\sqrt{2}}$

22. $\dfrac{5}{5-\sqrt{5}}$ 23. $\dfrac{3}{\sqrt{3}+\sqrt{2}}$ 24. $\dfrac{1}{2\sqrt{2}-\sqrt{3}}$

25. $\dfrac{5}{5\sqrt{3}-2\sqrt{5}}$ 26. $\dfrac{\sqrt{2}-\sqrt{3}}{\sqrt{2}+\sqrt{3}}$ 27. $\dfrac{\sqrt{5}-\sqrt{3}}{\sqrt{5}+2\sqrt{3}}$

Reduce the order of the given expressions.

28. $\sqrt[4]{4}$ 29. $\sqrt[6]{27}$ 30. $\sqrt[10]{32}$

31. $\sqrt[8]{16}$ 32. $\sqrt[4]{64}$ 33. $\sqrt[6]{a^3b^6c^9}$

34. $\sqrt[6]{a^4b^{10}}$ 35. $\sqrt[4]{4a^2b^6}$ 36. $\sqrt[n]{a^{n/2}b^{3n/2}}$

37. $\sqrt[3n]{a^3b^6}$

6-6 The number j

The square root of a positive number is defined as one of the two equal factors of that number. If $N = 9$, then the square root of N is $+3$, since $(+3)(+3) = 9$. Also, $(-3)(-3) = 9$, where the two equal factors of 9 are integers. The two equal factors of 6 are $\pm\sqrt{6}$, which is irrational. The two equal factors of -6 are $\pm\sqrt{-6}$, which has no *real* meaning and is therefore called *imaginary*.

The square root of -1 is the most elementary imaginary number. It is called the *imaginary unit* and is represented by the symbol j. (Many books use the symbol i; however, in technical mathematics the symbol i is reserved for electrical current.) Now

$$j = \sqrt{-1} \tag{25}$$

and

$$j^2 = -1.$$

Operations involving the j quantity will be discussed in detail in the chapter on complex numbers. We have presented here the definition of the j quantity (25). Whenever we have the square root of a negative number, we can use the j notation, as in

$$\sqrt{-9} = \sqrt{(-1)(9)} = \sqrt{-1}\cdot\sqrt{9} = 3j,$$
$$\sqrt{-20} = \sqrt{(4)(-1)(5)} = 2\sqrt{-1}\cdot\sqrt{5} = 2j\sqrt{5}.$$

Limited use of the j quantity is made in the following chapter on quadratics.

7

Quadratic equations

In earlier material we were concerned primarily with linear equations—that is, equations of first degree. In many applications, equations of degree greater than one appear; their study is essential in a program of technical mathematics. The simplest of the equations of degree greater than one is the quadratic equation, which we define and discuss in this chapter. The discussion also includes conic sections and simultaneous quadratics.

7-1 Definitions; the incomplete quadratic

We begin the study of quadratic equations by discussing the quadratic in one unknown. A *quadratic equation* is defined here as an equation of second degree. The highest power to which any of the terms is raised is the degree of the equation. For an equation with one variable, the degree of a term is the degree of the variable in the term. Based on the preceding definitions, the general quadratic in one unknown will be given here in the form

$$ax^2 + bx + c = 0 \qquad (a \neq 0), \tag{1}$$

where a, b, and c are constants and x is the variable. The condition $a \neq 0$ in Eq. (1) is essential, since (1) must be of second degree. If $a = 0$, the term of highest degree is the bx term, causing (1) to be a first- or lower-degree equation.

Note in (1) that no nonzero condition is mentioned regarding b or c. If $b = 0$ or $c = 0$ or $b = c = 0$, then (1) is called an *incomplete quadratic*. The following three cases illustrate methods of solving incomplete quadratics.

CASE 1 Let $b = 0$; then (1) becomes the incomplete quadratic

$$ax^2 + c = 0, \tag{2}$$

from which

$$x^2 = -\frac{c}{a},$$

and we have the two solutions

$$x = \pm\sqrt{-\frac{c}{a}}. \tag{3}$$

Note in Eq. (3) that if c and a are of the same sign, then x is the square root of a negative number and we say that we have no *real* solutions for x. If a and c are of different signs, two real solutions of x exist. These solutions are equal but opposite in sign.

Example 1 Solve $3x^2 - 12 = 0$.

Solution 1 Since

$$3x^2 - 12 = 0,$$

then

$$3x^2 = 12$$

and

$$x^2 = \frac{12}{3} = 4$$

from which

$$x = \pm\sqrt{4} = \pm 2. \tag{4}$$

Solution 2 The problem can also be solved by factoring. Removing the common monomial factor, we have

$$3x^2 - 12 = 3(x^2 - 4) = 0.$$

Next, $x^2 - 4$ is the difference of two squares; we can factor it, showing the fully factored, incomplete quadratic as

$$3(x - 2)(x + 2) = 0. \tag{5}$$

From (5) we have the product of three numbers equal to zero, where the numbers are 3, $x - 2$, and $x + 2$. Reflecting that when factors multiply to equal zero, at least one of the factors must be zero, then we have either

$$x - 2 = 0. \tag{6}$$

or

$$x + 2 = 0. \tag{7}$$

(Note that factor 3 cannot be zero.) From (6) and (7), $x = 2$ or $x = -2$, agreeing with the solutions in (4).

Example 2 A 12-ft ladder is inclined against a vertical wall as shown in Fig. 7-1. The inclination is such that the distance from the base of the wall to the top of the ladder is three times the distance from the base of the wall to the base of the ladder. What is the distance from the base of the wall to the top of the ladder?

Solution Referring to Fig. 7-1, we have, by the Pythagorean Theorem, equation

$$(3x)^2 + (x)^2 = (12)^2. \tag{8}$$

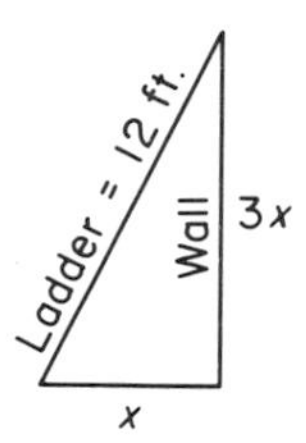

Figure 7-1

Squaring where indicated in (8) gives

$$9x^2 + x^2 = 144,$$

from which

$$10x^2 = 144$$

and

$$x = \pm\sqrt{14.4} = \pm 3.795.$$

The original problem requested $3x$; so we have the solution

$$3x = 3(3.795) = 11.385.$$

Even though $x = -3.795$ is a solution of (8), it is rejected here because physically it makes no sense.

CASE 2 Referring to (1), let $c = 0$; we have the incomplete quadratic

$$ax^2 + bx = 0. \tag{9}$$

Factoring (9) by removing the common monomial, we obtain

$$(x)(ax + b) = 0. \tag{10}$$

Then setting each factor of (10) equal to zero and solving for x, we have solutions

$$x = 0 \quad \text{and} \quad x = -\frac{b}{a}.$$

Example 3 Solve $3y^2 - 4y = 0$.

Solution Factoring, we have

$$y(3y - 4) = 0,$$

from which, on equating each factor to zero, we get

$$y = 0 \quad \text{and} \quad y = \frac{4}{3}.$$

Example 4 An object is thrown into the air with an initial velocity (upward) $v_0 = -128$ ft/sec. The distance-time equation is given as

$$s = v_0 t + \frac{1}{2} g t^2. \tag{11}$$

If $g = 32$ ft/sec/sec, at what times (t in seconds) does $s = 0$?

Solution Substituting the given data into (11), we have the incomplete quadratic

$$0 = -128t + \frac{1}{2}(32)t^2. \tag{12}$$

Clearing parentheses in (12) gives us

$$16t^2 - 128t = 0. \tag{13}$$

And factoring (13) results in

$$16t(t - 8) = 0$$

from which we have solutions $t = 0$ sec and $t = 8$ sec. The solutions here are physically explained as those times when the distance is zero or those times when the object is at the altitude of the launch point. At $t = 0$ sec the object is at the launch altitude heading upward; at $t = 8$ sec the object is at the launch altitude heading downward.

CASE 3 The third and last case of the incomplete quadratic has $b = c = 0$ in (1). This produces the trivial case

$$ax^2 = 0.$$

With a any nonzero constant, x must be zero, since the product of a and x^2 equals zero.

Later we will point out that the number of solutions of a quadratic of the form of (1) must be *two* solutions, not necessarily distinct. In Examples 1 to 4 two solutions were shown. In the generalization of Case 3 x has two values, both of which are zero.

Exercises 7-1

In Exercises 1–21, solve for the quantity indicated. In equations where only one literal quantity appears, solve for that quantity.

1. $2x^2 - 7 = 25$

2. $3(x^2 + 4) = x^2 + 16$

3. $-5x^2 + 2x = 0$

4. $36 - 4x^2 = 0$

5. $2(x^2 + 7) = 112$

6. $6(x^2 - 2) = 204$

7. $3(11 - 3x^2) - 15 = 0$

8. $4(3 - x^2) = 3x^2 - 16$

9. $2(3x^2 - 4) = 3x^2 + 19$

10. $6(x^2 - \frac{1}{2}) = 47 + 4x^2$

11. $(m + 4)^2 = 16$

12. $a^2 - y^2 = b^2$ (for y)

13. $r^2 - a^2 r = 0$ (for r)

14. $ar^2 + br^2 + a - b = 0$ (for r)

15. $(x + y)^2 - 5(x + y) = 0$ (for $x + y$)

16. $(\sin\phi)^2 - \frac{1}{4} = 0$ (for $\sin\phi$)

17. $9(\cos\theta)^2 = \cos\theta$ (for $\cos\theta$)

18. $\tan\phi\,(\tan\phi - 1) = 3\tan\phi$ (for $\tan\phi$)

19. $\dfrac{at + r}{t} = \dfrac{t}{at - r}$ (for r, t)

20. $\dfrac{3x + 4}{x} = \dfrac{x}{3x - 4}$

21. $0.0125t^2 - 2t = 0$

22. Given a square of side s, show that diagonal d of the square is $d = s\sqrt{2}$.

23. Given an equilateral triangle of side s, show that altitude h of the triangle is $h = s\sqrt{3}/2$.

24. The compound interest law is given as

$$A = P(1 + r)^n,$$

where A is the amount to which the principal P will accumulate in n years at an interest rate r. If a \$5000 investment accumulates by compound interest to \$5618 in 2 years, what is the interest rate?

25. A curve represented by the equation

$$y = 2x^3 - \tfrac{1}{2}x^2 + 4$$

has a slope y' for all x, where

$$y' = 6x^2 - x.$$

For what two values of x does the slope equal zero? For what two values of y does the slope equal zero?

26. A circle is inscribed in a square. The difference in areas of the circle and square is 10 in.^{2} What is the diameter of the circle?

27. In electricity, power p equals the product of intensity i and electromotive force e. Also, $e = ir$, where r is the resistance. Give an expression for i in terms of p and r only.

28. Given equation
$$s = v_0t + \tfrac{1}{2}gt^2,$$
show that $t = 0, -2v_0/g$ when $s = 0$.

29. A brick with six rectangular faces is such that the length exceeds the width by 2 and the height is 1 less than the width. The total surface area of the brick is 24 less than six times the width. Show that the width of the brick is 2.

7-2 Solution of the general quadratic by factoring

Many quadratics can be solved by a device called *factoring*. A trinomial equation like
$$ax^2 + bx + c = 0 \tag{14}$$
has two linear factors of form $x + r$ and $x + s$ and so (14) can be written
$$(x + r)(x + s) = 0. \tag{15}$$
From (15), since the product of two factors equals zero, either factor may be zero. Therefore from (15) we have the statements
$$x + r = 0 \quad \text{from which} \quad x = -r.$$
$$x + s = 0 \quad \text{from which} \quad x = -s.$$

Example 5 Solve for m if $m^2 - 2mn - 15n^2 = 0$.

Solution We recognize that the given expression may be factored as
$$(m - 5n)(m + 3n) = 0.$$
Setting each factor equal to zero and solving for m, we have solutions
$$m = 5n, \; -3n.$$

The factoring device is definitely a timesaver in the solution of a quadratic when the factors of the quadratic are easily recognized. However, many quadratics have irrational and even imaginary roots so that the factors may be difficult to recognize and the factoring device is of little use. In glancing over Examples 1 to 5, we note that the factoring device is useful for incomplete quadratics and complete quadratics; it is useful for quadratics with literal or numerical coefficients. It is also useful for solving equations of degree greater than two; once again, however, the obviousness of the factors of the expression dictates the use of the device.

7-3 Solution of the quadratic by completing the square

A general approach to the solution of a quadratic equation is a method called *completing the square*. This name accurately describes the procedure of the method because it involves the square of a binomial. Let us examine such a square.

Example 6 Expand quantities $(x - 5)^2$ and $(x + k)^2$.

Solution From earlier work we know that

$$(x - 5)^2 = x^2 - 10x + 25$$

and

$$(x + k)^2 = x^2 + 2kx + k^2.$$

Examination of these expansions reveals certain features. First, we might recognize one restriction: note that each is the square of a *binomial* and that each has a unity coefficient for x. In the expansions, three facts should be apparent.

1. The x^2 term has a unity coefficient.
2. The coefficient of the first-degree term is twice the value of the second term of the binomial; that is, -10 is twice -5 and $2k$ is twice k.
3. The last term of the expansion is the square of the second term of the binomial and is also the square of half the coefficient of the second term of the trinomial expansion.

We assert here that any quadratic possessing these three properties is the square of a binomial.

Example 7 Given the quadratic expression $x^2 - 6x + 9$. Examine the expression to determine whether it is the square of a binomial; if so, determine the binomial.

Solution We use two of the foregoing properties as a test for the square. First, the coefficient of x^2 is unity. Secondly, the last term, 9, is the square of one-half the coefficient of the center term, -6. With this simple test, the quadratic is shown to be the square of a binomial. The particular binomial is $x - 3$, where the -3 term is chosen by using either the second or third property above. The solution then is

$$x^2 - 6x + 9 = (x - 3)^2.$$

Any quadratic of form $x^2 + bx + c = 0$ can be solved by completing the square.

Example 8 Solve for x by completing the square if

$$x^2 - 6x - 16 = 0.$$

Solution First, rewrite the given expression as

$$x^2 - 6x = 16. \tag{16}$$

Then complete the square on the left side of (16). To obtain a square there, the constant, $+9$, must be added. The $+9$ is determined by taking one-half the coefficient of the first-degree term and squaring the result. Addition of $+9$ to the left side of the equation requires addition of the same quantity to the right side in order to retain the equality; therefore (16) becomes

$$x^2 - 6x + 9 = 16 + 9$$

or after factoring and combining terms

$$(x - 3)^2 = 25. \tag{17}$$

The addition of $+9$ to the left side of (16) was the critical step, and it completed the square of the binomial shown in (17). Taking the square root of both sides of (17), we have

$$x - 3 = \pm 5$$

or

$$x = 3 \pm 5,$$

from which we obtain the two solutions

$$x = 3 + 5 = 8$$

and

$$x = 3 - 5 = -2.$$

The reader may ask why this seemingly roundabout method of completing the square was used when the given expression in Example 8 could have been solved more readily by factoring. The reason is twofold: first, many quadratics are not easily factorable; secondly, the method of completing the square will be used in the next section to develop the quadratic formula, which is useful in solving the general quadratic.

Exercises 7-2 and 7-3

Solve Exercises 1–20 by factoring. When the factors are not readily apparent, solve by completing the square.

1. $x^2 - 4x - 21 = 0$

2. $2m^2 - 9m - 5 = 0$

3. $6t^2 - 5t - 6 = 0$

4. $12\phi^2 + \phi - 35 = 0$

5. $24k^2 - k - 44 = 0$

6. $x^2 + 4x - 20 = 0$

7. $m^2 + 6m + 6 = 0$

8. $2m^2 - 3m - 6 = 0$

9. $3b^2 + 4b + 1 = 0$

10. $4k^2 + 11k - 4 = 0$

11. $x^2 - 4x - 32 = 0$

12. $4k^2 - 4k - 3 = 0$

13. $m^2 - 9m + 14 = 0$

14. $b^2 - 8b + 12 = 0$

15. $t^2 + 2t - 3 = 0$

16. $2x^2 - 9x + 9 = 0$

17. $2k^2 - 5k - 12 = 0$

18. $b^2 - 13 = 6b - 6$

19. $t^2 + 4t + 2 = 0$

20. $t^2 - 6t + 4 = 0$

Solve Exercises 21–26 for the quantity indicated. Use factoring or completion of the square, whichever seems appropriate.

21. $a^2b^2 - 3ab - 4 = 0$ (for a)

22. $a^2 - 3ab - 4b^2 = 0$ (for a)

23. $(a + b)^2 + 5(a + b) - 6 = 0$ (for a)

24. $c^2(b - 1)^2 - c(b - 1) - 2 = 0$ (for c)

25. $6(\sin \phi)^2 - \sin \phi - 1 = 0$ (for $\sin \phi$)

26. $(a^2 - b^2)^2 + 2(a^2 - b^2) + 1 = 0$ (for b)

27. The standard form of a circle is given as

$$(x - h)^2 + (y - k)^2 = r^2.$$

By completing the square, assemble equation

$$x^2 + y^2 + 4x - 6y - 12 = 0$$

into standard form and identify h, k, and r.

28. Given equation

$$a^2 + 2ab + b^2 - 3ac - 3bc + 2c^2 = 0,$$

show that it may be factored into the form

$$(a + b - c)(a + b - 2c) = 0.$$

29. What is the equation of a quadratic whose solutions are $m = -4$ and $m = 3$?

30. Identify the equation that has solutions $x = 1, -1, 2$.

7-4 The quadratic formula

We are reminded that the complete quadratic equation in one unknown has the form

$$ax^2 + bx + c = 0. \tag{18}$$

Let us solve (18) for x by the method of completing the square as discussed in Sec. 7-3. First, the coefficient of the x^2 term must be unity; so we divide (18) by a, giving

$$x^2 + \frac{b}{a}x + \frac{c}{a} = 0. \tag{19}$$

Next, rewrite (19) as

$$x^2 + \frac{b}{a}x = -\frac{c}{a}. \tag{20}$$

Then complete the square in (20) by adding to both sides the square of one-half the coefficient of x. Here the coefficient of x is b/a; one-half the coefficient of x is $b/2a$ and the square of one-half the coefficient of x is $b^2/4a^2$. Now

$$x^2 + \frac{b}{a}x + \frac{b^2}{4a^2} = \frac{b^2}{4a^2} - \frac{c}{a} \tag{21}$$

and the square is completed.

Next, express the left side of (21) as the square of a binomial and place the right side over a least common denominator, or

$$\left(x + \frac{b}{2a}\right)^2 = \frac{b^2 - 4ac}{4a^2}. \tag{22}$$

Taking the square root of both sides of (22) gives

$$x + \frac{b}{2a} = \frac{\pm\sqrt{b^2 - 4ac}}{2a}. \tag{23}$$

Solving for x in (23), we have

$$x = \frac{-b \pm \sqrt{b^2 - 4ac}}{2a}, \tag{24}$$

where (24) is called the *quadratic formula* and is useful in solving any quadratic of form (18).

Certain features of (18) must be recognized before applying (24). First, the quantity that we call a is the coefficient of the square term. Second, the quantity that we call b is the coefficient of the first-degree term. Third, c is the constant. Fourth, all nonzero terms of (18) are on one side of the equal sign.

Example 9 Using the quadratic formula, solve for x if

$$x = 3x^2 - 2.$$

Solution First, we must rearrange the given equation into the form of (18) so that we can identify a, b, and c. Placing all the nonzero terms on one side of the equal sign, we have

$$3x^2 - x - 2 = 0. \tag{25}$$

Comparing (25) to (18), we see that $a = 3$, $b = -1$, and $c = -2$. Substituting these values into (24) gives

$$x = \frac{-(-1) \pm \sqrt{(-1)^2 - 4(3)(-2)}}{2(3)} = \frac{1 \pm \sqrt{25}}{6} = \frac{1 \pm 5}{6} = \frac{1}{6} \pm \frac{5}{6}$$

from which $$x = \frac{1}{6} + \frac{5}{6} = 1$$

or $$x = \frac{1}{6} - \frac{5}{6} = -\frac{2}{3}.$$

In Example 9 the coefficients and constants were numerical quantities; they needn't be in applying (24) because (24) is also useful when a, b, and c are literal quantities. Consider Example 10.

Example 10 Using the quadratic formula, solve for z if

$$m^2z^2 + 3mnz - 4n^2 = 0.$$

Solution Inspecting the given equation and comparing it to (18) show that $a = m^2$, $b = 3mn$, and $c = -4n^2$. Substituting into (24), we obtain

$$z = \frac{-3mn \pm \sqrt{(3mn)^2 - 4(m^2)(-4n^2)}}{2m^2}. \tag{26}$$

Simplification of (26) gives the solutions

$$z = \frac{n}{m}, -\frac{4n}{m}.$$

Further generality of the quadratic formula as it applies to (18) can be demonstrated. In Example 9 we saw that the quadratic formula was useful when the coefficients of (18) were numerical. In Example 10 it proved useful when the coefficients were literal. It is also useful when the unknown is something other than a monomial or some quantity other than a simple alphabetic letter.

Example 11 Solve for $\sin \phi$ if

$$6(\sin \phi)^2 + \sin \phi - 2 = 0.$$

Solution Here the unknown is the quantity $\sin \phi$, comparable to x in (18). Also, $a = 6$, $b = 1$, and $c = -2$. Substituting these data into (24),

$$\sin \phi = \frac{-1 \pm \sqrt{(1)^2 - 4(6)(-2)}}{2(6)} = \frac{-1 \pm \sqrt{49}}{12},$$

and $\sin \phi = 1/2, -2/3$ are the resulting solutions.

Example 12 Solve for m if

$$2m^2 + 4mn + 2n^2 - m - n - 15 = 0. \tag{27}$$

Solution Equation (27) may not only seem hopelessly complicated, it may also appear unlikely that it could be put into the form of (18). However, it can be written

$$2(m + n)^2 - (m + n) - 15 = 0. \tag{28}$$

Comparing (28) and (18), we have the variable $m + n$, with $a = 2$, $b = -1$, and $c = -15$. Applying (24), we get

$$m + n = \frac{-(-1) \pm \sqrt{(-1)^2 - 4(2)(-15)}}{2(2)} = \frac{1 \pm 11}{4} = 3, -\frac{5}{2}.$$

Now since

$$m + n = 3, \quad \text{then} \quad m = 3 - n,$$

and since

$$m + n = -\frac{5}{2}, \quad \text{then} \quad m = -\frac{5}{2} - n.$$

Examples 11 and 12 were introduced to show the broad usage of the quadratic formula.

Applications involving the quadratic equation seldom begin with a neatly organized equation of the form of (18), where simple substitution into the quadratic formula completes the solution. Instead the quadratic equation is often the byproduct of algebraic manipulation of some given data.

Example 13 A rectangular field 250 ft long and 180 ft wide has a sidewalk bordering it. If the sidewalk is of uniform width and its area is 2616 ft², what is the width of the sidewalk?

Solution Referring to Fig. 7-2, the sidewalk can be considered as composed of four rectangles. Two have the dimensions x by $250 + 2x$ and two have the dimensions x by 180, where x is the width of the sidewalk.

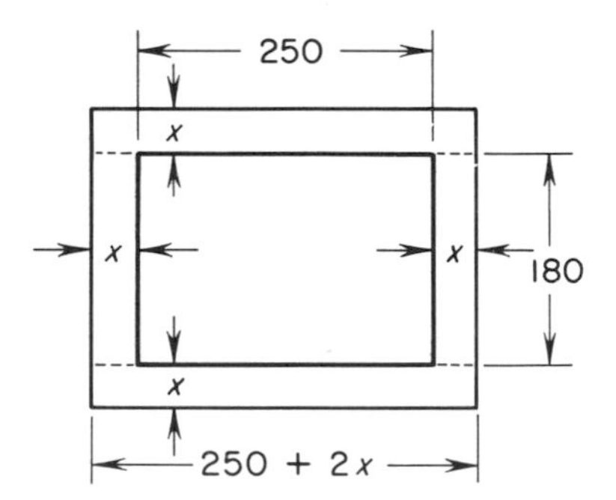

Figure 7-2

Adding the areas of the four rectangles, we obtain

$$2(x)(250 + 2x) + 2(x)(180) = 2616. \tag{29}$$

Equation (29) reduces to the quadratic

$$x^2 + 215x - 654 = 0,$$

which can be solved by formula or factored as

$$(x - 3)(x + 218) = 0$$

from which we obtain two solutions, $x = 3, -218$, where $x = 3$ is chosen as the acceptable solution.

Checking the solution, the dimensions of the four rectangles are now

$$x \text{ by } 250 + 2x \quad \text{or} \quad 3 \text{ by } 256$$

and

$$x \text{ by } 180 \quad \text{or} \quad 3 \text{ by } 180.$$

the total area is

$$2(3)(256) + 2(3)(180) = 1536 + 1080 = 2616.$$

Exercises 7-4

Solve Exercises 1–10 for the unknown indicated by using the quadratic formula except when factoring may prove an easier method.

1. $2m^2 + 5m - 25 = 0$

2. $m^2 + 5m - 15 = 0$

3. $bx^2 + 3x - 4 = 0$ (for x)

4. $x^2 + 3bx - 4 = 0$ (for x)

5. $\dfrac{1}{x+2} - \dfrac{1}{3} = \dfrac{2}{x+4}$

6. $\dfrac{9}{2y+6} + \dfrac{5}{2y-2} = 2$

7. $s = v_0 t + \frac{1}{2}gt^2$ (for t)

8. $A = 2\pi rh + 2\pi r^2$ (for r)

9. $x^2 + 2ax + a^2 = b^2$ (for x)

10. $a^2x^2 + a^2x + ab - b^2 = 0$ (for x)

11. The sum of the squares of two consecutive integers is 85. What are the integers?

12. The difference of the cubes of two consecutive integers is 61. What are the integers?

13. One leg of a right triangle exceeds the other leg by 4 in. The hypotenuse is 20 in. long. How long are the legs?

14. Given quadratic equation

$$x^2 + kx + 9 = 0,$$

what is the value of k if the solutions of the quadratic are equal?

15. Given quadratic equation

$$x^2 + 8x + k = 0,$$

for what values of k

(a) are the roots equal?
(b) is one root zero?
(c) is one root $+4$?

16. Find radius r of the circle shown in Fig. 7-3 if $\angle ABC = 90°$ and the dimensions shown are in inches.

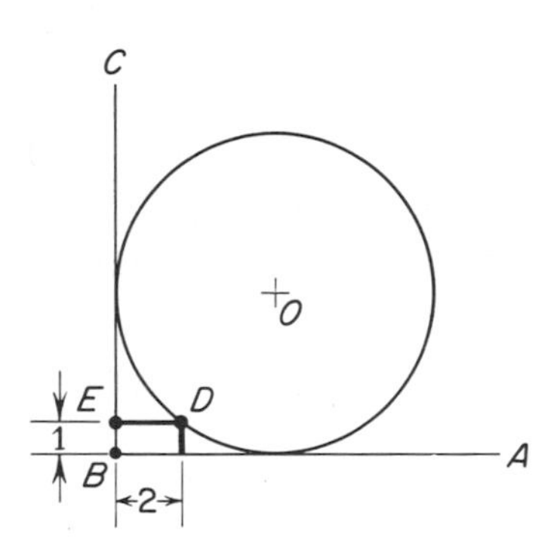

Figure 7-3

17. The quadratic equation

$$r^2 - 6r + 5 = 0$$

arises in Exercise 16; from it the solution $r = 5$ is found. The other solution of the quadratic is not a radius of the circle O shown in Fig. 7-3 but is the radius of another circle that, like O, is tangent to AB, tangent to BC, and passes through D. Describe the location of the center of the second circle and the radius of the circle.

18. The cosine law is given as

$$a^2 = b^2 + c^2 - 2bc \cos A,$$

where a, b, and c are the sides of a triangle. Find c if $a = 1.5$, $b = 1.4$, and $\cos A = 0.50$.

19. In Exercise 18, if all other conditions are the same, with $b = 1.8$, the solution of the quadratic in c results in a negative quantity under the radical sign, meaning that there is no real solution. With reference to the triangle, explain why no real solution is possible. (If $\cos A = 0.5$, then angle $A = 60°$.)

20. Two like magnetic poles of strengths m_1 and m_2 at a distance d apart are repelled from each other by a force F according to

$$F = \frac{m_1 m_2}{d^2}.$$

If one pole is of strength 40 and the other is of strength 30 and the force of attraction is 12, what is the distance between the poles?

21. The current in a circuit flows according to

$$i = 12 - 12t^2,$$

where i is the current and t is time in seconds.
(a) Find i when $t = 0$.
(b) Find t when $i = 0$.
(c) If t changes from $t = 0.1$ to $t = 0.6$, what is the change in i?

22. The maximum theoretical temperature starting from 0°C to which the products of combustion of hydrogen with oxygen are raised by the heat liberated is given approximately by

$$t = \frac{2600}{0.34 + 0.00015t}.$$

Solve for t, the maximum theoretical temperature.

23. The product of what two consecutive integers is 240?

24. A rectangle is three units longer than it is wide. If the width is doubled and the length is diminished by four units, the area is unchanged. What are the original dimensions?

25. The power output of a generator armature is given by

$$P_0 = E_g I - r_g I^2.$$

Find I, given $P_0 = 120$, $E_g = 16$, and $r_g = 0.5$.

26. Given the reactance of an electric circuit as

$$x = 2\pi f L - \frac{1}{2\pi f C},$$

solve for f.

27. Given the quadratic equation

$$y = x^2 - 2x - 3,$$

find x if $y = 0, -3, -4$.

7-5 Equations of the quadratic type

Earlier discussion of the quadratic in one unknown was limited to the form

$$ax^2 + bx + c = 0. \tag{30}$$

where the unknown x was present in degree 2 and was a monomial in most cases. We defined the quadratic as being of degree 2 and developed the method of solution accordingly.

Any equation of form

$$ax^{2n} + bx^n + c = 0 \tag{31}$$

can be solved by methods applicable to quadratic (30); for this reason, we can call (31) an *equation of quadratic type*. In (31), x is not restricted to being a first-degree monomial; it may be a polynomial, one or more trigonometric functions, a logarithmic quantity, or an exponential—to mention a few possibilities. The number n in (31) need not be a positive integer; it can be positive, negative, integral, or fractional. The condition that must be met regarding n in (31) is that the exponent of the variable in one of the terms be exactly twice that in another term.

Example 14 Solve for z, where

$$3z^{-1/2} - 2z^{-1/4} - 1 = 0. \tag{32}$$

Solution 1 Equation (32) is readily compared to (31), since $-1/2$ is twice $-1/4$, or one of the exponents of the variable is twice the other. Now (32) can be rewritten in the form of (30) as

$$3(z^{-1/4})^2 - 2(z^{-1/4}) - 1 = 0, \tag{33}$$

where $a = 3$, $b = -2$, $c = -1$, and the variable is not x but $z^{-1/4}$. If we apply the quadratic formula to (33), we solve not for x but for $z^{-1/4}$ instead. Solving, we have

$$z^{-1/4} = \frac{-(-2) \pm \sqrt{(-2)^2 - 4(3)(-1)}}{2(3)} = \frac{2 \pm 4}{6}$$

from which

$$z^{-1/4} = 1 \quad \text{or} \quad z = 1$$

and

$$z^{-1/4} = -\frac{1}{3} \quad \text{or} \quad z = 81.$$

Solution 2 We can choose to solve (32) by using a preliminary substitution that was suggested in Solution 1. Comparing (32) to (31), if we let

$$x = z^{-1/4}, \tag{34}$$

then

$$x^2 = z^{-1/2}$$

and, by substitution, (32) becomes $3x^2 - 2x - 1 = 0$, which has the solution $x = 1, -1/3$. These solutions in x have their equivalents in z through (34),

$$x = 1 = z^{-1/4} \quad \text{and} \quad z = 1$$

$$x = -\frac{1}{3} = z^{-1/4} \quad \text{and} \quad z = 81.$$

Solution 3 Equation (32) can be factored as

$$(3z^{-1/4} + 1)(z^{-1/4} - 1) = 0,$$

which, when each factor is equated to zero, has the solutions previously shown.

Example 15 Solve for $m + n$, where

$$(m + n)^4 - 13(m + n)^2 + 36 = 0. \tag{35}$$

Solution Equation (35) can be written

$$[(m + n)^2]^2 - 13[m + n]^2 + 36 = 0. \tag{36}$$

Comparing (36) to (30), we have $a = 1$, $b = -13$, $c = 36$, and the variable is the expression $(m + n)^2$. Now (36) can be solved by any of the methods shown in Example 14. We choose the substitution method shown in Solution 2. We let

$$x = (m + n)^2.$$

Then (36) becomes

$$x^2 - 13x + 36 = 0$$

which has the intermediate solutions $x = 4, 9$. Returning to $m + n$, we have

$$(m + n)^2 = 4 \quad \text{and} \quad (m + n)^2 = 9,$$

from which we obtain the desired solutions

$$m + n = \pm 2 \quad \text{and} \quad m + n = \pm 3.$$

It is worth noting that Eq. (35) in Example 15 was of fourth degree in $m + n$ and that four solutions were found for $m + n$. A theorem from the theory of equations asserts that an equation of form

$$a_0x^n + a_1x^{n-1} + a_2x^{n-2} + \cdots + a_{n-1}x + a_n = 0, \tag{37}$$

where all the a's are constants and n is a positive integer, has n solutions. Equations of form (37) that satisfy the conditions cited are called *polynomial equations*. These remarks are made in anticipation of equations with several solutions that will be found in subsequent exercises. Moreover, note that the solutions of (37) need not be real solutions; that is, some may be imaginary. Imaginary solutions of quadratic equations are discussed in Sec. 7-7.

Exercises 7-5

In the following exercises, if one literal quantity is present, solve the given equation for that quantity. If two or more literal quantities are present, solve for the quantity indicated in parentheses. The answers given show only the real solutions.

1. $x^4 - 2x^2 - 3 = 0$

2. $k^4 + 3k^2 - 4 = 0$

3. $y^4 - 5y^2 - 6 = 0$

4. $b^6 + 7b^3 - 8 = 0$

5. $t + 4t^{1/2} + 3 = 0$

6. $2b - 3b^{1/2} - 2 = 0$

7. $m^{-2/3} - m^{-1/3} - 2 = 0$

8. $p^{-2} - p^{-1} - 42 = 0$

9. $x^{-4} - 2x^{-2} - 3 = 0$

10. $2y^{-4} - 3y^{-2} - 2 = 0$

11. $\dfrac{1}{(m + \frac{1}{2})^2} + \dfrac{4}{m + \frac{1}{2}} + 4 = 0$

12. $\dfrac{1}{(a + b)^2} + \dfrac{6}{a + b} - 7 = 0$ (for $a + b$)

13. $\dfrac{2}{(x - \frac{1}{2})^2} + \dfrac{1}{(x - \frac{1}{2})} - 3 = 0$

14. $\dfrac{2}{(x + y)^2} - \dfrac{5}{(x + y)} - 3 = 0$ (for $x + y$)

15. $\frac{1}{x^2} - \frac{1}{4a^2} = 0$ (for x)

16. $x^2 - 6(x)(m - n)^{-1} + 8(m - n)^{-2} = 0$ (for $m - n$)

17. $2y^2 - 5y(a - b)^{-1} + 3(a - b)^{-2} = 0$ (for $a - b$)

18. $\frac{3}{k^2} - \frac{1}{3b^2} = 0$ (for k)

19. $(a + b)^2(x)^{-2/3} - (a + b)(y)(x)^{-1/3} = 12y^2$ (for x)

20. $(I_{max} \sin \phi)^4 - 2(I_{max} \sin \phi)^2 - 3 = 0$ (for $I_{max} \sin \phi$)

21. $(3 + \tan \phi)^6 - 9(3 + \tan \phi)^3 + 8 = 0$ (for $\tan \phi$)

7-6 Graphing the quadratic

Earlier we discussed solutions of the quadratic equation in one unknown. Here we extend our discussion of the quadratic to graphing a particular quadratic in two unknowns—namely,

$$y = ax^2 + bx + c, \tag{38}$$

where $a \neq 0$ and a, b, and c are real numbers.

One method of graphing equations of form (38) would be to assign arbitrarily selected values to x (called the *independent variable*) and to obtain the corresponding values that are forced on y (called the *dependent variable*). This procedure will produce several (x, y) pairs that can be plotted on an (x, y) plane. The method described here is somewhat uncertain in that many (x, y) pairs may be required before the shape of the curve is determined, and certain critical features of the curve may not be revealed.

We choose to graph equations of form (38) by determining four properties of the curve:

1. the roots or x intercepts (if they exist),
2. the y intercept,
3. the axis of symmetry,
4. the extreme point.

Let us discuss each briefly.

The roots or x intercepts

These are the points where the curve intercepts the x axis. Since these intercepts are on the x axis, the value of y at these points is zero. So it follows that the roots of (38) can be found by letting $y = 0$ and then solving for x.

This process involves solving the quadratic equation

$$ax^2 + bx + c = 0$$

as in previous sections.

The y intercept

This is the point where (38) crosses the y axis; here $x = 0$. So to find the y intercept of (38), let $x = 0$ and solve for y. In all cases, this solution will be $y = c$, which asserts that the y intercept of (38) is at $y = c$.

The axis of symmetry

The axis of symmetry is an "axis of reflection"; it divides the curve into two portions that are mirror images of each other. Figure 7-4 shows a curve that is symmetrical about the line $x = x_0$. Notice a particular feature of Fig. 7-4. An ordinate line $y = y_0$ intersects the curve in two points P and P', which are symmetrically located with respect to the axis of symmetry $x = x_0$.

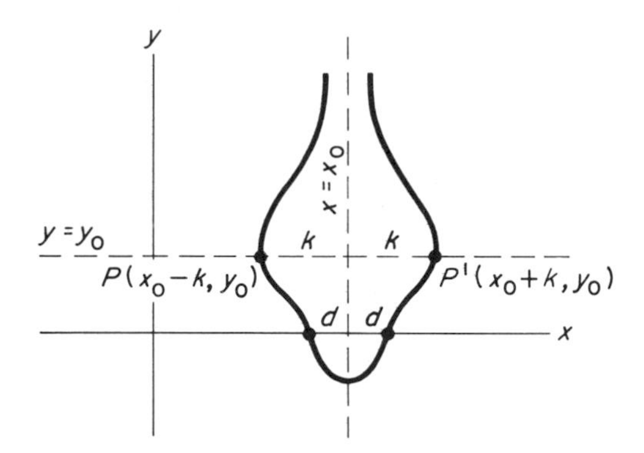

Figure 7-4

Stated another way, P is as far to the left of $x = x_0$ as P' is to the right of it; this distance is shown in Fig. 7-4 as k. Now since the abscissa of the axis of symmetry is $x = x_0$, then P has the abscissa $x_0 - k$ and P' has the abscissa $x_0 + k$. By the preceding discussion, we can see that the abscissas of the points where $y = y_0$ crosses the curve are $x = x_0 \pm k$. We also see that the curve has x intercepts (or roots) at $x = x_0 \pm d$, where the roots straddle the axis of symmetry.

Let us examine the roots of (38) next. If we let $y = 0$, we can write the roots of (38) as

$$x = -\frac{b}{2a} \pm \sqrt{\frac{b^2 - 4ac}{2a}}. \tag{39}$$

Examining (39) carefully, we see that the two roots straddle the value $x = -b/2a$ by an amount $\sqrt{b^2 - 4ac}/2a$. Should we choose to determine where the graph of (38) crosses any horizontal line $y = y_0$, we will have the equation

$$y_0 = ax^2 + bx + c$$

or

$$ax^2 + bx + c - y_0 = 0$$

which has solutions

$$x = -\frac{b}{2a} \pm \frac{\sqrt{b^2 - 4a(c - y_0)}}{2a}. \tag{40}$$

Then (40) shows that, for any preselected value y_0, the values of x straddle the value $x = -b/2a$, which shows that $x = -b/2a$ is the axis of symmetry of (38). One of the possible graphs of (38) is shown in Fig. 7-5; note the axis of symmetry and the disposition of the roots about that axis.

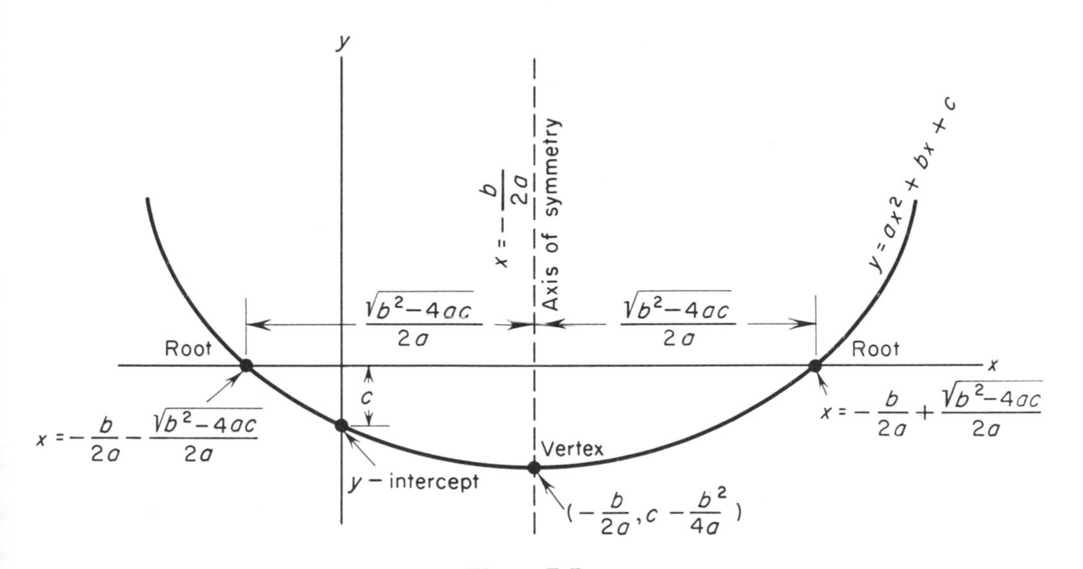

Figure 7-5

The extreme point

The extreme point on a curve is defined here as the highest or lowest point in the neighborhood. For (38), this point lies on the axis of symmetry and thus has an abscissa value $x = -b/2a$. Substituting $x = -b/2a$ into (38), we have

$$y = a\left(-\frac{b}{2a}\right)^2 + b\left(-\frac{b}{2a}\right) + c$$

from which

$$y = c - \frac{b^2}{4a},$$

and the coordinates of the extreme point of (38) are

$$x = -\frac{b}{2a} \qquad y = c - \frac{b^2}{4a}. \tag{41}$$

In many cases, the quadratic can be graphed adequately by plotting the y intercept, roots, and extreme point. It may be helpful to plot an extra point or two with the assistance of the axis of symmetry, to add detail. Consider an example.

Example 16 Plot the graph of the quadratic

$$y = x^2 - 2x - 3. \tag{42}$$

Solution Let us follow the four-step procedure described in preceding paragraphs.

Roots Let $y = 0$; then

$$x^2 - 2x - 3 = 0$$

from which

$$(x - 3)(x + 1) = 0$$

and we have the roots $x = 3, -1$. This means that the graph crosses the x axis at $x = 3$ and $x = -1$. (Refer to Fig. 7-6.)

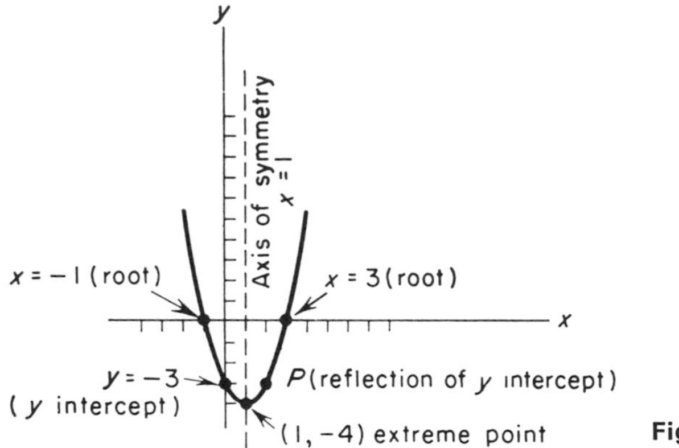

Figure 7-6

The y intercept Let $x = 0$ in (42), from which $y = -3$. This means that the graph crosses the y axis at $y = -3$.

Axis of symmetry Comparing (42) to (38), we have constants $a = 1$ and $b = -2$. Now the axis of symmetry is at

$$x = -\frac{b}{2a} = \frac{-(-2)}{2(1)} = +1.$$

We can find the axis of symmetry another way; it lies midway between the roots or midway between $x = -1$ and $x = +3$. This midvalue is found by adding the roots together and taking one-half the sum, or $(-1 + 3) \div 2 = +1$.

Extreme point The extreme point is on the axis of symmetry where $x = 1$. Substituting $x = 1$ into (42), we find that $y = -4$. This extreme point with coordinates $(1, -4)$ can also be located by using (41), where $a = 1$, $b = -2$, and $c = -3$.

We can easily add another point to the graph; this point is designated as P in Fig. 7-6 and is the reflection, across the axis of symmetry, of the y intercept. This applies to any other point that is plotted; it will have a companion across the axis of symmetry. In other words, only one-half the curve need be plotted, since the other half is a mirror copy.

Note that the quadratic graphed in Fig. 7-6 is shaped like the cross section of the reflector of a flashlight or an automobile headlight. This general shape is characteristic of all quadratics of the form of (38) and is called a *parabola*.

The following examples show that by modifying certain parts of (38), the parabola may be made to open downward, upward, left, or right.

Example 17 Plot the graph of the quadratic

$$y = -x^2 - 2x + 8. \tag{43}$$

Solution Let us look first for the roots and the axis of symmetry. Letting $y = 0$ and solving the remaining quadratic for x, we have

$$x = -\frac{2}{2} \pm \frac{\sqrt{(-2)^2 - 4(-1)(8)}}{2} = -1 \pm 3. \tag{44}$$

Inspecting (44) closely, we see that the axis of symmetry is at $x = -1$ and that the roots are three units to the right and left of the axis of symmetry or at $x = 2, -4$. See Fig. 7-7 for a plot of the roots. To find the extreme point, let $x = -1$ in (43), from which $y = 9$. All the points previously determined are shown in Fig. 7-7, along with the companion or reflection point of the y intercept. Note that the parabolas in Figs. 7-6 and 7-7 have the same basic shapes, but they open in different direc-

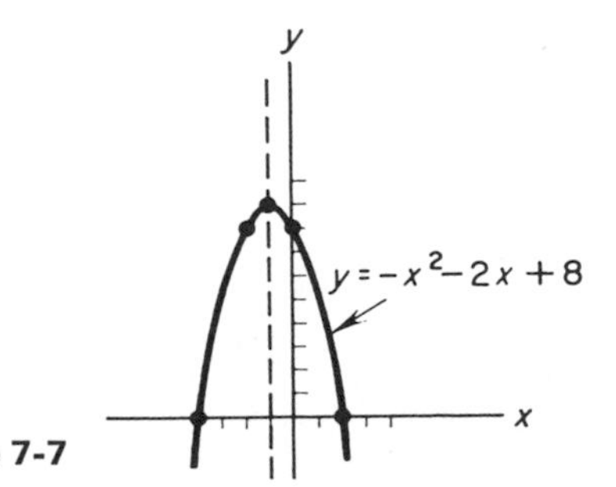

Figure 7-7

tions. This reversal of the direction of opening can be caused by changing the sign of the coefficient of the second-degree term.

Example 18 Plot the graph of equation

$$x = y^2 + y - 6. \tag{45}$$

Solution Comparing (45) with (38), we see that variables x and y are interchanged. This suggests that the properties that hold for x in (38) now hold for y in (45) and vice versa.

The roots are on the y axis for (45) and are found by letting $x = 0$; they are $y = -3, +2$ (see Fig. 7-8). The axis of symmetry is the line $y = -1/2$, which is midway between the two roots. The x intercept is found by letting $y = 0$, from which we obtain $x = -6$ as the horizontal intercept. The extreme point is now the leftmost point and is at $x = -6\frac{1}{4}$, which is found by letting $y = -1/2$ in (45).

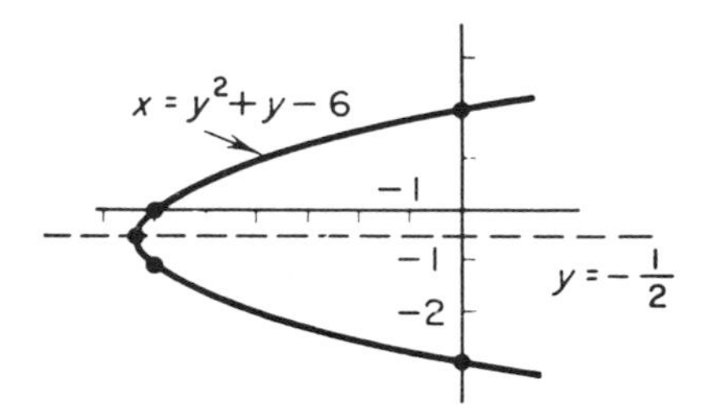

Figure 7-8

The graph of (45), as shown in Fig. 7-8, is shaped like a parabola but has a horizontal axis of symmetry. The horizontal position of the parabola is the result of interchanging the x and y variables in (38).

Exercises 7-6

Graph the quadratic given in Exercises 1–10. Plot the axis of symmetry, intercepts of both coordinate axes, and extreme point in all cases.

1. $y = x^2 - 4$
2. $y = x^2 - 2x$
3. $y = x^2 - 3x - 10$
4. $y = -x^2 + 4x$
5. $y = -x^2 + 3x + 4$
6. $y = 2x^2 - 3x - 2$
7. $x = y^2 + 3y$
8. $x = -y^2 + 8$
9. $x = -2y^2 - y + 3$
10. $x = y^2 - 3y - 4$
11. A free-falling body thrown vertically travels according to equation

$$s = \tfrac{1}{2}gt^2 + v_0t + s_0$$

where s is distance, t is time, and g, v_0, s_0 are constants. What are the values of s and t when s is an extreme value?

12. In Exercise 11, if t is the abscissa and s the ordinate, the (s, t) curve plots as a parabola opening upward or downward, depending on the sign of which constant? Changing which constant modifies the s intercept?

13. A ball thrown vertically from the ground with an initial velocity of 144 ft/sec has its distance from the ground measured according to equation

$$s = -16t^2 + 144t$$

where s is distance in feet and t is elapsed time in seconds. At what two times does $s = 0$? At what time does the ball reach maximum height? What is the maximum height?

14. A 40-volt generator with an internal resistance of 4 Ω delivers power to an external circuit amounting to $40i - 4i^2$ watts, where i is the current in amperes. For what current will the generator deliver maximum power?

15. A company successfully sells 500 units daily of a certain item at $1.20 per item. In order to increase sales, it is found that 50 additional items per day can be sold for each 10-cent decrease in the sale price of the item. What number of items at what unit price will provide maximum daily receipts?

7-7 More on graphs of quadratics; the discriminant

In the preceding section all the quadratics considered in examples and exercises had two real, distinct roots; that is, an equation of form

$$y = ax^2 + bx + c \tag{46}$$

was such that its graph crossed the x axis twice. However, this circumstance need not hold. Referring to Fig. 7-9, we see the graphs of

$$y = x^2 - 2x + c,$$

where $c = -3, +1$, and $+3$. When $c = -3$ as in Fig. 7-9(a), the curve crosses the x axis in two distinct points. In Fig. 7-9(b), when $c = +1$, the

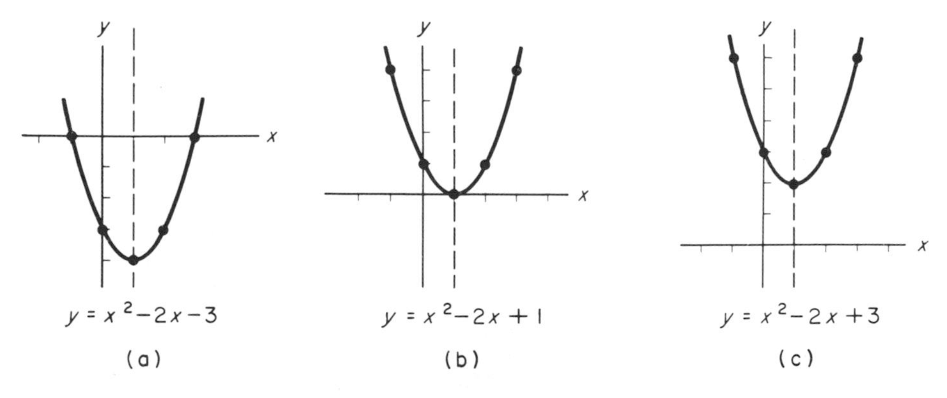

Figure 7-9

curve is tangent to the x axis. In Fig. 7-9(c), when $c = +3$, the curve fails to intercept the x axis.

If we accept an argument that an equation of form (46) has two algebraic solutions, then the solutions can be (from Fig. 7-9)

1. real, rational, and unequal, as in (a),
2. real, rational, and equal, as in (b),
3. imaginary, as in (c),
4. real, irrational, and unequal, not pictured in Fig. 7-9.

The preceding situations regarding the reality, equality, and rationality of the roots of (46) can be observed from the quadratic formula. Let $y = 0$ in (46), whence

$$x = -\frac{b}{2a} \pm \frac{\sqrt{b^2 - 4ac}}{2a}, \tag{47}$$

where we will call the portion $b^2 - 4ac$ the *discriminant*. Now inspect (47) carefully for the nature of the roots.

1. If the discriminant is positive ($b^2 - 4ac > 0$) and a perfect square, then $\sqrt{b^2 - 4ac}/2a$ is a rational number. Added to or subtracted from $-b/2a$ (which is also rational), we see that (47) has two roots, distinct (different) from each other and that both are real (as opposed to imaginary) and rational (as opposed to irrational). This case is pictured in Fig. 7-9(a), where
$$b^2 - 4ac = 16$$
and the roots are
$$x = -\frac{-2}{2} \pm \frac{\sqrt{16}}{2} = 1 \pm 2.$$
2. If the discriminant is positive ($b^2 - 4ac > 0$) but not a perfect square, then roots of (46) still exist (that is, they are real) and are distinct but irrational. Consider equation
$$y = x^2 - 2x - 2$$
from which
$$\sqrt{b^2 - 4ac} = \sqrt{12} = 2\sqrt{3}$$
and the solutions are
$$x = 1 \pm \sqrt{3},$$
both of which are real but irrational.
3. If the discriminant equals zero, then the roots of (46) are
$$x = -\frac{b}{2a} + 0 \quad \text{and} \quad x = -\frac{b}{2a} - 0,$$

which means that the roots are equal, rational, and real. This case is shown in Fig. 7-9(b), where equal roots indicate tangency.

4. In Fig. 7-9(c) it is clear that y cannot be zero, since the curve never crosses or is tangent to the x axis. If we try to solve (46) by the quadratic formula, we have automatically introduced the condition $y = 0$. In effect, we have imposed a condition that the equation cannot accept; therefore we cannot expect a real result. If

$$y = x^2 - 2x + 3$$

and we let $y = 0$, the roots are

$$x = -\frac{-2}{2} \pm \frac{\sqrt{-8}}{2} = 1 \pm j\sqrt{2},$$

where the roots are imaginary. Since the roots are imaginary, we cannot expect to plot them on the real (x, y) plane. When $b^2 - 4ac < 0$, the roots are always imaginary.

Table 7-1 summarizes the preceding information.

Table 7-1 Nature of Roots of a Quadratic for $y = ax^2 + bx + c$

Discriminant	*Nature of roots*	*Graph*
$b^2 - 4ac > 0$ and a perfect square	Real, distinct, rational	Crosses the x axis twice at rational values of x.
$b^2 - 4ac > 0$ and not a perfect square	Real, distinct, irrational	Crosses x axis twice at irrational values of x.
$b^2 - 4ac = 0$	Real, equal, rational	Is tangent to x axis.
$b^2 - 4ac < 0$	Imaginary	Does not cross x axis.

Example 19 Give the nature of the roots of

$$y = 2x^2 - 3x - 6.$$

Solution First, evaluate $b^2 - 4ac$. Referring to the given equation and comparing it with (46), we have $a = 2$, $b = -3$, and $c = -6$, from which

$$b^2 - 4ac = 9 + 48 = 57.$$

Now 57 is positive and so the roots are real; it is not a perfect square and so the roots are irrational; it is not zero and so the roots are unequal.

Example 20 Given equation

$$y = 2x^2 - kx + 8, \tag{48}$$

what condition regarding k is necessary if the roots of (48) are equal?

Solution If the roots of (48) are to be equal, the discriminant must be zero. Since $a = 2$, $b = k$, and $c = 8$, then

$$b^2 - 4ac = k^2 - 64$$

from which

$$k^2 - 64 = 0$$

and $k = \pm 8$, which are the requested solutions.

Exercises 7-7

In Exercises 1–10, describe the nature of the roots of the given equation. Also, plot the graph of the equation.

1. $y = x^2 - 5x$

2. $y = 3x^2 + 4x$

3. $y = 2x^2 - 6$

4. $y = x^2 - 8$

5. $y = x^2 - 6x + 9$

6. $y = x^2 + 4x + 4$

7. $y = 2x^2 + 4x + 6$

8. $y = 3x^2 + x + 1$

9. $y = 4x^2 - 5x - 9$

10. $y = 3x^2 + x - 1$

11. From equation $y = x^2 - 6x + c$ determine the value of c for which the roots of the equation are equal.

12. From equation $y = ax^2 + 3x + 2$, determine the value of a for which the roots of the equation are equal.

13. Quadratic $y = x^2 - 5x + c$ has roots such that one root is twice the other. Determine c.

14. For the quadratic equation $y = 3x^2 + bx + 3$, what range of values of b will permit the roots of the quadratic to be real?

Determine the nature of the roots in Exercises 15–18.

15. $y = x^4 - 13x^2 + 36$

16. $y = x^4 - 5x^2 - 36$

17. $y = x^4 + 13x^2 + 36$

18. $y = x^4$

19. Given equation

$$y = x^2 - 3x - 4,$$

substitute $x = w - 2$ and obtain an equation in y and w. Show that the roots of the (y, x) equation are each 2 less than the roots of the (y, w) equation. Note that the substitution merely shifts the location of the graph horizontally by 2 units.

20. For the quadratic equation $y = 2x^2 + bx + 8$, what range of values of b will permit the roots of the quadratic to be real?

7-8 Radical equations; extraneous roots

Applied problems in mathematics frequently involve equations in which one or more of the terms are a radical quantity; the most common radical involves a square root. Such equations are called *radical equations*; their solutions can involve quadratics in that a squaring operation intended to eliminate a radical may give rise to a second-degree expression. Then, too, it often happens that manipulation of a radical equation may give rise to solutions that are legitimate for some intermediate expression but not appropriate for the original expression; these solutions are called *extraneous*.

The process of solving radical equations requires arranging the radical expressions in such a way that successive squaring of the sides of the given equality will eliminate the radicals. This successive squaring assumes that the radicals present are square roots.

Example 21 Solve for x if

$$\sqrt{2x + 5} - x + 5 = 0. \tag{49}$$

Solution Squaring (49) in its given form will not eliminate the radical; however, if the radical quantity is isolated on one side of the equal sign and all other terms are transposed to the other side of the equal sign, then a single squaring operation will eliminate the radical. Thus

$$\sqrt{2x + 5} = x - 5.$$

If we square both sides, then

$$2x + 5 = x^2 - 10x + 25.$$

Collecting like terms, we have

$$x^2 - 12x + 20 = 0, \tag{50}$$

which has solutions $x = 10, 2$.

If we accept only the *positive* value of $\sqrt{2x + 5}$, then the only acceptable solution of (49) will be $x = 10$. The quantity $x = 2$, although an acceptable solution of intermediate equation (50), is *not* a solution of (49) and will therefore be regarded as extraneous.

Example 22 Solve for y if

$$\sqrt{6y + 7} = 6 + \sqrt{2y - 13}. \tag{51}$$

Solution No single squaring operation will eliminate both radicals in (51). Squaring both sides gives

$$6y + 7 = 36 + 12\sqrt{2y - 13} + 2y - 13. \tag{52}$$

Collecting like terms in (52) and dividing by 4, we have

$$y - 4 = 3\sqrt{2y - 13}.$$

If we square again, then

$$y^2 - 8y + 16 = 18y - 117,$$

which has solutions $y = 19, 7$, where, accepting again only the positive values of the square root, neither root is extraneous.

Example 23 Three circles are blended according to the dimensions shown in Fig. 7-10. Solve for the radius r of the center circle.

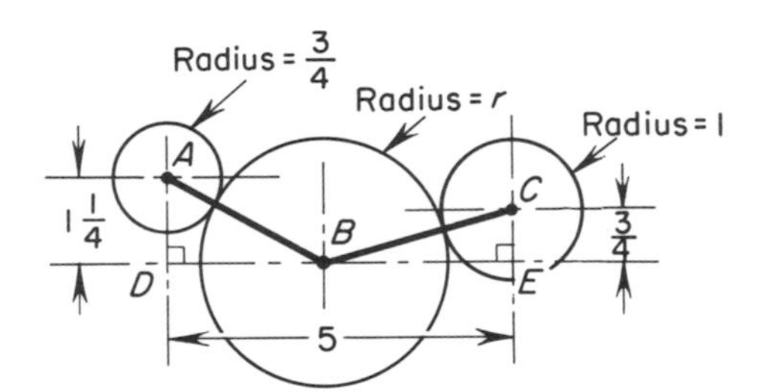

Figure 7-10

Solution Since radius r is the unknown quantity for which we are to solve, an equation relating r and the known data is required. The right triangles shown suggest a Pythagorean relationship. Now $AB = r + 3/4$ and $BC = r + 1$. From $\triangle ABD$

$$DB = \sqrt{(r + \tfrac{3}{4})^2 - (\tfrac{5}{4})^2},$$

and for $\triangle CBE$

$$BE = \sqrt{(r + 1)^2 - (\tfrac{3}{4})^2}.$$

Now $DB + BE = 5$ or

$$\sqrt{(r + \tfrac{3}{4})^2 - (\tfrac{5}{4})^2} + \sqrt{(r + 1)^2 - (\tfrac{3}{4})^2} = 5. \tag{53}$$

Solution of Eq. (53) for r now involves the solution of a radical equation. As in previous examples, radicals are eliminated by successive squaring operations. Transposing the second term of (53) to the right member, we have

$$\sqrt{(r + \tfrac{3}{4})^2 - (\tfrac{5}{4})^2} = 5 - \sqrt{(r + 1)^2 - (\tfrac{3}{4})^2}.$$

Removing parentheses, we obtain

$$\sqrt{r^2 + \tfrac{3}{2}r - 1} = 5 - \sqrt{r^2 + 2r + \tfrac{7}{16}}.$$

Then squaring both sides gives

$$r^2 + \tfrac{3}{2}r - 1 = 25 - 10\sqrt{r^2 + 2r + \tfrac{7}{16}} + r^2 + 2r + \tfrac{7}{16}.$$

Collecting like terms and clearing fractions result in

$$8r + 423 = 160\sqrt{r^2 + 2r + \tfrac{7}{16}}.$$

Finally, squaring both sides and collecting like terms, we have the quadratic

$$25{,}536r^2 + 44{,}432r - 167{,}729 = 0,$$

which has the positive solution $r = 1.84$ in.

Exercises 7-8

In Exercises 1–10, solve for the literal quantity. Disregard extraneous roots.

1. $\sqrt{m-3} = 6$
2. $12 = \sqrt{m+100}$
3. $\sqrt{3z+4} = z$
4. $2z - 1 = \sqrt{z^2 + 8z + 16}$
5. $\sqrt{y+13} - 3 = \sqrt{2y-5}$
6. $\sqrt{3m+28} = \sqrt{3m-5} + 3$
7. $-\sqrt{8k} + 15 - \sqrt{2k} = \dfrac{12}{\sqrt{2k}}$
8. $\sqrt{m+95} - 5 = \sqrt{m}$
9. $\sqrt{k^2+15} - 4 = \sqrt{k^2 - 4k - 5}$
10. $\sqrt{2r+5} = \sqrt{r} + \sqrt{5}$
11. Two numbers differ by 24. The square root of the larger number differs from the square root of the smaller number by 2. What are the numbers?
12. Determine the value of AB in Fig. 7-11 if $DC = 8$, $AC = 3$, and DB exceeds AB by 2. (Figure not to scale.)

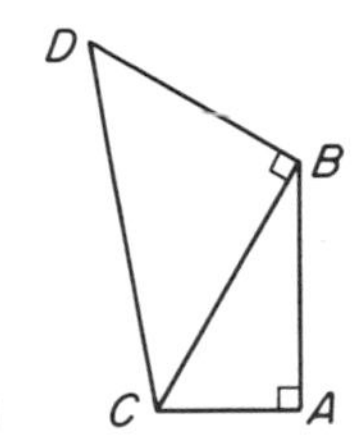

Figure 7-11

13. Using Fig. 7-12, find BD if given $AD = 12$, $AB = 8$, and $DC = 3$.

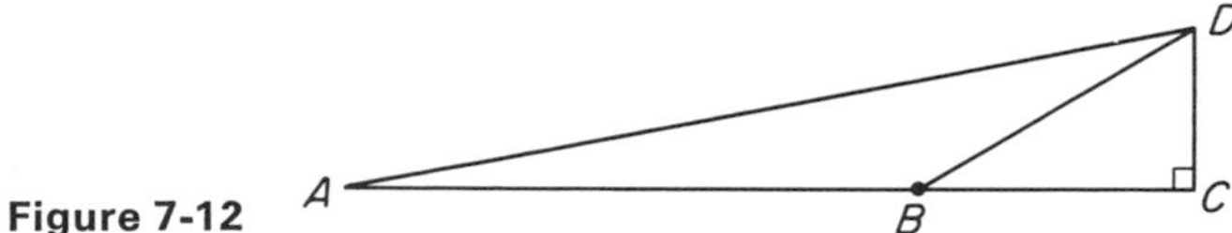

Figure 7-12

14. Using Fig. 7-13, find BE if given $AB = 5$, $BC = 9$, $ED = 3$, and $AE + DC = 12$.

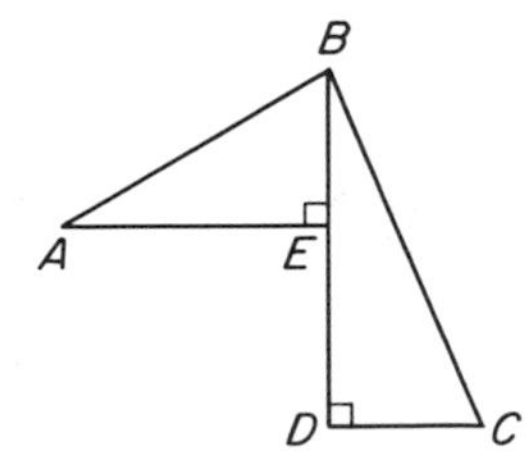

Figure 7-13

In Exercises 15–20, solve for the quantity indicated.

15. $v = \sqrt{\dfrac{g\alpha}{2\pi}}$ (for α)

16. $z = \sqrt{R^2 + (X_L - X_C)^2}$ (for R)

17. $m = \dfrac{m_0}{\sqrt{1 - (v/c)^2}}$ (for v)

18. $t = \dfrac{-v_0 \pm \sqrt{v_0^2 - 2sg}}{g}$ (for s)

19. $A = \pi(R^2 - r^2)$ (for R)

20. $r = \dfrac{-\pi h \pm \sqrt{\pi^2 h^2 + A\pi}}{\pi}$ (for A)

21. Show that expression $x + a = \sqrt{y^2 + (x - a)^2}$ can be reduced to expression $y^2 = 4ax$.

22. Show that expression

$$\sqrt{(x + f)^2 + y^2} + \sqrt{(f - x)^2 + y^2} = 2a$$

can be reduced to expression

$$\frac{x^2}{a^2} + \frac{y^2}{b^2} = 1,$$

where $b^2 = a^2 - f^2$.

Additional exercises 7-8

1. Find the values of a, b, and c for which the roots of the quadratic are equal.
(a) $5x^2 + 4x + c = 0$
(b) $3x^2 + bx + 8 = 0$
(c) $ax^2 + x - 6 = 0$
(d) $ax^2 - 4x + a = 0$

2. Given quadratic equation

$$2x^2 - 3x + k = 0.$$

(a) For what value of k are the roots such that one root exceeds the other by unity?
(b) For what value of k are the roots such that one root is twice the other?

3. A runner of width 3 ft is placed over a 9 by 12-ft rectangular rug as shown in Fig. 7-14(a). Find the area of the runner.

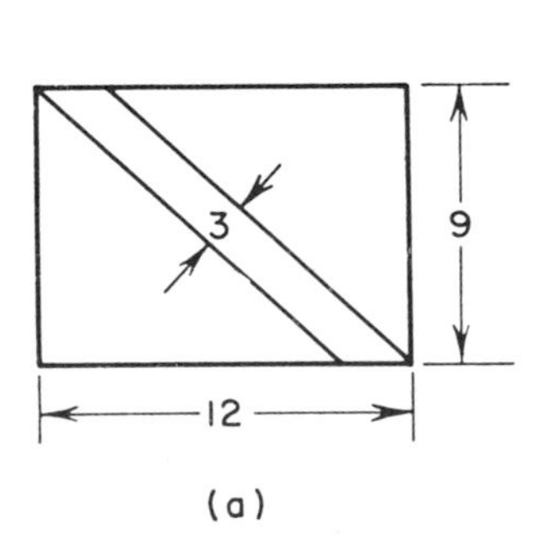

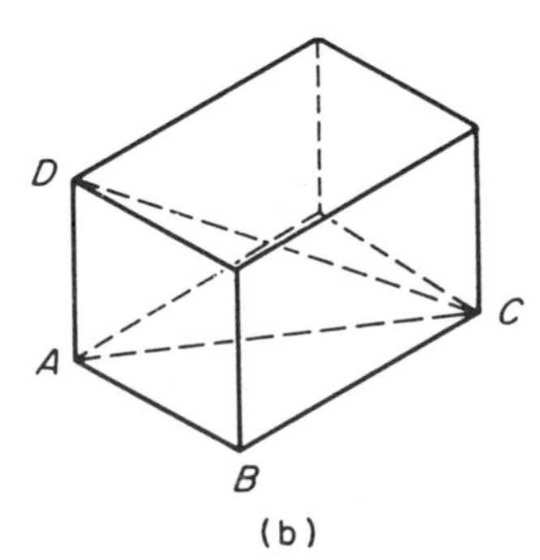

Figure 7-14

4. Two circles are such that the sum of their circumferences is 24π ft and the sum of their areas is 84π ft^2. What are the radii of the circles?

5. Solve for L if

$$n = \sqrt{\frac{R^2}{4L^2} - \frac{1}{LC}}.$$

6. Referring to Fig. 7-15,
   (a) Find the lengths AC, AE, BD, DC, and DE.
   (b) There is a point F on AD such that $FB = 1.2FE$. Show that F is either 2.61 or 21.0 in. from A.

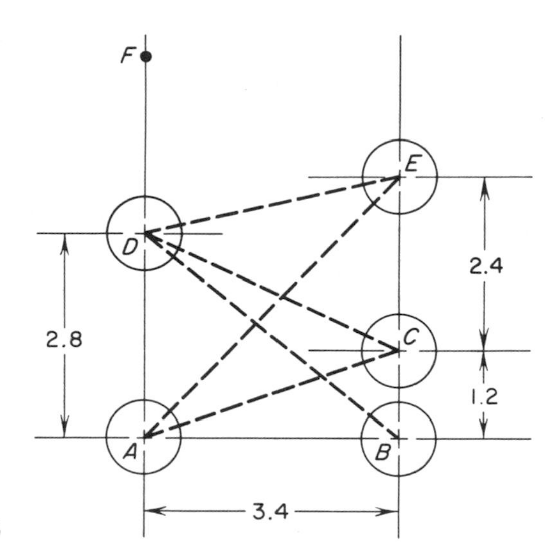

Figure 7-15

7. Figure 7-14(b) shows a prism with six rectangular faces. Note by the Pythagorean Theorem that

$$(AB)^2 + (BC)^2 = (AC)^2$$

and that

$$(AC)^2 + (AD)^2 = (DC)^2$$

from which

$$(DC)^2 = (AB)^2 + (BC)^2 + (AD)^2.$$

   (a) Find DC if $AD = AB = 5.2$ and $BC = 2AB$.
   (b) Find AB if $BC = AD$, DC exceeds BC by 2, and BC exceeds AB by 1.
   (c) If the prism is a perfect cube, show that $DC = AB\sqrt{3}$.

8. It can be shown from Graham's law of gaseous diffusion that, under similar conditions of temperature and pressure, the ratio of the rates of diffusion of two gases is the reciprocal of the ratio of the square roots of the molecular weights. That is,

$$\frac{D_a}{D_b} = \frac{\sqrt{M_b}}{\sqrt{M_a}}$$

where D is the rate of diffusion and M is molecular weight.

(a) If $D_a = 1.12D_b$, show that $M_a = 0.796M_b$.
(b) Find D_a/D_b, where gas a is oxygen (molecular weight 32) and gas b is nitrogen (molecular weight 28).

9. A best estimate of standard error of the difference between the means of two random samples is given as

$$\sigma = \sqrt{\frac{S_1^2}{N_1 - 1} + \frac{S_2^2}{N_2 - 1}}.$$

Show that

$$S_1 = \sqrt{\frac{N_1 - 1}{N_2 - 1}(\sigma^2 N_2 - \sigma^2 - S_2^2)}.$$

10. The equivalent resistance of two resistors in parallel is

$$R = \frac{R_1 R_2}{R_1 + R_2}.$$

A constant resistance R_1 is placed in parallel with a thermistor R_2 that has a resistance of 27 ohms (Ω) at 20°C and 4 Ω at 80°C. Find the value of R_1 that will cause the equivalent resistance of the cold circuit to be 0.8 Ω greater than that of the hot circuit.

11. An object with initial velocity v undergoes an acceleration a for time t. Displacement s of the object for this time is given by equation

$$s = vt + \tfrac{1}{2}at^2.$$

(a) Solve for t.
(b) Find t if $s = 100$ ft, $v = 20$ ft/sec, and $a = 8$ ft/sec^2.

12. A force F in Fig. 7-16 is attempting to shear a rivet across its circular cross-sectional area $\pi D^2/4$, where diameter D is unknown. The maximum force that the rivet can endure safely is 15,000 times the rivet's cross-sectional area, where

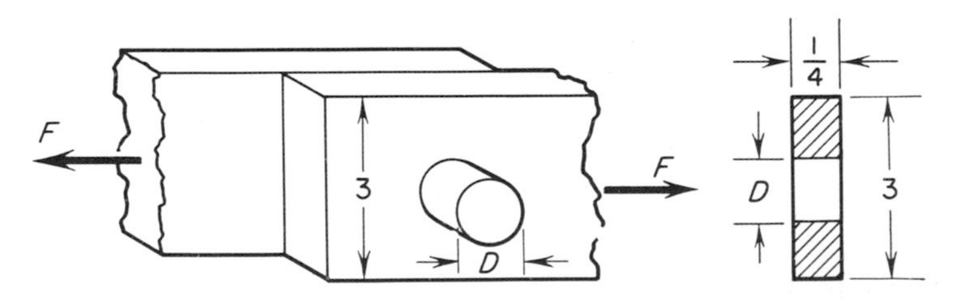

Figure 7-16

the area is expressed in square inches. The force F is also attempting to pull the bar apart (tension failure) at its weakest cross-sectional area, located at the rivet hole. The maximum force that the bar can endure safely is 20,000 times the bar's cross-sectional area, taken at the weakest section. Find the rivet diameter required to make the load-carrying ability of the rivet equal to that of the bar.

13. In Eqs. (54a) and (54b), S_x and S_y represent, respectively, the horizontal and vertical components of displacement of a projectile.

$$S_x = (V \cos \theta)(t) \tag{54a}$$

$$S_y = (V \sin \theta)(t) - \tfrac{1}{2}gt^2 \tag{54b}$$

(a) Eliminate t (time in flight) from the equation for S_y.

(b) From the answer in part (a), let $\theta = 30°$, $\sin \theta = 1/2$, $\cos \theta = \sqrt{3}/2$, $g = 32$ ft/sec², and $V = 400$ ft/sec. Solve for S_x when $S_y = 300$ ft.

(c) When the assumed values for part (b) are substituted into the answer for part (a), equation

$$S_y = (0.577)S_x - (0.000133)S_x^2$$

is obtained. Solve for S_x when $S_y = 0$. Graph the equation with S_x on the horizontal axis and S_y on the vertical axis.

(d) Calculate the maximum value of S_y in part (c) and compare the result with that shown on the graph.

14. A circuit is designed with a switching device that places two resistors either in parallel or in series. The parallel resistance $R_p = R_1R_2/(R_1 + R_2)$ is 10 Ω and the series resistance $R_s = R_1 + R_2$ is 48 Ω. Find R_1 and R_2.

15. A company has N machines of equal capacity that produce a total of 180 pieces each workday. If two machines break down, the work load of the remaining machines is increased by three pieces each per day to maintain production. Find N.

16. The total power developed in a 100-volt electric generator is the product of the generated voltage and current, or $100I$, where I is in amperes. Part of the total power and voltage is lost in the generator, due to the current flow through the internal resistance R_i of 5 Ω. The internal power loss is I^2R_i, or $5I^2$ watts. Consequently, the output power $P = 100I - 5I^2$. Find the current that gives the maximum output power. Find the maximum power.

17. In Exercise 16, show that the output power is maximum when $I =$ (generated voltage)/$2R_i$ and that this power equals the quotient of the generated voltage squared and $4R_i$. Show that the output power is maximum when it equals one-half the total power generated.

18. A nozzle directs a jet of water tangent to a hydraulic turbine wheel to hit turbine buckets and drive a generator. The jet speed is J ft/sec and the peripheral speed of the buckets is B ft/sec. The turbine power P is proportional to the product of the bucket speed and the difference between the jet and bucket speeds, or $P = KB(J - B)$. When $J = 100$ and $B = 70$, the power is 210 kW. If the jet speed is constant, what bucket speed gives maximum power and what is the maximum power?

19. A beam of length L shown in Fig. 7-17 is supported by reaction forces R_1 and R_2 that vary while two wheels of a traveling crane roll across the beam from left to right. Forces P_1 and P_2 are the constant wheel loads, and x is the varying distance from the left end of the beam to the load P_1. As the wheels move

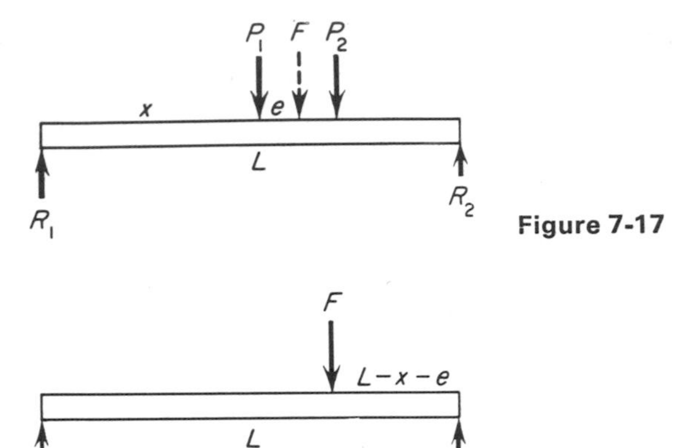

Figure 7-17

Figure 7-18

across the beam, there is for each value of x a value of the bending effect (bending moment) inside the beam directly under P_1. The general expression for this bending moment is $M = R_1x$. Since R_1 changes with x, a function of x may be substituted for R_1. Referring to Fig. 7-18, it can be shown from mechanics that

$$R_1 = \frac{F(L - x - e)}{L}$$

where F is the resultant load effect of P_1 and P_2 and is located some constant distance e from P_1. F, L, and e remain constant. Show that the maximum value of M, under load P_1, occurs when $x = (L - e)/2$.

7-9 The general quadratic; conics

The general quadratic is represented by the second-degree equation

$$Ax^2 + Bxy + Cy^2 + Dx + Ey + F = 0, \tag{55}$$

where, in order to remain second degree, A, B, and C cannot all be zero. The quadratic familiar to students of more elementary mathematics was second degree in one variable only and was generally of the form $Ax^2 + Dx + F = 0$, where $B = C = E = 0$, or of the form $Cy^2 + Ey + F = 0$, where $A = B = D = 0$. Modification of constants A to F in (55) results in equations that, when graphed, describe a group of curves called the *conic sections*. Several of these modifications are discussed in subsequent sections from the point of view of inspecting for mathematical types and graphing the curves.

Circle

A *circle* is defined as the locus of points equidistant from a given point. If a circle of radius r has its center on the origin [see Fig. 7-19(a)], then

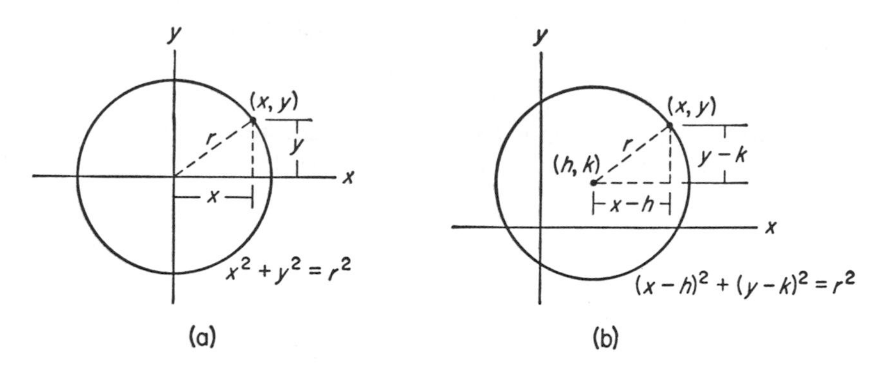

Figure 7-19

$x^2 + y^2 = r^2$ is the equation of the circle by the Pythagorean Theorem. This is a modification of Eq. (55) such that $Ax^2 + Cy^2 + F = 0$, where $A = C$ and $B = D = E = 0$. The equation of a circle with center on the origin can, therefore, be readily identified by the presence of the sum of two squares with equal coefficients. Thus

$$x^2 + y^2 = 49$$

is an origin-centered circle of radius 7 and

$$2x^2 + 2y^2 = 72$$

is an origin-centered circle of radius 6.

If the center of the circle is at a point other than the origin, say (h, k), as in Fig. 7-19(b), the equation of the circle, again by the Pythagorean Theorem, is

$$(x - h)^2 + (y - k)^2 = r^2. \tag{56}$$

This is the *standard form* of the circle. If Eq. (56) is expanded, we have

$$x^2 + y^2 - 2hx - 2ky + h^2 + k^2 - r^2 = 0, \tag{57}$$

where, by comparison with Eq. (55), we have

$$B = 0, \quad A = C, \quad D = -2h, \quad E = -2k, \quad F = h^2 + k^2 - r^2;$$

therefore

$$Ax^2 + Ay^2 + Dx + Ey + F = 0 \tag{58}$$

is an expression equivalent to (56) and is called the *general form* of the circle.

Example 24 Give the equation of the circle with radius 3 and center at $(4, -2)$.

Solution Substituting into Eq. (56) the values $r = 3$, $h = 4$, and $k = -2$, we have

$$(x - 4)^2 + (y + 2)^2 = 3^2.$$

Example 25 Describe the circle whose equation is

$$x^2 + y^2 - 10x + 2y = -10.$$

Solution This equation is given in the general form and can be made more meaningful by reverting to the standard form. Completing the square, we have

$$(x - 5)^2 + (y + 1)^2 = 4^2,$$

which is a circle of radius 4 and center at $(5, -1)$.

Exercises 7-9A

Assemble Exercises 1–8 into the standard form. Identify the center and radius and sketch the graph.

1. $2x^2 + 2y^2 = 162$

2. $3x^2 + 3y^2 = 27$

3. $x^2 + y^2 + 8x = 0$

4. $x^2 + y^2 - 8y = 0$

5. $x^2 + y^2 + 6x + 4y = 0$

6. $x^2 + y^2 + 3x - 7y = \frac{1}{4}$

7. $36x^2 + 36y^2 + 108x - 96y + 1 = 0$

8. $9x^2 + 9y^2 + 6y - 24 = 0$

Find the equations of the following circles.

9. With center on origin and radius 10.

10. With center at $(3, -8)$ and radius $\sqrt{5}$.

11. With center $(10, 0)$ and tangent to $y = x$.

12. With center $(-5, 12)$ and passing through the origin.

Ellipse

An ellipse is defined as the locus of points, the sum of whose distances to two fixed points is constant. These fixed points are called *foci* and are pictured in Fig. 7-20 as F and F'. From Fig. 7-20, the horizontal and vertical

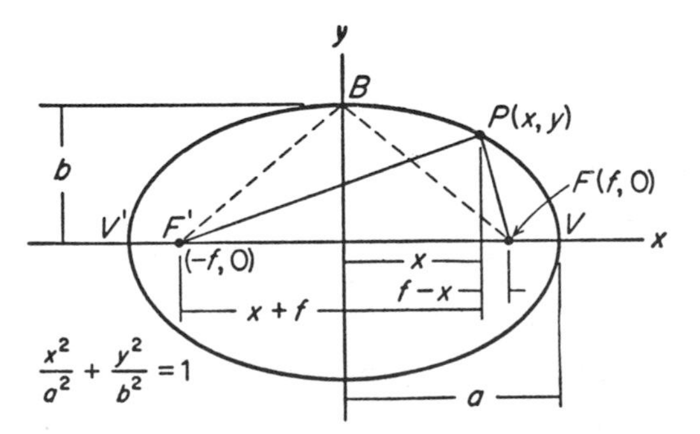

Figure 7-20

distances between P and F' are $x + f$ and y, respectively, whereas the corresponding distances between P and F are $f - x$ and y. The sum of the slanting distances PF' and PF must equal $2a$, for if P rests at V, then

$$PF = a - f \quad \text{and} \quad PF' = a + f.$$

Consequently,

$$PF + PF' = 2a.$$

Also, by the locus definition of an ellipse, $PF + PF'$ must remain fixed. With P anywhere on the curve, then, we have the equation

$$\sqrt{(x + f)^2 + y^2} + \sqrt{(f - x)^2 + y^2} = 2a$$

from which

$$\frac{x^2}{a^2} + \frac{y^2}{a^2 - f^2} = 1.$$

Since the vertical axis of the ellipse in Fig. 7-20 is the perpendicular bisector of FF', point B is equidistant from F and F' with the result that $BF = a$ and hence the Pythagorean relationship $b^2 = a^2 - f^2$. Consequently, the *standard form* of the ellipse is either

$$\frac{x^2}{a^2} + \frac{y^2}{b^2} = 1 \qquad (a^2 > b^2)$$

or

$$\frac{y^2}{a^2} + \frac{x^2}{b^2} = 1 \qquad (a^2 > b^2). \tag{59}$$

The second equation is the case where the major axis is vertical rather than the horizontal case pictured in Fig. 7-20. The distance a is always measured from the center to the vertex, and, from the relationship $b^2 = a^2 - f^2$, the requirement that $a^2 > b^2$ is apparent. Note that Eqs. (59) are modifications of the general quadratic equation (55) with $B = D = E = 0$.

It can also be shown that an ellipse with center (h, k) will be of the form

$$\frac{(x - h)^2}{a^2} + \frac{(y - k)^2}{b^2} = 1 \qquad (a^2 > b^2)$$

or

$$\frac{(y - k)^2}{a^2} + \frac{(x - h)^2}{b^2} = 1 \qquad (a^2 > b^2). \tag{60}$$

It should be noted that if Eqs. (60) are expanded to the form of (55), the characteristic of the equation of the ellipse is seen to be the sum of two squares with the coefficients of the square terms unlike.

The following examples use some of the properties of the ellipse and the standard forms.

Example 26 Give the equation of the ellipse with horizontal axis 6 units, vertical axis 12 units, and center at $(-3, 4)$.

Solution Because the long axis is vertical, we use the second of Eqs. (60) with $h = -3$, $k = 4$, $a^2 = 36$, and $b^2 = 9$. Substituting, we have

$$\frac{(y-4)^2}{36} + \frac{(x+3)^2}{9} = 1.$$

Example 27 Describe the ellipse represented by equation

$$25x^2 + 4y^2 - 100x + 24y + 36 = 0.$$

Solution Completing the square,

$$25(x^2 - 4x + 4) + 4(y^2 + 6y + 9) = -36 + 100 + 36$$

from which

$$\frac{(x-2)^2}{4} + \frac{(y+3)^2}{25} = 1.$$

This is recognized as an ellipse with center (2, −3). The graph is shown in Fig. 7-21. The major axis is vertical and is 10 units long, placing the

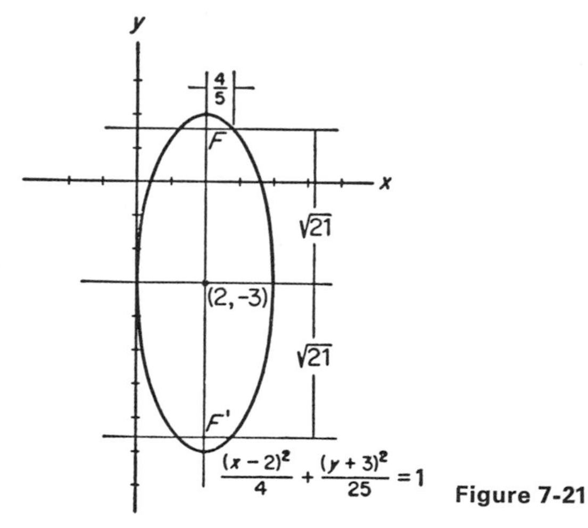

Figure 7-21

vertices at (2, 2) and (2, −8). The minor axis is 4 units long. The foci are (from $f^2 = a^2 - b^2 = 25 - 4 = 21$) located $\sqrt{21}$ units above and below the center and thus have coordinates $(2, -3 + \sqrt{21})$ and $(2, -3 - \sqrt{21})$. The alert reader will recognize that ellipses

$$\frac{(x-2)^2}{4} + \frac{(y+3)^2}{25} = 1 \quad \text{and} \quad \frac{x^2}{4} + \frac{y^2}{25} = 1$$

differ only in one respect—the locations of their centers. Their dimensions are identical. Since the latter form is easier to graph, it would be advisable in this example to assume the ellipse to be origin centered, graph it accordingly, and then move the origin away from the center of the ellipse to its proper relative position.

Exercises 7-9B

Collect the ellipses in Exercises 1–7 into standard form. Plot the center, vertices, and foci of each and sketch the graph.

1. $4x^2 + 9y^2 = 36$

2. $4x^2 + y^2 = 16$

3. $5x^2 + y^2 - 50x + 100 = 0$

4. $x^2 + 4y^2 + 16y = 0$

5. $2x^2 + 3y^2 + 12x - 30y + 81 = 0$

6. $5x^2 + 4y^2 - 60x + 72y + 464 = 0$

7. $6x^2 + 3y^2 - 72x + 24y + 222 = 0$

Determine the equation of the ellipse from the given properties.

8. Center at $(1, -5)$, horizontal major axis 12 units long and minor axis 4 units.

9. Center at $(\frac{1}{2}, 1)$, vertical major axis of length 6 and minor axis of length 3.

10. Center at $(5, 0)$, one focus at $(0, 0)$, major axis 12 units long.

11. Vertices at $(-3, 2)$ and $(3, 2)$, minor axis half the major axis.

Hyperbola

The *hyperbola* is defined as the locus of points, the difference of whose distances to two fixed points is constant.

The equation of the hyperbola is derived in a manner somewhat similar to that of the ellipse; the *standard form* of a hyperbola with center (h, k) is given as either

$$\frac{(x-h)^2}{a^2} - \frac{(y-k)^2}{b^2} = 1$$

or

$$\frac{(y-k)^2}{a^2} - \frac{(x-h)^2}{b^2} = 1. \tag{61}$$

From Eqs. (61), if $h = k = 0$, we have the origin-centered equations shown in Fig. 7-22. We choose the origin-centered hyperbola for a more detailed study primarily because it is less cumbersome, and, as in the case of the ellipse, the relocation of the center in no way affects the shape of the conic.

The reader will notice these features from Fig. 7-22.

1. Only one of the coordinate axes intercepts the origin-centered hyperbola. This axis becomes the *transverse axis* as opposed to the *conjugate axis*.
2. The vertices are $\pm a$ from the center along the transverse axis.
3. The foci are also on the transverse axis. In deriving the distance from the center to the focus, we obtain $f = \pm\sqrt{a^2 + b^2}$. (The derivation is left as a problem.)

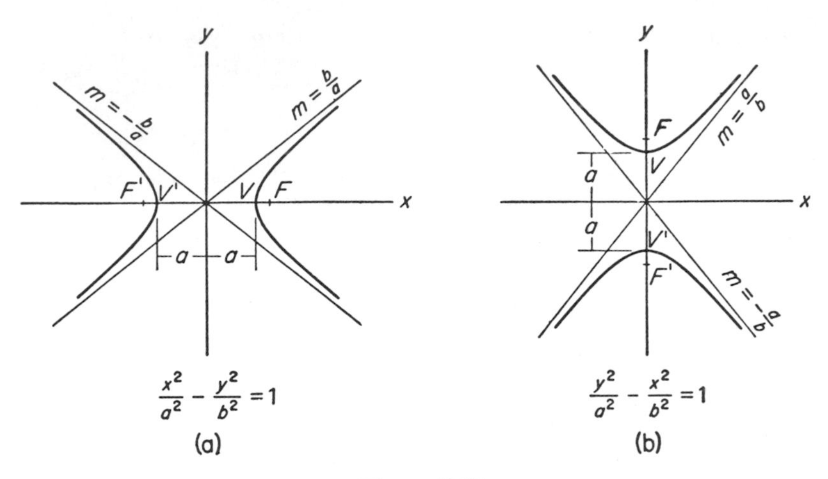

Figure 7-22

4. The hyperbola approaches limiting lines whose slopes are $\pm a/b$ or $\pm b/a$, depending on the direction of the transverse axis.

The limiting lines suggested above are called *asymptotes*; they are lines that a curve approaches but never reaches.

The equations of these asymptotes and other properties of the hyperbola are more clearly illustrated by an example.

Example 28 Discuss and graph the hyperbola

$$\frac{x^2}{4} - \frac{y^2}{9} = 1.$$

Solution Referring to Fig. 7-23, in order to find the intercepts, we let $y = 0$ and obtain $x = \pm 2$ as the x intercepts. If we let $x = 0$, we discover that the y intercepts do not exist.

If the equation is solved for y, we have

$$y = \pm\sqrt{\tfrac{9}{4}x^2 - 9}.$$

Two interesting points are noted here.

1. If $9x^2/4 - 9$ is negative, y is imaginary. In other words, in order to have a defined y, $(9/4)x^2 \geqq 9$ and the absolute value of x equals or exceeds 2—that is, $|x| \geqq 2$. (See Fig. 7-23.) This is the domain of x for which y is defined and is accordingly called the *domain of definition* of y.
2. As x becomes very large, $9x^2/4$ dominates the radicand, and the constant -9 has but a trifling effect on the value of y. For large abscissas,

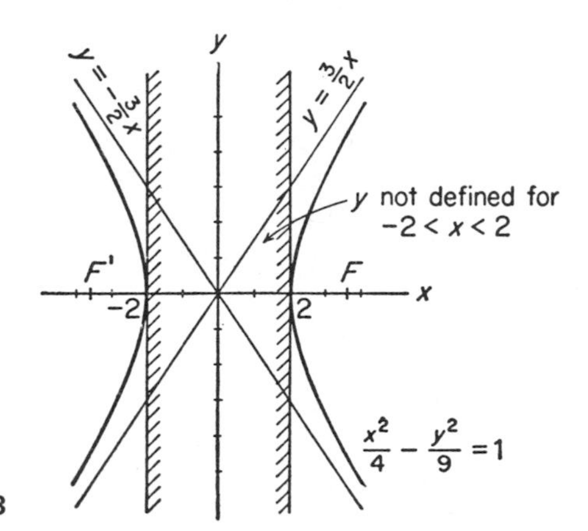

Figure 7-23

then, $y = \pm 3x/2$ is a near-correct description of the hyperbola, becoming even more correct as x continues to enlarge. It is by this reasoning that $y = \pm 3x/2$ are the limiting lines of the hyperbola, or asymptotes.

If Eqs. (61) are expanded, the result is of the general form

$$Ax^2 + Cy^2 + Dx + Ey + F = 0,$$

where, in every case, A and C differ in sign. The characteristic of the equation of a hyperbola is, then, for purposes of rapid identification, the difference of the squares of the variables.

Example 29 Graph equation $xy = 12$.

Solution This equation asserts that the product of the abscissa and ordinate equals 12. Immediately suggested are such coordinate pairs as

$$(12, 1), (3, 4), (-6, -2), (-4, -3), \text{etc.},$$

where it is apparent that the ordinate and abscissa must be of the same sign, forcing the graph to exist only in the first and third quadrants. This information, combined with the condition that the coordinate axes are the asymptotes, produces the graph in Fig. 7-24.

Equation $xy = c$ assumes the form

$$(x - h)(y - k) = c$$

by shifting the center to the point (h, k). When expanded, this takes on the form

$$xy + Dx + Ey + F = 0,$$

which is a modification of the general quadratic equation.

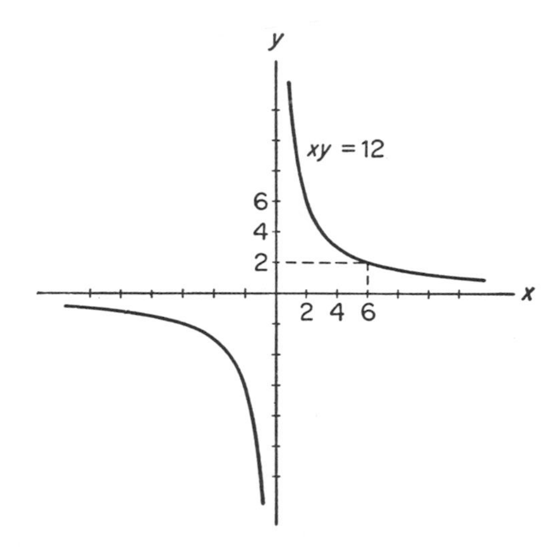

Figure 7-24

Exercises 7-9C

Assemble Exercises 1–7 into standard form. Locate the center, foci, and vertices. Sketch the asymptotes and the curve.

1. $x^2 - 4y^2 = 36$

2. $y^2 - 6x^2 = 24$

3. $y^2 - 9x^2 + 8y - 20 = 0$

4. $4x^2 - 9y^2 - 24x - 108y - 432 = 0$

5. $x^2 - y^2 - 4x - 12 = 0$

6. $xy = -16$

7. $xy - 3y + 4x = 0$

Determine the equations of the following hyperbolas with horizontal transverse axes.

8. Asymptotes of slopes ± 1, center (0, 0), distance between vertices 6.

9. Center (0, 0), vertex $(\frac{3}{2}, 0)$, asymptotes of slopes ± 2.

10. Center (0, 0), passing through (4, 1), $a = \sqrt{15}$.

Simultaneous systems involving conics

In Sec. 5-7 the solution of simultaneous linear systems was discussed. It was demonstrated that two straight lines in a plane are parallel, intersecting, or coincident, yielding, respectively, no, one, or an indefinite number of solutions. Here we consider the solutions of systems involving (a) a straight line and a conic and (b) two conics.

A straight line and a conic can intersect in no, one, or two distinct points as shown in Fig. 7-25. The number of real solutions depends on whether the line fails to intersect the conic, is tangent to it, or cuts it as a secant. A theorem from the theory of equations states that the number of

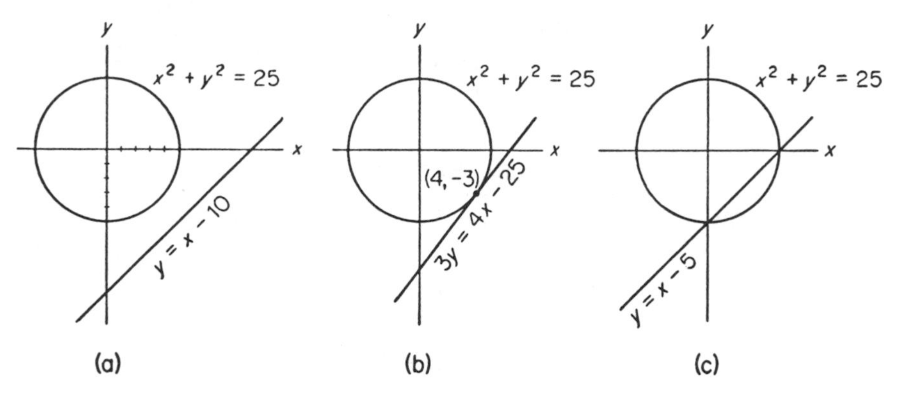

Figure 7-25

solutions of two equations equals the products of their degrees. Figure 7-25(a) shows no *real* solutions; any effort to obtain the solutions algebraically would produce an (x, y) set of which either x or y or both are imaginary because the quadratic resulting from eliminating one of the unknowns will contain a negative discriminant. An algebraic example of this situation follows.

Example 30 Solve the system containing

$$x^2 + y^2 = 25 \text{ and } y = x - 10.$$

Solution By substitution,

$$x^2 + (x - 10)^2 = 25.$$

Expanding,

$$2x^2 - 20x + 75 = 0.$$

This quadratic in x produces coordinates

$$x = \frac{10 \pm 5\sqrt{-2}}{2} \quad \text{and} \quad y = \frac{-10 \pm 5\sqrt{-2}}{2}.$$

This shows the solutions to be imaginary. A graph of the system is approximated in Fig. 7-25(a), where the line and circle fail to intersect.

When the line is tangent to the circle as in Fig. 7-25(b), an attempt to solve the system algebraically would result in a quadratic in one unknown whose discriminant is zero. An example of this situation follows.

Example 31 Solve the system containing

$$x^2 + y^2 = 25 \quad \text{and} \quad y = \frac{4x}{3} - \frac{25}{3}.$$

Solution By substitution,

$$x^2 + \left(\frac{4x}{3} - \frac{25}{3}\right)^2 = 25$$

from which

$$x^2 - 8x + 16 = 0,$$

which has the roots 4, 4 and produces the points of intersection

$$\begin{pmatrix} x = & 4 \\ y = & -3 \end{pmatrix}, \begin{pmatrix} x = & 4 \\ y = & -3 \end{pmatrix}.$$

The multiple roots indicate a tangency situation.

When the line intersects the circle in two distinct points, the reader would properly anticipate two distinct solutions when solving algebraically. The quadratic resulting in the elimination of one unknown will have a positive discriminant, indicating real and distinct roots. An example of this situation follows.

Example 32 Solve the system containing

$$x^2 + y^2 = 25 \quad \text{and} \quad y = x - 5.$$

Solution By substitution,

$$x^2 + (x - 5)^2 = 25$$

or

$$2x^2 - 10x = 0,$$

producing roots

$$\begin{pmatrix} x = & 0 \\ y = & -5 \end{pmatrix}, \begin{pmatrix} x = 5 \\ y = 0 \end{pmatrix}.$$

These roots show that the intersections are real and distinct. Figure 7-25(c) illustrates this example.

The preceding situations are duplicated and expanded for systems involving two noncoincident conics, that can intersect in 0, 1, 2, 3, or 4 distinct points. A theorem from the theory of equations states that imaginary roots occur in pairs. In eliminating one of the unknowns from a system of simultaneous quadratics, four solutions will always result; none, two, or four of them can be imaginary. The others will be real or infinite and, in the case of tangencies, will be multiple. The separate situations involving a circle and a parabola are pictured in Fig. 7-26.

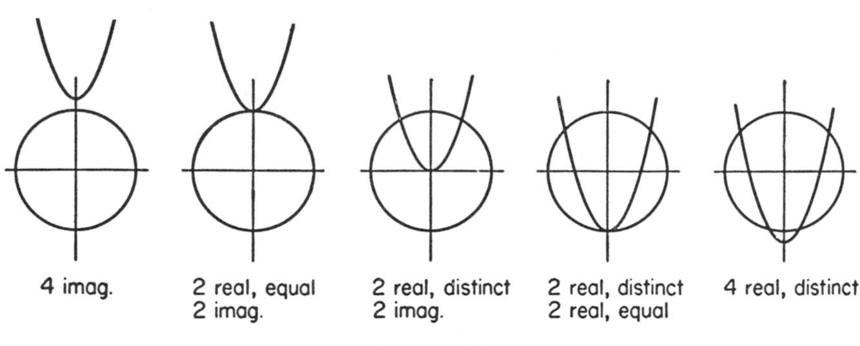

Figure 7-26

Example 33 Solve the system containing

$$x^2 + y^2 = 25 \quad \text{and} \quad y = \frac{4x^2}{9}.$$

Solution Substituting,

$$y^2 + \frac{9y}{4} - 25 = 0$$

from which

$$\begin{pmatrix} x = \pm 3 \\ y = 4 \end{pmatrix}, \begin{pmatrix} x = \pm \dfrac{15}{4} j \\ y = -\dfrac{25}{4} \end{pmatrix}$$

This is the situation of four roots, two of which are imaginary and two real and distinct, as pictured in the middle sketch in Fig. 7-26.

Exercises 7-9D

Solve Exercises 1–8, both graphically and algebraically.

1. $9x^2 + 4y^2 = 36$
$2y - 3x = 6$

2. $x^2 - y^2 = 12$
$2y - x = 0$

3. $4y^2 - x^2 = 36$
$x^2 + y^2 = 9$

4. $4x - 8 = y^2$
$x^2 + 9y^2 = 4$

5. $9x = 4y^2 - 40y + 100$
$x^2 + y^2 - 10y = 0$

6. $x^2 - 2x + y^2 = 24$
$x^2 - y^2 = 16$

7. $9x^2 + 16y^2 - 36x - 64y = 0$
$3x^2 - 12x - 2y = 0$

8. $x^2 - 4x + 4y^2 = 0$
$x^2 - 4x = y$

Solve the following systems algebraically.

9. $x^2 + xy = 3$
$4xy + y^2 = 12$

10. $xy + y = 10$
$x + xy = 6$

11. $x^2 + 2xy + y^2 = 64$
$x^2 + 4xy + 4y^2 = 100$

12. $x^3 + y^3 = 26$
$x + y = 2$

13. $\dfrac{1}{x} + \dfrac{1}{y} = 5$
$\dfrac{1}{2x} - \dfrac{1}{6y} = -1$

14. $x + \dfrac{1}{x} = y$
$x - \dfrac{2}{x} = 2y$

8

Identification and approximation of roots

In the preceding chapter we discovered that quadratic equations often have roots that are irrational in nature. The irrational roots of a quadratic are inexact when expressed in decimal form. This same inexactness may occur in logarithmic, trigonometric, and exponential forms; it also may occur in many algebraic forms.

Third-degree and fourth-degree polynomial equations, like quadratic equations, have radical solutions. These radical solutions are often very involved and might better be obtained in decimal form by approximate methods. No general radical solution is available for equations of fifth degree and higher.

In this chapter we introduce methods of identifying roots, isolating roots, and approximating roots.

8-1 Introduction

In considering approximate solutions, it is convenient to use the term *function*. We have already gained sufficient background to enable us to define a function. We recall from previous chapters that, for equations such as

$$y = x - 5, \tag{1}$$

$$y = \tfrac{1}{2}x^2 + x - 6, \tag{2}$$

the assignment of a numerical value to x imposes a unique value on y. In each case, we arrive at an (x, y) pair that satisfies the equation in question. In (1) and (2) we can state that y is the value of a function of x; we can call the right-hand member of (1) the f function of x and of (2) the g function of x, giving us new symbols:

$$y = f(x) = x - 5,$$

$$y = g(x) = \tfrac{1}{2}x^2 + x - 6.$$

The symbols $f(x)$ and $g(x)$ simply mean f function of x and g function of x and do not imply multiplication. Going back to the (x, y) pairs, we can define a *function* as *a set of ordered pairs of numbers* (x, y) *such that to each value of the first variable* (x) *there corresponds a unique value of the second variable* (y).

The composite of all (x, y) pairs that satisfy (1) can be shown as a straight line on a graph, whereas the composite of (x, y) pairs that satisfy (2) can be shown as a parabola (Fig. 8-1). We note a particular property of the two functions. First, we observe that for *all* real values of x there is a real value for y. Secondly, as a result, the graphs in Fig. 8-1 are smooth and *continuous*. We will describe a continuous function here in terms of the graphical representation of that function: *if the graph of a function possesses*

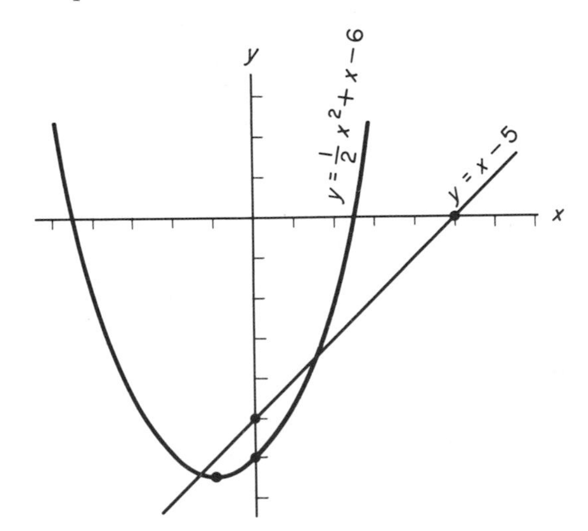

Figure 8-1

no unusual "leaps" or "breaks," the function represented by that graph is continuous. Figure 8-2 pictures some graphs that contain the "leaps" or "breaks" cited and gives the accompanying equations. Note in Fig. 8-2(a) that $f(x)$ is infinitely large for $x = 0$; in Fig. 8-2(b), $f(x)$ is infinitely large for $x = 1$. In each case, $f(x)$ "leaps" to "infinity" for certain values of x. In Fig. 8-2(c), $f(x)$ does not exist when x is exactly an integer and the graph "leaps" in steps.

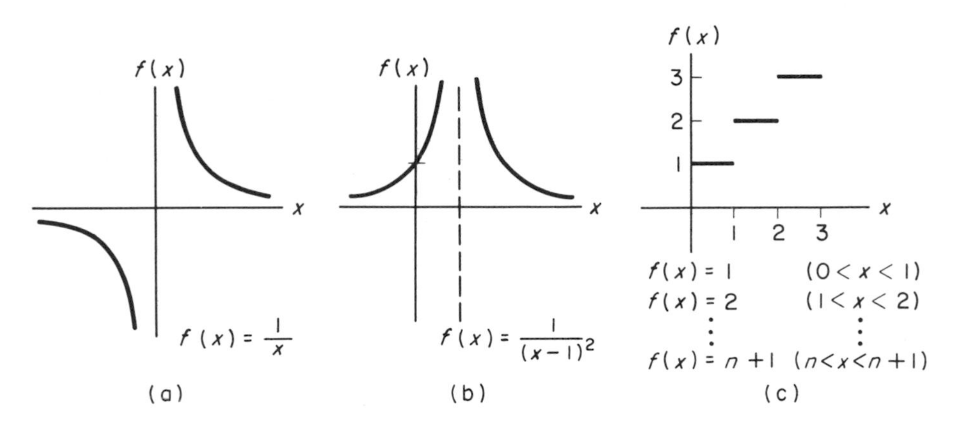

Figure 8-2

8-2 Polynomial equations; integral roots; synthetic division

We turn next to the special function

$$f(x) = x^n + a_1x^{n-1} + a_2x^{n-2} + \cdots + a_n. \tag{3}$$

We will impose certain conditions on (3); first, n is a positive integer and $a_1, a_2, a_3, \ldots, a_n$ are constants. If we set the right side of (3) equal to zero, then

$$x^n + a_1x^{n-1} + a_2x^{n-2} + \cdots + a_n = 0, \tag{4}$$

where (4) is *an integral rational equation of* nth *degree in* x and $f(x)$ from (3) is called *a polynomial of* nth *degree in* x. We assert here that (3) is continuous for all x.

Let us discuss a certain simple integral rational equation for the purpose of obtaining its roots. Given

$$y = f(x) = x^2 - 2x - 3, \tag{5}$$

we have an integral rational equation of degree 2 in x. Let us find the roots of (5); we recall that the roots can be defined as the x intercepts and that y has the value zero at the x intercept. So if we want the roots of (5), we let

$y = 0$ and we have

$$x^2 - 2x - 3 = 0, \tag{6}$$

which is now a polynomial of second degree in x.

Recalling the factoring method of solving (6), we have

$$x^2 - 2x - 3 = (x - 3)(x + 1) = 0.$$

Then since the product of $(x - 3)$ and $(x + 1)$ is zero, either

$$x - 3 = 0 \quad \text{or} \quad x + 1 = 0$$

and we have the roots $x = 3$, $x = -1$.

Note that the root $x = 3$ is obtained from the factor $x - 3$; similarly, $x = -1$ is obtained from the factor $x + 1$. Similarly, a root $x = r$ would be obtained from the factor $x - r$. Note also from the original polynomial $x^2 - 2x - 3$ that roots 3 and -1 are factors of the constant term -3.

In the general case of a quadratic, if we have roots $x = r_1$ and $x = r_2$, then $x - r_1$ and $x - r_2$ are factors of the original quadratic and the original quadratic is

$$(x - r_1)(x - r_2) = x^2 - (r_1 + r_2)x + r_1 r_2,$$

where it is again noted that roots r_1 and r_2 are factors of the constant $r_1 r_2$.

This situation is broadened to include the cubic, quartic, and the general equation (4) so that we can say that if (4) has integral roots, then these roots are factors of a_n.

Example 1 Determine by inspection whether

$$y = x^3 - 2x^2 + 2x - 4 \tag{7}$$

has any integral roots.

Solution From the preceding material we recall first that a number is a root of (7) if it satisfies the equation

$$x^3 - 2x^2 + 2x - 4 = 0. \tag{8}$$

Secondly, if (8) has an *integral* root, that root is a factor of the constant term -4. This limits our choice to $x = \pm 4, \pm 2, \pm 1$. Substituting these six values of x into (7), we have

x	$+4$	-4	$+2$	-2	$+1$	-1
y	$\neq 0$	$\neq 0$	0	$\neq 0$	$\neq 0$	$\neq 0$

,

where our interest is in determining the exact value of y only if $y = 0$. If $y \neq 0$, we have no concern about its exact value.

Our conclusion here is that $x = 2$ is a root of (7) because $y = 0$ when $x = 2$.

Reverting again to (4), the nth-degree polynomial has n roots $r_1, r_2, r_3, \ldots, r_n$. This asserts that it has n factors so that (4) can be written

$$x^n + a_1x^{n-1} + a_2x^{n-2} + \cdots + a_n$$
$$= (x - r_1)(x - r_2) \cdots (x - r_n) = 0. \tag{9}$$

The right member in (9) is the product of n factors, and we assert that if we divide both members of (9) by any one of these n factors (say $x - r_1$), we have the new polynomial

$$x^{n-1} + b_1x^{n-2} + b_2x^{n-3} + \cdots + b_{n-1}$$
$$= (x - r_2)(x - r_3) \cdots (x - r_n) = 0, \tag{10}$$

where the left member of (10) equated to zero is called a *depressed equation,* since it is Eq. (4) with one factor removed. Note that removal of the linear factor $x - r_1$ reduces the degree of (4) by unity. This suggests a ready method of solving certain third-degree equations.

Example 2 Inspect equation

$$y = x^3 + 6x^2 + 7x - 2 \tag{11}$$

for integral roots. If an integral root is found, depress the equation to a quadratic and then solve the quadratic, thus obtaining all the roots of (11).

Solution Referring to (11), if $y = 0$, then

$$x^3 + 6x^2 + 7x - 2 = 0. \tag{12}$$

The only integral roots of (12) are factors of -2, suggesting $x = \pm 1$, $x = \pm 2$. Trying all of them, we have

x	1	-1	2	-2
y	$\neq 0$	$\neq 0$	$\neq 0$	0

.

We have ascertained that $x = -2$ is a root of (12) and $x + 2$ is a factor of $x^3 + 6x^2 + 7x - 2$. Dividing the third-degree expression by $x + 2$, we have

$$\begin{array}{r|l}
 & x^2 + 4x - 1 \\
\hline
x + 2 & x^3 + 6x^2 + 7x - 2 \\
 & x^3 + 2x^2 \\
\cline{2-2}
 & \quad 4x^2 + 7x \\
 & \quad 4x^2 + 8x \\
\cline{2-2}
 & \qquad -x - 2 \\
 & \qquad -x - 2 \\
\cline{2-2}
\end{array}$$

where the quotient $x^2 + 4x - 1$ is the other factor of the third-degree expression. Thus

$$x^3 + 6x^2 + 7x - 2 = (x + 2)(x^2 + 4x - 1) = 0. \tag{13}$$

Each factor of (13) may be zero so that

$$x^2 + 4x - 1 = 0$$

from which

$$x = -2 \pm \sqrt{5},$$

and the three roots of (11) are

$$\begin{aligned} x &= -2, \\ x &= -2 + \sqrt{5} = 0.236, \\ x &= -2 - \sqrt{5} = -4.236. \end{aligned}$$

By successively depressing a polynomial equation, all the integral roots can be determined. If the equation can be depressed to a quadratic, the last two roots can easily be obtained through the quadratic formula, regardless of the character of those roots—that is, whether the remaining roots are real or imaginary, equal or unequal, rational or irrational.

Synthetic division

One fast and convenient way of dividing a polynomial by a binomial is by a process called *synthetic division*. As an introduction, we see that

$$\frac{P(x)}{x - r} = q(x) + \frac{R}{x - r}$$

or

$$P(x) = (x - r)q(x) + R$$

where these equations assert that division of a polynomial $P(x)$ by a binomial $x - r$ results in a quotient $q(x)$ and a remainder R. With R independent of x, $q(x)$ will be a depressed polynomial of degree one less than that of $P(x)$. Note from the latter equation that $P(x) = R$ when $x = r$, meaning that the remainder is the value of the function $P(r)$.

We will show how synthetic division is derived from the long division process outlined in Sec. 2-7. A model of that process is given here.

$$\begin{array}{r|rrrrr l}
 & 3x^3 & + 10x^2 & + 20x & + 5 & & \\ \cline{2-6}
1x - 2 & 3x^4 & + 4x^3 & + 0x^2 & - 35x & - 7 & \\
 & 3x^4 & - 6x^3 & & & & (3) \\ \cline{2-3}
 & & 10x^3 & + 0x^2 & & & (0) \\
 & & 10x^3 & - 20x^2 & & & (10) \\ \cline{3-4}
 & & & 20x^2 & - 35x & & (-35) \\
 & & & 20x^2 & - 40x & & (20) \\ \cline{4-5}
 & & & & 5x & - 7 & (-7) \\
 & & & & 5x & - 10 & (5) \\ \cline{5-6}
 & & & & & 3 &
\end{array}$$

This demonstrates division of the polynomial $P(x) = 3x^4 + 4x^3 - 35x - 7$ by the binomial $x - 2$ to obtain the quotient $q(x) = 3x^3 + 10x^2 + 20x + 5$ and remainder $R = 3$.

Let us make three changes in this division display. First, omit all powers of x. Then avoid the duplication of showing the "brought down" numbers 0, -35, and -7 shown in parentheses at the right. Finally, avoid duplicating the subproduct numbers 3, 10, 20, and 5. Compress into three lines the numbers that remain and we have the new array

$$\begin{array}{r|rrrrr} 1-2 & 3 & 4 & 0 & -35 & -7 \\ & & -6 & -20 & -40 & -10 \\ \hline & & 10 & 20 & 5 & 3 \end{array}$$

We drop coefficient 1 from the divisor (it is always 1 due to our condition of dividing by $1x - r$) and, for convenience, duplicate the 3 of the first line onto the third line.

$$\begin{array}{r|rrrrr} -2 & 3 & 4 & 0 & -35 & -7 \\ & & -6 & -20 & -40 & -10 \\ \hline & 3 & 10 & 20 & 5 & 3 \end{array}$$

From this array we see that we can *multiply* any number on the third line by -2 of the divisor in order to obtain the next number of the second line. Also, we obtain any number in the third line by *subtracting* the second line number from the first line number in the column above it. Since *multiplication by* -2 and *subtraction* is the equivalent of *multiplication by* $+2$ and *addition*, we can make final modifications in the array, providing the model for *synthetic division*:

$$\begin{array}{r|rrrrr} 2 & 3 & 4 & 0 & -35 & -7 \\ & & 6 & 20 & 40 & 10 \\ \hline & 3 & 10 & 20 & 5 & 3 \end{array}$$

Note the components of the last array when compared to important portions of the original problem. The dividend $P(x)$ is expressed without powers of x; however, those powers of x assisted in ordering the sequence of the coefficients. The quotient $q(x)$ appears in the last line, as does the remainder $R = 3$. Observe that all columns are added and that all second-row entries are obtained by multiplying the previous third-row entry by the divisor. Let us try two examples.

Example 3 Using synthetic division, divide

$$(x^3 + 6x^2 + 7x - 2) \div (x + 2).$$

Solution Using the simplified array suggested, duplicate the coefficients of the dividend and show the divisor as -2.

$$\begin{array}{r|rrrr} -2 & 1 & 6 & 7 & -2 \\ & & & & \\ \hline & 1 & & & \end{array}$$

Draw a line under what will become the second line and "bring down" the lead coefficient 1 as the first coefficient on the third line. Then multiply the third-line coefficient 1 by the divisor -2 and place the product -2 under the first-line coefficient 6.

$$\begin{array}{r|rrrr} -2 & 1 & 6 & 7 & -2 \\ & & -2 & & \\ \hline & 1 & 4 & & \end{array}$$

Add the 6 and -2 to obtain the second coefficient, 4, in the third line. Multiply 4 by the divisor -2 and place the product -8 under the first-line coefficient 7 and add to obtain the third-line coefficient -1.

$$\begin{array}{r|rrrr} -2 & 1 & 6 & 7 & -2 \\ & & -2 & -8 & \\ \hline & 1 & 4 & -1 & \end{array}$$

Multiply -1 by -2 and place the product 2 under the first-line coefficient -2 and add to obtain 0.

$$\begin{array}{r|rrrr} -2 & 1 & 6 & 7 & -2 \\ & & -2 & -8 & 2 \\ \hline & 1 & 4 & -1 & 0 \end{array}$$

This finishes the synthetic division process and leaves only an interpretation of the results. We observe that the quotient is $q(x) = x^2 + 4x - 1$ and the remainder is $R = 0$; all this information is obtained from the third line.

Example 4 Is $x = 5$ a root of $x^3 + 2x^2 - 23x + 60 = 0$?

Solution If $x = 5$ is a root, then $x - 5$ is an exact divisor of the cubic, meaning that the remainder will be $R = 0$. Applying synthetic division, we have

$$\begin{array}{r|rrrr} 5 & 1 & 2 & -23 & 60 \\ & & 5 & 35 & 60 \\ \hline & 1 & 7 & 12 & 120 \end{array}$$

From this operation the quotient is $q(x) = x^2 + 7x + 12$ and the remainder is $R = 120$, showing that $x = 5$ is not a root of the cubic.

Exercises 8-2

In Exercises 1–10, divide the polynomial by the binomial. Express both the quotient and the remainder. Use synthetic division.

1. $x^2 - 2x - 3$ by $x + 1$

2. $x^2 + 4x - 5$ by $x - 1$

3. $3x^3 + 8x^2 + 3x - 1$ by $x + 2$

4. $2x^3 - 5x^2 + 7x - 4$ by $x + 3$

5. $63m + 4m^3 - 6$ by $m - 4$

6. $5p + 9 - p^3$ by $p + 6$

7. $k^4 - 1$ by $k + 2$

8. $y^4 + 8$ by $y - 1$

9. $-6x^3 + 2x^2 - 3x^3 + 4x^4 - 5$ by $x + 1$

10. $9y^4 - 6 + 5y^2 - 3y$ by $y - 1$

11. Show that $x = -3$ is a root of $x^3 + 5x^2 + 8x + 6 = 0$.

12. Is $x - 1$ a factor of $x^5 - 2x + 1$?

13. Given $P(x) = 4x^3 + x - 110$, show that $R > 0$ for $P(3)$ and $R < 0$ for $P(2)$.

14. Given $P(x) = x^3 + ax^2 - 3x + 4$, for what value of a is $P(x)$ exactly divisible by $x - 1$?

15. Using synthetic division, find the value of $P(x) = 3x^4 + 4x^3 + x - 18$ for $x = -1$. Compare this remainder to $P(-1)$.

In Exercises 16–28, identify the integral roots of the given polynomial equations (if they possess integral roots). When an integral root is identified, depress the equation. If the equation is depressible to a quadratic, find all roots, regardless of their character.

16. $y = x^3 - 2x + 1$

17. $y = x^3 + 2x + 3$

18. $y = x^3 - x^2 - 4x - 6$

19. $y = x^3 + 5x^2 + 8x + 6$

20. $y = x^4 - x^3 - 4x^2 - 5x - 3$

21. $y = x^4 + 3x^3 - x - 3$

22. $y = x^3 - 1$

23. $y = x^3 + 1$

24. $y = x^4 - 1$

25. $y = x^4 + 1$

26. $y = x^3 + 3x^2 - 5x + 12$

27. $y = x^4 - 3x^3 - 4$

28. $y = x^3 + 4x^2 - 5x$

29. The rectangular sheet of tin in Fig. 8-3 has equal squares cut from its corners. The resulting sheet is folded along the dotted lines to form a topless box. What three different dimensions can each square have so that the volume of the box is 16 in.^{3}?

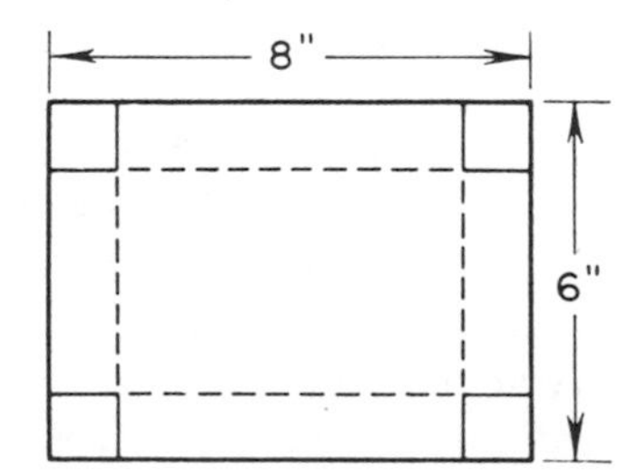

Figure 8-3

30. The length, width, and height of the large rectangular bin shown in Fig. 8-4 are each increased by the same amount. The increases change the volume of the bin to 864 ft^3. What are the new dimensions of the bin?

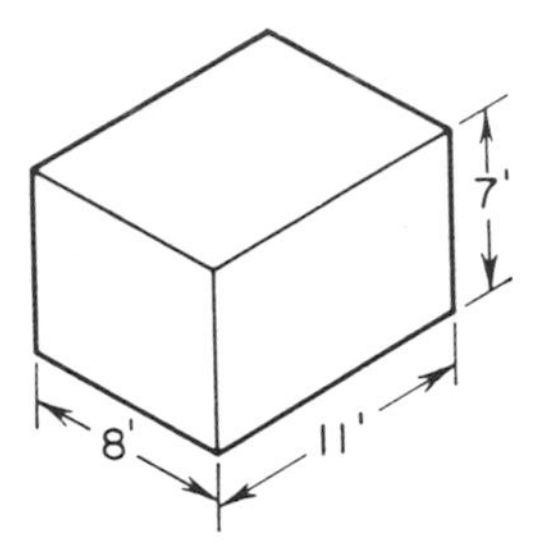

Figure 8-4

31. The sum of the volume and total surface area of a certain cube is 275. What are the dimensions of the cube?

32. The dimensions of a box are such that it is one inch taller than it is wide and one inch longer than it is tall. The sum of the volume and surface area is 628. What are the dimensions of the box?

33. The amount of inertia about the x axis of the thin, hollow, rectangular plate shown in Fig. 8-5 is given as

$$I_x = \frac{bh^3}{3} - \frac{b_1h_1(3h^2 + h_1^2)}{12}.$$

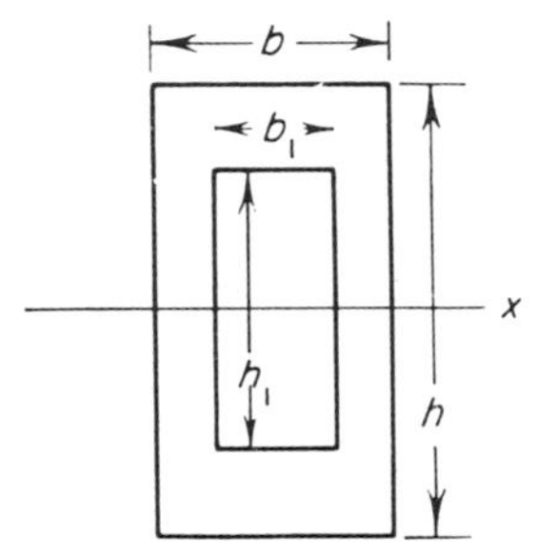

Figure 8-5

What height h must a designer choose if he or she wishes to have a moment of inertia $I_x = 400/3$ with $b_1 = 2$, $b = 3$, and $h_1 = 4$?

8-3 Isolation of roots between consecutive integers

Earlier we discussed integral roots of a polynomial equation. Special care was used to avoid examples and exercises in which a polynomial equation had no integral roots. We extend our discussion now to include the case of polynomial equations with no integral roots and eventually extend it to include equations (whether polynomial or not) that possess no integral roots and that may have discontinuities but discontinuities that are not sufficiently close to a root to disturb the methods of solution that we describe.

A completely continuous function $y = f(x)$ may have a root $x = r$ where r is irrational and none of its other roots is rational. Let us assume that the root r is between two consecutive integral values of x such that $k < r < k + 1$. Referring to Fig. 8-6, we see three graphs of different continuous functions $y = f(x)$. In each case, a root at $x = r$ is shown and $k < r < k + 1$, where k is an integer. Note in Fig. 8-6(a) that when $x = k$, $y < 0$; when $x = r$, $y = 0$; and when $x = k + 1$, $y > 0$. Note in Fig. 8-6(b) that

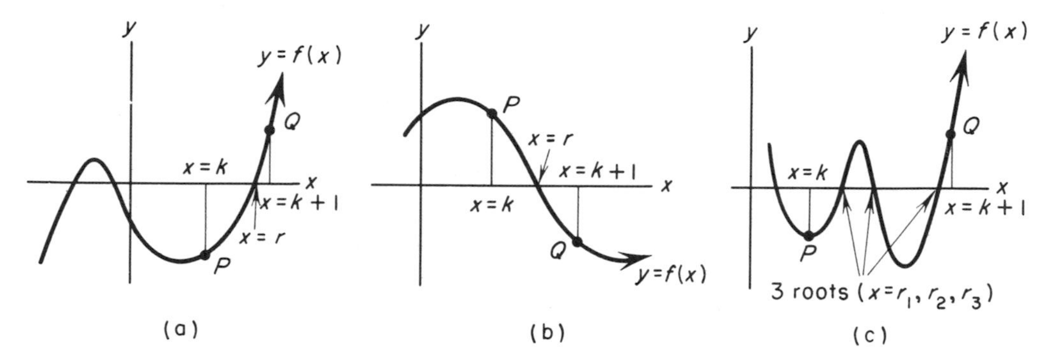

Figure 8-6

when $x = k$, $y > 0$; when $x = r$, $y = 0$; and when $x = k + 1$, $y < 0$. What is suggested here is the following fact. *If a function $y = f(x)$ is continuous in the interval from $x = k$ to $x = k + 1$ and the sign of y is different at $x = k$ from the sign of y at $x = k + 1$, then at least one root of $y = f(x)$ exists between $x = k$ and $x = k + 1$.* When there is a sign change, we say that we have *isolated* a root between $x = k$ and $x = k + 1$.

The reason for the assertion that *at least one root* exists between $x = k$ and $x = k + 1$ is pictured in Fig. 8-6(c). There may be *more than one* root. Inspection will bear out the assertion that the number of roots will be an *odd* number.

In Fig. 8-7 we see that the signs of y at $x = k$ and $x = k + 1$ are unchanged. The *lack* of a sign change does not prove that there is no root

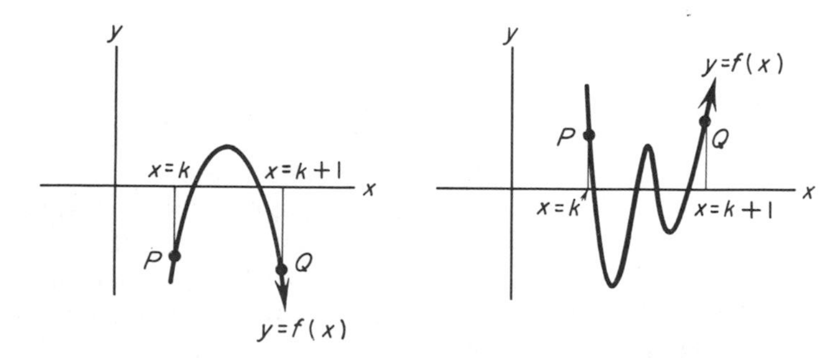

Figure 8-7

between $x = k$ and $x = k + 1$ because there may be an even number of roots in the interval.

Let us demonstrate the preceding observation with an example.

Example 5 The equation

$$y = x^3 - 3x^2 - 19x - 15 \tag{14}$$

possesses one positive root. Isolate the root between consecutive integers.

Solution We begin with arbitrarily chosen integral values of x and determine the sign of y. We emphasize that the magnitude of y is not important; only the *sign* is important. We see that y is negative for $0 < x < 6$

x	0	1	2	3	4	5	6	7
y	−	−	−	−	−	−	−	+

but positive for $x = 7$. This asserts that there is at least one root where $6 < r < 7$. The graph of the important portion of (14) is shown in Fig. 8-8. The actual numerical value of the root is $x = 2 + \sqrt{19}$ or $x = 6.359$.

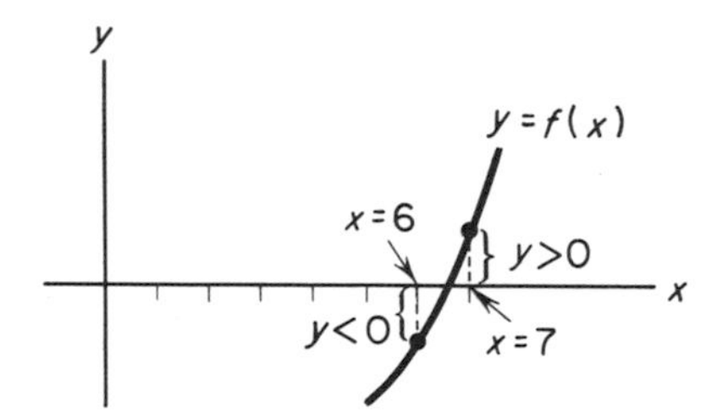

Figure 8-8

8-4 Approximation of roots

In Sec. 8-3 we discussed a method of isolating a root between consecutive integers. It may be desirable to describe the root more accurately, perhaps correct to the nearest tenth or hundredth or thousandth. We can broaden the method in Sec. 8-3 to embrace further refinement; in that section we stated that for $y = f(x)$, if $f(k)$ and $f(k + 1)$ differ in sign, then there is a root $x = r$ such that $k < r < k + 1$. We restricted k to integral values, and the x interval was one unit wide with k on the left edge of the interval and $k + 1$ on the right edge.

Actually, we need not be so restrictive about the edges of the interval. We may say that *a continuous function* $y = f(x)$ *has a root* $x = r$ *in the interval* $a < x < b$ *if the signs of* $f(a)$ *and* $f(b)$ *are different.* In Fig. 8-9 we show the interval $a < x < b$, where $f(a) < 0$ and $f(b) > 0$. It is important to note that a and b need not be integers; the quantity $b - a$ may be small or large. If we wish to obtain r correct to hundredths or thousandths, it is important that $b - a$ be a small quantity, perhaps on the order of tenths or hundredths.

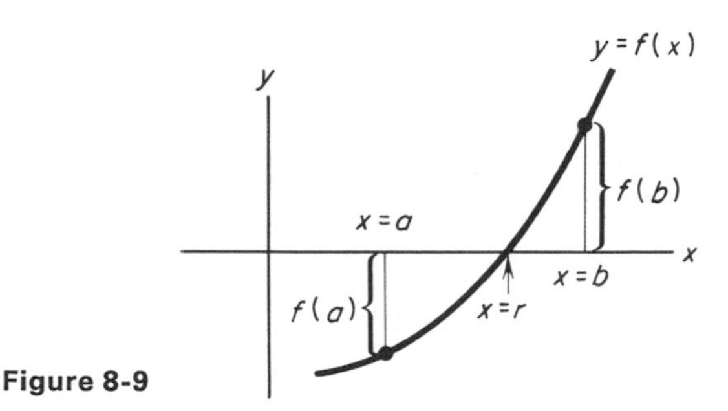

Figure 8-9

A method of refining the estimate of the value of a root is suggested here. It presumes that there is only one root in the interval $a < x < b$; it also presumes continuity in the interval. We shall call it the *method of linear interpolation* and discuss it with reference to Fig. 8-10. Suppose that we have isolated the root between $x = a$ and $x = b$ so that $f(a)$ and $f(b)$ are of opposite sign. The curve $y = f(x)$ is shown as a curved line. Draw the chord AB that will intersect the x axis between a and b at H. Ignoring signs, we

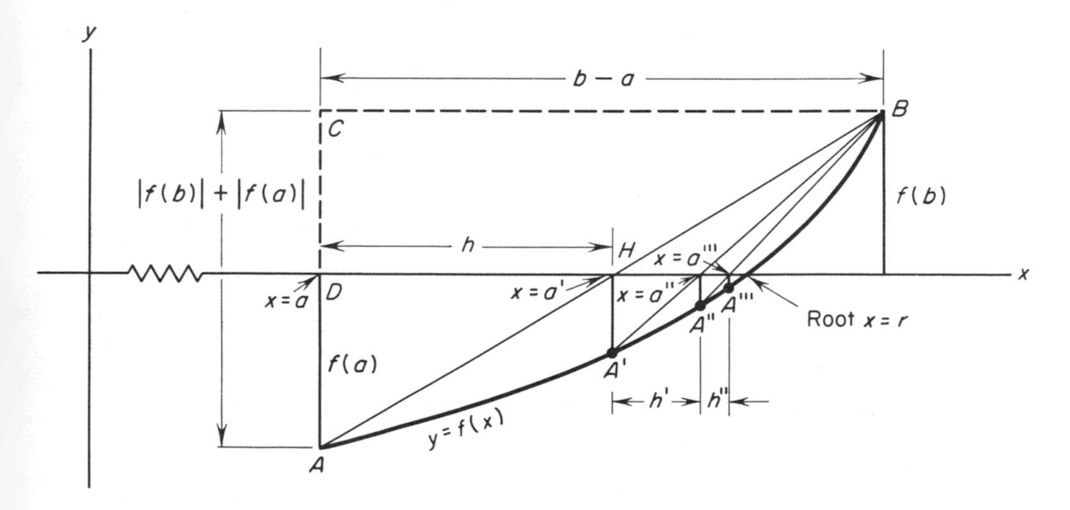

Figure 8-10

have $\triangle ADH$ similar to $\triangle ACB$, from which

$$\frac{DH}{DA} = \frac{CB}{AC} \quad \text{or} \quad \frac{h}{|f(a)|} = \frac{|b-a|}{|f(b)| + |f(a)|}$$

and so

$$h = \frac{|b-a||f(a)|}{|f(b)| + |f(a)|}, \tag{15a}$$

where h is an *adjustment* away from a toward b. Absolute values are used in (15a) because of possible modifications of the signs of a, b, $f(a)$, and $f(b)$.

Expression (15a) can be used repeatedly to refine the estimate of the root. After one application of (15), the root is a', where

$$a' = a + h.$$

For the next refinement, we consider the interval from a' to b where (15a) remains unchanged except that we are obtaining a second adjustment h', using a' as a reference so that (15a) is

$$h' = \frac{|b-a'||f(a')|}{|f(b)| + |f(a')|} \tag{15b}$$

and the second estimate of the root is

$$a'' = a' + h' = a + h + h'.$$

Let us use the method of linear interpolation in the next example.

Example 6 Use the method of linear interpolation to determine the positive root of

$$y = x^3 - 4x^2 - 12x - 6 = f(x) \tag{16}$$

correct to thousandths.

Solution First, we isolate the root of (16) between consecutive integers by substituting consecutive integral values of x into (16) and looking for a change in the sign of y.

x	0	1	2	3	4	6	7
y	$-$	$-$	$-$	$-$	$-$	-6	$+57$

(17)

We note that a root exists such that $+6 < r < +7$ because $f(6) < 0$ and $f(7) > 0$.

Next, we observe from Fig. 8-11′ (drawn out of scale) that possibly the root favors the left side of the interval bounded on the left by $x = 6$ and on the right by $x = 7$. We have not yet assigned the a and b of Eq. (15a). Let us first narrow the interval by estimating a value $x = b$ with $6 < b < 7$ such that $f(b) > 0$ but $f(b)$ is nearer zero than $+57$. It

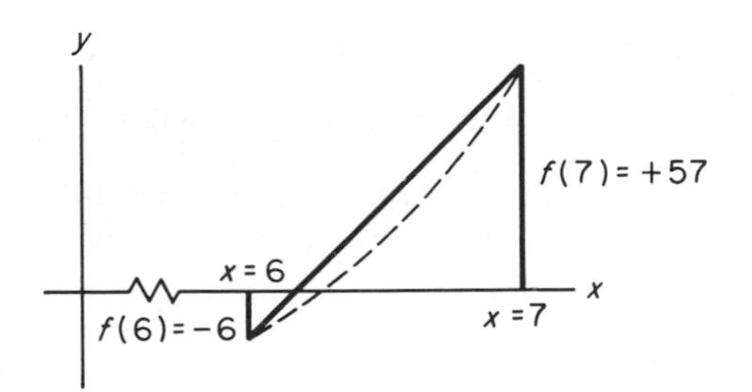

Figure 8-11

seems logical to try $b = 6.2$ so that

$$f(b) = f(6.2) = (6.2)^3 - 4(6.2)^2 - 12(6.2) - 6 = +4.168. \qquad (18)$$

Next, we make our assignments appropriate to (15a), letting $a = +6$, with $f(a) = f(6) = -6$ from (17), and letting $b = 6.2$ with $f(b) = f(6.2) = +4.168$ from (18). Substituting these facts into (15a),

$$h = \frac{|6.2 - 6||-6|}{|4.168| + |-6|} = \frac{(0.2)(6)}{10.168} = 0.12.$$

Refer to Fig. 8-12 for a graph involving a, b, $f(a)$, $f(b)$, and h.

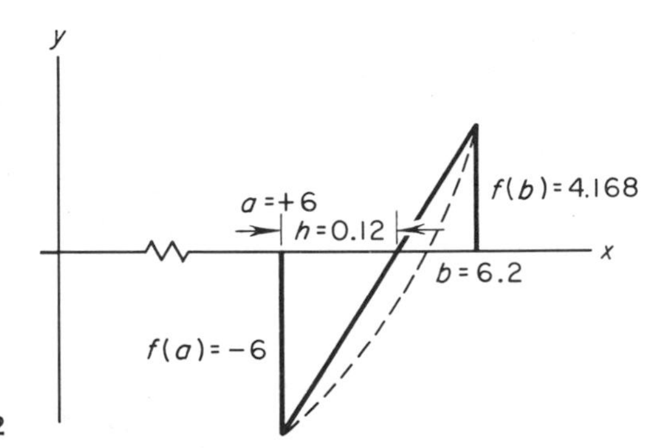

Figure 8-12

The new estimate of the root is

$$a' = a + h = 6 + 0.12 = 6.12.$$

To find the second estimate, we apply (15b) with $b = 6.2$, $f(b) = +4.168$, $a' = 6.12$, and

$$f(a') = f(6.12) = (6.12)^3 - 4(6.12)^2 - 12(6.12) - 6 = -0.036672,$$

and so

$$h' = \frac{|6.2 - 6.12||-0.036672|}{|4.168| + |-0.036672|} = \frac{(0.08)(0.036672)}{4.204672}$$

$$= \frac{0.00293376}{4.204672} = 0.0007.$$

Our new estimate of the root is

$$a'' = a' + h' = 6.12 + 0.0007 = 6.1207$$

and we conclude that the desired root of (16), correct to thousandths, is 6.121. Refer to Fig. 8-13 for a graph involving $a', f(a'), b, f(b)$, and h'.

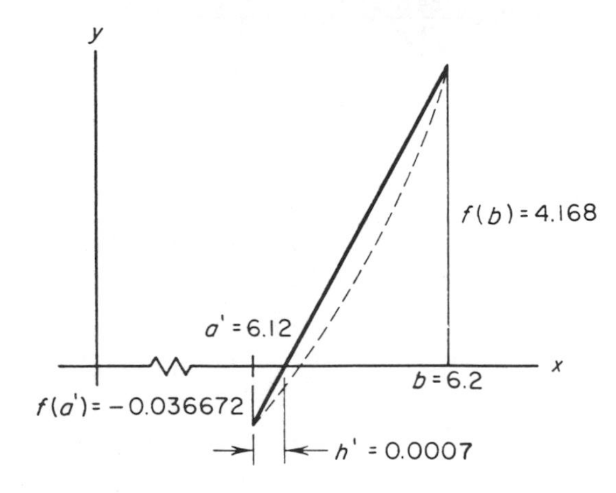

Figure 8-13

The method of linear interpolation need not converge on a root as rapidly as in Example 6. It will always converge on the root as long as the function is continuous for $a < x < b$ and $f(a)$, $f(a')$, $f(a'')$, and so on are different in sign from $f(b)$ but will converge rather slowly if the curve nearly parallels the x axis near a or b. See Fig. 8-14.

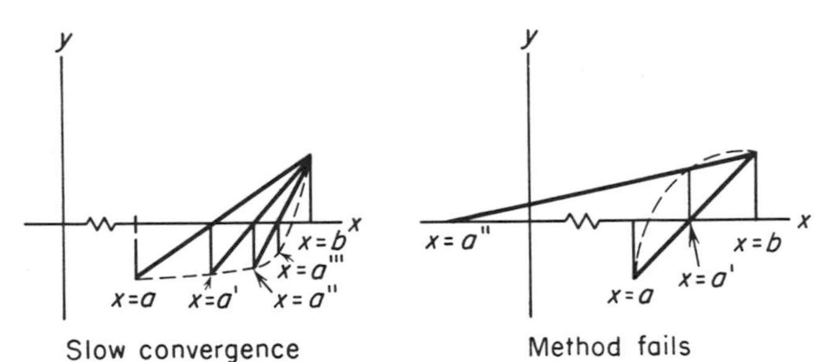

Figure 8-14

Exercises 8-3 and 8-4

In Exercises 1–15, isolate the indicated root by using procedures given in the previous section. After isolation, refine the approximation of the root correct to thousandths.

1. $y = x^3 - 4x^2 - 3x - 29$, the real root

2. $y = x^3 - 4x^2 - 3x - 28$, the real root

3. $y = x^3 - 4x^2 - 3x - 7$, the real root

4. $y = x^3 - 4x^2 - 3x - 6$, the real root

5. $y = x^3 - 4x^2 - 3x + 1$, the negative root

6. $y = x^3 - 4x^2 - 3x + 3$, the negative root

7. $y = x^3 - 4x^2 - 3x + 10$, the smaller positive root

8. $y = x^3 - 4x^2 - 3x + 11$, the larger positive root

9. $y = x^3 - 4x^2 - 3x + 20$, the real root

10. $y = x^3 - 4x^2 - 3x + 22$, the real root

11. $y = 3x^4 + x - 66$, the positive root

12. $y = 4x^3 + x - 112$, the real root

13. $x^2y = x^3 - 45$, the positive root

14. $x^2y = x^3 - 73$, the positive root

15. $y = x^5 - 100x^3 + 42{,}200$, the negative root

In Exercises 16–20, find the root indicated by using the method of linear interpolation.

16. Cube root of 12

17. Cube root of 23

18. Two fourth roots of 36

19. Two fourth roots of 50

20. Fifth root of 100

21. By adding one foot to each side of a cube, the volume of the cube is doubled. How long were the sides of the original cube?

22. The right cylindrical container shown in Fig. 8-15 has a volume of 25 gal where one gallon equals 231 in.³. Find the dimensions of the container to the nearest tenth of an inch if the height exceeds the radius by 3 in.

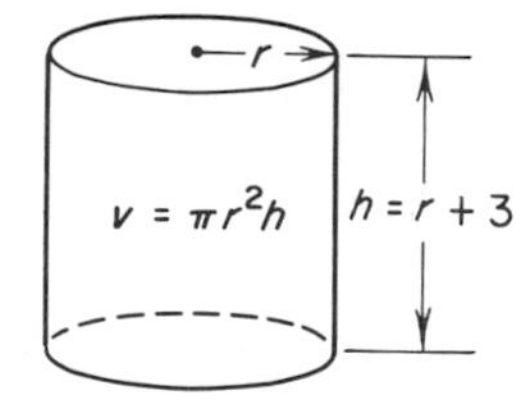

Figure 8-15

23. The deflection at P, σ_x, of a simple beam with a concentrated load at the center is given by equation

$$\sigma_x = \frac{Px}{48EI}(3l^2 - 4x^2),$$

where the situation described is pictured in Fig. 8-16. Find l if $l = 2x$, $P =$ 4000 lb, $E = 3 \times 10^7$, $I = 10.2$, and $\sigma_x = 0.02$.

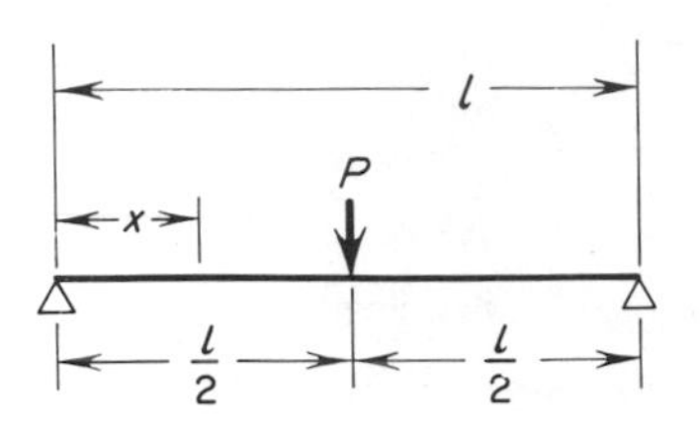

Figure 8-16

24. In Fig. 8-17 a simple beam is loaded with a load that increases uniformly from one end to the other. The deflection σ_x at some distance x from the end A is given by

$$\sigma_x = \frac{Wx}{180EIl^2}(3x^4 - 10l^2x^2 + 7l^4).$$

Find x if $W = 5000$ lb, $E = 3 \times 10^7$, $I = 11.2$, $\sigma_x = 0.03$, and $l = 100$.

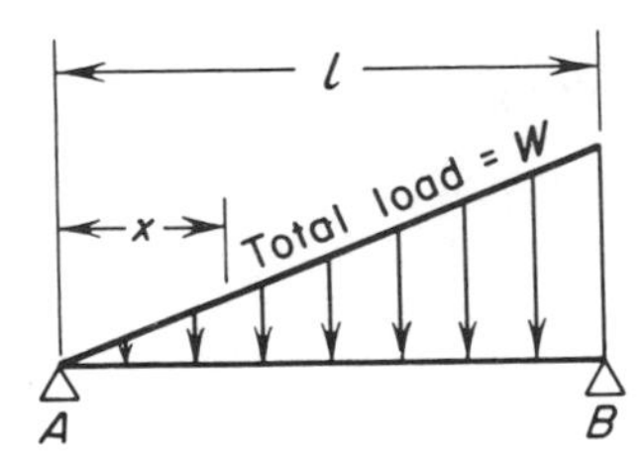

Figure 8-17

25. The value V of a certain quantity is given by the expression

$$V = 1 + x + \frac{x^2}{2!} + \frac{x^3}{3!} + \frac{x^4}{4!} + \cdots + \frac{x^n}{n!} + \cdots,$$

where $n!$ is the product of all integers from 1 through n. Using the first seven terms of the expansion on the right, find an approximate value for x if $V = 0.15$.

26. The volume of a spherical segment shown in Fig. 8-18 is given as

$$V = \pi h^2\left(r - \frac{h}{3}\right),$$

where r is the radius of the sphere and h is the thickness of the segment. A sphere of radius 10 ft contains 390 gal of water (231 in.^{3} = 1 gal). What is the greatest depth of the water, correct to hundredths?

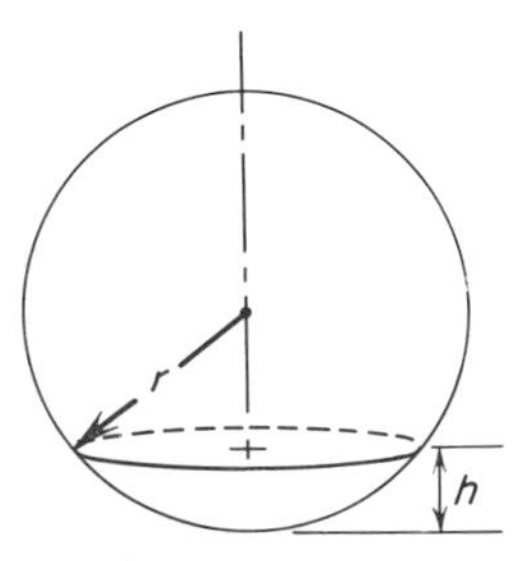

Figure 8-18

8-5 Imaginary and multiple roots

In Chapter 5 we introduced the simultaneous solutions of linear equations; we showed that a simultaneous solution of two straight lines can be interpreted as the point where the two lines intersect. In determining an x intercept (or root) of a linear equation, we can interpret this as the simultaneous solution of the given line and the x axis; in other words, the root of $ax + by = c$ can be interpreted as the solution of the system

$$ax + by = c \quad \text{and} \quad y = 0$$

where $y = 0$ is the equation of the x axis.

This same interpretation can be made regarding the roots of a quadratic; we can find the roots (x intercepts) of a quadratic by solving the system

$$y = ax^2 + bx + c \quad \text{and} \quad y = 0.$$

An important point arises here; perhaps, for the given quadratic, y cannot be zero. Yet we are, in attempting the simultaneous solution, imposing the condition that y is zero! So we must obtain no real solutions. In this case, we will obtain imaginary roots.

A theorem from the theory of equations asserts that two equations of degree m and n will have $m \cdot n$ solutions. Thus if $y = f(x)$ is of degree n and is solved simultaneously with the first-degree equation $y = 0$, there will be $n \cdot 1 = n$ solutions. We attempt to interpret all n of these solutions as being points of intersection of $y = f(x)$ with the x axis (or x intercepts), but the curve need not cross the x axis n times. Consider the three equations

$$y = x^3 - 3x^2 - x + 3, \tag{19}$$

$$y = x^3 - 3x^2 + 4, \tag{20}$$

$$y = x^3 - 3x^2 + x + 5, \tag{21}$$

which graph as shown in Fig. 8-19. In the figure we see the intersections of (19), (20), and (21) with the x axis. In Fig. 8-19(a) observe that (19) intersects the x axis in three distinct points; therefore we say that (19) has three real, distinct roots. In Fig. 8-19(b) observe that (20) intersects the x axis at $x = -1$ and is tangent to the x axis at $x = 2$; we say that (20) has three real roots with a root of multiplicity 2 at $x = 2$. In Fig. 8-19(c) we see that (21) intersects the x axis only once; therefore (21) has one real root and two imaginary roots; any effort to locate these imaginary roots on the x axis would be in vain because they do not exist.

The roots of (21) are shown by factoring as

$$x^3 - 3x^2 + x + 5 = (x + 1)(x^2 - 4x + 5) = 0,$$

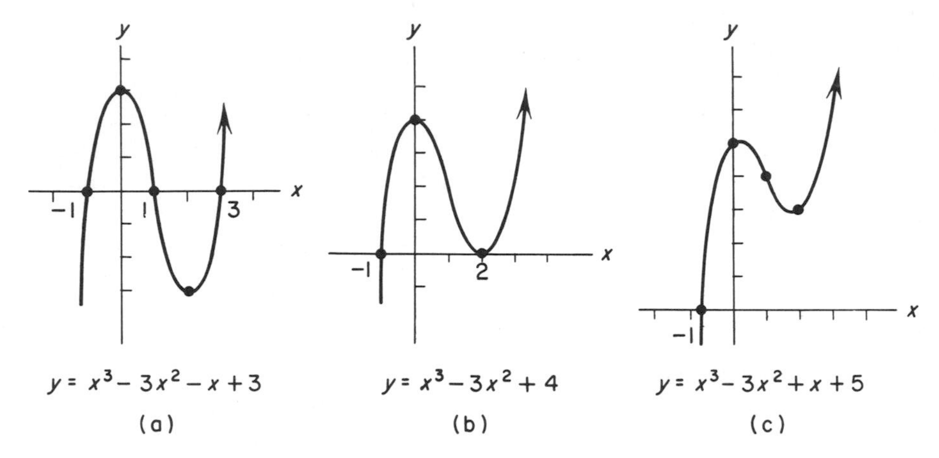

Figure 8-19

from which $x = -1$ and $x = 2 \pm j$, where the roots $x = 2 + j$ and $x = 2 - j$ are called *complex conjugates.* General complex conjugates are of the form $a + jb$ and $a - jb$. A theorem from the theory of equations asserts that imaginary roots occur in complex conjugate pairs.

The general shape of a quartic (fourth-degree polynomial equation) is shown in Fig. 8-20. Various positions of the quartic display some combinations of imaginary and real roots. Not listed is the possibility of four roots at one point; this is a distorted quartic whose shape is like a parabola.

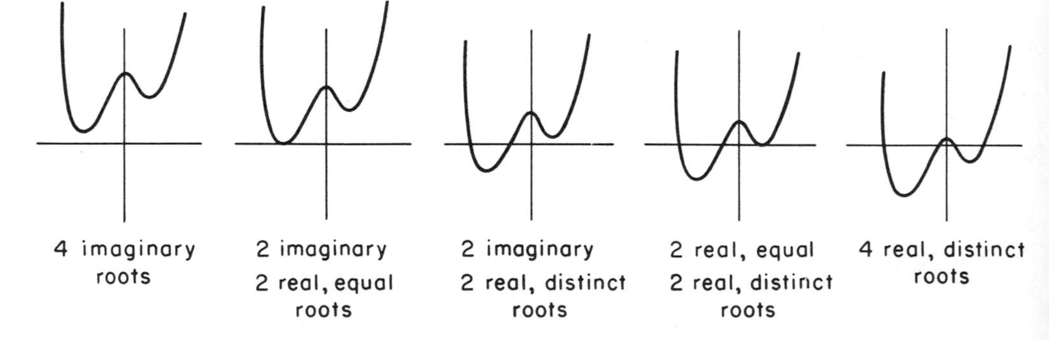

Figure 8-20

Exercises 8-5

Describe the nature of the roots of the given equations in Exercises 1–21. Graph the equation in each case.

1. $y = x^2$

2. $y = x^2 - 4$

3. $y = x^2 + 4$

4. $y = x^3$

5. $y = x^3 - 8$

6. $y = x^3 + 8$

7. $y = x^4 - 16$

8. $y = x^4 + 16$

9. $y = x^3 + 2x - 3$

10. $y = x^3 - 4x^2 - 3x + 18$

11. $y = (x - 1)^3$

12. $y = x^2 + 3x - 7$

13. $y = 5x^2 - 24x + 17$

14. $y = 3x^2 + 10$

15. $y = x^3 + x^2 - 8x + 6$

16. $y = x^2 + 25x - 25$

17. $y = x + \frac{1}{x} - 1$

18. $y = x + \frac{1}{x} - 3$

19. $y = x + \frac{1}{x} - 2$

20. $y = x^2 + 6x + 10$

21. $y = x^2 + 6x + 8$

Give the polynomial equation of form $Z = f(s)$ that has the roots given in Exercises 22–28.

22. $s = 2, -2$

23. $s = 2, -2 \pm j$

24. $s = 0, 0, 0, 0$

25. $s = \pm 3j$

26. $s = 6, 3, \pm j$

27. $s = \pm 2, 3$

28. $s = 1 \pm j, 2 \pm 3j$

9

Exponentials and logarithms

In preceding sections our discussion was limited to algebraic expressions and groups of equations that might be classified as algebraic equations. Our ultimate goal is to describe the elementary functions, which include algebraic, exponential, logarithmic, and trigonometric functions. In this chapter we describe the second and third of these functions—namely, the exponential and logarithmic functions. Exponentials are involved in, among others, such areas as investment, bacterial growth, radioactivity, and sound. They are intimately related to logarithms through a simple device that is used to convert an expression from the exponential form to the logarithm form.

9-1 Definitions; how exponential expressions arise

Frequently the question is asked "Which would you rather have, a million dollars or the amount obtained by starting on the first of the month with one cent and doubling your holdings each day for the entire month?" The answer lies in the amount obtained in the doubling operation; call the amount at any time A. If we double at the end of each day,

$$A_1 = 2 = (2)^1 \text{ at the end of the first day,}$$

$$A_2 = 4 = (2)^2 \text{ at the end of the second day,}$$

$$A_3 = 8 = (2)^3 \text{ at the end of the third day.}$$

We see that the power of 2 is the same as the number of the day; therefore

$$A_x = (2)^x \text{ at the end of the } x\text{th day.}$$

So dropping the subscript, we have the equation

$$A = (2)^x, \tag{1}$$

where A is the amount at the end of a given day and exponent x is the day number. Expression (1) is called an *exponential equation.* In answer to our original problem, if the month is considered to have 30 days in it, then $x = 30$ and (1) becomes

$$A = (2)^{30} = 1{,}073{,}741{,}824¢ = \$10{,}737{,}418.24. \tag{2}$$

The choice between the offered million and the amount shown in (2) is easy to make.

In Eq. (1) A is called the *amount* or *number*, 2 is called the *base*, and x is the *exponent*. A general expression for the *exponential equation* is

$$y = (b)^x, \tag{3}$$

where y is the number, b is the base, and x is the exponent. We note in (3) that x and y are variables and the base is constant. The fact that we have a variable exponent with a constant base is the identifying feature of the *exponential form*. This situation is in opposition to the *power form*, which has a variable base and a constant exponent, such as $y = x^2$.

The base in (1) can be any constant, including both negative and positive choices. If b is positive and other than 1, then exponent x can take on an alternative name, *logarithm*, which is studied in Sec. 9-4.

Another illustration that shows how exponentials arise is the case of compound interest. Let us start with the assumption that \$5000 is invested in a financial program that guarantees 6% interest compounded annually. This means that the amount in the account at the end of any year is 1.06

times the amount at the beginning of that year. Thus amount A is

$$
\begin{aligned}
A &= 5000(1.06) \text{ at the end of year 1,} \\
A &= 5000(1.06)^2 \text{ at the end of year 2,} \\
A &= 5000(1.06)^3 \text{ at the end of year 3,} \\
A &= 5000(1.06)^n \text{ at the end of year } n.
\end{aligned}
\tag{4}
$$

So the expression in (4) is a modification of (3) with the base a decimal quantity. In general, for compound interest, if a principal P is compounded annually at a yearly interest rate i, then amount A is given by

$$A = P(1 + i)^n \tag{5}$$

where n is the number of years.

If the interest rate in (5) is considered negative, we have *depreciation*. Thus if the value of machinery is depreciating at 10% per year, this means that the value of the machinery at the end of a given year is 90% of its value at the beginning of that year and (5) becomes

$$A = P(1 - i)^n.$$

Being able to formulate exponentials is important. Exercises on formulation and definitions follow.

Exercises 9-1

1. Evaluate 3^x for $x = 2, 3, 4$.
2. Evaluate $(\frac{1}{3})^x$ for $x = 2, 3, 4$.
3. Evaluate 3^{-x} for $x = 1, 2, 3$.
4. Evaluate $(\frac{1}{3})^{-x}$ for $x = 1, 2, 3$.
5. Evaluate $(2^x)^2$ for $x = 3$.
6. Evaluate 2^{x^2} for $x = 3$.
7. Evaluate 0.5^x for $x = 2, 3, 4$.
8. Evaluate $(1 - 0.1)^x$ for $x = 1, 2, 3$.
9. Evaluate $(1 + 0.1)^x$ for $x = 3$.
10. Evaluate 9^x for $x = 1.5, 2.5$.
11. A savings bank offers a compound interest rate of $3\frac{1}{2}$% yearly with the compounding taking place at the end of each year. Give an expression showing the amount to which an initial account of $150 accumulates in 6 years.
12. A radioactive material decays in such a way that it loses one-half its radioactive component every 7.8 years. How many years are required for 32 g to decay to 2 g?

13. Assume that the population of a bacteria culture doubles every 50 hours. A culture of 5000 units will be how large at the end of 350 hours?

14. Assume that the atmospheric pressure at sea level measures 30 in. of mercury. If each increase in altitude of one kilometer is accompanied by a 14% fall in the pressure, what is the predicted pressure at a height of 3 km?

15. Show that $a^{-x} = b^x$, where $a = 1/b$. This asserts that reciprocating the base changes the sign of the exponent.

16. Which of the given forms are exponential as opposed to the power form?

(a) $W = \eta\beta^{1.6}$ (b) $k = (I/A)^{1/2}$
(c) $A = 100(1.05)^i$ (d) $\delta = Wl^3/15EI$
(e) $i = I(1 - e^{-kt})$ (f) $y = (3.05)^{-2x}$

17. A man starts stepping toward a wall that is initially 10 ft away. If his steps are such that each step covers one-half the distance remaining to the wall, what is the size of his fifth step? What is the equation relating d, the length of a given step, and n, the step number?

9-2 Graphs of exponential equations

Graphs of exponentials display growth or decay of the dependent variable. From the general exponential form

$$y = ab^x, \tag{6}$$

x is the *independent* variable and y is the *dependent* variable. They are so named because assignment of arbitrarily chosen values of x imposes values on y. Let us graph an equation of the form of (6) by plotting arbitrarily chosen points.

Example 1 Graph equation

$$y = 10\left(\frac{1}{2}\right)^x. \tag{7}$$

Solution Let us choose several values of the independent variable x and compute the corresponding values of y, arranging them as follows:

x	-3	-2	-1	0	1	2	3	4	5
y	80	40	20	10	5	$\frac{5}{2}$	$\frac{5}{4}$	$\frac{5}{8}$	$\frac{5}{16}$

Note from the chart that when x increases by unity, the value of y is multiplied by 1/2; this factor is the base in (7) and is often called the *decrement* or rate of decay for a decreasing exponential like (7). The graph of (7) is shown in Fig. 9-1, where the axes are intentionally drawn

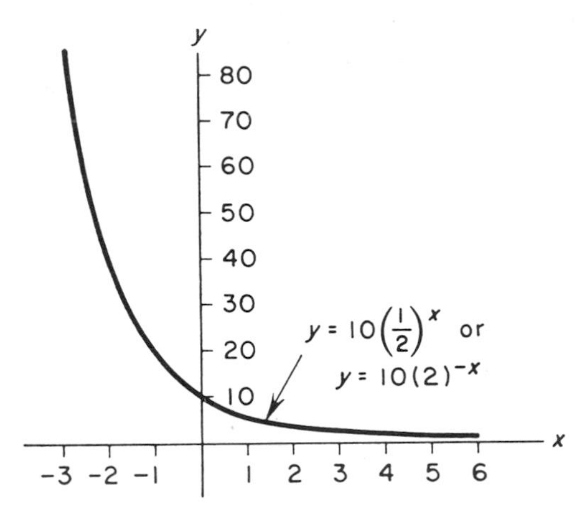

Figure 9-1

to different scales to show the complete range of values drawn from the chart. Note that as x increases to large positive values, y becomes very small, suggesting that the curve approaches the x axis but will never reach it. This means that the x axis is a limiting line or an *asymptote*.

Example 2 Graph equation

$$y = 10(2)^{-x}. \tag{8}$$

Solution Before attempting to graph (8), let us manipulate the expression. First, we know that

$$2 = \left(\frac{1}{2}\right)^{-1}.$$

Therefore

$$(2)^{-x} = \left[\left(\frac{1}{2}\right)^{-1}\right]^{-x} = \left(\frac{1}{2}\right)^{x}$$

by a law of exponents. Then, by substitution, (8) can be rewritten

$$y = 10(2)^{-x} = 10\left(\frac{1}{2}\right)^{x}.$$

Now (8) and (7) are identical and the graph of (8) is exactly as shown in Fig. 9-1.

We have demonstrated that reciprocating the base of an exponential form changes the sign of the exponent. So for the exponential (6),

$$y = a(b)^{x} = a\left(\frac{1}{b}\right)^{-x}.$$

Examining (6) for its properties, we see that

1. If $0 < b < 1$, y decreases as x increases and the curve is shaped as (a) in Fig. 9-2.
2. If $b > 1$, y increases as x increases and the curve is shaped as (b) in Fig. 9-2.
3. If $b = 1$, $y = a$ for all x as in (c) in Fig. 9-2.
4. In all cases, the y intercept is $y = a$.

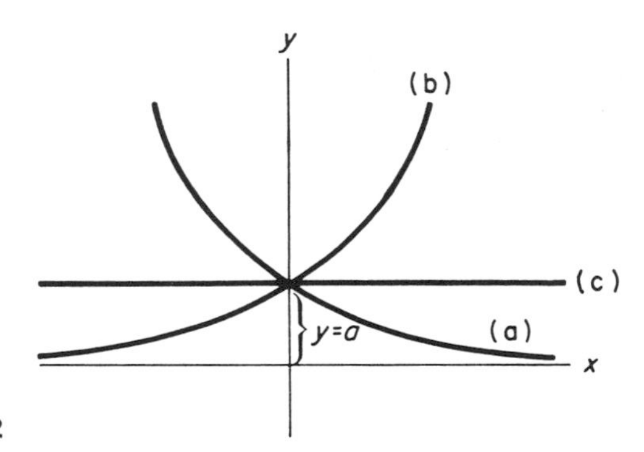

Figure 9-2

Consider another graph that is particularly useful in electricity. Refer to Example 3.

Example 3 Graph equation

$$y = k(1 - b^{-x}) \qquad (b > 1). \tag{9}$$

Solution Equation (9) can be written

$$y = k - kb^{-x}. \tag{10}$$

If we graph $y_1 = k$ and $y_2 = kb^{-x}$ as shown by the dotted lines in Fig. 9-3, we see that the desired y in (10) is the difference $y_1 - y_2$. Here y is

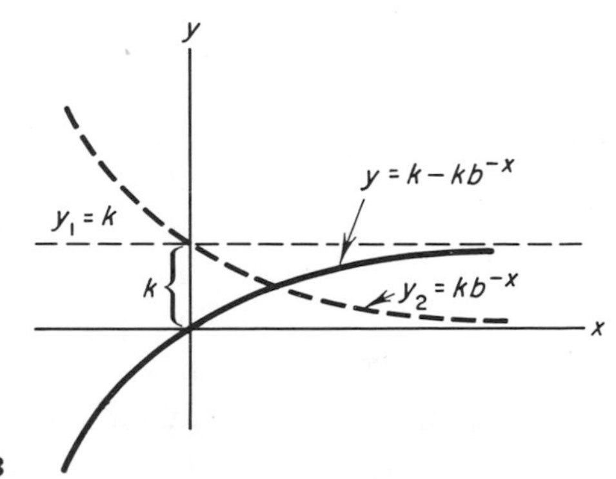

Figure 9-3

obtained by subtracting from a fixed $y_1 = k$ an ever-diminishing quantity obtained from $y_2 = kb^{-x}$. Thus the greater the value of x, the closer the value of y in (10) is to the fixed value $y_1 = k$.

Exponential equations are easily formulated from given data. It is useful (for simplicity) if the given values of one of the variables are uniformly spaced. Consider an example.

Example 4 From the data given in the accompanying chart, formulate an equation whose graph will pass through the given points.

x	-2	-1	0	1	2	3	4
y	81	27	9	3	1	$\frac{1}{3}$	$\frac{1}{9}$

Solution On inspecting the given data, we see that a unit increase in x is accompanied by a decrease in y that is a factor 1/3. So continual multiples by 1/3 are appropriate and an exponential with base 1/3 is suggested. Also, when $x = 0$, $y = 9$, suggesting a constant factor 9. Combining this information, we have equation

$$y = 9\left(\frac{1}{3}\right)^x. \tag{11}$$

Applying values of x drawn from the given chart, we see that each (x, y) pair satisfies (11).

In the more general form, if we are given (x, y) pairs (x_1, y_1), (x_2, y_2), (x_3, y_3), . . . , (x_i, y_i), and the x values are uniformly spaced, then

$$\frac{y_2}{y_1} = \frac{y_3}{y_2} = \frac{y_4}{y_3} = \cdots = \frac{y_i}{y_{i-1}} = b,$$

where b is the base in the form

$$y = ab^x.$$

Exercises 9-2

In Exercises 1–15, show the graphs of the given equations. When more than one literal quantity is present, the variables are indicated at the right.

1. $y = 2^x$

2. $y = 2^{-x}$

3. $y = -2^x$

4. $y = -2^{-x}$

5. $x = 2^y$

6. $x = 2^{-y}$

7. $x = -2^y$

8. $x = -2^{-y}$

9. $y = 1 - (3)^{-x}$

10. $y = 4(1 - 2^{-x})$

11. $p = p_0(0.86)^h$ (p, h)

12. $A = P(1.06)^n$ (A, n)

13. $A = P(0.94)^n$ (A, n)

14. $i = I(1 - e^{-t})$ $(i, t; e = 2.7)$

15. $y = \sqrt{64}$

In Exercises 16–20, develop the equation of an exponential passing through the given (x, y) points.

16.

x	0	1	2	3	4	5
y	4.4	2.2	1.1	0.55	0.275	0.1375

17.

x	0	$\frac{1}{3}$	$\frac{2}{3}$	1	$\frac{4}{3}$	$\frac{5}{3}$
y	4.4	2.2	1.1	0.55	0.275	0.1375

18.

x	0	−2	−4	−6	−8	−10
y	4.4	2.2	1.1	0.55	0.275	0.1375

19.

x	1.5	2.5	3.5	4.5	5.5	6.5
y	4.4	2.2	1.1	0.55	0.275	0.1375

20.

x	−3	−1	1	3	5	7
y	4.4	2.2	1.1	0.55	0.275	0.1375

21. A financial amount A is such that at the end of each year it is 96% of its amount at the beginning of the same year. Give an expression for A if the initial amount was P and the number of years is n. Give an expression for the amount by which P has been depleted in n years.

22. In compound interest problems the amount is usually given by equation

$$A = P(1 + i)^n, \tag{12}$$

where i is the yearly interest rate and n is the number of years. In many cases, compounding is done more than once yearly, say m times per year. In these cases, the interest rate is i/m and the exponent is mn so that (12) becomes

$$A = P\left(1 + \frac{i}{m}\right)^{mn}.$$

Given $P = 1$, $i = 0.04$, and $n = 1$, approximate values for A with $m = 1, 2, 4, 80, 1600$. Does it seem that as m becomes large, A approaches 0 or 1 or some other number?

23. In simple interest problems interest is computed yearly on only the *original* principal. Thus in 5 years at rate 3%, with a principal of \$100, the interest accumulated is

$$(\$100)(0.03)(5) = \$15$$

and the amount is

$$A = \$100 + \$15 = \$115.$$

Formulate an expression for amount A derived from principal P if the simple interest rate is i and is applied for n years. Compare this expression to (12). Graph both on the same set of axes. Is the simple interest formula an exponential?

9-3 Changing bases, base *e*, base 10

In preceding sections we applied the exponential to a variety of bases, including integral and fractional numbers. By using a table of logarithms or a slide rule containing LL scales, bases can easily be interchanged. We will postpone to a later section those changes that require tables; in this section we discuss, with certain exceptions, only those base changes that require no special tables. Change of bases is especially easy if the old base is some simple power of the new base; we say some *simple* power because any positive number can be expressed as a power of another positive number; however, the powers involved are often difficult to obtain.

Example 5 Express (a) $(\frac{1}{2})^{3x}$, (b) $4^{-0.6x}$, (c) $64^{-x/4}$ as exponentials, using base 8.

Solution In all cases, we must determine the power of 8 that is the equivalent of the given base.

(a) We want to determine what power of 8 is of value $\frac{1}{2}$. Therefore we are asking to find k, where

$$8^k = \frac{1}{2}. \tag{13}$$

We can rewrite (13) in such a way as to express both sides as powers of the same base. The convenient new base is base 2; thus

$$8^k = (2^3)^k = (2)^{3k} \tag{14}$$

and

$$\frac{1}{2} = (2)^{-1}. \tag{15}$$

Equating the last two members of (14) and (15), we have

$$2^{3k} = 2^{-1}. \tag{16}$$

Then, with the bases in (16) equal, the exponents are also equal and

$$3k = -1$$

from which

$$k = -\frac{1}{3}.$$

Substituting $k = -\frac{1}{3}$ into (13) gives

$$(8)^{-1/3} = \frac{1}{2}.$$

Replacing base $\frac{1}{2}$ by the number $(8)^{-1/3}$ in the given expression $(\frac{1}{2})^{3x}$, we have

$$\left(\frac{1}{2}\right)^{3x} = (8^{-1/3})^{3x} = 8^{-x}, \tag{17}$$

where (17) shows the conversion of the given expression from base $\frac{1}{2}$ to base 8.

(b) If we proceed as in part (a), we must determine what power of 8 the number 4 is. Thus

$$8^k = 4$$

from which $2^{3k} = 2^2$ and $k = \frac{2}{3}$. Now $4 = 8^{2/3}$ and

$$4^{-0.6x} = (8^{2/3})^{-0.6x} = 8^{-0.4x}.$$

(c) To find the power of 8 to which $64^{-x/4}$ corresponds, we need not go through the intermediate base 2 as in parts (a) and (b). Since $64 = 8^2$, then

$$64^{-(1/4)x} = (8^2)^{-(1/4)x} = 8^{-(1/2)x}.$$

Example 6 Given

$$2 = 10^{0.3010}, \qquad 3 = 10^{0.4771}, \qquad 5 = 10^{0.6990},$$

express the following numbers as powers of 10: (a) 16, (b) 15, (c) $\sqrt{6}$, (d) $\sqrt[3]{30}$.

Solution In each case, we will convert the given number to a power of 10 through use of one or more of the powers of 10 given.

(a) $16 = 2^4 = (10^{0.3010})^4 = 10^{1.2040}$

(b) $15 = 3 \cdot 5 = (10^{0.4771})(10^{0.6990}) = 10^{1.1761}$

(c) $\sqrt{6} = (2 \cdot 3)^{1/2} = (10^{0.3010} \cdot 10^{0.4771})^{1/2} = 10^{0.7781/2} = 10^{0.3891}$

(d) $\sqrt[3]{30} = (2 \cdot 3 \cdot 5)^{1/3} = (10^{0.3010} \cdot 10^{0.4771} \cdot 10^{0.6990})^{1/3}$
$= 10^{1.4771/3} = 10^{0.4924}$

In Example 6 a group of numbers was converted to base 10. In computations with common logarithms, base 10 is used throughout. Tabular entries in the table of common logarithms simply indicate the power of 10 that represents the number in question. The number 10, familiar to all of us, is called the *common base*. Computations are also made using base e, where e is called the *natural base* and is a rather strange number.

The number e is defined as a limit; it is the limit of the quantity $(1 + x)^{1/x}$ as x goes to zero, which is written

$$e = \lim_{x \to 0} (1 + x)^{1/x}. \tag{18}$$

Inspection of (18) often results in two incorrect points of view. One person views the base $1 + x$, remarks that the base goes to unity as x goes to zero and that unity raised to any power is unity, suggesting that e has the value unity. Another person views the exponent $1/x$ and remarks that as $x \longrightarrow 0$, $1/x \longrightarrow \infty$, suggesting that e is infinitely large. Neither point of view is correct. If we substitute successively smaller values of x into (18), say $x = 0.1$, $x = 0.001$, $x = 0.00001$, and so forth, e appears to converge on the value $e = 2.718\ldots$; this is the value of e that we will accept without proof.

Exercises 9-3

In Exercises 1–10, convert the given expression into another expression in the base indicated in parentheses.

1. 9^{2x} (3) **2.** $16^{3.5x}$ (2) **3.** 2^{-8x} (16)

4. $3^{-0.3x}$ (27) **5.** 32^{-2y} (8) **6.** $64^{-0.1y}$ (2)

7. $0.3^{7.2y}$ (0.09) **8.** $2500^{-1.26t}$ (50) **9.** 8^{2t-3} (16)

10. 16^{t+4} (8)

In Exercises 11–30, given the values

$$2 = 10^{0.3010}, \quad 5 = 10^{0.6990}, \quad 10 = e^{2.3026}$$

$$3 = 10^{0.4771}, \quad e = 10^{0.4343}, \quad \pi = 10^{0.4971}$$

express the given numbers as powers of the base indicated in parentheses.

11. 2π (10) **12.** 5π (10) **13.** 50 (10)

14. 15 (10) **15.** 9 (10) **16.** 25 (10)

17. 36 (10) **18.** 8 (10) **19.** 15,000 (10)

20. 810 (10) **21.** $\frac{1}{3}$ (10) **22.** $\frac{1}{2}$ (10)

23. 0.6 (10) **24.** $\frac{2}{3}$ (10) **25.** 100 (e)

26. 1000 (e) **27.** $\sqrt[3]{10}$ (e) **28.** $\sqrt{10}$ (e)

29. $\dfrac{10\pi}{e}$ (10) **30.** $12\pi^2$ (10)

In Exercises 11–30, the operations involved adding, subtracting, doubling, halving, etc., the given exponents. In view of those operations, provide answers to the questions in Exercises 31–35.

31. If we multiply two numbers a and b that are in the power form, how is the exponent of the product ab related to the exponents of a and b?

32. If we divide a by b, how is the logarithm of $a \div b$ related to the logarithms of a and b? Assume base 10.

33. If we raise a to the nth power, how is the logarithm of a^n related to the logarithm of a?

34. If we take the nth root of a, how is the logarithm of $\sqrt[n]{a}$ related to the logarithm of a?

35. Given the number $N = a^2b^3/5c$, give an expression for $\log_{10} N$.

9-4 Logarithms

In preceding sections we discussed exponentials; particularly appropriate were those discussions that suggested that exponents can be called *logarithms* as long as the base being considered is a positive base other than one. Any number can be expressed as a power of a base; for example, the number 64 can be expressed as $(2)^6$, or

$$64 = (2)^6. \tag{19}$$

Here 64 is the *number*, 2 is the *base*, and 6 is the exponent or *logarithm*. Stated in the inverse form, "6 is the logarithm to base 2 of number 64", and we write

$$6 = \log_2 64. \tag{20}$$

In general, a number y can be considered the xth power of a base b. The general form of (19) can be written

$$y = b^x. \tag{21}$$

Next, (21) can be written in logarithmic form (20) where the general logarithmic form of (21) is

$$x = \log_b y. \tag{22}$$

In (21) base b is assumed to be constant, and both the logarithm x and the number y are variables. Once again the description distinguishes the exponential form from the power form.

Expressions (22) and (20) assert that a quantity is a logarithm of a second quantity with a third quantity as a base. A given number may be the logarithm of many different numbers, depending on the choice of base. Consider the case of several squares.

$$9 = 3^2 \quad \text{and} \quad 2 = \log_3 9$$
$$100 = 10^2 \quad \text{and} \quad 2 = \log_{10} 100$$
$$10 = 3.162^2 \quad \text{and} \quad 2 = \log_{3.162} 10$$

In each instance, the logarithm of the given number is 2. The numbers differ and the bases also differ.

Two bases are commonly used—base 10 and base e. Our initial discussion here concerns the *common* or *Briggsian* base 10. In each case cited, it is the base used. Table C in the back of the book is a table of common logarithms of numbers. It lists a number and its logarithm, using base 10. We emphasize here that Table C does not give the *entire* logarithm but only a part of it called the *mantissa*.

Let us examine a logarithm more closely with the objective of using Table C and analyzing the complete logarithm. First, logarithms of integral powers of 10 are integers. We have

$$\begin{aligned} 1 &= 10^0 \quad \text{and} \quad 0 = \log_{10} 1, \\ 10 &= 10^1 \quad \text{and} \quad 1 = \log_{10} 10, \\ 100 &= 10^2 \quad \text{and} \quad 2 = \log_{10} 100, \\ 1000 &= 10^3 \quad \text{and} \quad 3 = \log_{10} 1000, \\ 10{,}000 &= 10^4 \quad \text{and} \quad 4 = \log_{10} 10{,}000. \end{aligned} \tag{23}$$

From expressions (23) we see that the logarithms of positive powers of 10 are positive integers (using base 10). Similarly, from expressions (24) we note that the logarithms of negative powers of 10 are negative integers.

$$\begin{aligned} 0.1 &= \frac{1}{10} = 10^{-1} \quad \text{and} \quad -1 = \log_{10} 0.1, \\ 0.01 &= \frac{1}{100} = 10^{-2} \quad \text{and} \quad -2 = \log_{10} 0.01, \\ 0.001 &= \frac{1}{1000} = 10^{-3} \quad \text{and} \quad -3 = \log_{10} 0.001, \\ 0.0001 &= \frac{1}{10000} = 10^{-4} \quad \text{and} \quad -4 = \log_{10} 0.0001. \end{aligned} \tag{24}$$

Certain general facts are clear from (23) and (24).

1. If a number N is greater than unity, then $\log_{10} N$ is greater than zero; that is, if
$$N > 1, \quad \text{then} \quad \log_{10} N > 0.$$
2. If a number N is between 0 and 1, then $\log_{10} N$ is negative; that is, if
$$0 < N < 1, \quad \text{then} \quad \log_{10} N < 0. \tag{25}$$
3. If a number $N = 1$, then $\log_{10} N = 0$.

In the next part of our discussion we draw particular attention to (25). In expressions (23) and (24) we considered only those numbers that are integral powers of 10. Now let us consider the logarithms of some numbers that are not integral powers of base 10.

Example 7 Find $\log_{10} N$ if (a) $N = 2$, (b) $N = 20{,}000$, (c) $N = 0.002$.

Solution Referring to Table C, we find no entry in the N column listed as 2, or 20,000, or 0.002. There is a listing for a number designated by digits 200 (where the digits are *not* intended to mean the number "two hundred"). The tabular entry under 200 is 3010. This means simply that there is a number whose one significant digit, 2, is identified with a logarithm with significant digits 3010. As a result of (23) and (25), we can be more explicit.

(a) Since $N = 2$, then $0 < \log_{10} N < 1$. Actually, from Table C, we then have

$$2 = 10^{0.3010} \quad \text{or} \quad 0.3010 = \log_{10} N.$$

(b) Here $N = 20{,}000$. We can write 20,000 in standard notation as

$$20{,}000 = 2 \times 10^4.$$

In part (a) and from Table C we have

$$2 = 10^{0.3010}$$

so that

$$20{,}000 = 2 \times 10^4 = 10^{0.3010} \times 10^4 = 10^{0.3010+4}. \qquad (26)$$

The exponent (logarithm) in (26) is written in a seemingly unusual fashion. From (26) we have

$$\log_{10} 20{,}000 = 0.3010 + 4, \qquad (27)$$

where the portion 0.3010 of the logarithm was obtained from the table and is called the *mantissa* of the logarithm, and the +4 portion was obtained through standard notation. The +4 portion is called the *characteristic* of the logarithm; it was *not* obtained from the table and serves simply to indicate the position of the decimal point.

(c) Here $N = 0.002$. Written in standard notation,

$$N = 0.002 = 2 \times 10^{-3}.$$

Then

$$0.002 = 2 \times 10^{-3} = 10^{0.3010} \times 10^{-3} = 10^{0.3010-3}$$

from which

$$\log_{10} 0.002 = 0.3010 - 3.$$

Here the characteristic is the negative number -3.

It is noteworthy that any number in decimal form can be written in scientific notation. If we accept the fact that proper scientific notation requires that a number be written as the product of two parts, the first part being a quantity between 1 and 10 and the second part being a power of 10 that properly locates the decimal point, then we can make these observations:

1. The table of logarithms will provide the mantissa.
2. The characteristic *is* the exponent in the power of 10 shown in scientific notation.

Consider another example.

Example 8 Find $\log_{10} N$ if (a) $N = 987{,}000$, (b) $N = 0.00155$.

Solution

(a) First, express the number in scientific notation.

$$N = 987{,}000 = 9.87 \times 10^5$$

Then from the table of logarithms,

$$\log_{10} 9.87 = 0.9943 \quad \text{or} \quad 9.87 = 10^{0.9943}$$

and $\qquad N = 9.87 \times 10^5 = 10^{0.9943} \times 10^5 = 10^{0.9943+5}$

and we have the statement

$$\log_{10} 987000 = 0.9943 + 5$$

(b) Proceeding more rapidly with the problem of finding the logarithm, we have

$$N = 0.00155 = 1.55 \times 10^{-3} = 10^{0.1903-3}$$

and $\qquad \log_{10} 0.00155 = 0.1903 - 3.$

where the mantissa, 0.1903, was found in Table C.

Observation: In part (a) *both* the characteristic and the mantissa are positive for

$$\log_{10} 987000 = 0.9943 + 5. \tag{28}$$

It would not be confusing or improper to write (28) as

$$\log_{10} 987000 = 5.9943, \tag{29}$$

where the characteristic in (29) *precedes* the mantissa, as opposed to (28) where the characteristic follows the mantissa. We assert that interchanging the positions of the characteristic and mantissa in (28) is not confusing; however, in part (b) it may be confusing to make the interchange. Thus if

$$\log_{10} 0.00155 = 0.1903 - 3, \tag{30}$$

we have a *positive* mantissa and a *negative* characteristic. If we wrote (30) by interchanging the positions of the characteristic and mantissa, we would have

$$\log_{10} 0.00155 = -3.1903, \tag{31}$$

where in (31) it is difficult to understand that the mantissa is positive and the characteristic is negative. Therefore other conventions are adopted. Some references use the standard

$$\log_{10} 0.00155 = \bar{3}.1903$$

to show that the negative sign belongs *only* with the characteristic. A more common convention writes the characteristic in two parts, one before the mantissa and the other trailing, with the trailing portion usually the number -10. Thus since $-3 = 7 - 10$, we can write (30) as

$$\log_{10} 0.00155 = 7.1903 - 10. \tag{32}$$

Note that the leading and trailing portions of the characteristic in (32) can be *any* two numbers that add to -3; for example, $97 - 100$, $41 - 44$, and so on. Flexibility in the choice of parts will be useful later.

Exercises 9-4

In Exercises 1–20, find the logarithm of the given number. Provide both the characteristic and the mantissa. Split the characteristic into two parts with the trailing part the number -10. Use the table of common logarithms in Table C.

1. 3 **2.** 30 **3.** 3000

4. 0.3 **5.** 0.00003 **6.** 48

7. 0.48 **8.** 4850 **9.** 48.5

10. 485.0 **11.** 4.85 **12.** 9270

13. 0.00927 **14.** 7.25×10^{-5} **15.** 6.22×10^{-7}

16. 622×10^{-5} **17.** 7250×10^{-3} **18.** 10^{-5}

19. 10^{-9} **20.** 10^{3}

In Exercises 21–40, find the number whose logarithm is given. Express the number in scientific notation.

21. 0.8451 **22.** 0.7404 **23.** 3.8451

24. 4.7404 **25.** $6.8451 - 10$ **26.** $7.7404 - 10$

27. $14.8451 - 8$ **28.** $12.7404 - 7$ **29.** $0.8451 - 27$

30. $0.7404 - 21$ **31.** 4.8513 **32.** 3.1761

33. 0.8555 **34.** $0.1917 - 6$ **35.** $19.8561 - 20$

36. $9.1987 - 10$ **37.** $19.8603 - 10$ **38.** $12.2810 - 10$

39. 0.8704 **40.** 3.3979

9-5 Interpolation

In Sec. 5-3 entitled Division of a Line Segment we presented expressions (7) and (8), which give coordinates of specified points of division of a line segment. We further mentioned that division of a line segment, as presented there, is called interpolation.

We define *interpolation* here as the act of supplying an intermediate term in a series of terms. More specifically, we relate interpolation to the process of supplying numbers that are intermediate to numbers entered in a table. Since we are concerned in this chapter with logarithms, our reference is to the table of common logarithms in Table C.

This table actually presents a thousand number pairs satisfying the equation

$$N = 10^x \quad \text{or} \quad x = \log_{10} N \tag{33}$$

where N is a *number* listed to two significant figures in the N column (with the third significant figure given in an additional column) and x is the mantissa of the logarithm of the number, with x tabular; that is, it is read from the body of the table. A number N given to *four* significant figures is not included in the limited column headings provided in Table C: however, such a four-digit number can be considered to exist a fractional portion of the way through an interval whose endpoints are in the table and bracket the number in question. Similarly, the logarithm of that number is the same fractional portion of the way through the logarithm interval defined. With this notion of partial movement through an interval, we are duplicating the idea advanced in (7) and (8) of Sec. 5-3.

Equation (33) is graphed in Fig. 9-4, and Table C can be considered a table of coordinate pairs from which (33) is graphed. Two consecutive coordinate pairs taken from Table C plot as points very close to each other on the graph in Fig. 9-4. In *linear interpolation* we consider that the two nearby points in question are joined by a *straight line*. Details of the interpolation process are shown below.

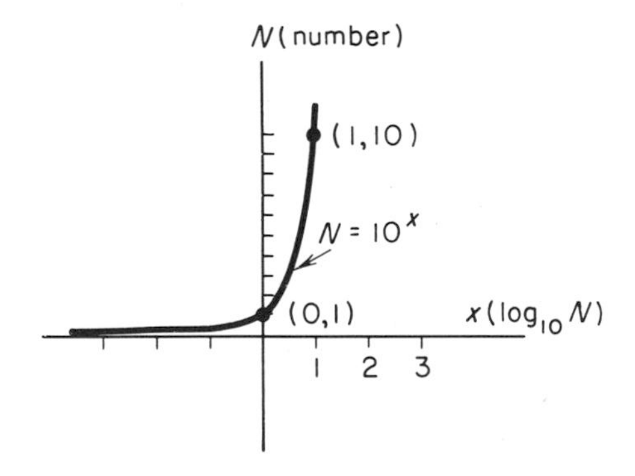

Figure 9-4

Example 9 Find log 3.526, using the table of logarithms in Table C.

Solution From Table C we see that there is no tabular entry for $N = 3.526$; however, we can read $N = 3.52$ as $N = 3.520$ and $N = 3.53$ as $N = 3.530$ for purposes of establishing the entries that immediately bracket $N = 3.526$.

Then $\log 3.520 = 0.5465$ and $\log 3.530 = 0.5478$; so

$$0.5465 < \log 3.526 < 0.5478.$$

Assuming linearity, we assert "Because 3.526 is six-tenths of the way through the 3.520–3.530 number interval, then log 3.526 is six-tenths of the way through the 0.5465–0.5478 logarithm interval." This assertion is detailed in Fig. 9-5. With the log interval being 0.0013 unit wide, six-tenths of the interval is $(0.6)(0.0013) = 0.00078$ or 0.0008 unit beyond the lower (0.5465) end of the interval, leading us to state that

$$\begin{aligned} \log 3.526 &= \log 3.520 + 0.6(\log 3.530 - \log 3.520) \\ &= 0.5465 + 0.6(0.0013) = 0.5465 + 0.00078 \\ &= 0.5465 + 0.0008 = 0.5473. \end{aligned}$$

The reasoning in the preceding paragraph and Fig. 9-5 is displayed in the block arrangement shown.

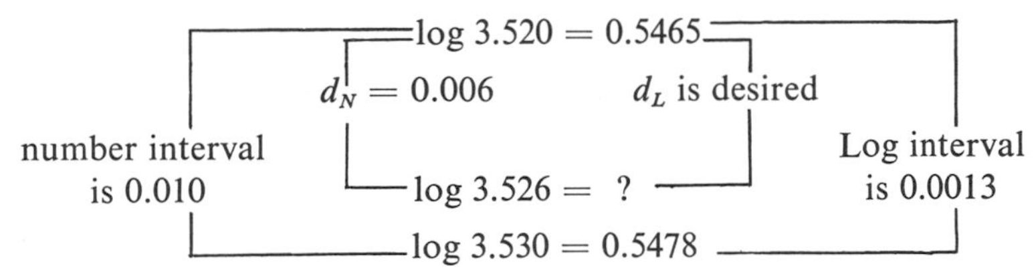

The block arrangement suggests the proportion

$$\frac{0.006}{0.010} = \frac{d_L}{0.0013}$$

from which $d_L = 0.0008$ and

$$\log 3.526 = 0.5465 + d_L = 0.5465 + 0.0008 = 0.5473.$$

Interpolation can also be used to find a number whose logarithm is given. This number is called the *antilogarithm* of the given logarithm. The method suggested is identical with that described in Example 9.

Example 10 Find N if $\log N = 3.4127$.

Solution From Table C we list the two logarithms immediately straddling the given logarithm, along with the antilogarithms. The characteristic of the logarithms is also provided in the block arrangement shown.

$\log 2590 = 3.4133$

d_N is desired $\qquad d_L = 0.0006$

Number interval is 10 $\qquad \log N = 3.4127 \qquad$ Log interval is 0.0017

$\log 2580 = 3.4116$

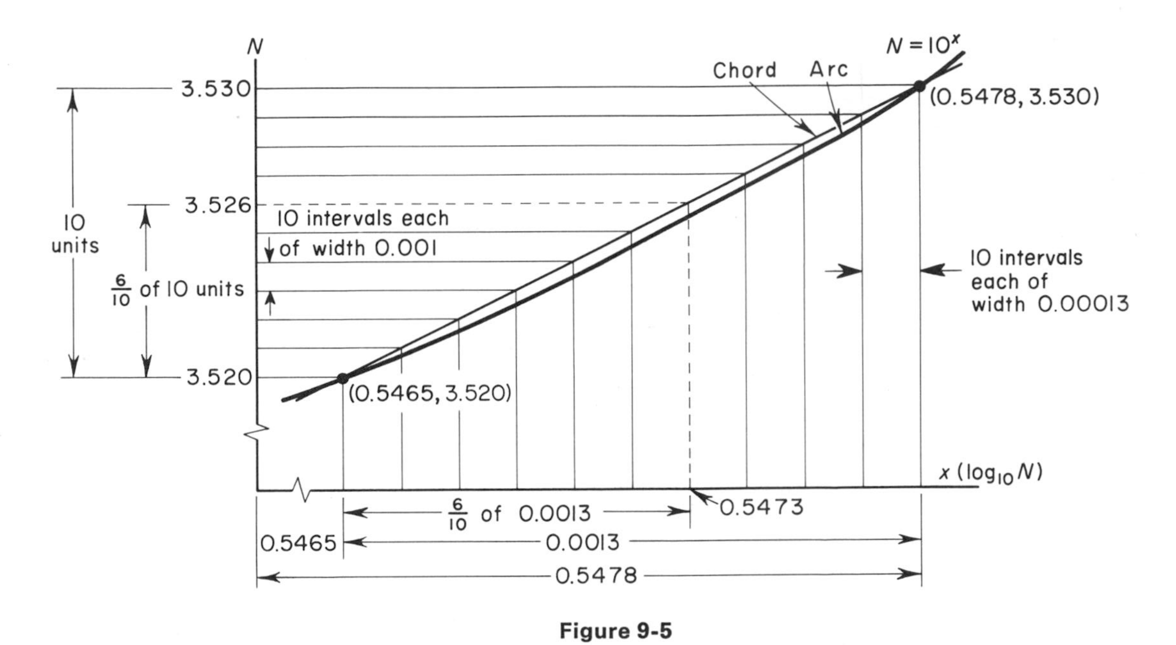
N
$N = 10^x$
Chord
Arc
3.530
(0.5478, 3.530)
3.526
10 intervals each
of width 0.001
10 units
$\frac{6}{10}$ of 10 units
3.520
(0.5465, 3.520)
10 intervals
each of
width 0.00013
x ($\log_{10} N$)
$\frac{6}{10}$ of 0.0013
0.5473
0.5465
0.0013
0.5478

Figure 9-5

From the block arrangement we have the proportion

$$\frac{d_N}{10} = \frac{0.0006}{0.0017}$$

from which $d_N = 3.53$ or 4 and antilog $3.4127 = 2590 - 4 = 2586$.

Conventions with logarithm symbols We note that base 10 was not included in the logarithm notation in Examples 9 and 10. This exclusion is due to certain popular conventions:

1. $\log N$ indicates the *common logarithm*, base 10.
2. $\ln N$ indicates the *natural logarithm*, base e.
3. $\log_b N$ indicates logarithms using the specified base b.

Summary of table usage The customary table of common logarithms gives only mantissas and restricts them to positive mantissas. The characteristic is obtained through scientific notation. Errors in the characteristic produce errors in decimal point location. Errors in the mantissa produce errors in the significant figures of the number. Although tables give only positive mantissas, many brands of calculators will give both positive and negative mantissas.

Exercises 9-5

In Exercises 1–10, find the logarithm of the given number by interpolation. Use the table of common logarithms in Table C.

1. $N = 2.504$
2. $N = 5.413$
3. $N = 30.04$
4. $N = 72.44$
5. $N = 0.03091$
6. $N = 0.2249$
7. $N = 16890$
8. $N = 84240$
9. $N = 0.0007249$
10. $N = 0.005929$

In Exercises 11–20, find the antilogarithm of the given logarithm by interpolation. Use Table C.

11. $\log N = 0.3640$
12. $\log N = 0.4825$
13. $\log N = 2.2596$
14. $\log N = 3.3728$
15. $\log N = 8.8102 - 10$
16. $\log N = 7.9222 - 10$
17. $\log N = 4.9222$
18. $\log N = 1.1003$
19. $\log N = 16.1616$
20. $\log N = 10.0003$

9-6 Multiplication and division with logarithms

From the laws of exponents we have

$$a^m \cdot a^n = a^{m+n}, \tag{34}$$

$$a^m \div a^n = a^{m-n}, \tag{35}$$

where (34) asserts that powers of like bases can be multiplied by adding the exponents and (35) asserts that powers of like bases can be divided by subtracting the exponent of the divisor from the exponent of the dividend. The table of logarithms in Table C is constructed using base 10; therefore if we are to use the table of logarithms, it would be helpful to modify (34) and (35) to read

$$10^m \cdot 10^n = 10^{m+n}, \tag{36}$$

$$10^m \div 10^n = 10^{m-n}. \tag{37}$$

In (36) and (37) the exponents m and n are logarithms that can be obtained through the table of logarithms. Consider an example.

Example 11 Find $N = 2 \times 3$ by use of logarithms.

Solution 1 Using the table of logarithms to convert numbers 2 and 3 to powers of base 10, we have

$$\begin{aligned} N = 2 \times 3 &= 10^{0.3010} \times 10^{0.4771} \\ &= 10^{0.3010+0.4771} = 10^{0.7781}. \end{aligned} \tag{38}$$

In (38) we have a statement that $N = 10^{0.7781}$. If we enter the table to find the number whose logarithm is 0.7781, we find that $N = 6$.

Solution 2 In practice, displaying the given factors as powers of 10 as in (38) is not common. We did so in (38) to show that the multiplication operation is performed by adding logarithms. In general, we realize that the base is 10 and that we are dealing with logarithms so that we simply reduce the operation to the following format:

$$\begin{array}{rll} \log 2 &= 0.3010 & \text{(from table)} \\ + \log 3 &= 0.4771 & \text{(from table)} \\ \hline \log N &= 0.7781 & \text{(addition)} \\ N &= 6.000 & \text{(antilogarithm)} \end{array}$$

Example 12 Find N if $N = 6.832 \div 0.4626$.

Solution Here we have division, which calls for the subtraction of logarithms. Using the format shown in Example 11, we have

$$\begin{array}{rll} \log 6.832 &= 10.8345 - 10 & \text{(table, interpolation)} \\ - \log 0.4626 &= 9.6652 - 10 & \text{(table, interpolation)} \\ \hline \log N &= 1.1693 & \text{(subtraction)} \\ N &= 14.77 & \end{array}$$

It is important to note in Example 12 that we resorted to the device of breaking the characteristic into two parts. This step is especially necessary if the divisor is greater than the dividend. We may have to resort to some unusual breakdowns of the characteristic when necessary.

Combined operations involving multiplication and division require expansion of our format somewhat. Such operations are mere extensions of simple multiplications or divisions. The following example calls for combined operations and some juggling of the characteristic.

Example 13 Find N if

$$N = \frac{0.325 \times 4.695}{15.46 \times 0.673 \times 8.882}.$$

Solution Here we have the quotient of products; by usual division we have

$$\log N = \log S - \log T,$$

where S is the numerator and T is the denominator, or

$$\log N = (\log 0.325 + \log 4.695) - (\log 15.46 + \log 0.673 + \log 8.882).$$

Introducing a format, we have

$$\begin{array}{rl} \log 0.325 & = 29.5119 - 30 \\ + \log 4.695 & = 10.6716 - 10 \\ \hline \log S & = 40.1835 - 40 \\ - \log T & = 31.9657 - 30 \\ \hline \log N & = 8.2178 - 10 \\ N & = 0.01651 \end{array}$$

$$\begin{array}{rl} \log 15.46 & = 11.1892 - 10 \\ + \log 0.673 & = 9.8280 - 10 \\ + \log 8.882 & = 10.9485 - 10 \\ \hline \log T & = 31.9657 - 30 \end{array}$$

Note in Example 13 that the characteristic of log 0.325 was chosen as 29 minus 30. This made it possible to have the leading part of the characteristic of the logarithm of the numerator greater than the leading part of the characteristic of the denominator, enabling the subtraction of the logarithms of the numerator and denominator. We chose to do the juggling with log 0.325; actually, this juggling could have been done at other points in the solution without loss of accuracy.

At this point we introduce two laws of logarithms:

$$\log_b MN = \log_b M + \log_b N, \tag{39}$$

$$\log_b \frac{M}{N} = \log_b M - \log_b N. \tag{40}$$

We applied laws (39) and (40) in Examples 11, 12, and 13. The laws are shown here for generality.

Exercises 9-6

Evaluate the expressions in Exercises 1–25 by using the table of common logarithms in Table C. Express results correct to accuracy that can be achieved through interpolation.

1. 23.42×653.2

2. 14.29×185.3

3. 692×0.004276

4. 312.2×0.009293

5. 0.09295×0.006293

6. 0.008325×0.008526

7. $613.5 \div 4.952$

8. $900 \div 2.600$

9. $\pi \div 57.3$

10. $3.098 \div 292.1$

11. $0.004942 \div \pi$

12. $0.008295 \div 0.07967$

13. $\dfrac{315.2 \times 404.2}{1859}$

14. $\dfrac{418.1 \times 3.462}{17.15 \times 255.6}$

15. $\dfrac{96.66 \times 4.247}{3.919 \times 3.142}$

16. $1 \div 792.6$

17. $1 \div 3.982$

18. 6344×0.1234

19. $920.2 \div 2.476$

20. 0.006265×12.27

21. $\dfrac{157.6 \times 60.63}{92.96}$

22. $\dfrac{727.5 \times 1.469}{20.26}$

23. $0.9884 \div \pi$

24. $4.682 \div 0.03139$

25. 979.8×46.85

26. The density of mercury is given as 13.546 grams per cubic centimeter. There are 16.38 cubic centimeters per cubic inch. What is the weight of a column of mercury filling a right circular cylinder with a base of area 0.04255 in.^{2} and height 3.295 in.?

27. Give the logarithm for the reciprocal of π.

28. There is one dyne of gravity force acting on 1.0197×10^{-3} gram of mass at the earth's surface. Find the number of grams that has a weight of one poundal if the poundal is 1.3825×10^{-4} dyne.

29. Find the area of a trapezoid with bases of lengths $b_1 = 3.2952$ in. and $b_2 = 5.6271$ in. and altitude $h = 2.7919$ in. if area A is

$$A = \frac{h}{2}(b_1 + b_2).$$

30. The location of the centroid of a trapezoid is given by equation

$$\bar{h} = \frac{h(2b_1 + b_2)}{3(b_1 + b_2)},$$

where $\bar{h}$ is measured from b_2. Find $\bar{h}$ for the trapezoid described in Exercise 29.

31. A point is on the rim of a wheel 14.37 in. from the center of the wheel. If the wheel rotates at the rate of 218 revolutions each 32.5 sec, find the speed of the point in inches per second.

32. By using logarithms, find the approximate value of 10!, where 10! (ten factorial) is the product of all the integers from 1 through 10.

33. The pH value of a substance is defined as the negative of the logarithm of the hydrogen ion concentration. If a substance is such that only one part in one hundred million is ionized hydrogen, what is the pH value?

9-7 Powers and roots

The definition of a power of a number combines with the law of multiplication by use of logarithms to present a method of computing powers of numbers by logarithms. If

$$N = b^k \qquad (k \text{ an integer}, b \neq 1),$$

then by the definition of a power,

$$N = b \cdot b \cdot b \cdot \cdots \cdot b \qquad (k \text{ factors})$$

and by the law of multiplication by logarithms in the preceding section,

$$\log N = \log b + \log b + \log b + \cdots + \log b \qquad (k \text{ addends})$$

or $\quad \log N = k \log b.$

Thus we have a law of logarithms of powers

$$\log_b M^n = n \log_b M. \tag{41}$$

Law (41) can be worded in a specific example as "the logarithm of the cube of a number is three times the logarithm of the number." Examples are

$$\log_{10} 5^2 = 2 \log_{10} 5,$$
$$\log_{10} (0.0034)^{1.25} = 1.25 \log_{10} 0.0034,$$
$$\log_7 (4.25)^{-2.3} = -2.3 \log_7 4.25.$$

Example 14 Find N if $N = (0.8325)^4$.

Solution Using (41) to express the given quantity as an equation of logarithms, we have

$$\log N = 4 \log 0.8325.$$

Using a format for computations,

$$\begin{aligned} \log 0.8325 &= 9.9204 - 10 \\ & \underline{\phantom{39.6816 -{}} \times\ 4} \\ \log N &= 39.6816 - 40 \\ N &= 0.4803. \end{aligned}$$

The preceding example illustrated a case in which the exponent was integral. A slight bit of juggling of the characteristic is useful if the exponent is not integral. The trailing portion of the characteristic is wisely chosen in such a way that the product of that portion of the characteristic and the exponent is an integer; although not necessary, it often has a simplifying effect. Consider Example 15.

Example 15 Evaluate N if $N = (0.1265)^{3.12}$.

Solution Since the exponent contains a decimal to hundredths, the trailing portion of the characteristic is wisely chosen as 100; so from

$$\log N = 3.12 \log 0.1265,$$

we have the format

$$\begin{aligned} \log 0.1265 &= 99.1021 - 100 \\ & \qquad\qquad \times\ 3.12 \\ \hline \log N &= 309.1985 - 312 \\ N &= 1.580 \times 10^{-3} \end{aligned}$$

It is interesting to note, at the risk of confusion, that the product of 99.1021 by 3.12 in Example 15 could have been obtained by the use of logarithms, adding log 99.1021 and log 3.12. The sum would not have been $\log N$ but would have been log 309.1985 (shortened by the limitations of the table of logarithms). The antilogarithm of log 309.1985 would be $\log N$, which, in turn, is such that the antilogarithm of $\log N$ is N. In the preceding discussion the location of the decimal point in N is ignored.

Roots by logarithms can be computed in the same manner as powers if desired. The indicated root can be expressed as a power. Thus if

$$N = \sqrt[4]{3.26},$$

then

$$N = (3.26)^{1/4} = (3.26)^{0.25}$$

and we can apply (41). As an alternative, we can use the law

$$\log_b \sqrt[n]{M} = \frac{1}{n} \log_b M. \tag{42}$$

In applying (42), it is useful to have the trailing portion of the characteristic divisible by n with no remainder.

Example 16 Evaluate $N = \sqrt[6]{0.009642}$.

Solution By (42), we have

$$\log N = \frac{1}{6} \log 0.009642$$

$$= \frac{1}{6}(3.9842 - 6)$$

$$= 0.66403 - 1,$$

$$N = 0.4613.$$

Combined operations involving roots and powers are performed by selecting appropriate laws among (39), (40), (41), and (42). Thus if

$$N = \sqrt[3]{\frac{a^2b}{5cd^4}},$$

then

$$\log N = \log \left(\frac{a^2b}{5cd^4}\right)^{1/3}$$

and

$$\begin{aligned} \log N &= \frac{1}{3}(\log a^2b - \log 5cd^4) \\ &= \frac{1}{3}(2 \log a + \log b - \log 5 - \log c - 4 \log d). \end{aligned} \tag{43}$$

It is often useful to express a problem in logarithmic form (43). It is also useful to establish a format that will provide an organized plan for the sequence of operations, showing the intended steps of addition, subtraction, and so on. The advantages of a format are orderliness and the fact that once the format is completely built, there is no longer reason to switch back and forth between the problem and the table of logarithms. Such switching is often confusing and tends toward disorganization.

Exercises 9-7

Evaluate the expressions in Exercises 1–26. Use the table of logarithms in Table C.

1. $(9.826)^2$

2. $(4.735)^2$

3. $(0.1629)^3$

4. $(0.02404)^3$

5. $(4.913)^2$

6. $(8.219)^3$

7. $(3.276 \times 0.0907)^2$

8. $(0.01345 \times 0.0615)^2$

9. $(3.258)^3$

10. $(5.417 \times 0.196)^2$

11. $\sqrt{819.4}$

12. $\sqrt{1006.4}$

13. $\sqrt[3]{1.842}$

14. $\sqrt[3]{76.68}$

15. $\sqrt[4]{0.0003294}$

16. $\sqrt[5]{0.08275}$

17. $\sqrt{204.85}$

18. $\sqrt[6]{4096}$

19. $\sqrt{\dfrac{94.05}{275.47}}$

20. $\left(\dfrac{0.0615}{57.53}\right)^2$

21. $(2.166\pi)^2$

22. $\left(\dfrac{57.53}{0.0615}\right)^2$

23. $\dfrac{31.35}{826.4}$

24. $\sqrt{\dfrac{9.254}{10.985}}$

25. $\left(\dfrac{43.26 \times 125.67}{2.14 \times 6.285}\right)^2$

26. $\left(\dfrac{99.04}{34.62 \times 0.2163}\right)^3$

27. Given a circle of radius 6.245, find
(a) circumference C.
(b) area A.
(c) moment of inertia I_A about one of the diameters if $I_A = \pi r^4/4$.

28. Using logarithms, find $\sqrt{7}$ to four figures.

29. The moment of inertia I_A of an annulus (ring) about a diameter is given as

$$I_A = \pi \frac{r_1^4 - r_2^4}{4},$$

where r_2 is the inside radius and r_1 is the outside radius. Find I_A if $r_2 = 1.356$ and $r_1 = 2.138$.

30. Using the compound interest law $A = P(1 + i)^n$, find A if $P = \$9500$, $n = 20$, and $i = 0.055$.

31. Find the volume of a sphere of radius $r = 1.254$ in.

32. What is the characteristic of $\log N$ if (a) $N = 1 \times 10^{-27}$? and (b) $N = 1.62 \times 10^{-27}$?

33. By using logarithms, find the hypotenuse of the right triangle with legs 3.406 in. and 5.866 in.

34. Find CB in Fig. 9-6 if $AB = 1.495$, $EB = 0.8925$, and $DC = 1.666$.

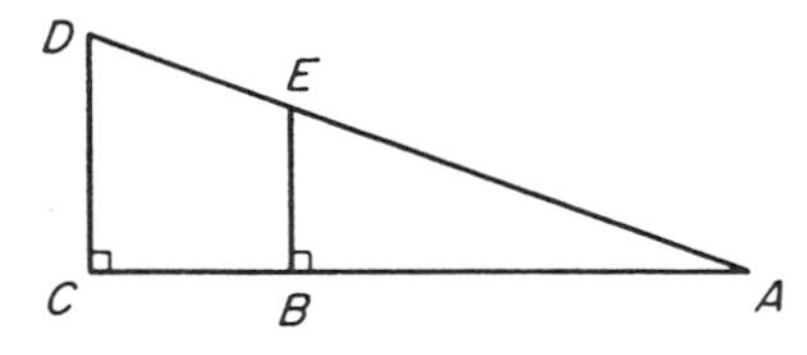

Figure 9-6

35. Triangle ABC in Fig. 9-7 is inscribed in a semicircle. If CD is perpendicular to diameter AB with $AD = 4.182$ in. and $DB = 1.369$ in., find CD.

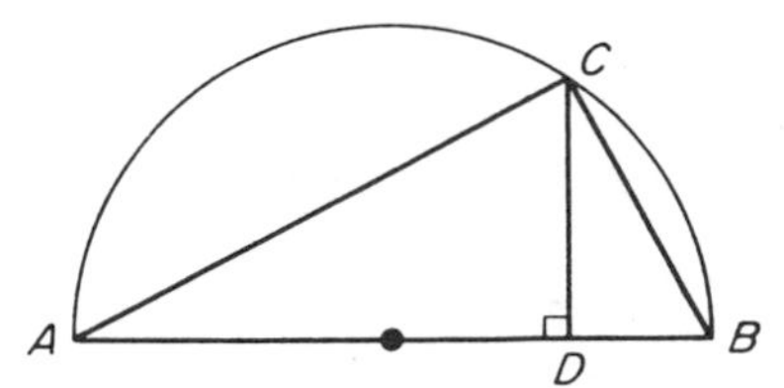

Figure 9-7

36. If R dollars per year are placed into a fund for n years and interest accumulates at rate i, the amount K is given as

$$K = Rs_{\overline{n}|},$$

where $s_{\overline{n}|}$ is

$$s_{\overline{n}|} = \frac{(1 + i)^n - 1}{i}.$$

This yearly payment into a fund is called an annuity. Find K if \$200 per year is deposited for 6 years at 4%.

37. If a man sets up an account today and wishes to withdraw R dollars annually from it for n years, the initial value of the account must be

$$A = Ra_{\overline{n}|},$$

where

$$a_{\overline{n}|} = \frac{1 - (1 + i)^{-n}}{i}.$$

Here A is the present amount of an annuity. Find A if \$1000 is to be taken from the account yearly for 10 years, with the account yielding 3.5%.

10

Miscellaneous topics in exponentials and logarithms

In this chapter we introduce several special topics involving exponentials and logarithms. Among them are special topics that will facilitate certain operations with logarithms, solution of exponential and logarithmic equations, use of the electronic calculator for special solutions, and applications and curve rectification. The basic notions advanced here rely on methods and laws established in Chapter 9.

10-1 Cologarithms

A *cologarithm* of a number is defined here as the logarithm of the reciprocal of the number; it is abbreviated colog. Thus

$$\text{colog}\left(\frac{1}{N}\right) = \log N$$

or

$$\text{colog}\, N = \log\left(\frac{1}{N}\right). \tag{1}$$

From (1) we can determine another relationship between colog N and log N, since (1) can be written

$$\text{colog}\, N = \log 1 - \log N$$

but log 1 = 0; therefore

$$\text{colog}\, N = -\log N, \tag{2}$$

where (2) asserts that the colog of a number is the negative of the logarithm of the number. Cologs are useful in computations where use of logarithms requires extensive manipulation of the characteristic. Let's consider two examples; the first involves finding a colog and the second uses the colog in a computation.

Example 1 Find (a) colog 76 and (b) colog 0.000372.

Solution

(a) From (2)

$$\text{colog}\, 76 = -\log 76.$$

From the table of logarithms

$$\log 76 = 1.8808$$

so that

$$\text{colog}\, 76 = -(1.8808). \tag{3}$$

Expression (3) is written in such a way as to indicate that both the characteristic and the mantissa are negative. Doing so is permissible and sometimes useful. Expression (3) can be written in the more popular form by adding and subtracting 10 such that

$$\text{colog}\, 76 = 10 - (1.8808) - 10 = 8.1192 - 10 \tag{4}$$

where (4) is the more popular form.

(b) From (2)

$$\begin{aligned}\text{colog}\, 0.000372 &= -\log 0.000372\\ &= -(6.5705 - 10) = 10 - 6.5705 = 3.4295.\end{aligned}$$

It is again asserted here that negative mantissas are permissible in logarithms. So if

$$N = \frac{1}{\sqrt{10}} = (10)^{-1/2} = 10^{-0.5000}$$

then $\log_{10} N = -0.5000$, where 5000 is the mantissa and is negative. It is also permissible to defy former usage and have a positive characteristic with a negative mantissa. Thus if

$$N = \frac{10^3}{\sqrt{10}} = 10^3 \cdot 10^{-0.5000},$$

then $\log_{10} N = 3 - 0.5000$, where $+3$ is the characteristic and -0.5000 is the mantissa.

Example 2 Find N if $N = (0.00493)^{-2.14}$.

Solution By the laws of exponents, we can write the given expression as

$$N = \left(\frac{1}{0.00493}\right)^{+2.14}. \tag{5}$$

A reason for rewriting in the form of (5) is that the original problem has a negative exponent and the logarithm of the base is negative. Multiplication of these two can be confusing because of the minus signs. Now from (5)

$$\log N = 2.14 \log \left(\frac{1}{0.00493}\right), \tag{6}$$

and using (2), we can write (6)

$$\log N = 2.14(-\log 0.00493).$$

Next,

$$-\log 0.00493 = -(7.6928 - 10) = +2.3072$$

and

$$\log N = (2.14)(2.3072) = 4.9374$$

from which $N = 86580$.

10-2 Solutions by logarithms where some quantities are negative

In the preceding chapter we established the requirement that we will not consider the logarithm of a negative number. However, in computations, since numbers used are negative, preliminary manipulation to determine the final outcome of the sign can assist us. Consider Example 3.

Example 3 Evaluate N if $N = \sqrt[3]{\dfrac{(-4.17)^2}{-8.65}}$.

Solution By examining the given expression, we see that $N = \sqrt[3]{a}$, where a is negative; therefore N is negative. We could rewrite the given expression as

$$-N = \sqrt[3]{\frac{(4.17)^2}{8.65}}, \tag{7}$$

where preliminary manipulation eliminates the negative signs in the radicand and we know that $-N$ is positive because N is negative. Now we can proceed, knowing that we will not be taking the logarithms of negative numbers. Thus from (7)

$$\log(-N) = \frac{1}{3}[2 \log 4.17 - \log 8.65] = \frac{1}{3}[2(0.6201) - 0.9370]$$

$$= \frac{1}{3}[1.2402 - 0.9370] = \frac{1}{3}(0.3032) = 0.1011$$

from which $-N = 1.262$ and $N = -1.262$.

Extending this discussion somewhat, if we have the form

$$N = \sqrt[k]{a},$$

where k is an odd integer and a is positive, no special manipulation is necessary. If a is negative when k is odd, manipulation such as that shown in Example 3 is needed. If k is even and a is positive, again no manipulation is necessary. If k is even and a is negative, we are faced with an even root of a negative number; this will involve removing a factor $\sqrt{-1} = j$ and then proceeding as usual.

10-3 Expressions involving sums and differences

Logarithms can be used to compute certain quantities that involve sums and differences. Their use, with certain exceptions, is not direct and so is often quite laborious. Here we introduce some manipulations involving sums and differences and certain shortcuts.

Among the common applications involving differences is the Pythagorean Theorem. From that theorem

$$b = \sqrt{c^2 - a^2}, \tag{8}$$

where c is the hypotenuse and a and b are the legs of a right triangle. Then from (8)

$$b = \sqrt{c^2 - a^2} = \sqrt{(c + a)(c - a)}$$

and

$$\log b = \frac{1}{2}[\log(c + a) + \log(c - a)], \tag{9}$$

where (9) is direct compared with other methods of finding log b. If we did not use (9), we would have to square c by logarithms, then square a by logarithms, subtract the squares, and, finally, take the square root by logarithms. A logarithmic representation of the procedure mentioned is

$$\log b = \frac{1}{2} \log [\text{antilog} (2 \log c) - \text{antilog} (2 \log a)].$$

Example 4 Referring to Fig. 10-1, find b.

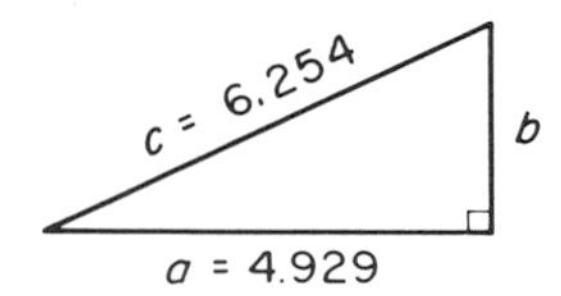

Figure 10-1

Solution From the figure

$$b = \sqrt{(6.254)^2 - (4.929)^2}. \tag{10}$$

Factoring the radicand, (10) becomes

$$b = \sqrt{(6.254 + 4.929)(6.254 - 4.929)} = \sqrt{(11.18)(1.325)}$$

$$\log b = \frac{1}{2}(\log 11.18 + \log 1.325)$$

$$= \frac{1}{2}(1.0486 + 0.1222) = \frac{1}{2}(1.1708) = 0.5854$$

from which $b = 3.849$.

Whenever possible, it is desirable to factor the radicand in order to facilitate direct use of logarithms and avoid intermediate antilogairthms. The process of factoring is not always possible; some simplicity, however, can be gained by other devices.

Example 5 Find log c if $c = \sqrt{a^2 + b^2}$.

Solution From $c = \sqrt{a^2 + b^2}$ we have

$$c = \sqrt{\frac{a^2(a^2 + b^2)}{a^2}} = a\sqrt{1 + \left(\frac{b}{a}\right)^2}$$

and $$\log c = \log a + \frac{1}{2} \log [1 + \text{antilog} (2 \log b - 2 \log a)].$$

Here we have reduced the number of antilogarithms to one, a useful reduction in the amount of work required for computation.

Exercises 10-1, 10-2, and 10-3

In Exercises 1–6, find the colog of the given number.

1. 0.286 **2.** 0.619 **3.** 0.000355

4. 0.0000572 **5.** 14.06 **6.** 27.18

In Exercises 7–16, evaluate the given expressions by use of logarithms.

7. $\left(\frac{-4.35}{6.19}\right)^2$ **8.** $\left(\frac{4.98}{-3\pi}\right)^2$ **9.** $\sqrt[3]{\frac{-285}{1.16 \times 3.29}}$

10. $\sqrt{\frac{(-14.9)(27.6)}{1.385}}$ **11.** $(+0.0032)^{-1.25}$ **12.** $(-12.36)^{2/3}$

13. $(14.6)^{-3}(-0.045)^{-5}$ **14.** $(0.0969)^{-32.9}$ **15.** $(\frac{1}{8})^{-4.51}$

16. $(\frac{1}{2})^{-0.05}$

In Exercises 17–30, evaluate the given expressions by use of logarithms. Use the values $a = 5.62$, $b = 4.99$, $c = 1.62$.

17. $\sqrt{a^2 - b^2}$ **18.** $\sqrt{b^2 - c^2}$ **19.** $a^2 + c^2 - b^2$

20. $a^2 + b^2 - c^2$ **21.** $\sqrt{2ab - a^2}$

22. $\sqrt{s(s-a)(s-b)(s-c)}$, where $s = \frac{1}{2}(a + b + c)$

23. $ab + bc$ **24.** $\sqrt{3a(a + 2b)}$ **25.** $a^2 + 2ab - 3b^2$

26. $a^2 - 3ac + 2c^2$ **27.** $a^2 - 4c^2$ **28.** $a^3 - 3a^2c + 2ac^2$

29. $\frac{a^2 + c^2 - 2ac}{2ab}$ **30.** $\sqrt[3]{c^2 - 3ac}$

10-4 Miscellaneous applications of logarithms

pH value

In chemistry the pH value of a solution is defined as the negative of the logarithm of the hydrogen ion concentration (using base 10). Thus if a_H is the hydrogen ion concentration, then the pH of the solution is

$$\text{pH} = -\log_{10} a_H.$$

So if the hydrogen ion concentration of a solution is given as $a_H = 3 \times 10^{-3}$, then

$$\text{pH} = -\log(3 \times 10^{-3}) = -\log(0.003)$$
$$= -(7.48 - 10) = -(-2.52) = 2.52.$$

The greater the hydrogen ion concentration, the more acid the solution. A solution of pH = 7 is considered neutral.

If a pH value is known, the hydrogen ion concentration can be determined. Thus if a substance has pH = 9.8, then from the definition of pH,

$$-\log_{10} a_H = 9.8 \quad \text{and} \quad \log_{10} a_H = -9.8.$$

Since 10 is the base and -9.8 is the logarithm, we have

$$a_{\mathrm{H}} = 10^{-9.8} = 10^{-10} \times 10^{+0.2}.$$

Using the table of logarithms, $10^{+0.2} = 1.58$; then

$$a_{\mathrm{H}} = 1.58 \times 10^{-10}.$$

We note here that the number of significant figures involved in pH and concentrations is usually not beyond three; so the entries in Table C are more than sufficient.

Decibels

M is the sound initially produced by a source, and N is a new sound produced by the same source after a change. If P_M and P_N are the power levels of the two sounds, the difference in loudness between M and N is measured in decibels by the expression

$$\text{decibel gain or loss} = 10 \log_{10} \frac{P_N}{P_M}. \tag{11}$$

If the ratio $0 < P_N/P_M < 1$, then $\log_{10} P_N/P_M < 0$ and we have a decibel loss. This situation occurs when the power has been diminished. In the case of a gain, the later power is greater than the initial power.

Example 6 A loundspeaker is supplied by 2.5 watts producing a certain sound volume in an enclosure. If the power is increased to 3.5 watts, what is the gain in decibels?

Solution Here $P_M = 2.5$ and $P_N = 3.5$. Substituting into (11), decibel gain (db) is

$$\text{db} = 10 \log_{10} \frac{3.5}{2.5} = 10 \log_{10} 1.4 = 10(0.146) = 1.46.$$

The power in sound-level studies can be described indirectly; that is, $P = I^2R$ in electricity, where I is intensity and R resistance. Also, $E = IR$, where E is electromotive force. Combinations of E, I, and R can be given from which P is derived.

Radioactive decay

The intensity of radioactivity is measured at intervals referred to as *half-life* intervals. The half-life is the time period required for the material's activity to decay by 50%. Thus if a material has a level of activity today of L, then the time period lapsing between now and the time when the activity level is $\frac{1}{2}L$ is called its half-life. The radioactive level of a substance is a measure of the amount of radioactive material present; therefore 10 mg of radioactive material will decompose to 5 mg in a half-life.

An example of radioactive material is radium, which has a half-life of 1580 years. This does not presume that full decay occurs in 2(1580) years, for such would be an assumption that the decay is linear as opposed to exponential. Actually, the amount at the end of 2(1580) year is $(\frac{1}{2})(\frac{1}{2})$ of the original amount. The amount after t years is given as

$$A = A_0\left(\frac{1}{2}\right)^{t/1580},$$

where A is the instantaneous amount, A_0 is the amount at $t = 0$, and t is the time in years.

Example 7 How many years are required for 10 mg of radium to decay to 4 mg?

Solution Using information from the preceding paragraph, $A = 4$ and $A_0 = 10$; so

$$4 = 10\left(\frac{1}{2}\right)^{t/1580}$$

from which

$$0.4 = \left(\frac{1}{2}\right)^{t/1580} \quad \text{and} \quad \frac{t}{1580}\log\left(\frac{1}{2}\right) = \log 0.4.$$

Using base 10,

$$\frac{t}{1580}(-0.301) = -0.398$$

from which we have the solution $t = 2090$ years.

Exercises 10-4

1. The pH value of a solution is given in parts (a) through (d). What are the hydrogen ion concentrations?
   (a) 8.5 (b) 9.6 (c) 3.2 (d) 4.6
2. The hydrogen ion concentration of a solution is given in parts (a) through (d). What is the pH value in each case?
   (a) 1.65×10^{-12} (b) 3.86×10^{-10}
   (c) 6.19×10^{-9} (d) 8.37×10^{-7}
3. In parts (a) through (d), a loudspeaker is initially supplied with a power of P watts and the power is later changed to R watts. What is the gain (or loss) of decibels for the values of P and R given?
   (a) $P = 4.3$, $R = 6.5$ (b) $P = 1.3$, $R = 4.2$
   (c) $P = 5.2$, $R = 2.5$ (d) $P = 8.4$, $R = 2.1$
4. From electricity we have $P = I^2R$, where P is power, I is intensity (amperes, A), and R is resistance (ohms, Ω). Also, $E = IR$, where E is electromotive force

(volts, V). If an amplifying system has a resistance of 10 Ω and the voltage is changed from E_1 to E_2, find the decibel change if

(a) $E_1 = 50, E_2 = 100$ (b) $E_1 = 100, E_2 = 50$
(c) $E_1 = 75, E_2 = 52$ (d) $E_1 = 108, E_2 = 126$

5. If actinium has a half-life of 0.002 sec, 3 mg of actinium will decay to what amount after 0.10 sec?

6. If a radioactive material has a half-life of 6 min, in how many seconds will it decay by 10%?

10-5 Exponential and logarithmic equations

An *exponential equation* is an equation in which the variable is present in the exponent. Such equations can easily be solved by use of logarithms. It is generally less confusing to use logarithms to base e rather than base 10, but once again manipulation of the characteristic in base 10 can permit solutions. We will obtain our solutions in base 10.

Two devices are useful in solving exponential equations. (a) If feasible, take the logarithm of both sides of the equation or (b) retain the exponential form, changing all bases to the same base. These two devices actually are the same device if the logarithms in (a) are of the same base as the base used in (b). Consider an example.

Example 8 Solve for x if

$$2^{3x-4} = 7^x.$$

Solution 1 If we take the logarithm of both sides, we have

$$(3x - 4) \log 2 = x \log 7$$

from which $(3x - 4)(0.3010) = x(0.8451)$.
We have a linear expression in x from which $0.0579x = 1.2040$ and

$$x = \frac{1.2040}{0.0579} = 20.76.$$

It is interesting to note that the division of 1.204 by 0.0579 can be done by logarithms despite the fact that these numbers are obtained from logarithms.

Solution 2 Given $2^{3x-4} = 7^x$, we can convert each base (2 and 7) to powers of the same base. If we use the table of logarithms in Table C, the base is 10. We could use base e or any other base. Thus $2 = 10^{0.3010}$ and $7 = 10^{0.8451}$ so that

$$[(10)^{0.3010}]^{3x-4} = [(10)^{0.8451}]^x.$$

With the bases alike, the exponents can be equated and

$$0.3010(3x - 4) = 0.8451x.$$

The remainder of the solution is identical to Solution 1.

Sometimes the bases involved can be converted to the same base, which is neither 10 nor e.

Example 9 Find x if $(2)^{x-1} = 8$.

Solution Here we see that both bases are powers of 2 and so we have $(2)^{x-1} = (2)^3$. Equating exponents because the bases are alike, we have $x - 1 = 3$ and $x = 4$.

10-6 Logarithmic equations

A *logarithmic equation* is an equation in which at least one term involves a logarithmic expression of one or more variables. We confine our work to expressions involving only one variable. Solutions depend on fundamental properties of logarithms and often involve conversion to the exponential form.

Example 10 Find y if

$$\log (y + 3) + \log (y + 1) = \log 2 + \log (y + 3).$$

Solution By the laws of logarithms, the given expression can be converted to

$$\log (y + 3)(y + 1) = \log 2(y + 3),$$

since the sum of the logarithms is the logarithm of the product. Next, we take the antilogarithm of both sides, giving

$$(y + 3)(y + 1) = 2(y + 3),$$

which can be solved as a quadratic to yield solutions $y = -3$, $y = +1$. Only the solution $y = 1$ is permissible, for $y = -3$ would involve the logarithm of a negative number.

Example 11 Find x if $\log (5x + 20) = 2$.

Solution Convert the given expression to the exponential form by using base 10. Here 2 is the exponent and 10 is the base so that $10^2 = 5x + 20$, from which $5x + 20 = 100$, $x = 16$.

Exercises 10-5 and 10-6

In Exercises 1–20, solve the given exponential equation for the variable present. Use whatever base seems to be most convenient.

1. $2^x = 8$

2. $3^x = 81$

3. $2^x = \frac{1}{16}$

4. $4^x = \frac{1}{16}$

5. $3^{-2x} = 27$

6. $2^{-3x} = 16$

7. $4^{x-1} = 2^{3x-1}$

8. $3^{2x+4} = 9^{-x-1}$

9. $81^{(x/2)+1} = 27^{4x-1}$

10. $25^{Z+4} = 125^{3Z-2}$

11. $4^{2Z+2} = 3^{Z-1}$

12. $2^{x-2} \cdot 3^{2x+1} = 1$

13. $(0.4)^{0.5Z} = (1.9)^{2Z+3}$

24. $\sqrt[K]{25} = 5$

15. $\sqrt[3K]{81} = 3$

16. $\sqrt[K]{3} \times \sqrt[2K]{6} = 4$

17. $\sqrt[m]{5} \times \sqrt[3m]{63} = 12$

18. $\sqrt[2]{0.004} = 3^m$

19. $\sqrt[3]{0.0065} = 4^n$

20. $\sqrt[n]{4} = (4)^n$

In Exercises 21–27, we are given the compound interest formula

$$A = P(1 + i)^n,$$

where A is amount, P is principal, i is interest rate per year, and n is the number of years.

21. What is the interest rate if an amount of $968 results from compounding a principal of $850 for 4 years?

22. What is the interest rate if an amount of $255 results from compounding a principal of $225 for 3 years?

23. What principal is required to accumulate to an amount of $10,000 in 10 years at 5% interest rate?

24. What principal is required to accumulate to an amount of $3000 in 5 years at 5% interest rate?

25. In how many years at 3.5% will a principal of $7000 accumulate to $10,000?

26. In how many years at 5% will a principal double?

27. If a man wishes to double his money in 12 years, what would be the interest rate required?

In Exercises 28–30, apply the formula

$$i = \frac{V}{R}(1 - e^{-Rt/L}),$$

where V is a steady voltage applied to an inductance L in series with a resistance R with i = current and t = time.

28. At what time does $i = 3$ amperes if $V = 120$, $R = 12$, and $L = 0.3$? ($e = 2.718 = 10^{0.4343}$.)

29. What voltage V is required if $i = 5$ when $t = 0.006$ with $R = 10$ and $L = 0.5$?

30. Determine L for a system where $V = 100$, $R = 2$, and $i = 5$ when $t = 0.01$.

In Exercises 31–38, solve the given logarithmic equation for the variable present.

31. $\log_{10} x^2 = 4$

32. $\log_{10}(x - 4) = 1$

33. $\log_2 A + \log_2 (A + 6) = 4$

34. $1 + \log_{10} x = \log_{10}(x + 1)$

35. $2 + \log_2 k = \log_2 (k + 5)$

36. $\log A - \log P = n \log(1 + i)$ (Solve for P.)

37. $\log x = \dfrac{1}{n} \log A.$

38. $bn \log x = \log y - \log a$ (Solve for y.)

10-7 Conversions between bases

Assume that we are given an exponential in one base (base a) and wish to convert this expression to an exponential in another base (base b). We start with the expression

$$y = a^x \tag{12}$$

and convert it to another expression

$$y = b^z. \tag{13}$$

If we take the logarithm (to any base, say base c) of (12) and (13), we have

$$\log_c y = x \log_c a \quad \text{and} \quad \log_c y = z \log_c b. \tag{14}$$

Then by substitution, we obtain

$$x \log_c a = z \log_c b$$

so that

$$z = \frac{x(\log_c a)}{(\log_c b)}$$

and the result is

$$a^x = b^{x(\log_c a)/(\log_c b)}. \tag{15}$$

In the preceding paragraph it is usually unnecessary to introduce a third base c. The logarithm taken to obtain (14) might more wisely have been to base a or b, from which (15) becomes either

$$a^x = b^{x(\log_a a)/(\log_a b)} = b^{x/\log_a b} \tag{16}$$

or

$$a^x = b^{x(\log_b a)/(\log_b b)} = b^{x \log_b a}. \tag{17}$$

For conversions between base 10 and base e, using (16), let $a = 10$ and $b = e$ so that

$$10^x = e^{x/\log_{10} e}. \tag{18}$$

Then in (18) $\log_{10} e = \log_{10} 2.718 = 0.4343$, and (18) becomes

$$10^x = e^{x/0.4343}, \tag{19}$$

where (19) asserts that *to change a logarithm from base 10 to base e, divide the logarithm by 0.4343*. Also, from (19), if we raise both sides to the 0.4343 power, we have

$$10^{0.4343x} = e^x, \tag{20}$$

where (20) asserts that *to change a logarithm from base e to base 10, multiply the logarithm by 0.4343.*

Example 12 Using common logs, express 12 as a power of base e.

Solution Using common logs, we start with

$$12 = 10^{1.0792}. \tag{21}$$

Then from (19)

$$10^{1.0792} = e^{1.0792/0.4343} = e^{2.4849}.$$

$$\log_e 12 = \ln 12 = 2.4849. \tag{22}$$

Convention: It is a common convention to express $\log_e N$ *by another notation* ln N, *where* ln *is used to indicate natural* (*or base* e) *logarithms.*

Example 13 If $\ln N = 2.0000$, find N by using Table C.

Solution From $\ln N = 2.0000$ we have the exponential $e^2 = N$, but from (20)

$$e^2 = 10^{2(0.4343)} = 10^{0.8686} = 7.389.$$

Exercises 10-7

Solve the given equation for the variable present by taking the logarithm to base 10 of both sides.

1. $10^2 = 3^y$	**2.** $10^3 = 6^y$	**3.** $10^{2.5} = e^y$
4. $10^{1.8} = e^y$	**5.** $6^3 = 10^y$	**6.** $2^{3.6} = 10^y$
7. $e^4 = 10^x$	**8.** $e^{2.5} = 10^x$	**9.** $3^{10} = 6^y$
10. $2^{1.1} = 8^x$	**11.** $6^{-2} = 3^y$	**12.** $15^{-.55} = 2^y$
13. $y = \log_4 100$	**14.** $x = \log_{0.5} 10$	**15.** $x = \log_2 3^{1.5}$

10-8 The equation $y^x = N$, a summary

The equation $y^x = N$ consists of a base y, an exponent x, and a number N. If two of the three elements x, y, and N are known, the third is found rather easily. Obviously three different cases of solution result, since any one of the three elements may be the unknown.

General solution The general solution of $y^x = N$ begins by taking the logarithm of both sides of the equation. Thus if

$$y^x = N \tag{23}$$

then

$$x \log y = \log N \tag{24}$$

and solution for the single unknown follows readily. The logarithm taken in (24) may involve any base: if $y = e$, base e is a good choice; if $y = 10$, then base 10 is a good choice. Consider the three cases presented in Example 13.

Example 13 Determine the value of the literal quantity in (a) $(3.1)^{1.2} = N$, (b) $y^{2.4} = 16.2$, and (c) $1.5^x = 7.4$.

Solution

(a) Comparing $(3.1)^{1.2} = N$ to (23), we have $y = 3.1$ and $x = 1.2$ and N is the unknown. Taking the common logarithm of both sides of (a) as in (24), we have $1.2 \log 3.1 = \log N$ and with $\log 3.1 = 0.4914$ from Table C, then

$$\log N = (1.2)(0.4914) = 0.5897$$

from which (by antilogarithm) $N = 3.887$.

Note: The solution of (a) is direct on a hand calculator with y^x capability. The steps are (1) Enter 3.1, (2) Depress y^x, (3) Enter 1.2, (4) Depress =.

(b) Comparing $y^{2.4} = 16.2$ to (23), base y is unknown, whereas exponent $x = 2.4$ and number $N = 16.2$ are known. Taking the common logarithm of both sides of (b) as in (24), we have $2.4 \log y = \log 16.2$ from which

$$\log y = (\log 16.2) \div 2.4 = 1.2096 \div 2.4 = 0.5040$$

from which (by antilogarithm) $y = 3.191$.

Note: The solution of (b) is possible on a hand calculator with a y^x and inverse capacity. The steps are (1) Enter 16.2, (2) Depress INV, (3) Depress y^x, (4) Enter 2.4 (5) Depress =.

(c) Comparing $1.5^x = 7.4$ to (23), we have $y = 1.5$ and $N = 7.4$ with exponent x being the unknown. Taking the logarithm of both sides of (c) as in (24), $x \log 1.5 = \log 7.4$ from which

$$x = \frac{\log 7.4}{\log 1.5} = \frac{0.8692}{0.1761} = 4.936.$$

Note: The solution of (c) on most hand calculators requires the same steps as the algebraic solution. Steps include (1) Enter 7.4, (2) Depress log (or ln), (3) Depress ÷, (4) Enter 1.5, (5) Depress log (or ln), (6) Depress =.

Exercises 10-8

Find the literal quantity.

1. $e^{4.2} = N$ **2.** $e^{-8.6} = N$ **3.** $e^{0.047} = N$

4. $e^{-0.37} = N$ **5.** $e^{-0.096} = N$ **6.** $(1.04)^{4.2} = N$

7. $(0.95)^{-3.6} = N$ **8.** $(0.02)^{-0.02} = N$ **9.** $\sqrt[25]{4216} = N$

10. $\sqrt[10]{0.00385} = N$ **11.** $e^x = 4.5$ **12.** $e^x = 0.025$

13. $(0.04)^x = 27$ **14.** $(0.982)^x = 0.02$ **15.** $b^{2.7} = 48.2$

16. $b^{-0.09} = 1.06$ **15.** $(0.619)^{1.95} = N$ **18.** $(2 \times 10^{-5})^{0.033} = N$

19. $(0.775)^{-3.4} = N$ **20.** $e^{14.3} = N$ **21.** $e^{-12.7} = N$

22. $e^{22.7} = N$ **23.** $(12.4)^{9.8} = N$ **24.** $(0.092)^{0.00042} = N$

25. $\sqrt[5]{12{,}800} = N$ **26.** $(0.985)^{15.6} = N$

Additional exercises 10-8

1. When a switch is closed to apply a constant voltage E to a circuit containing an inductance L and a resistance R, the current i increases from zero to a maximum value. The instantaneous value of current at any time t sec after the switch is closed is given by

$$i = \frac{E}{R}(1 - e^{-Rt/L}).$$

If L is 0.8 henrys (H), R is 16 Ω, and E is 32 volts, find i when t is 0.03 sec.

2. Referring to the equation of Exercise 1, show that $i = 0$ when $t = 0$ and that i approaches E/R as a maximum value when t is increased. (Try $t = 1$ sec.)

3. Find t in Exercise 1 when $i = 90\%$ of its maximum value.

4. A convenient measure of the rapidity of current increase in the circuit of Exercise 1 is the value of time t numerically equal to the ratio L/R, known as the time constant. Show, for any such circuit, that the current has increased to 63.2% of its maximum value when t equals L/R sec.

5. From the relation $\ln N = \ln N_0 - kt$, show that $N = N_0 e^{-kt}$.

6. The atoms of a radioactive material disintegrate, forming a new substance and reducing the amount of active material present. In the equation below N_0 is the original amount of radioactive material in a sample at some time. N is the amount remaining t sec later.

$$N = N_0 e^{-kt}.$$

If N_0 is 3.5 g and the decay constant k for the material is $3.2(10)^{-6}$, find the amount remaining 2 days later.

7. The half-life of a radioactive material is the time required for half the atoms to disintegrate. From the equation of Exercise 6, show that the half-life equals $0.693/k$.

8. In 10 days, 5% of a radioactive sample has disintegrated. Using the equation

of Exercise 6, find the total time in days required for disintegration of 80% of the original amount. The decay constant and the time may be solved by using days for time units.

9. An annual payment P, invested at the end of each year for n years and earning interest at i% compounded annually, will build a "sinking fund" S. The annual payment required is

$$P = \frac{Si}{(1 + i)^n - 1}.$$

If a company has just purchased a $30,000 machine and wishes to have a sinking fund available in 15 years to replace it, what annual payment must be made? Interest is at 5%, or $i = 0.05$.

10. The expression for sound level in decibels is dB = 10 log (I/I_0), where I is the intensity of sound power is watts/cm² at the point of interest. I_0 is 10^{-16} watts/cm², a reference value approximating a barely audible sound. If the sound level is 80 dB at a certain distance R from a small source, what would the sound level be at a distance five times as great? The intensity of sound power varies inversely as the square of distance from the source, or $I_1/I_2 = R_2^2/R_1^2$.

11. If people begin life as a single biological cell and by cell division become fully grown with about ten billion cells, how many generations of cell division are required? Neglect the effect of cell deaths and assume that all cells require the same time from one division to the next.

12. Electrical resistance R of a thermistor decreases as its temperature T is increased, according to the relation

$$R = (R_a)e^{B(1/T - 1/T_a)}.$$

where R_a is the resistance at temperature T_a. Find R when T is 370°K if R_a is 25 Ω at 300°K. The constant B, which depends on the type of thermistor, has the value 3530.

13. At what temperature T would the thermistor in Exercise 12 have a resistance R of 10 Ω?

14. From the equation of Exercise 12, make a graph of R versus T for values of T from 300 to 370°K. Four points may be sufficient. Use the given values for B, R_a, and T_a.

15. If the work put into compressing a gas is done at the same rate at which energy is removed from the gas by a cooling system, there is no change in gas temperature and the process is called *isothermal compression.* The work required is

$$W = P_1 V_1 \ln \frac{V_2}{V_1}.$$

Find the foot-pounds of work required if initial pressure P_1 is 2100 lb/ft², the initial volume V_1 is 5 ft³, and the final volume V_2 is 1.2 ft³. A negative answer indicates work is done on the gas.

16. Referring to Exercise 15, what initial volume V_1 can be isothermally compressed by twice the amount of work in Exercise 15? Use the same initial pressure and final volume. A trial-and-error procedure may be required.

17. If a principal P is invested at interest rate $i = 0.04$ compounded annually, in how many years will the investor's money double? In how many years will it triple? Use the compound interest formula

$$A = P(1 + i)^t.$$

18. Given $i = 120$, solve for t in equation

$$i = 240e^{t^2 - 2t}.$$

19. The population of a certain city doubles every 40 years. If the population in 1900 was 10,000, what was the population in 1960?

20. A sequence of terms is given as

$$100, 90, 81, 72.9, \ldots$$

where each term is 90% of the previous term. Calculate the value of the tenth term of the sequence; the 51st term. Use logarithms for the computations.

21. Solve for t in Exercise 1.

22. The temperature of the atmosphere in a certain room is 70°F. A vat of liquid at temperature 180°F is brought into the room. The liquid cools in such a way that its temperature falls, during any 30-minute interval, to a reading halfway between the room temperature and the liquid temperature at the beginning of the 30-minute time period. That is, during the first 30 minutes the liquid temperature falls $(180° - 70°)/2 = 55°$ to a reading of 125°F; during the second 30 minutes the temperature falls $(125° - 70°)/2 = 27.5°$ to a reading of $125° - 27.5° = 97.5°$F, and so on. In how many minutes will the temperature of the liquid fall from 180°F to 75°F, assuming that the room temperature remains at 70°F?

11

Trigonometry of right triangles

In this chapter trigonometric functions are introduced and definitions of the trigonometric functions, reciprocal and cofunctional relationships, are described. The approach is vectorial in nature, with vectors placed on the cartesian coordinate plane. Emphasis is placed on the determination of horizontal and vertical components of vectors, with a view to applications in mechanics, physics, surveying, and electricity.

11-1 Vectors on the cartesian coordinate plane

Such quantities as light, mass, area, and volume, which are characterized by magnitude only, are given the name *scalars*. Scalars are usually discussed in conjunction with *vectors*, quantities characterized by both magnitude and direction. A vector is generally represented by a straight line; the length of the line is intended to represent the magnitude of the quantity involved, and a direction is assigned. Physical quantities, like velocity, acceleration, and force are examples of vectors.

In Fig. 11-1 conventions regulating the description of vectors on the cartesian coordinate plane are displayed. In all cases, the vectors originate at the origin O and terminate at P_1, P_2, or P_3. In this text the name given to a vector is the same as the terminal point, so the directed line segment OP_1 is called vector $\mathbf{P}_1$, or, more simply, $\mathbf{P}_1$. Length units on the vector, consistent with those on the x axis and the y axis, are shown. The arrowhead or barb at the terminal end of the vector conveys direction or, perhaps more explicitly, *sense*. Note that $\mathbf{P}_1$ is of magnitude 5, $\mathbf{P}_2$ is of magnitude 4, and $\mathbf{P}_3$ is of magnitude approximately 4.5.

Directions of the vectors in Fig. 11-1 are shown in angular units (degrees in this case) and are consistently measured from the right-hand horizontal, with the counterclockwise direction being called positive and the clockwise direction negative. The angle between the right-hand horizontal and the vector is called the *reference angle*. Thus $\mathbf{P}_1$ is of magnitude 5 and has a reference angle of 45°. Note that $\mathbf{P}_3$ has a magnitude of approximately 4.5 and a reference angle of either +285° or −75°.

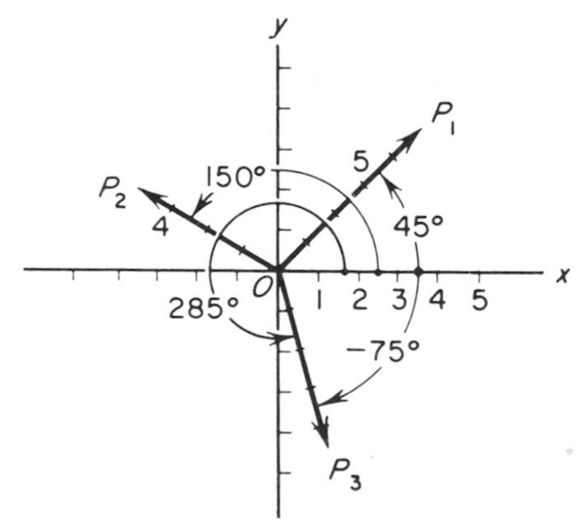

Figure 11-1

For our purposes, we will consider the magnitudes of all vectors as positive; this is merely a convention used here and is not universally adopted. We will depart from the convention in a later chapter on the j operator. The general designation that we will assign to the magnitude of the vector is ρ (Greek letter rho) and the general reference angle is θ (Greek letter theta).

In Fig. 11-2 we see a vector **P** with initial end at O and terminal end at P. The point P is at a position designated by coordinates (x, y). If a vertical line MP is drawn from P perpendicular to the x axis, two sides of a right triangle OPM are formed. MP is called the *y component* or *vertical component* **P**. The length of MP is exactly the ordinate of P—namely, y. OM is called the *x component* or *horizontal component*; its length is exactly the abscissa of P—namely, x.

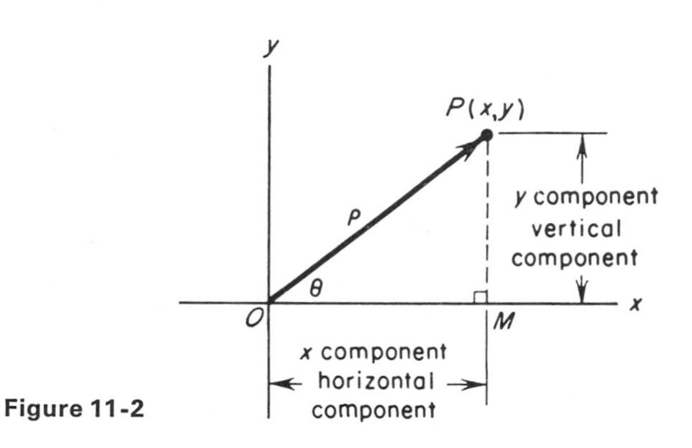

Figure 11-2

It is important to recognize that even though the magnitude ρ is, by our convention, always positive, the horizontal and vertical components can be either negative or positive, depending on the size of θ. Referring to Fig 11-3(a), we note the following. If OP terminates in Quadrant II ($90° < \theta < 180°$), then the vertical component is positive and the horizontal component is negative. If OP terminates in Quadrant IV ($270° < \theta < 360°$), the vertical component is negative and the horizontal is positive. From Fig. 11-3(b) we

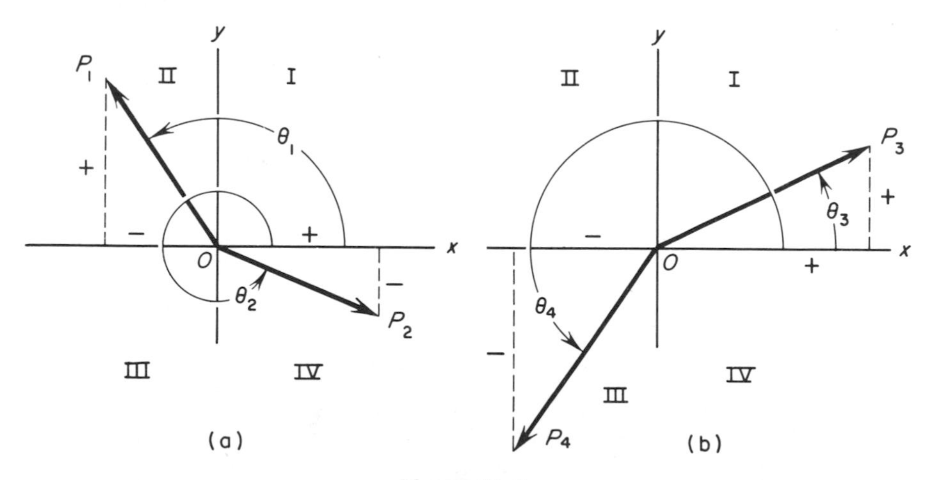

Figure 11-3

see that both components are positive in Quadrant I ($0° < \theta < 90°$) and negative in quadrant III ($180° < \theta < 270°$).

Exercises 11-1

1. Calling the magnitude ρ (always positive) of a vector, the horizontal component x, and the vertical component y, we have six possible ratios involving the division of one of these quantities by another:

$$\frac{y}{\rho}, \quad \frac{x}{\rho}, \quad \frac{y}{x}, \quad \frac{\rho}{y}, \quad \frac{\rho}{x}, \quad \frac{x}{y}.$$

   Determine the sign of each of these six ratios for a vector that initiates at the origin and terminates in the
   (a) first quadrant.
   (b) second quadrant.
   (c) third quadrant.
   (d) fourth quadrant.

2. If two vectors are exactly 180° apart, compare the signs of their horizontal components, vertical components, magnitudes, ratio of horizontal-to-vertical components.

What positive reference angle, less than 360°, is the equivalent of the angle given in Exercises 3–8?

3. 400° 4. −80° 5. 725°

6. 1000° 7. −885° 8. −718°

11-2 Definitions of the trigonometric functions

We consider six trigonometric functions. Each is a ratio of two sides of the triangle sketched in Fig. 11-4 and each is a function of angle θ. These functions are purely definitions and must therefore be memorized. Taking our symbols from Fig. 11-4, we define the trigonometric functions as

$$\text{sine of } \theta = \sin\theta = \frac{y}{\rho}, \tag{1}$$

$$\text{cosine of } \theta = \cos\theta = \frac{x}{\rho}, \tag{2}$$

$$\text{tangent of } \theta = \tan\theta = \frac{y}{x}, \tag{3}$$

$$\text{cosecant of } \theta = \csc\theta = \frac{\rho}{y}, \tag{4}$$

$$\text{secant of } \theta = \sec\theta = \frac{\rho}{x}, \tag{5}$$

$$\text{cotangent of } \theta = \text{ctn}\,\theta = \frac{x}{y}. \tag{6}$$

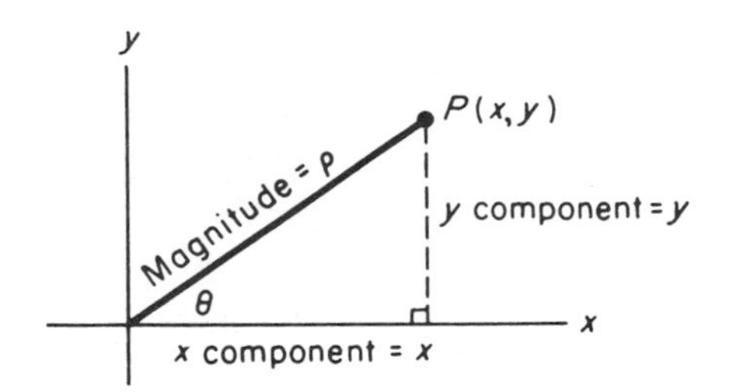

Figure 11-4

We note that the six definitions given made reference to a vector drawn on a cartesian coordinate plane and the components of the vector. Many readers are acquainted with geometrically slanted definitions that use the terms *side opposite*, *side adjacent*, and *hypotenuse*; these terms are shown in Fig. 11-5 and have their equivalents in Fig. 11-4. It is worthwhile to draw comparisons here. In Fig. 11-5(a) we are given a right triangle with the hypotenuse, the side opposite the angle θ, and the side adjacent to the angle θ shown. In Fig. 11-5(b) the same triangle is placed on a cartesian coordinate plane, showing the equivalencies of the hypotenuse, side opposite, side adjacent to the magnitude, y component, and x component, respectively.

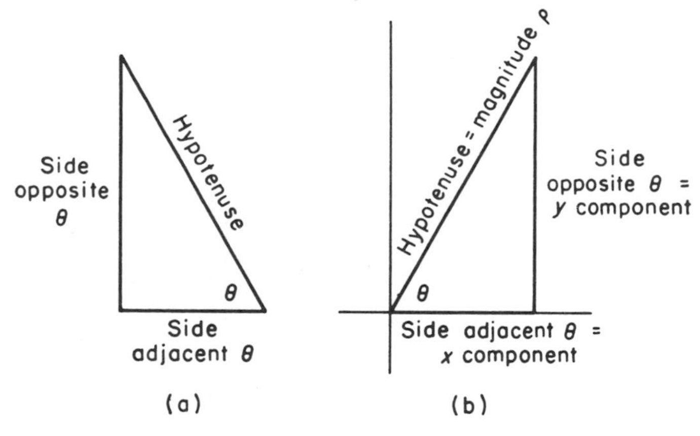

Figure 11-5

In view of the comparative nomenclature drawn from Fig. 11-5, we can redefine the trigonometric functions in terms of the hypotenuse, side opposite, and side adjacent as

$$\sin\theta = \frac{\text{side opposite } \theta}{\text{hypotenuse}}, \tag{7}$$

$$\cos\theta = \frac{\text{side adjacent } \theta}{\text{hypotenuse}}, \tag{8}$$

$$\tan\theta = \frac{\text{side opposite } \theta}{\text{side adjacent } \theta}, \tag{9}$$

$$\csc\theta = \frac{\text{hypotenuse}}{\text{side opposite }\theta}, \tag{10}$$

$$\sec\theta = \frac{\text{hypotenuse}}{\text{side adjacent }\theta}, \tag{11}$$

$$\text{ctn}\,\theta = \frac{\text{side adjacent }\theta}{\text{side opposite }\theta}, \tag{12}$$

Example 1 Give the numerical values of the six trigonometric functions of θ in Fig. 11-6(a).

Solution Using notation consistent with Fig. 11-4 and definitions (1) through (6), we place the given triangle on a cartesian coordinate plane as in Fig. 11-6(b). Then $x = 4$, $y = 3$, and $\rho = 5$ so that

$$\sin\theta = \frac{y}{\rho} = \frac{3}{5} = 0.600 \qquad \csc\theta = \frac{\rho}{y} = \frac{5}{3} = 1.667,$$

$$\cos\theta = \frac{x}{\rho} = \frac{4}{5} = 0.800 \qquad \sec\theta = \frac{\rho}{x} = \frac{5}{4} = 1.250,$$

$$\tan\theta = \frac{y}{x} = \frac{3}{4} = 0.750 \qquad \text{ctn}\,\theta = \frac{x}{y} = \frac{4}{3} = 1.333.$$

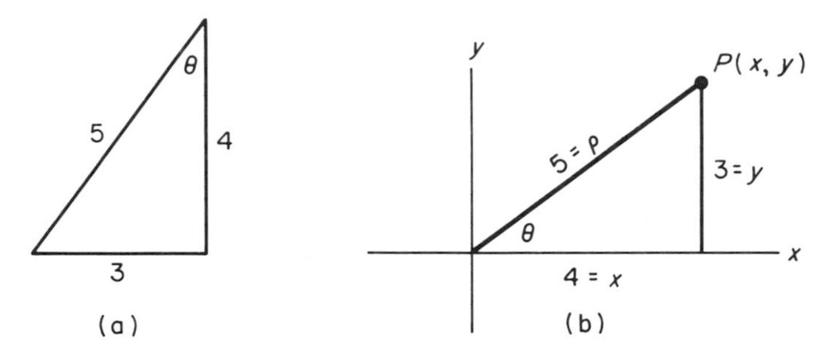

Figure 11-6

In Example 1 we could have said that the side opposite θ is 3, the side adjacent θ is 4, and the hypotenuse is 5 and then applied definitions (7) through (12).

Let us examine the components of vectors whose reference angles are 30, 45, and 60°. These special angles will provide opportunity for additional exercises involving the trigonometric functions, without reference to the tables of trigonometric functions.

Functions of 30° and 60°

An equilateral triangle can be assumed to have sides of length $2s$; its angles are all 60°. If a perpendicular is drawn to one side from the opposing

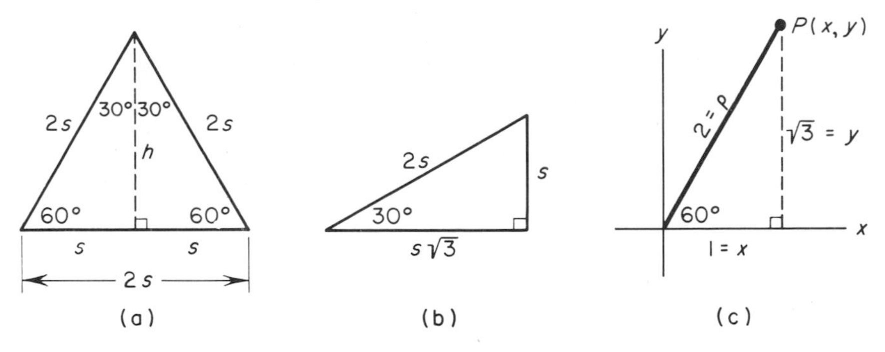

Figure 11-7

vertex, as the dotted line in Fig. 11-7(a), it bisects that side. If we call the altitude h, we have the Pythagorean relationship

$$(h)^2 + (s)^2 = (2s)^2$$

from which

$$h^2 = 4s^2 - s^2 = 3s^2$$

and

$$h = s\sqrt{3}. \tag{13}$$

Expression (13) shows that either triangle formed by the bisector can be shown as the triangle in Fig. 11-7(b), where the ratio of the sides of a 30°-60° right triangle is s: $2s$: $s\sqrt{3}$, or $1:2:\sqrt{3}$, where the smallest side (s) is opposite the 30° angle, the largest side ($2s$) is the hypotenuse, and the remaining side ($s\sqrt{3}$) is opposite the 60° angle.

Example 2 Find tan 60°, sec 60°, and csc 60°.

Solution Sketch a vector with reference angle 60° measured from the right-hand horizontal in a counterclockwise direction. From the terminal end of the vector drop a perpendicular to the x axis as shown in Fig. 11-7(c). Since the resulting right triangle is a 30°-60° right triangle, assign values in the proper $1:2:\sqrt{3}$ ratio as shown. Then

$$\tan 60^\circ = \frac{y}{x} = \frac{\sqrt{3}}{1} = \sqrt{3},$$

$$\sec 60^\circ = \frac{\rho}{x} = \frac{2}{1} = 2,$$

$$\csc 60^\circ = \frac{\rho}{y} = \frac{2}{\sqrt{3}} = \frac{2\sqrt{3}}{3}.$$

Functions of 45°

If a diagonal is drawn in a square of side s, it divides the square into two right triangles with angles 45, 45, and 90°. The sides of the 45° right triangle can be shown to be in the ratio $1:1:\sqrt{2}$ as in Fig. 11-8(a). (The proof is left as an exercise.) Sides s and s (or 1 and 1) are opposite the 45° angles and sides $s\sqrt{2}$ is opposite the 90° angle.

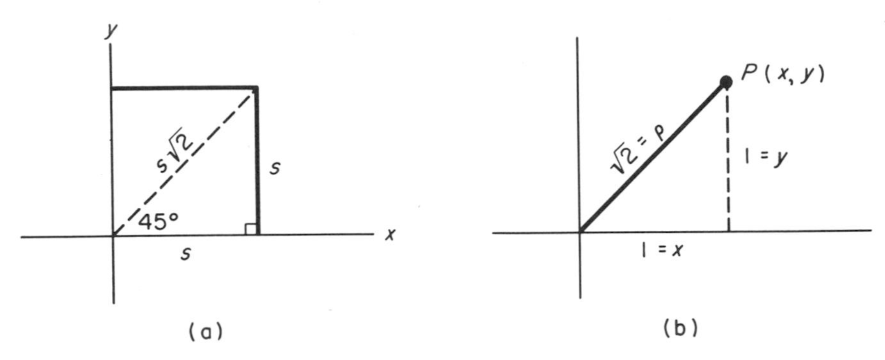

Figure 11-8

Example 3 Find cos 45°, ctn 45°, and sin 45°.

Solution From Fig. 11-8 we see components $x = 1$, $y = 1$, and magnitude $\rho = \sqrt{2}$. Then

$$\cos 45^\circ = \frac{x}{\rho} = \frac{1}{\sqrt{2}} = \frac{\sqrt{2}}{2},$$

$$\text{ctn } 45^\circ = \frac{x}{y} = \frac{1}{1} = 1,$$

$$\sin 45^\circ = \frac{y}{\rho} = \frac{1}{\sqrt{2}} = \frac{\sqrt{2}}{2}.$$

Exercises 11-2

1. Complete the chart shown. Rationalize denominators. Recall the ratios given in Figs. 11-7 and 11-8.

θ	$\sin\theta$	$\cos\theta$	$\tan\theta$	$\csc\theta$	$\sec\theta$	$\text{ctn}\,\theta$
30°	$\frac{1}{2}$					
45°		$\frac{\sqrt{2}}{2}$		$\sqrt{2}$		
60°					2	

2. Given the triangle in Fig. 11-9, give all six of the trigonometric functions of each θ and ϕ. Since θ and ϕ are *complementary* angles, determine which func-

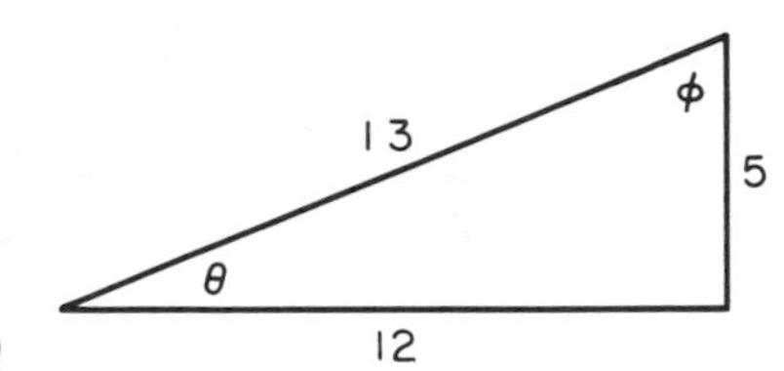

Figure 11-9

tions of θ and ϕ are equal; this might provide a lead to the prefix *co-* on three of the trigonometric fuctions.

3. Given two points on a cartesian coordinate plane with descriptions $P_1(-3, -4)$ and $P_2(2, 3)$, what is the tangent of the angle θ formed by a horizontal line drawn to the right through P_1 and by the line P_1P_2?

4. Given Fig. 11-10 with $AB = 1$, $BC = 2$, and $CD = 2$, find sec ϕ and cos θ.

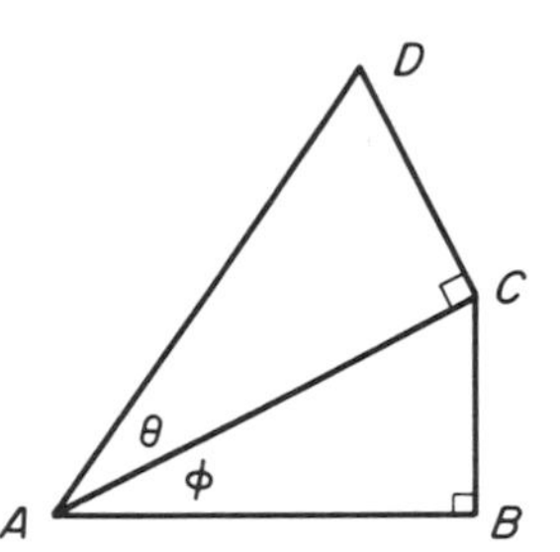

Figure 11-10

5. Given a 30°–60° right triangle with hypotenuse 10, find the lengths of the legs of the triangle, using the definitions of the sine and cosine functions.

6. In Fig. 11-4 we have the Pythagorean relationship

$$x^2 + y^2 = \rho^2. \tag{14}$$

If we divide through by ρ^2, we have

$$\frac{x^2}{\rho^2} + \frac{y^2}{\rho^2} = 1$$

or

$$\left(\frac{x}{\rho}\right)^2 + \left(\frac{y}{\rho}\right)^2 = 1, \tag{15}$$

where, by (2), $\cos\theta = x/\rho$ and, by (1), $\sin\theta = y/\rho$. Substituting them into (14) or (15), we have the identity

$$(\cos\theta)^2 + (\sin\theta)^2 = 1,$$

which asserts that the sum of the squares of the sine and cosine functions by any angle equals unity. By using similar techniques in dividing (14) by x^2, establish the identity

$$1 + (\tan\theta)^2 = (\sec\theta)^2.$$

Also, divide (14) by y^2 and obtain the identity

$$1 + (\text{ctn}\,\theta)^2 = (\csc\theta)^2.$$

7. Three forces F_1, F_2, and F_3 are applied to an object A in the directions shown in Fig. 11-11. The magnitudes of the forces are $F_1 = 10$ lb, $F_2 = 12$ lb, and $F_3 = 15$ lb. If AF_1 is parallel to the x axis, find the sum of the horizontal components contributed by the three forces; the vertical sum.

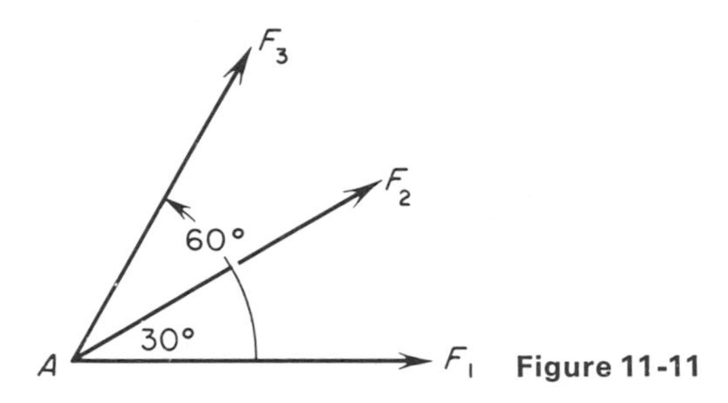

Figure 11-11

8. In Exercise 7, resolve the three forces into a single force whose horizontal and vertical components are the sums given as answers. Find the cosine function of the reference angle of the new vector.

11-3 Functions of angles reducible to first-quadrant angles

In preceding sections our discussion was restricted primarily to first-quadrant angles—that is, to angles between 0° and 90°. In this section we enlarge our discussion to include angles in all quadrants. We will use reference angles that are multiples of 30° and 45°.

In Sec. 11-1 we established the convention that reference angles are measured from the x axis. In Fig. 11-12 we see a vector **P** with reference angle θ. If θ is sufficiently large (as in Fig. 11-12) to place the vector in the second quadrant, then right triangle OPM has an acute angle $180° - \theta$, with a vertical component that is positive and a negative horizontal component. Since $180° - \theta$ is now less than 90°, we have reduced the second-quadrant angle θ to a first-quadrant angle $180° - \theta$, and we can proceed to describe the trigonometric functions of θ by using the proper signs of the components.

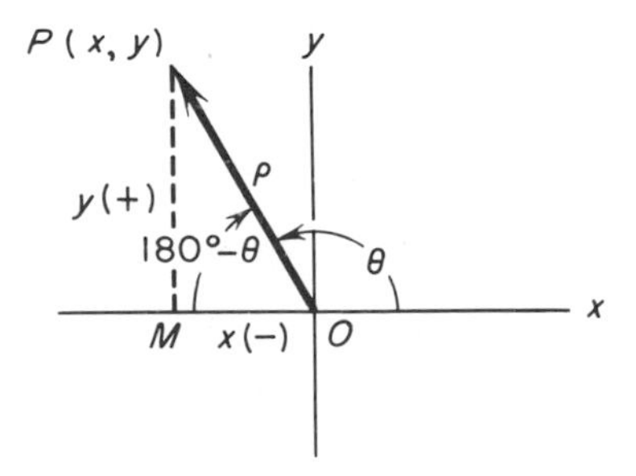

Figure 11-12

Example 4 Find (a) tan 135°, (b) sin 210°, and (c) sec 300°.

Solution Let us follow a four-step procedure.

1. Sketch the vector in the proper position.
2. Drop a perpendicular from the end of the vector to the x axis.
3. Determine the signs of the components of the vector.
4. Use the definitions to determine the values of the trigonometric functions of the angle involved.

(a) To find tan 135°, sketch vector **P** with reference angle 135° as in Fig. 11-13(a). The perpendicular drawn from P to the x axis forms a 45° right triangle with horizontal component $x = -1$, vertical component $y = +1$, and magnitude $\rho = \sqrt{2}$ (this assumes ratio $s : s : s\sqrt{2}$ and proper signs). Then by definition (3),

$$\tan 135° = \frac{y}{x} = \frac{+1}{-1} = -1.$$

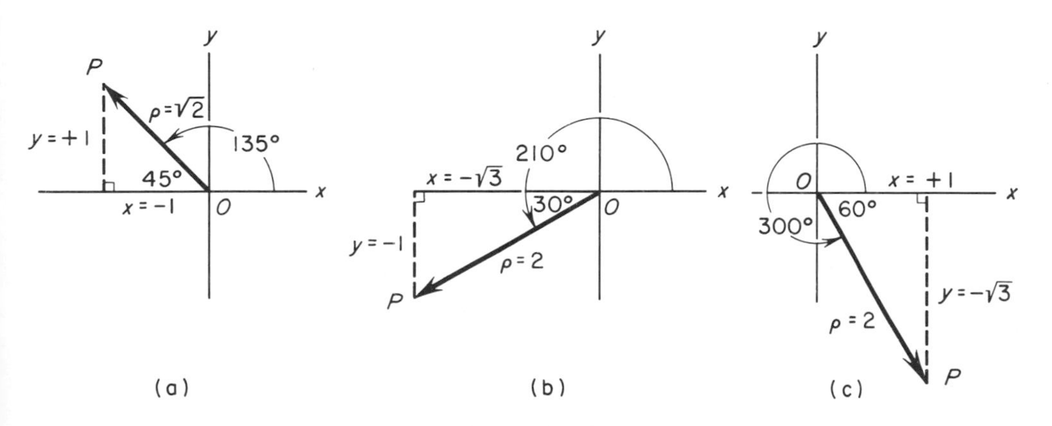

Figure 11-13

(b) To find sin 210°, sketch vector **P** as in Fig. 11-13(b). The right triangle formed by dropping a perpendicular from P to the x axis is a 30°–60° right triangle, and we consider the angle reduced to the first-quadrant angle 30° with components $x = -\sqrt{3}$, $y = -1$ and magnitude $\rho = 2$. By definition (1),

$$\sin 210° = \frac{y}{\rho} = \frac{-1}{2} = -\frac{1}{2}.$$

(c) To find sec 300°, sketch vector **P** as in Fig. 11-13(c). Assign signs and numbers to the components and magnitude, as dictated by the fourth-quadrant position of the 60° angle. Then by definition (5),

$$\sec 300° = \frac{\rho}{x} = \frac{2}{+1} = 2.$$

For angles greater than 360° or for negative angles, we proceed in the fashion discussed in Example 4. Sketch the vector in its appropriate position, drop a perpendicular from the end of the vector to the x axis, determine the signs of the components, and then apply the definitions.

Example 5 Find (a) ctn 930° and (b) csc (−420°).

Solution

(a) In Fig. 11-14(a) we see that the vector with reference angle 930° is two complete revolutions plus 210° more removed from the zero position. This places the vector in the third quadrant, 30° past the 180° position. Resolving the components and applying the proper definition, we have

$$\text{ctn } 930° = \text{ctn } 210° = \frac{x}{y} = \frac{-\sqrt{3}}{-1} = \sqrt{3}.$$

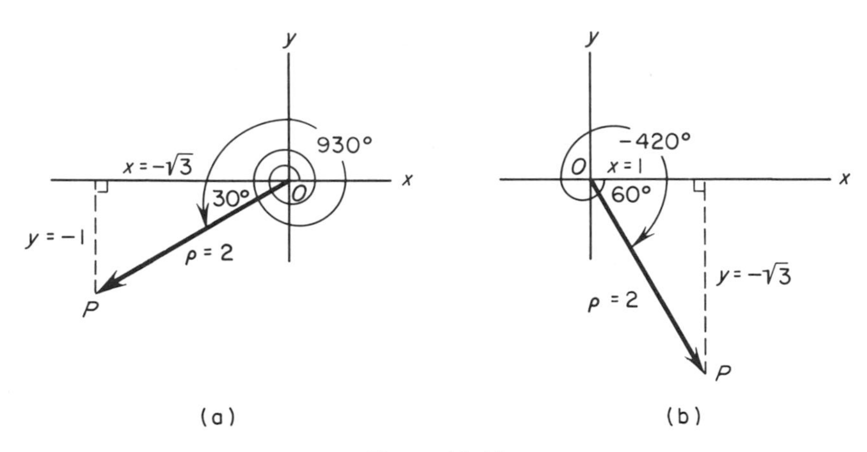

Figure 11-14

(b) From Fig. 11-14(b) the reference angle sends us −420° or one full revolution in the clockwise direction, plus 60° more. This places the vector at −60° (or +300°) in the fourth quadrant. Applying previous procedures, we have

$$\csc(-420°) = \csc(-60°) = \csc 300° = \frac{\rho}{y} = \frac{2}{-\sqrt{3}} = -\frac{2\sqrt{3}}{3}.$$

We are now ready to apply the reverse of Examples 4 and 5. We may be given the components of a vector and asked to find the reference angle. For this discussion we confine ourselves to positive values of θ less than 360°. Actually, an infinite number of values of θ will be available; discussion of all values follows in a later section.

Example 6 Find θ if (a) $\tan\theta = -1/+1$ and (b) $\sec\theta = -(2/\sqrt{3})$.

Solution

(a) From the given information and definition (3)

$$\tan\theta = \frac{-1}{+1} = \frac{y}{x},$$

which suggests components $y = -1$ and $x = +1$. Sketching these components, we have only one position available for **P** as shown in Fig. 11-15(a). So when $\tan\theta = -1/+1$, then $\theta = 315°$.

(b) From the given information and definition (5),

$$\sec\theta = -\frac{2}{\sqrt{3}} = \frac{\rho}{x}.$$

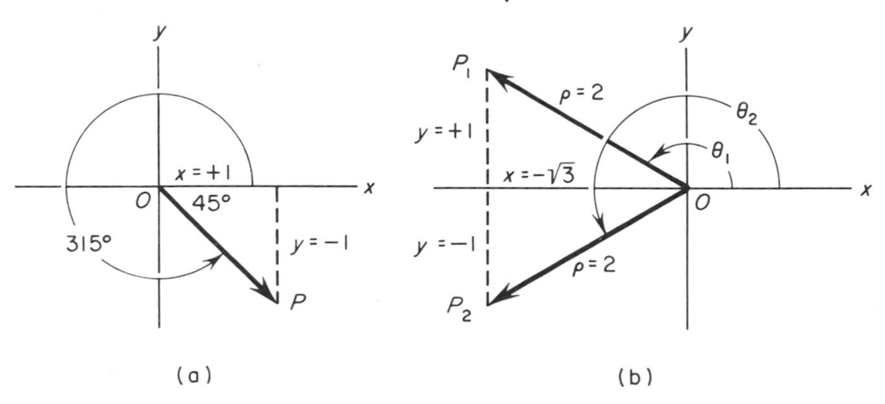

Figure 11-15

By convention, ρ is always positive; so $\rho = +2$. This result requires that $x = -\sqrt{3}$. Sketching the vector as shown in Fig. 11-15(b), two different vectors can be drawn with $\rho = +2$ and $x = -\sqrt{3}$. Consequently, we have two solutions:

$$\theta_1 = 180° - 30° = 150°$$

and

$$\theta_2 = 180° + 30° = 210°.$$

It is worth noting that if we had been given $\tan\theta = -1$ in part (a), we would have been obliged to show two solutions there $\cdots\theta = 135°, 315°\cdots$ because the signs of the components are not specified.

Exercises 11-3

1. Give all six trigonometric functions of each of the following angles: 30°, 45°, 60°, 120°, 135°, 150°, 210°, 225°, 240°, 300°, 315°, 330°. Since there will be 72 answers, it may be helpful to arrange them in a chart.

Determine whether the expressions given in Exercises 2–15 are true or false. We suggest that you do so by sketching vectors representing each member of the expression, where the magnitudes are identical and a general value of θ is assumed.

2. $\cos\theta = \cos(-\theta)$

3. $\sin\theta = \sin(-\theta)$

4. $\tan\theta = \tan(\theta + 180°)$

5. $\tan\theta = -\tan(\theta + 90°)$

6. $\sec\theta = \sec(\theta + 90°)$

7. $\sin\theta = \sin(180° - \theta)$

8. $\csc\theta = \csc(180° - \theta)$

9. $\text{ctn}\,\theta = -\text{ctn}(180° - \theta)$

10. $\cos\theta = \cos(180° - \theta)$

11. $\cos\theta = \cos(360° - \theta)$

12. $\sin\theta = \sin(360° - \theta)$

13. $\sin(180° + \theta) = \sin(180° - \theta)$

14. $\tan(180° + \theta) = -\tan(180° - \theta)$

15. $\cos(180° + \theta) = \cos(180° - \theta)$

16. Complete the table below, where the entries indicate the sign of a trigonometric function for reference angles in the quadrant indicated.

Quadrant	sin	cos	tan	csc	sec	ctn
I		+				+
II	+		−			
III				−		
IV					+	

Give values of θ (there may be more than one solution) for $0° < \theta < 360°$ that are solutions of the expressions given in Exercises 17–26.

17. $\tan\theta = \sqrt{3}$

18. $\sec\theta = -\sqrt{2}$

19. $\csc\theta = -\sqrt{2}$

20. $\sin\theta = \frac{1}{2}$

21. $\cos\theta = -\frac{1}{2}$

22. $\tan\theta = \frac{-\sqrt{3}}{+1}$

23. $\text{ctn}\,\theta = \frac{-1}{-1}$

24. $\sec\theta = \frac{2}{\sqrt{3}}$

25. $\sin\theta = -\frac{1}{2}$

26. $\sin\theta = \cos\theta$

Find the value of each of the given expressions.

27. $\sin 930°$

28. $\cos 1470°$

29. $\tan 1125°$

30. $\csc(-330°)$

31. $\sec(-390°)$

32. $\text{ctn}(-675°)$

11-4 Trigonometric functions of 0°, 90°, 180°, 270°

Earlier we carefully avoided reference to trigonometric functions of angles coincident with the axes of the cartesian coordinate plane—that is, angles of magnitude 0, 90, 180, and 270°. One reason is the requirement of division by zero. Let us consider division by zero or, more exactly, division of a nonzero quantity by a number approaching zero.

Given $10/k$, let us assign values to k, where successive values of k approach zero.

Let $k = 1$; then $\frac{10}{k} = 10.$

Let $k = 0.01$; then $\frac{10}{k} = \frac{10}{0.01} = 1000.$

Let $k = 0.0001$; then $\frac{10}{k} = \frac{10}{0.0001} = 100{,}000.$

Let $k = 1 \times 10^{-10}$; then $\frac{10}{1 \times 10^{-10}} = 1 \times 10^{11}.$

We see that as k goes to zero ($k \longrightarrow 0$), $10/k$ becomes very large. If k is zero, then $10/k$ is immense; this immense quantity is symbolized ∞ and we call it infinitely large. Note that ∞ is no real number; it cannot be manipulated by the usual laws of algebra. We note, too, that if k had been negative and approaching zero, we would eventually arrive at the same division by zero, suggesting that there is no real distinction between ∞ and $-\infty$.

Let us consider the functions of 90°. Here we sketch vector **P** in several positions with reference angles approaching 90° as in Fig. 11-16(a) Notice the behavior of the magnitude and the components. First, the magnitude of the vector is assumed to remain fixed. In Fig. 11-16(a) the magnitude is the radius of an origin-centered circle. Next, as P passes through P_1, P_2, and P_3;—that is, as θ increases toward 90°—the y component increases as at y_1, y_2, and y_3. Similarly, the x component diminishes. By inspection, we agree that as $\theta \longrightarrow 90°$, the y component approaches as its limit the length OP, which is the magnitude ρ; also, the x component approaches zero. In mathematical symbols, we have:

$$\text{as } \theta \longrightarrow 90°, \quad \text{then} \quad y \longrightarrow \rho \quad \text{and} \quad x \longrightarrow 0.$$

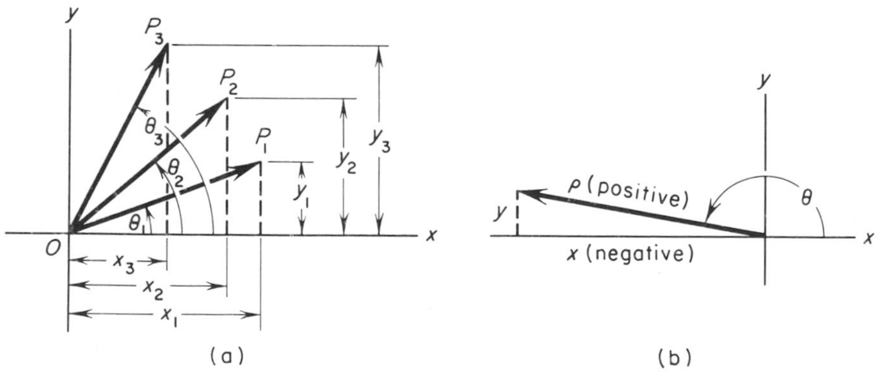

Figure 11-16

Consider the functions of 90°.

$$\sin 90° = \frac{y}{\rho} = \frac{\rho}{\rho} = 1 \qquad \csc 90° = \frac{\rho}{y} = \frac{\rho}{\rho} = 1,$$

$$\cos 90° = \frac{x}{\rho} = \frac{0}{\rho} = 0 \qquad \sec 90° = \frac{\rho}{x} = \frac{\rho}{0} = \infty$$

$$\tan 90° = \frac{y}{x} = \frac{y}{0} = \infty \qquad \text{ctn}\ 90° = \frac{x}{y} = \frac{0}{y} = 0$$

The functions of 0, 180, and 270° can be determined as in preceding paragraphs. First, determine what happens to the components as the reference angle approaches the angle desired and then apply the definitions. A note of caution; consider Fig. 11-16(b). As $\theta \longrightarrow 180°$, the x component approaches a length equal to the magnitude ρ; at least it seems to. However, we have accepted the convention that magnitude ρ is always positive. As $\theta \longrightarrow 180°$, the x component, being negative, approaches $-\rho$. Ratios involving the magnitude and the x component at 180° will therefore be negative. Similarly, as $\theta \longrightarrow 270°$, the y component approaches $-\rho$ and functions of 270° involving the magnitude and the y component will be negative.

Exercises 11-4

1. Give all six trigonometric functions of each of angles 0, 90, 180, and 270°. Since 24 answers are involved, we suggest that they be arranged in a chart.
2. Using the results in Exercise 1 and the discussion in Sec. 11-4, answer the following questions.
   (a) What is the range of value of $\sin \theta$? That is, as θ increases from 0 to 360°, $\sin \theta$ passes through a set of values. All these values lie between two certain numbers; what are these two numbers?
   (b) What is the range of values of $\cos \theta$?
   (c) What is the range or values of $\tan \theta$?
   (d) What is the range of values of $\sec \theta$? $\csc \theta$? $\text{ctn}\ \theta$?
3. Using sketches of vectors, their ratios of components to magnitudes, and the definitions of the functions, explain why $\sec \theta$ and $\csc \theta$ can never have values between -1 and $+1$.
4. If a vector has a horizontal component of zero and a vertical component equal to $-\rho$, what is the reference angle?
5. If a vector has a vertical component of zero and a horizontal component equal to ρ, what is the reference angle?
6. Consider a reference angle between 0 and 90°. Which trigonometric function always has the larger value, $\sin \theta$ or $\tan \theta$? (*Suggestion:* Sketch a vector in various positions between 0 and 90°, determine the components, and set up appropriate ratios from the definitions and compare.)

11-5 Cofunctions and reciprocal functions

From definitions (1) through (6) it is clear that certain trigonometric functions are reciprocals of others. Restating these definitions with reference to Fig. 11-17, we have

$$\sin\theta = \frac{y}{\rho}, \tag{16}$$

$$\cos\theta = \frac{x}{\rho}, \tag{17}$$

$$\tan\theta = \frac{y}{x}, \tag{18}$$

$$\csc\theta = \frac{\rho}{y}, \tag{19}$$

$$\sec\theta = \frac{\rho}{x}, \tag{20}$$

$$\operatorname{ctn}\theta = \frac{x}{y}. \tag{21}$$

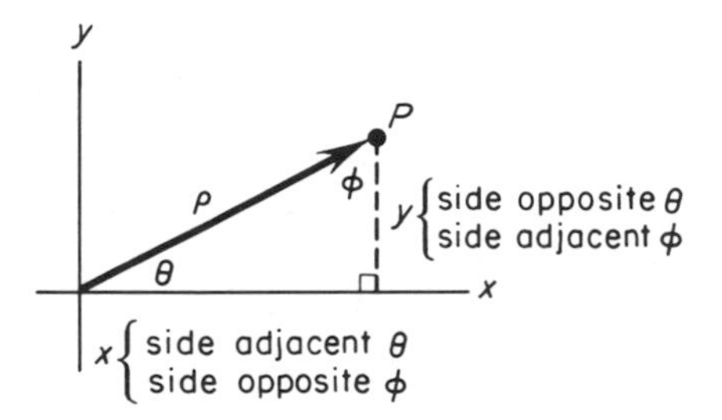

Figure 11-17

Referring to definitions (16) and (19), we see that $\sin\theta = y/\rho$ and $\csc\theta = \rho/y$. Considering that $y/\rho = 1/(\rho/y)$, then

$$\sin\theta = \frac{1}{\csc\theta} \quad \text{or} \quad \csc\theta = \frac{1}{\sin\theta}. \tag{22}$$

Similarly, from (17) and (20)

$$\cos\theta = \frac{1}{\sec\theta} \quad \text{or} \quad \sec\theta = \frac{1}{\cos\theta}, \tag{23}$$

and from (18) and (21)

$$\tan\theta = \frac{1}{\operatorname{ctn}\theta} \quad \text{or} \quad \operatorname{ctn}\theta = \frac{1}{\tan\theta}. \tag{24}$$

Expressions (22), (23), and (24) are called the *reciprocal relationships* between the trigonometric functions. They are particularly useful in simplifying trigonometric expressions, in computations where they will permit multiplication instead of division, and in calculator operations.

Referring again to Fig. 11-17, we can determine another group of relationships between trigonometric functions. First, observe that three of the six trigonometric functions possess the prefix *co-* —namely, the cosine, cosecant, and cotangent. This situation suggests that there might be some relationship between the sine and cosine, the secant and cosecant, and the tangent and cotangent.

In Fig. 11-17 we see two angles θ and ϕ; the triangle shown is a right triangle so that $\theta + \phi = 90°$. In other words, θ and ϕ are *complementary* angles. Next, observe any function of θ, say $\sin\theta$. From previous definitions

$$\sin\theta = \frac{y}{\rho} = \frac{\text{side opposite } \theta}{\rho}. \tag{25}$$

Then notice the same ratio, y/ρ, as a function of angle ϕ. We see that the side opposite θ is the side adjacent ϕ and

$$\cos\phi = \frac{\text{side adjacent } \phi}{\rho}. \tag{26}$$

From (25) and (26) we have

$$\text{sine } \theta = \textit{co}\text{sine } \phi \qquad (\theta + \phi = 90°). \tag{27}$$

Then since θ and ϕ are complementary angles, we see from (27) that the *sine of an angle equals the cosine of the complementary angle or the cosine of an angle equals the sine of the complementary angle.* Apparently the prefix *co-* is associated with the *co*mplementary angle.

By reference to Fig. 11-17, we find that

$$\tan\theta = \text{ctn}\,\phi \tag{28}$$

and

$$\sec\theta = \csc\phi, \tag{29}$$

where θ and ϕ are complementary angles and (27), (28), and (29) are the *cofunction relationships*. They are useful in many of the same ways that the reciprocal relationships are useful.

Example 7 Find θ for $\theta \leqq 90°$ if (a) $\tan\theta = \text{ctn}\,30°$, (b) $\csc\theta = 1/\sin 22°$, and (c) $\sec\theta = 1/\sin 15°$.

Solution

(a) Given $\tan\theta = \text{ctn}\,30°$. By (28), the tangent of an angle equals the cotangent of the complementary angle. Therefore

$$\tan 60^\circ = \text{ctn } 30^\circ \quad \text{and} \quad \theta = 60^\circ.$$

(b) Given $\csc \theta = 1/\sin 22^\circ$. By (22), $\csc \theta = 1/\sin \theta$ and so

$$\csc 22^\circ = \frac{1}{\sin 22^\circ} \quad \text{and} \quad \theta = 22^\circ.$$

(c) Given $\sec \theta = 1/\sin 15^\circ$. By (23), $\sec \theta = 1/\cos \theta$. Also, by (27), $\sin \theta = \cos (90^\circ - \theta)$. Then

$$\frac{1}{\sin 15^\circ} = \frac{1}{\cos 75^\circ} = \sec 75^\circ$$

and we have solution $\theta = 75^\circ$.

Exercises 11-5

In Exercises 1–12, find the first quadrant value of θ for the given expressions.

1. $\sin 47^\circ = \cos \theta$
2. $\sin 62^\circ = \dfrac{1}{\csc \theta}$
3. $\tan \theta = \text{ctn } 14^\circ$
4. $\tan \theta = \dfrac{1}{\text{ctn } 25^\circ}$
5. $\cos \theta = \dfrac{1}{\sec 12^\circ}$
6. $\cos \theta = \sin 82^\circ$
7. $\cos \theta = \dfrac{1}{\csc 32^\circ}$
8. $\text{ctn } \theta = \tan 85^\circ$
9. $\sin \theta \csc \theta = 1$
10. $\sec \theta \cos \theta = 1$
11. $\sin \theta \sec \theta = \dfrac{1}{\csc \theta}$
12. $\csc 72^\circ = \dfrac{1}{\sin \theta}$
13. In Fig. 11-18 we are given a right triangle with acute angle θ and a unity hypotenuse. Using the definitions of the trigonometric functions, show that (a) $y = \sin \theta$, (b) $x = \cos \theta$, and (c) $\sin \theta / \cos \theta = \tan \theta$. Using the Pythagorean Theorem and the conclusions in parts (a) and (b), show that

$$(\sin \theta)^2 + (\cos \theta)^2 = 1.$$

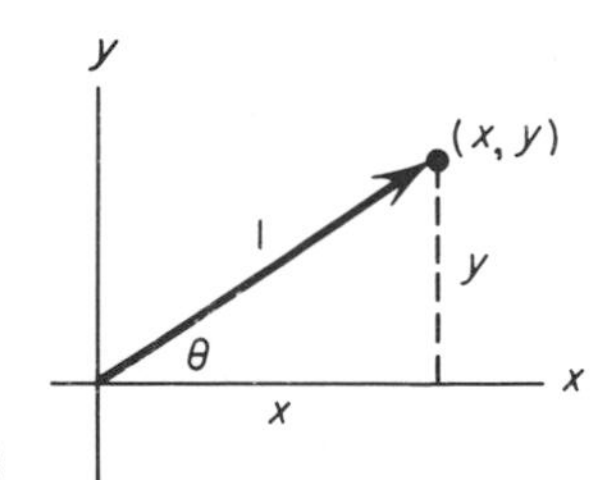

Figure 11-18

14. Using reciprocal and cofunction relationships, show that any trigonometric function of an acute angle can be expressed in terms of the sine or tangent of the angle or its complement. The conclusion here is useful in calculator operations.

15. In Fig. 11-19 we are given a right triangle with acute angle θ and a unity x component. Using the definitions of the trigonometric functions, show that (a) $y = \tan\theta$, (b) $\rho = \sec\theta$, and (c) $\tan\theta/\sec\theta = \sin\theta$. Using the Pythagorean Theorem and the conclusions in parts (a) and (b), show that

$$1 + (\tan\theta)^2 = (\sec\theta)^2.$$

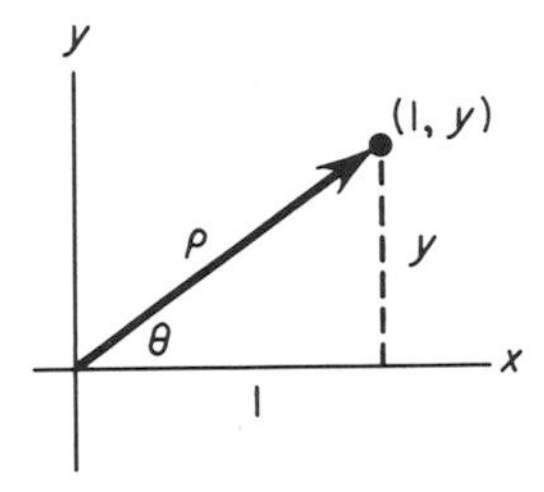

Figure 11-19

16. Using a triangle as in Fig. 11-19 with an acute angle θ and a unity x component, show that

(a) y component $= \tan\theta$.
(b) $\rho = \sec\theta$.
(c) $\cos\theta = \dfrac{\operatorname{ctn}\theta}{\csc\theta}$.
(d) $1 + (\operatorname{ctn}\theta)^2 = (\csc\theta)^2$.

17. Using methods similar to those in Exercises 13, 15, and 16, show in Fig. 11-20 that $u = \sec\theta$, $v = \tan\theta$, $z = \sec\theta\tan\phi$, and $w = \sec\theta\sec\phi$.

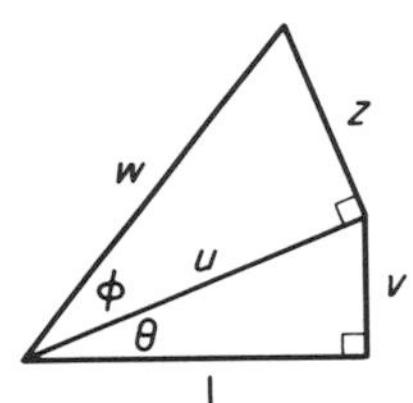

Figure 11-20

18. Given Fig. 11-21 with $ABCD$ a rectangle with diagonal AC forming an acute angle θ with side AB. Angle $EDC = 90°$. $EFAD$ is a square. Edge $CB = 1$.

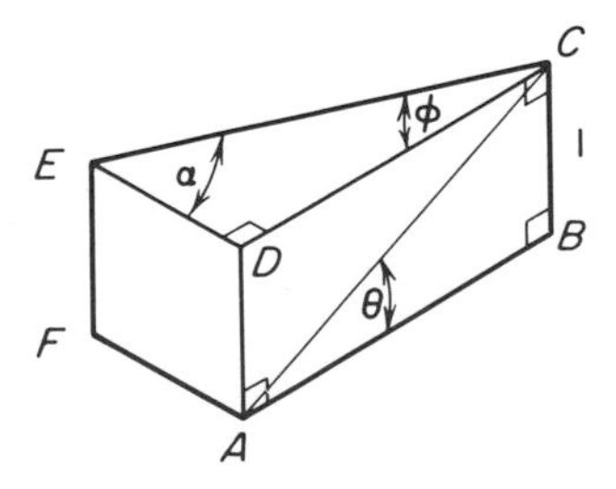

Figure 11-21

Show that
(a) $ED = \tan \phi \operatorname{ctn} \theta$.
(b) $ED = \operatorname{ctn} \alpha \operatorname{ctn} \theta$.

19. Given Fig. 11-22 with $AB = 1$, show that $EC = (\cos \alpha)(\sin \beta \tan \theta + \cos \beta)$.

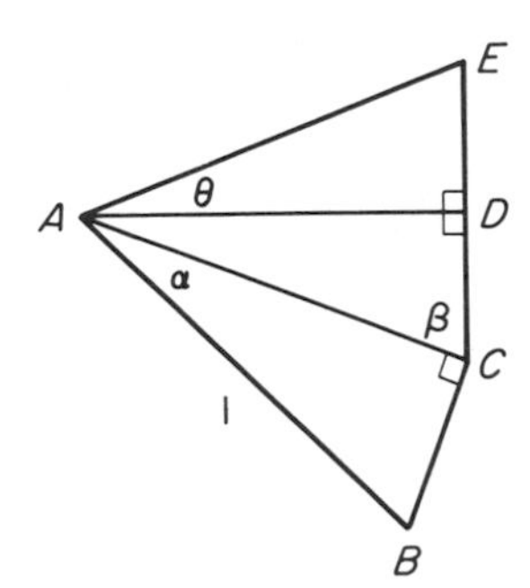

Figure 11-22

11-6 Vector quantities

In many applications, such quantities as force, velocity, and acceleration are involved. These quantities have magnitude, direction, and sense and are called *vector quantities*. Earlier we defined the vector quantity and the scalar quantity. We did not clearly identify the third property of the vector—*sense*. A vehicle traveling at 30 mph in a specified direction is said to have a velocity of 30mph; velocity here is a vector quantity. At times we may assume that the direction was totally specified when, in fact, it was not. Suppose that we state that a vehicle is traveling at 30 mph in a direction parallel to the Equator. In this case, magnitude is described fully, direction is described, but we do not know whether the vehicle is eastbound or westbound; this factor concerns *sense*. The barb at the end of a vector gives sense to the vector.

The following paragraphs contain a limited amount of repetition of material from Sec. 11-1. This material is directed toward the solution of vector problems.

Example 8 Three forces are applied to a body. Force F_1 is 6 lb in a 40° direction, force F_2 is 8 lb in a 180° direction, and force F_2 is 4 lb in a 300° direction. Sketch the system in a cartesian coordinate plane.

Solution Using methods discussed in Sec. 11-1, we sketch the system as shown in Fig. 11-23.

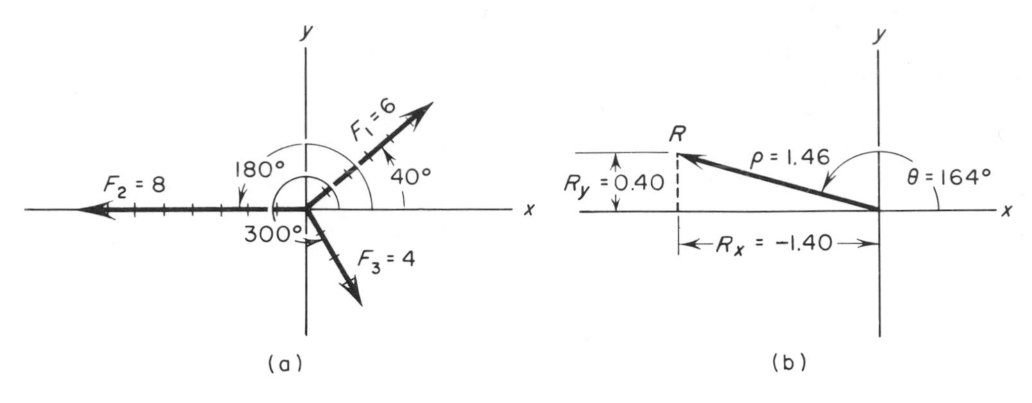

Figure 11-23

Components of vectors

Operations with vectors (adding and subtracting) are often performed by determining the rectangular components of the vectors. These components are most wisely chosen as the horizontal and vertical components of the vectors. Thus in Fig. 11-24(a) the horizontal component of OP is

$$OM = OP \cos \theta \tag{30}$$

and the vertical component of OP is

$$ON = PM = OP \sin \theta. \tag{31}$$

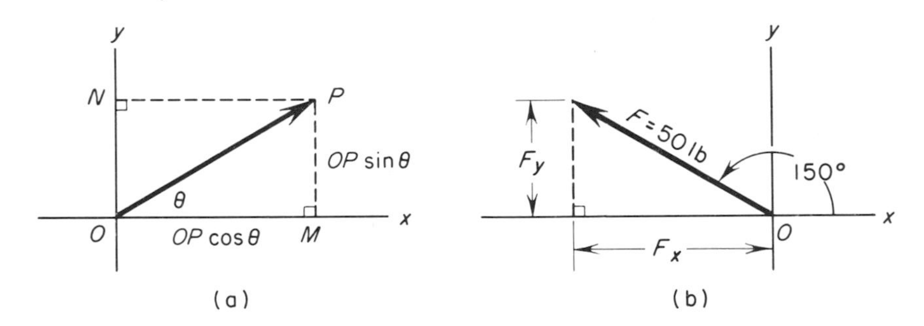

Figure 11-24

These rectangular components are the horizontal and vertical projections of the magnitude of the vector on the coordinate axes. The process of breaking down a vector into its components is called *resolution* or *resolution of components*.

Example 9 A force $F = 50$ lb is applied to a body in a direction of 150°. Find the horizontal component F_x and the vertical component F_y of this force.

Solution By (30) and (31) and Fig. 11-24(b),

$$F_x = 50 \cos 150° = -43.3 \text{ lb}$$

and

$$F_y = 50 \sin 150° = +25 \text{ lb}.$$

Resultant vectors

If two forces are applied to a body from different directions or the same direction, then there is a single force (with its own direction) that could be applied to obtain the same result on the body. This single force is called the *resultant force* and, in the more general case of vectors, a single vector that can provide the displacement of two or more vectors is called the *resultant vector*. From Example 9 and Fig. 11-24(b) two forces (25 lb at 90° and 43.3 lb at 180°) have a resultant of 50 lb at 150°.

A method commonly used to find the resultant of several vectors is to resolve the vectors into their rectangular components; the resultant vector will have as its horizontal component the sum of the horizontal components of the several vectors and as its vertical component the sum of the vertical components of the several vectors. Thus resultant **R** of vectors **A**, **B**, and **C** has

$$R_x = A_x + B_x + C_x,$$

$$R_y = A_y + B_y + C_y.$$

Example 10 Find the resultant of the forces described in Example 8 and pictured in Fig. 11-23(a).

Solution For force F_1,

$$F_{1x} = 6 \cos 40° = 4.60,$$

$$F_{1y} = 6 \sin 40° = 3.86,$$

meaning that the horizontal component of F_1 is of magnitude 4.6 lb and the vertical component is 3.86 lb.

For F_2, none of the force is applied in the vertical direction and all of it is applied to the left in the horizontal direction so that

$$F_{2x} = 8 \cos 180° = -8.00,$$

$$F_{2y} = 8 \sin 180° = 0.$$

For F_3,

$$F_{3x} = 4 \cos 300° = 2.00,$$

$$F_{3y} = 4 \sin 300° = -3.46.$$

The horizontal component R_x of resultant **R** is

$$R_x = F_{1x} + F_{2x} + F_{3x} = 4.60 - 8.00 + 2.00 = -1.40$$

and the vertical component R_y of resultant **R** is

$$R_y = F_{1y} + F_{2y} + F_{3y} = 3.86 + 0 - 3.46 = +0.40.$$

Figure 11-23(b) shows the resultant **R** drawn to a larger scale for clarity. There

$$\rho = \sqrt{R_x^2 + R_y^2} = \sqrt{(-1.40)^2 + (0.40)^2} = 1.46$$

and $$\tan \theta = \frac{R_y}{R_x} = \frac{0.40}{-1.40} = -0.286$$

from which $\theta = 164°$. We conclude that the resultant of the system shown in Fig. 11-23(a) is a force of magnitude 1.46 and direction $\theta = 164°$.

In Example 10 the reader may not yet be prepared to find θ by use of tables where $\tan \theta = 0.40/-1.40$. In this case, θ may be found by use of a protractor where a vector with components 0.40 and -1.40 is laid off as in Fig. 11-23(b) and θ is found by measurement. The same applies to Example 8, where a 40° angle is given and the components of a vector of magnitude 6 are needed.

The *equilibrant* of a force system is the single force that would balance the system—that is, the force that would exactly nullify the resultant. In Example 10 the resultant is of magnitude 1.46 at an angle of 164°. The equilibrant is of equal magnitude but oppositely directed, or of magnitude 1.46 at 344°.

Parallelogram method applied to vector problems

In previous examples we considered vector problems in which all vectors originated in a common point. Often, however, we wish to consider a group of vectors in series; that is, the first vector may originate at the origin in a cartesian coordinate plane, the second where the first terminates, the third where the second terminates, and so forth. The method of solution of such a series system can be similar to that shown in Example 10. We could also use the parallelogram method, which is described in Example 11.

Example 11 The progress of an aircraft is plotted on a cartesian coordinate plane. It flies at a heading of 45° at a velocity of 200 mph. A crosswind of 40 mph blows steadily on a heading of 300°. How many ground miles does the aircraft fly in one hour and what is its resultant direction?

Solution In Fig. 11-25 we have pictured the vectors involved. Note that vector **A** represents the progress of the aircraft, neglecting the crosswind. Its direction, 45°, is measured from the right-hand horizontal

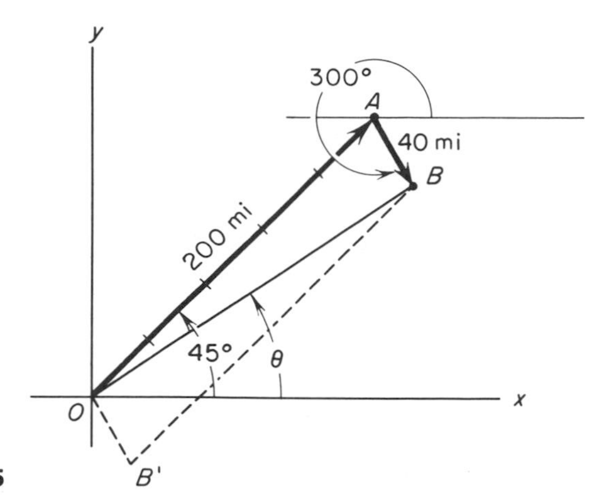

Figure 11-25

in a counterclockwise direction. Vector **B** represents the crosswind; its direction is measured in the same manner. OB represents the resultant with magnitude OB and direction θ.

A parallelogram $OABB'$ can be sketched. From the properties of the parallelogram $OB' = AB$ and OB' is in the same direction as AB; therefore, OB may be considered the resultant of A and B'. The system is the same one described in previous sections where all the vectors originate from a common point. Solving, we have the horizontal component of OB as

$$(OB)_x = 200 \cos 45^\circ + 40 \cos 300^\circ = 161.4$$

and the vertical component of OB as

$$(OB)_y = 200 \sin 45^\circ + 40 \sin 300^\circ = 106.8.$$

Then

$$OB = \sqrt{(OB)_x^2 + (OB)_y^2} = \sqrt{(161.4)^2 + (106.8)^2} = 193.4$$

and

$$\tan \theta = \frac{(OB)_y}{(OB)_x} = \frac{106.8}{161.4} = 0.661$$

from which $\theta = 33.4^\circ$. Once again the reader may not be prepared to evaluate θ where $\tan \theta = 0.661$ is known. However, by carefully sketching the vector with components 106.8 and 161.4 and measuring the reference angle with a protractor, θ can be determined.

We conclude that the actual distance traveled by the aircraft is $OB = 193.4$ ground miles and the direction is $\theta = 33.4^\circ$.

Exercises 11-6

Provide solutions as directed. when the reference angle is not a multiple of 30 or 45°, draw the vector to scale, using a straightedge and protractor. Where the components are known, find the angle with a protractor.

1. An automobile travels at 45 mph in a direction 30° east of north. What are its northerly and easterly components of velocity?

2. A projectile fired from the ground at a distant target is traveling, at a certain instant, at 1800 fps and is falling toward the earth such that the angle between its path and the horizontal is 15°. What are its horizontal and vertical components of velocity?

3. An object on a cartesian coordinate plane starts at a point A. It moves from A to B, where the direction is 225° and the straight-line distance from A to B is 12 in. Next, it proceeds from B to C, where the direction is 300° and the distance is 15 in. Find the straight-line distance AC and the direction from A to C.

4. Find the magnitude and direction of the system of forces shown in Fig. 11-26 if
(a) $F_1 = 95$ lb, $F_2 = 210$ lb, $F_3 = 0$; $\theta = 25°$, $\phi = 130°$.
(b) $F_1 = 95$ lb, $F_2 = 210$ lb, $F_3 = 95$ lb; $\theta = 25°$, $\phi = 130°$, $\gamma = 235°$.
(c) $F_1 = F_2 = F_3$; $\phi = \theta + 120°$, $\gamma = \phi + 120°$.

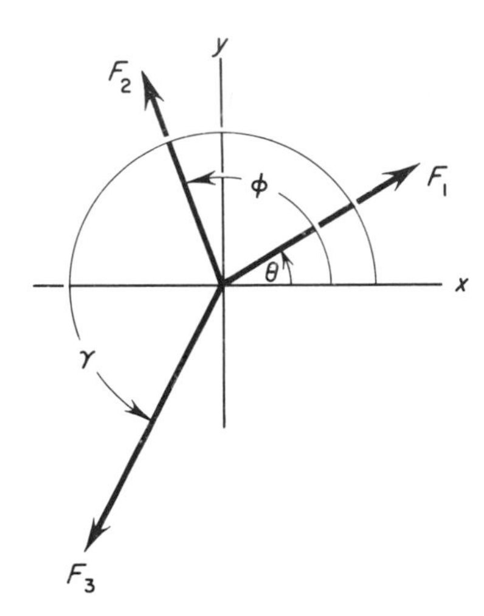

Figure 11-26

5. Find the magnitude and direction of the equilibrant of the forces in Fig. 11-27.

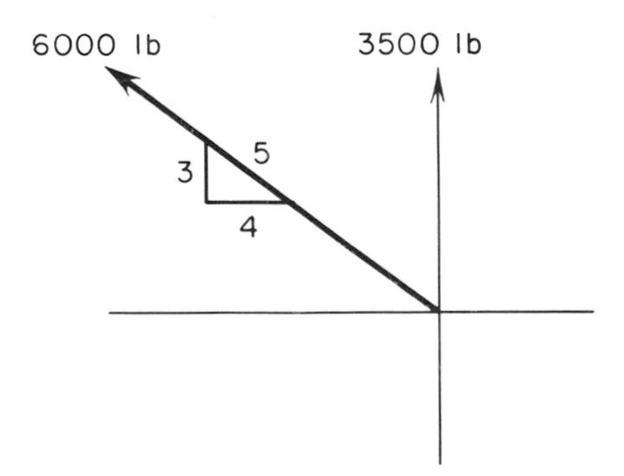

Figure 11-27

12

Computations involving right-triangle trigonometry

In Chapter 11 we defined the trigonometric functions and discussed cofunctions and reciprocal functions. Little attention was given to computations involving trigonometric functions. Since such computations are an important part of technical mathematics, they are described here.

One use of trigonometry is in indirect measurements; that is, some known quantities are used to compute another magnitude or distance or force that may not be readily measurable because of obstructions, inaccessibility, and so forth. Another use is in resolution or composition of forces. We will concentrate on these applications.

Since computation is a primary goal, use of the table of natural trigonometric functions, use of the electronic calculator, interpolation, and angular measurements are discussed.

12-1 Angular measurement, degree system

The most commonly used unit for measuring angles is the *degree*. We define a degree by referring to Fig. 12-1. A circle O has a circumference that is the length of the circular line measured along arc $A_0, A_1, A_2, \ldots,$ starting at any point on the arc and terminating at that point and passing through all points on the arc once and only once. Figure 12-1(a) shows a circle O with the circumference divided into eight equal parts; this situation creates eight equal angles like $\angle A_0OA_1$ with vertices at O, called *central angles*. If the circumference of O were divided into 360 equal arcs, there would be 360 equal central angles like $\angle A_0OA_1$ in Fig. 12-1(b). These 360 equal central angles are of magnitude 1° by definition. Figure 12-1 (b) is drawn out of scale, because of the difficulty in representing a 1° angle.

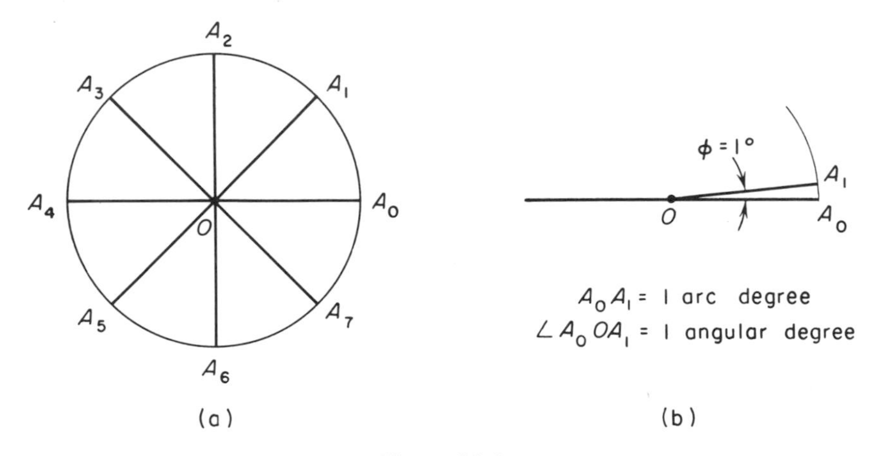

Figure 12-1

The term *degree* is not restricted to angles. Instead we refer to $\angle A_0OA$ in Fig. 12-1(b) as being of magnitude *one angular degree*. The arc A_0A_1 is called *one arc degree*. So in a circle there are 360 arc degrees and 360 angular degrees.

For accuracy in many technical situations, the degree, despite its small magnitude, must be subdivided. The subdivisions are either tenths, hundredths, thousandths, and so on (decimal) or one-sixtieth and one-sixtieth of one-sixtieth. Let us clarify the last statement. One-tenth of one degree is called simply "one-tenth of one degree" and is symbolized 0.1°. One-hundredth of one degree is 0.01° and so on. In contrast, one-sixtieth of one degree is called a *minute* and is symbolized

$$1' = \left(\frac{1}{60}\right)^\circ.$$

Similarly, one-sixtieth of one minute is called a *second* and is symbolized

$$1'' = \left(\frac{1}{60}\right)' = \left(\frac{1}{60}\right)\left(\frac{1}{60}\right)^\circ = \left(\frac{1}{3600}\right)^\circ \tag{1}$$

and

$$60' = 1^\circ, \qquad 60'' = 1', \qquad 3600'' = 1^\circ. \tag{2}$$

Conversions between decimal degrees and minutes and seconds is reasonably simple. Let us consider some examples.

Example 1 Convert 22.384° to degrees, minutes, and seconds.

Solution We have 22° and 0.384 of another degree. Two options are available. Under option 1 we can convert 0.384° to seconds and then determine minutes. Under option 2 we can find minutes and then seconds.

Option 1 by (2), $1^\circ = 3600''$. Therefore

$$0.384^\circ = (0.384 \text{ deg})(3600 \text{ sec/deg}) = 1382.4''.$$

Since $60'' = 1'$, then by (1),

$$1382.4'' = (1382.4 \text{ sec})(1/60 \text{ min/sec}) = 23'$$

with remainder 2.4″ and we have the solution $22.384^\circ = 22^\circ 23' 02.4''$.

Option 2 By (2), $1^\circ = 60'$. Therefore

$$0.384^\circ = (0.384 \text{ deg})(60 \text{ min/deg}) = 23.04'.$$

Also, by (2),

$$0.04' = (0.04 \text{ min})(60 \text{ sec/min}) = 2.4''$$

and we have $22.384^\circ = 22^\circ 23' 02.4''$.

Note: The second option is easily performed on an electronic calculator. The process involves identification of integral degrees, removing these degrees by subtraction, and being left with the fractional degree. Multiplying this fraction by 60 identifies the number of minutes; removing the integral minutes by subtraction leaves the fractional minutes. Multiplying this fractional portion of a minute by 60 gives us the seconds. Sequence of steps on an electronic calculator would be as follows.

1. Enter 22.384
2. − (minus)
3. 22 (degrees)
4. = (Register shows 0.384 degrees)
5. × (times)

6. 60
7. = (Register shows 23.04 minutes)
8. — (minus)
9. 23 (minutes)
10. = (Register shows 0.04 minute)
11. × (times)
12. 60
13. = (Register shows 2.4 seconds)

In the preceding example we converted from decimal notation to notation in minutes and degrees. In our next example the conversion is in the opposite direction.

Example 2 Convert 12°14′26″ to decimal notation accurate to thousandths.

Solution If we convert 14′26″ to seconds and divide by 3600, we will have the decimal equivalent.

$$14'26'' = (14 \text{ min})(60 \text{ sec/min}) + 26 \text{ sec} = 866''$$

From (1)

$$866' = (866 \text{ sec})(1 \text{ deg}/3600 \text{ sec}) = 0.240^\circ$$

and we have the solution 12°14′26″ = 12.240°.

Note: Conversion of degrees, minutes, and seconds to decimal degrees in this example proceeds rapidly on an electronic calculator. The process involves conversion of the partial degree to seconds and dividing these seconds by 3600 to obtain the fractional degree equivalent. Here is the sequence of steps.

1. 14
2. × (times)
3. 60
4. = (Register shows 840 seconds)
5. + (add)
6. 26
7. = (Register shows 866 seconds)
8. ÷ (division by)
9. 3600
10. = (Register shows 0.240 degree)

An idea of the magnitude of a minute or a second of angle is easily developed and is useful in considerations of accuracy. The nautical mile is a length defined by an interception of the Equator by the sides of an angle of 1′ with its vertex at the North Pole. Thus the arc length $\widehat{AB}$ in Fig. 12-2 is a nautical mile. Since there are 360° along the Equator and 60′ per degree, then $\widehat{AB}$ is 1/21,600 of the length of the Equator and is 6080 ft approximately. In other words, in distance OA, which is a quarter of the way around the earth, an angle of 1′ intercepts an arc of 6080 ft. Similarly, an angle of 1″ with vertex at the North Pole would intercept an equatorial arc of approximately (1/60)(6080) = 101.3 ft.

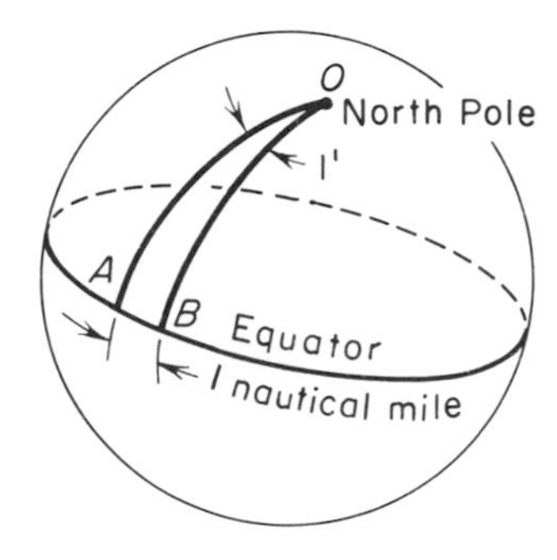

Figure 12-2

Exercises 12-1

In Exercises 1–8, convert the given decimal expressions to expressions in degrees, minutes, and seconds, correct to the nearest second.

1. 43.287° **2.** 16.006° **3.** 12.009° **4.** 97.842°

5. 108.972° **6.** 14.5° **7.** 13.6° **8.** 185.35°

In Exercises 9–16, convert the given expressions to decimal notation, correct to thousandths.

9. 12°26′45″ **10.** 82°15′15″ **11.** 76°42′00″ **12.** 118°05′00″

13. 327°00′40″ **14.** 225°09′25″ **15.** 112°03′06″ **16.** 62°59′59″

17. Referring to Fig. 12-2, find AB in statute miles (1 statute mile = 5280 ft) if $\angle AOB = 1°$.

18. According to Fig. 12-2 and the accompanying discussion, a 1′ angle "runs out" 1 nautical mile in a distance of 6250 miles. How far would the same angle run out in a distance to the moon, given as 235,000 miles?

19. Express a conversion unit between one-thousandth of a degree and seconds. That is, find x if $0.001° = x''$.

20. What is the decimal degree equivalent of one second of angle?

21. We are given that the circumference of a circle is $C = 2\pi r$ or circumference $= 2\pi$ radii. This means that if the radius were bent into a curve to fit the arc, 2π such radii would be required to fit the entire circumference. How many degrees are in a central angle that intercepts an arc equal to the radius?

12-2 Angular measurement, radian system

In Sec. 12-1 we discussed the degree system for measuring angles. Another system, commonly used in technical practice because of its properties, is the radian system. We refer to Fig. 12-3 to picture the radian, which is the basic

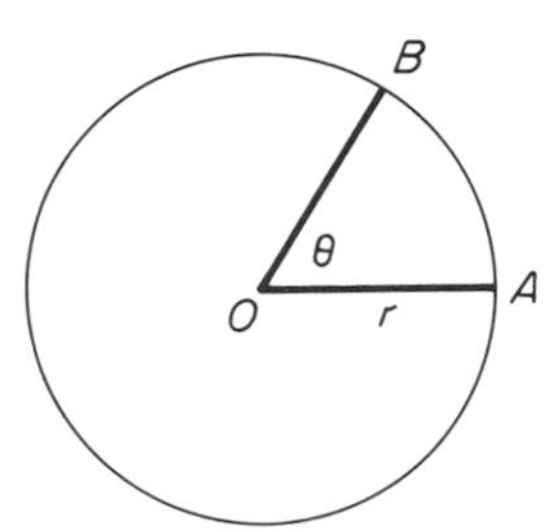

Figure 12-3

unit in the radian system. Here we have a central angle θ intercepting an arc $\widehat{AB}$. The length of arc $\widehat{AB}$ is exactly the length of radius OA and we have $OA = r = \widehat{AB}$; under these restrictions, θ is defined as one radian. We have then the definition

An angle of one radian is the central angle whose intercepted arc is equal in length to one radius.

Let's consider the relationship between radians and degrees. The circumference of a circle is given as

$$C = 2\pi r, \tag{3}$$

where we can read (3) as "the circumference is 2π radii." According to the definition of a radian, one radian of angle intercepts an arc equal in length to one radius; so 2π radians of angle will intercept 2π radii of arc. Since 2π radii of arc constitute a circumference by (3), then 2π radians of angle constitute a complete circular angle of 360°; thus we have the conversion equation

$$2\pi \text{ radians} = 360° \tag{4}$$

from which

$$\pi \text{ radian} = 180° \tag{5}$$

and

$$1 \text{ radian} = \frac{180°}{\pi} = 57.3°. \tag{6}$$

From (4), if we divide both sides by 360, we have

$$1° = \frac{2\pi}{360} = 0.01745 \text{ radian}. \tag{7}$$

An interesting property of the radian is worthwhile discussing. Since an *angle* of one radian intercepts an *arc* of length one radius, there is a one-to-one correspondence between radians and radii. We cannot strictly say that a radian equals a radius, for it would be an attempt to call an angular unit equal to a length unit. We can, however, correctly state that the *number* of radians of angle equals the *number* of radii of arc. This means that we can shift between linear and angular measurements at will if the angular measurements are in radians.

Example 3 How many degrees are in (a) 3 radians and (b) $2\pi/3$ radians?

Solution

(a) Since 1 radian $= 57.3°$ by (6), then

$$3 \text{ radians} = 3(57.3°) = 171.9°.$$

(b) Since, by (5), π radians $= 180°$, then

$$\frac{2\pi}{3} \text{ radians} = \frac{2}{3}(180)° = 120°.$$

Example 4 A wheel of radius 12 in. rolls without slipping along a straight line. If it rolls through five revolutions, (a) how many radians of angle does it rotate through and (b) how far does it travel?

Solution

(a) Since one revolution $= 2\pi$ radians, then five revolutions $= 10\pi$ radians.

(b) Since, by (a), 10π radians of angle are generated, then 10π radii of arc are traversed and, with one radius equal to 12 in. then the distance is

$$10\pi \text{ radii} = 10\pi(12) \text{ in.} = 120\pi \text{ in.}$$

Exercises 12-2

In Exercises 1–10, the angle given is in radians. Convert the angle to degrees.

1. $\theta = \frac{\pi}{3}$ **2.** $\theta = 3\pi$ **3.** $\theta = \frac{5\pi}{12}$ **4.** $\theta = \frac{7\pi}{18}$

5. $\theta = \frac{37\pi}{18}$ **6.** $\theta = 2$ **7.** $\theta = 51$ **8.** $\theta = 6.7$

9. $\theta = 0.02$ **10.** $\theta = 10.5$

Convert the given angle in Exercises 11–20 into radians.

11. $\theta = 90°$ **12.** $\theta = 45°$ **13.** $\theta = 150°$

14. $\theta = 198°$ **15.** $\theta = 342°$ **16.** $\theta = 27.6°$

17. $\theta = 825°$ **18.** $\theta = 125°12'26''$ **19.** $\theta = 209°15'06''$

20. $\theta = 925°47'12''$

21. Given the angular velocity of a rotating body as 10 rad/sec, express this velocity in (a) degrees per minute and (b) revolutions per second.

22. If a body is rotating at an angular velocity k revolutions per second, what is its angular velocity in degrees per minute?

23. A wheel of radius 28 in. is rolling along a flat surface in a straight line at 75 mph. What is the angular velocity of a point on the wheel?

24. If a wheel of radius 12 in. rotates at 75 rpm, what is the linear velocity (a) of a point on the rim of the wheel and (b) of a point on a spoke of the wheel one-half of a radius from the center?

25. If a vector **OP** of magnitude ρ is rotated with an angular velocity ω radians per second, what is the linear velocity of **P**?

12-3 Fundamental operations on angles in degrees, minutes, and seconds

Adding and subtracting angles and multiplying or dividing an angle by an integer will, for our purposes, constitute the fundamental operations on angles. We restrict our discussion to angles expressed in degrees, minutes, and seconds (as opposed to angles in decimal form) for operations on these angles in decimal form are like the fundamental operations on decimal quantities studied in elementary mathematics.

Addition of angles is performed by columnizing like quantities and adding by normal procedures. If the sum of the seconds exceeds 60″, multiples of 60″ in the sum are converted to minutes and added to the minutes column. Excesses of 60′ are treated similarly in converting to degrees. Consider an example.

Example 5 Given $A = 14°25'49''$, $B = 38°46'51''$, and $C = 42°37'46''$, find $A + B + C$.

Solution Obeying the rules of algebra, which assert that only like quantities can be combined by addition, we regard all quantities in degree units alike, all quantities in minutes alike, and all quantities in seconds alike. So we columnize the quantities as

$$\begin{aligned} A &= 14°25'49'' \\ B &= 38°46'51'' \\ C &= \underline{42°37'46''} \\ A + B + C &= 94°108'146''. \end{aligned} \tag{8}$$

The sum (8) bears inspection. We have 146″, which exceeds 60″; our problem is to determine the number of complete minutes that are contained in 146″ and transfer these minutes to the minute column.

$$146 \text{ sec} \div 60 \text{ sec/min} = 2' + 26'' \text{ remainder}$$

or $146'' = 2'26''$. So (8) becomes

$$94°108'146'' = 94°108' + 2' + 26'' = 94°110'26''.$$

Next, 110′ are converted through division by 60 min/deg to $110' = 1° + 50'$. So

$$94°110'26'' = 94° + 1° + 50' + 26'' = 95°50'26''.$$

In subtraction of angles, the usual procedure is to require the number of minutes of the minuend to exceed the number of minutes of the subtrahend; the same requirement applies to seconds. (This requirement assumes that negative minutes and seconds are not permitted.) We still obey the rule of algebra that only like quantities can be subtracted and so we columnize again.

Example 6 Find $A - B$ if $A = 75°15'35''$ and $B = 36°45'42''$.

Solution Columnizing, we have

$$\begin{aligned} A &= 75°15'35'' \quad &&\text{(minuend)} \\ B &= 36°45'42'' \quad &&\text{(subtrahend).} \end{aligned}$$

Referring to earlier comments, the number of seconds in the subtrahend here exceeds the number of seconds in the minuend. Subtraction would require a negative number of seconds in the difference. This situation can be remedied by borrowing $1' = 60''$ from the minutes column in A and applying the 60″ to the seconds column; thus

$$\begin{aligned} A &= 75°15'35'' = 75°14'95'' \\ B &= 36°45'42'' = 36°45'42'' \end{aligned} \tag{9}$$

In (9) the number of minutes in A is less than the number of minutes in B. Borrowing $1° = 60'$ from the degree column in A, we have

$$\begin{aligned} A &= 75°14'95'' = 74°74'95'' \\ B &= 36°45'42'' = 36°45'42'' \\ \hline A - B & = 38°29'53''. \end{aligned} \tag{10}$$

Multiplication of an angle by an integer involves little difficulty; the primary concern is that of aggregating more than 60″ or 60′ in the respective columns. Note Example 7.

Example 7 Multiply 15°35′42″ by 5.

Solution We have

$$(15°35'42'') \times 5 = 75°175'210'' \tag{11}$$

but

$$75°175'210'' = 75°178'30'' \tag{12}$$

and

$$75°178'30'' = 77°58'30'', \tag{13}$$

where (12) is obtained from (11) by transferring 180″ of the 210″ to the minutes column as 3′ and (13) is obtained from (12) by transferring 120′ of the 178′ to the degree column as 2°.

Division of an angle by a constant is complicated by the possibility of obtaining quotients that contain fractions of degrees, minutes, or seconds. This possibility can be avoided by manipulating the dividend in a way that will require the degrees and minutes of the quotient to be integral and by allowing the seconds part of the quotient to contain a fraction or decimal.

Example 8 Perform the operation (96°24′16″) ÷ 5.

Solution Here we change the degrees to 95° (the largest multiple of 5 contained in 96°). Next, transfer the excess degree to the minutes as 60′ so that

$$96°24'16'' = 95°84'16''.$$

Then change the minutes to 80′ (the largest multiple of 5 contained in 84′). Next, transfer the excess 4′ = 240″ to the seconds, obtaining

$$96°24'16'' = 95°84'16'' = 95°80'256''.$$

Now we are prepared to divide:

$$(95°80'256'') \div 5 = 19°16'51.2''.$$

We could have divided the original expression by 5 so that

$$(96°24'16'') \div 5 = 19.2°4.8'3.2''.$$

Looking at the answer, 19.2° = 19°12″ and

$$19.2°4.8'3.2'' = 19°16.8'3.2''.$$

Finally, 16.8′ = 16′48″ and

$$19°16.8'3.2'' = 19°16'51.2''.$$

The latter solution converted fractional degrees to minutes and fractional minutes to seconds. This solution is preferred if the divisor is not integral.

Exercises 12-3

In Exercises 1–15, we are given $A = 95°15'12''$, $B = 53°56'29''$, $C = 37°42'05''$, and $D = 18°03'58''$. Find the following.

1. $A + B$

2. $B + D$

3. $A - B$

4. $C - D$

5. $A + C - D$

6. $B - (C + D)$

7. $3B$

8. $4C$

9. $3B - 4C$

10. $A - 5D$

11. $\frac{1}{2}A$

12. $\frac{2}{3}B$

13. $\frac{3}{4}C$

14. $\frac{1}{2}(A + B) - \frac{1}{2}(A - B)$

15. $\frac{1}{2}(C + D) - \frac{1}{2}(C - D)$

Refer to Fig. 12-4 in Examples 16–20.

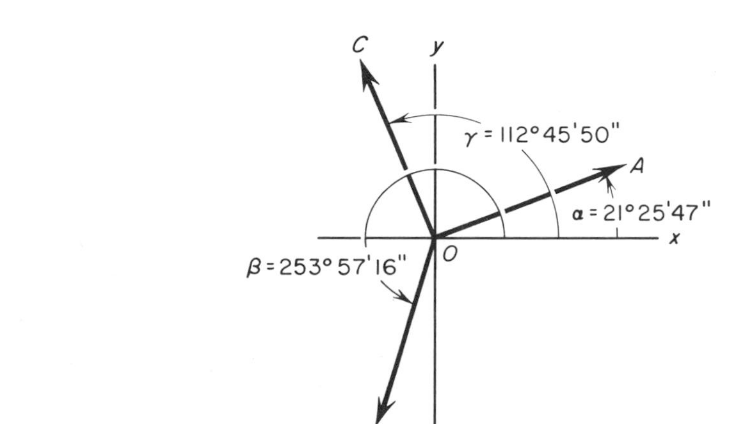

Figure 12-4

16. Find the reference angle of the bisector of $\angle AOC$.

17. Express the direction of OB as an angle measured in a clockwise direction from Ox.

18. Find the direction of the vector bisecting $\angle BOC$.

19. The angle BOA is trisected. Find the directions of the trisectors. Note that there can be two answers in the fourth quadrant and two more in the second quadrant 180° removed from their fourth-quadrant companions. Accept the fourth-quadrant answers.

20. What is the average angle represented?

12-4 Use of the natural trigonometric tables

In Chapter 11 we defined the natural trigonometric functions and performed some elementary operations with them, including the solution of a restricted set of triangles. We used all six of the functions but applied them only to those angles that are multiples of 30 or 45°. In restricting our usage to the angles mentioned, we avoided using a set of values of the trigonometric functions of many angles. Our purpose here is to broaden our use of the trigonometric functions to include the functions of angles described in accuracy in degrees and minutes. We know, for instance, from Chapter 11 the functions of 30°, but what about the functions of 30°25′? In order to obtain an answer, we must refer to the table of natural trigonometric functions in Table B at the back of the book. Let us examine a very small portion of that table by creating Table 12-1.

In Table 12-1 we reproduced that portion of Table B from which we can read the six trigonometric functions of angles between 18 and 27° and also between 63 and 72°, spaced at 10′ intervals. The functions of 0 to 45° in

Table 12-1 Trigonometric functions

Degrees	Radians	Sin	Cos	Tan	Cot	Sec	Csc		
18°00′	.3142	.3090	.9511	.3249	3.078	1.051	3.236	1.2566	72°00′
10′	171	118	502	281	047	052	207	537	50′
20′	200	145	492	314	018	053	179	508	40′
30′	.3229	.3173	.9483	.3346	2.989	1.054	3.152	1.2479	30′
40′	258	201	474	378	960	056	124	450	20′
50′	287	228	465	411	932	057	098	421	10′
19°00′	.3316	.3256	.9455	.3443	2.904	1.058	3.072	1.2392	71°00′
.									.
.									.
.									.
26°00′	.4538	.4384	.8988	.4877	2.050	1.113	2.281	1.1170	64°00′
10′	567	410	975	913	035	114	268	141	50′
20′	596	436	962	950	020	116	254	112	40′
30′	.4625	.4462	.8949	.4986	2.006	1.117	2.241	1.1083	30′
40′	654	488	936	022	991	119	228	054	20′
50′	683	514	923	057	977	121	215	025	10′
27°00′	.4712	.4540	.8910	.5095	1.963	1.122	2.203	1.0996	63°00′
		Cos	Sin	Cot	Tan	Csc	Sec	Radians	Degrees

Table B read "down-left"—that is, sin 18°40′ = 0.3201 by reading 18° at the left and then down to 40′. Also, cot 26°20′ = 2.020 by reading the same way. The functions of 45 to 90° are read "up-right"—that is, tan 63°30′ = 2.006 and sec 71°10′ = 3.098.

Because of efforts to save space and avoid cluttering, certain features of Table B need clarifying.

1. Because of cofunctional relationships discussed in Sec. 11-5, all angles from 0 to 45° have cofunctional companions in the interval from 45 to 90°. *Example*: sin 19°00′ = cos 71°00′ = 0.3256, where these two functional entries are found in precisely the same place in Table B. In effect, the table "folds back" on itself at 45°.
2. Because of (1), the six functions at the *top* of any page in Table B pair with the angle at the *left* when the angle is between 0 and 45°. However, the six functions at the *bottom* pair with the angles at the *right* when the angle is between 45 and 90°.
3. Tabular entries in Table B are given to four figures, with all four figures printed in every *third* location. Intermediate entries omit the repetitive lead figure; however, *it is understood* that the lead figure is supplied by the reader. Thus tan 18°10′ = 0.3281, where 281 is printed in the table, but it is understood that 0.3281 is the actual entry, with 0.3 omitted to avoid overcrowding and perhaps limit typesetting costs.

To find an angle, when knowing a function of that angle, requires an awareness of the location of the function (at the head versus the foot of the column). Thus if $\sin A = 0.9465$ and we wish to find A, we scan the sine column in search of the tabular entry 0.9465. If we scan the columns with "Sin" at the top, we never find 0.9465, but switching to columns with "Sin" at the bottom, we find 0.9465 opposite 71°10′ on the *right*. Thus if $\sin A = 0.9465$, then $A = 71°10′$.

Example 9 Find (a) tan 36°20′, (b) sec 75°30′, and (c) angle A if $\csc A = 1.090$.

Solution From Table B we have (a) tan 36°20′ = 0.7355, (b) sec 75°30′ = 3.994, and (c) if $\csc A = 1.090$, then $A = 66°30′$.

Exercises 12-4

Using the table of natural trigonometric functions (Table B), find the function of the angle given in Exercises 1–12. Also find angle θ in Exercises 13–24, where $0° < \theta < 90°$.

1. sin 12°25′ **2.** cos 21°35′ **3.** sin 32°27′

4. ctn 41°15′ **5.** sin 65°56′ **6.** cos 81°12′

7. ctn 76°05′ **8.** tan 49°59′ **9.** cos 127°15′

10. sin 227°27′	**11.** tan 333°21′	**12.** ctn 462°15′
13. $\sin\theta = 0.3461$	**14.** $\cos\theta = 0.9776$	**15.** $\tan\theta = 0.8002$
16. $\text{ctn}\,\theta = 1.0392$	**17.** $\cos\theta = 0.6111$	**18.** $\sin\theta = 0.8090$
19. $\text{ctn}\,\theta = 0.4010$	**20.** $\tan\theta = 4.705$	**21.** $\sin\theta = 0.9930$
22. $\text{ctn}\,\theta = 10.78$	**23.** $\tan\theta = 10.78$	**24.** $\text{ctn}\,\theta = 0.0160$

12-5 Solution of a right triangle

The general right triangle shown in Fig. 12-8 contains six parts, three angles and three sides. One of the angles is a right or 90° angle. The side opposite the right angle is often called the *hypotenuse*. To solve the triangle, all six parts must be found. Certain facts about triangles are useful when solving them.

1. The sum of the angles of a triangle is 180°. In a right triangle the two acute angles are complementary; that is, they add to 90°.
2. In any triangle the largest side is opposite the largest angle and the smallest side is opposite the smallest angle. This fact is useful in checking for gross errors in solutions of triangles.
3. A triangle can be solved only if at least three of its six parts are known; at least one of those known parts must be a side.

In most trigonometry problems solution of the entire triangle is not necessary. However, finding one unknown part requires three known parts as described in (3).

General procedure for solution In solving for an unknown part of a right triangle, we know in advance that one angle is 90°. We also must know two more parts, either two sides or one side and one more angle. The desired unknown part may be a side or an angle. There are three cases.

1. If two angles (including the right angle) are known, the fact that the sum of the angles of a triangle is 180° leads us easily to the third angle.
2. If two sides are known and the third side of the right triangle is needed, the Pythagorean Theorem may lead us easily to the solution. If two sides are known and an angle is needed, a trigonometeric ratio is used to find the angle.
3. If a side and an angle of a right triangle are known and a second side is requested, a trigonometric ratio will be used to find the side.

What trigonometric ratio is used in a solution? Because a trigonometric function of an *angle* equals the ratio of *two sides*, using a ratio relates *two sides* and an *angle*; one of these *three* items may be the unknown in a problem if the other two are known. Choice of the trigonometric function depends entirely on the relative positions of the known and unknown items. Consider Examples 10 and 11.

Example 10 Find z in Fig. 12-5.

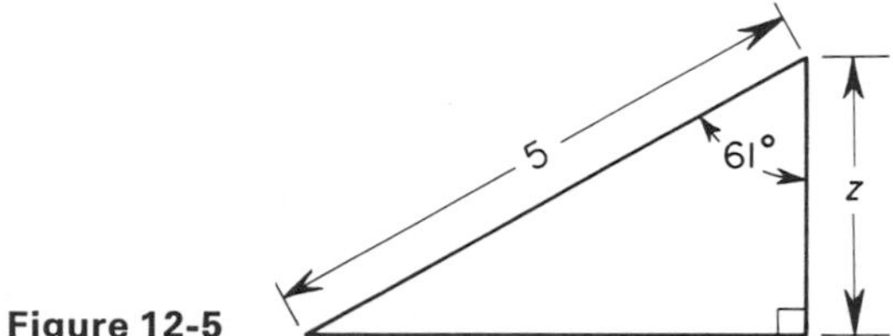

Figure 12-5

Solution Considering the right triangle, the known information, and the unknown z in Fig. 12-5, we have a *known angle* (61°), we know the *hypotenuse*, and we wish to find z, which is the *side adjacent* to the known angle. The trigonometric ratios of an angle that involve the *side adjacent* to that angle and the *hypotenuse*, per Sec. 11-2, are *either*

$$\cos 61^\circ = \frac{z}{5} \quad \textit{or} \quad \sec 61^\circ = \frac{5}{z}.$$

Side z can be found from either one; using the first, we have

$$z = 5 \cos 61^\circ = 5(0.4848) = 2.424.$$

Example 11 Find θ in Fig. 12-6.

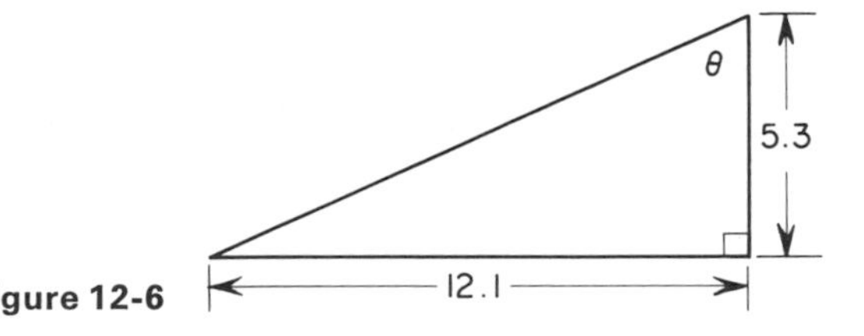

Figure 12-6

Solution Here we have angle θ as the unknown and we know the *side opposite* θ (12.1) and the *side adjacent* θ (5.3). The trigonometric ratios of an angle that involve the *sides opposite* and *adjacent* to that angle are *either*

$$\tan \theta = \frac{12.1}{5.3} \quad \textit{or} \quad \text{ctn}\, \theta = \frac{5.3}{12.1}.$$

Now θ may be found by using either ratio. Using the first, we have

$$\tan \theta = \frac{12.1}{5.3} = 2.283$$

from which $\theta = 66^\circ 20'$ to the nearest $10'$.

Solution using an electronic calculator Solution of a right triangle presents a strong case for using an electronic calculator as a way to reduce the time and effort involved in arithmetic operations and use of tables. On the other hand, the calculator is useless until the problem solution is set up—that is, until we know which trigonometric function is used and how. Since most calculators have only three trigonometric function buttons (usually sin, cos, and tan), it is useful to know the reciprocal functions (Sec. 11-5). If a problem is set up using a function other than the customary sin, cos, or tan, we can use one of the given function buttons and then reciprocate the number. That is, $\csc 30° = 1 \div \sin 30°$; therefore if we wish to use $\csc 30°$, we call out $\sin 30°$ and then reciprocate with the $1/x$ button. Consider the solution of Example 10 with an electronic calculator of standard scientific capacity.

For $\cos 61° = z/5$, we have $z = 5 \cos 61°$ and the keystrokes are 61, cos, ×, 5, =. For $\sec 61° = 5/z$, $z = 5 \div \sec 61°$ and the keystrokes are 5, ÷, (, 61, cos, $1/x$,), = or, as an option, 61, cos, $1/x$, ÷, 5, =, $1/x$. Note that other options are available.

Example 12 Find $\angle B$ in Fig. 12-7.

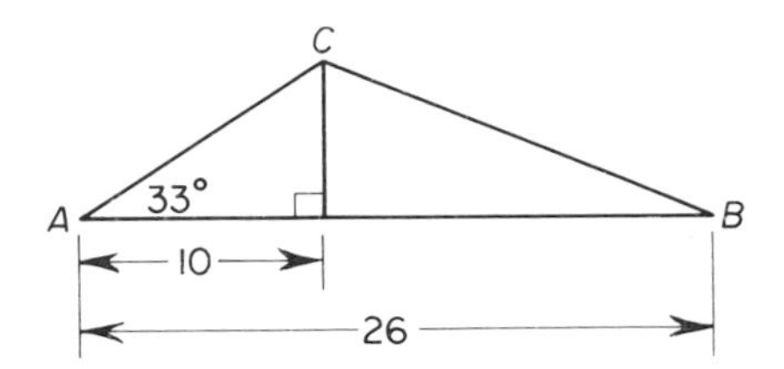

Figure 12-7

Solution Clearly $\angle B$ cannot be found without some intermediate computations. In right triangle BCD we can compute BD (*adjacent* to $\angle B$) as $BD = 26 - 10 = 16$. From triangle ACD we can compute CD:

$$\tan 33° = \frac{CD}{10} \quad \text{or} \quad CD = 10 \tan 33°.$$

Then

$$\tan B = \frac{CD}{BD} = \frac{10 \tan 33°}{16} = \frac{10}{16}(0.6494) = 0.4059$$

from which $\angle B = 22°10'$ to the nearest $10'$.

Interpolation We apply interpolation to Table B for greater accuracy by locating readings between tabular entries. The method of *linear interpolation* is discussed with reference to logarithms in Sec. 9-5. In this section, by use of example, we discuss this same linear interpolation as it relates to trigonometry tables.

Example 13 Using Table B, find tan 29°36′ by linear interpolation.

Solution Table B gives trigonometric functions of angles spaced at intervals of 10′. There are entries for tan 29°30′ and tan 29°40′. Linear interpolation asserts that "Because 29°36′ is 6′ through the 10′ interval, which starts at 29°30′ and ends at 29°40′, then the *angle* is 6/10 of the way through the *angle interval* and tan 29°36′ will be 6/10 of the way through the *tangent interval.*" Looking up the tabular values, we lay out the block arrangement shown

tan 29°40′ = 0.5696

10′ — tan interval = 0.0038

tan 29°36′ =

6′ or $\frac{6}{10}$ of way through angle interval — .6 of way through tan interval

tan 29°30′ = 0.5658

and see that

$$\tan 29°36' = 0.5658 + \tfrac{6}{10}(0.0038) = 0.5658 + 0.0023 = 0.5681.$$

Example 14 Find arc sin 0.8300.

Solution There is no 0.8300 in a Sin column in Table B. However, the two entries that immediately bracket 0.8300 in the Sin column are sin 56°10′ = 0.8307 and sin 56°00′ = 0.8290. Now our value 0.8300 is 10/17 of the way through the *sine interval* and, by linear interpolation, is considered 10/17 of the way through the *angle interval* (going in the same direction). But the angle interval is 10′ wide; so our angle is 10/17 × 10 = 100/17 = 5.88 = 6′ through the interval. Therefore we assert that arcsin 0.8300 = 56°00′ + 06′ = 56°06′.

Exercises 12-5

In Exercises 1–12 reference is to Fig. 12-8. Find the quantity indicated for the values of the other given quantities. When an angle is requested, find that angle to the closest minute.

1. $a = 3, b = 4$, find θ.

2. $a = 3.2, b = 5.6$, find ϕ.

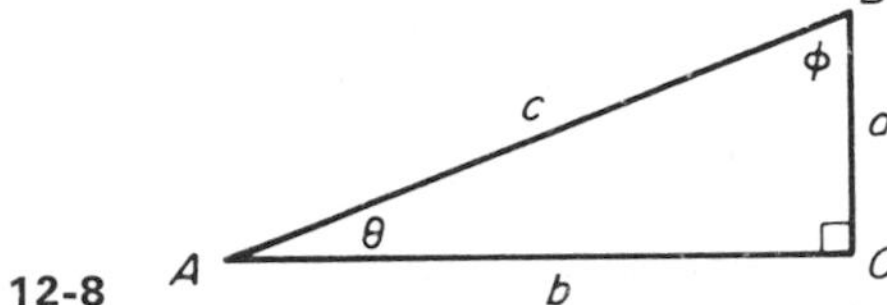

Figure 12-8

3. $c = 12.5$, $a = 7.29$, find θ.

4. $c = 4.85$, $a = 2.16$, find ϕ.

5. $b = 8.18$, $c = 10.12$, find ϕ.

6. $b = 13.11$, $c = 17.75$, find θ.

7. $A = 15°15'$, $a = 4$, find b.

8. $A = 32°15'$, $a = 3.15$, find c.

9. $A = 47°17'$, $b = 12.18$, find a.

10. $B = 62°36'$, $b = 6.5$, find c.

11. $B = 75°57'$, $c = 18.0$, find a.

12. $B = 75°17'$, $c = 18.0$, find b.

13. A surveyor proceeds from a point O to a point B in a direction θ measured off a base line Ox as in Fig. 12-9 and then in a direction ϕ to D with BM parallel to Ox. If $\theta = 15°40'$, $OB = 308$ ft, and $\phi = 47°30'$ with $BD = 427$ ft.
(a) How far is D from the line BM?
(b) How far is D from the alternate line Ox?

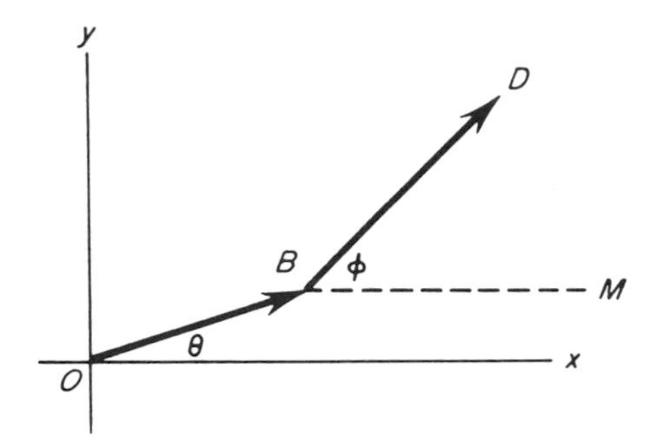

Figure 12-9

14. In Exercise 37, what is the direction of OD as measured from Ox?

15. A man starts at point A, walks due east 8 miles, turns due north for 6 miles, goes west for 2 miles, and then south for 3 miles to a point B. What is the angle between AB and a north–south line through A correct to the nearest minute?

Using Fig. 12-8, solve for the indicated unknown in Exercises 16–22.

16. Given $a = 14.4$, $c = 26.7$, find A, b.

17. Given $c = 0.0486$, $A = 23.5°$, find a, b.

18. Given $b = 16.85$, $c = 38.6$, find A, a.

19. Given $a = 3.65$, $b = 8.82$, find A.

20. Given $a = 4.72$, $b = 4.07$, find A.

21. Given $A = 17.4°$, $a = 12.25$, find b.

22. Given $B = 64.5°$, $a = 0.746$, find b.

In Exercises 23–25, refer to Fig. 12-10 and apply the oblique triangle law:

$$\frac{a}{\sin A} = \frac{b}{\sin B} = \frac{c}{\sin C}.$$

From the given information find the requested parts in Fig. 12-10.

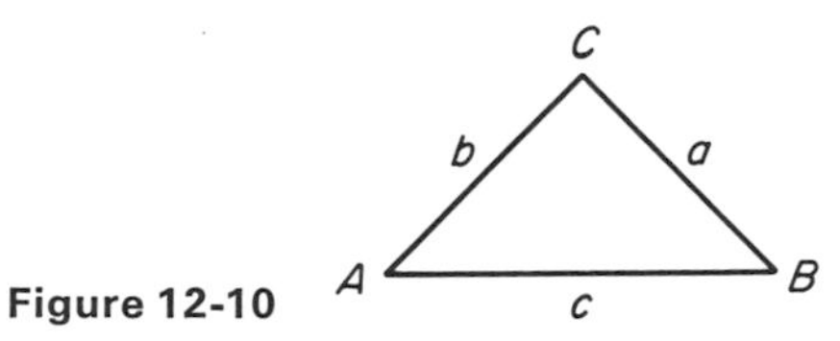

Figure 12-10

23. Given $a = 4.25$, $A = 36.5°$, $B = 21.5°$, find b.

24. Given $a = 82.6$, $A = 61°$, $b = 75.5$, find c.

25. Given $C = 106°$, $B = 45°$, $a = 12.25$, find c.

In Exercises 26–31, solve for the indicated unknown.

26. Solve for θ in Fig. 12–11.

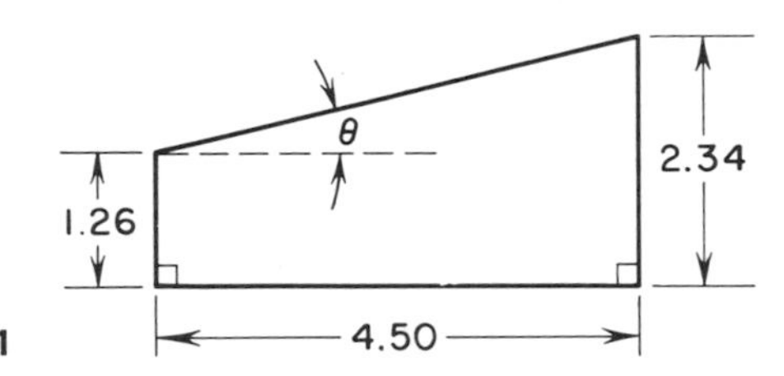

Figure 12-11

27. Solve for θ in Fig. 12-12.

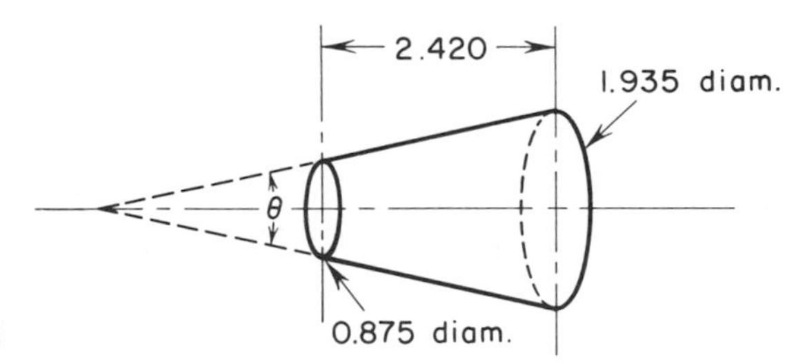

Figure 12-12

28. Solve for θ in Fig. 12-13.

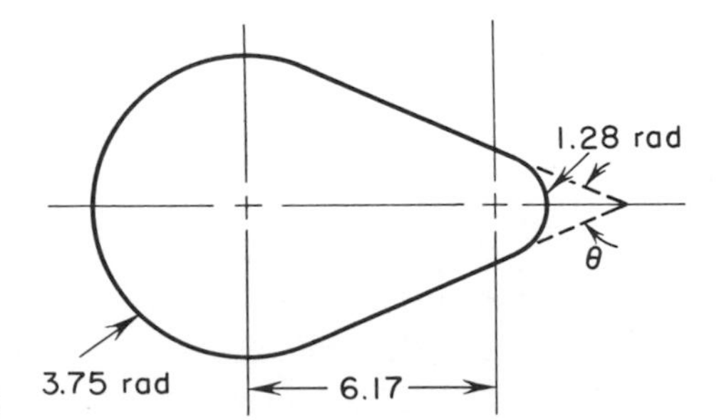

Figure 12-13

29. Solve for θ in Fig. 12-14.

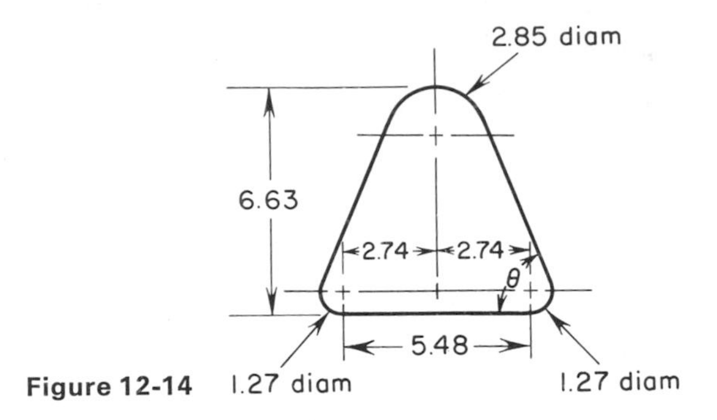

Figure 12-14

30. Solve for x in Fig. 12-15.

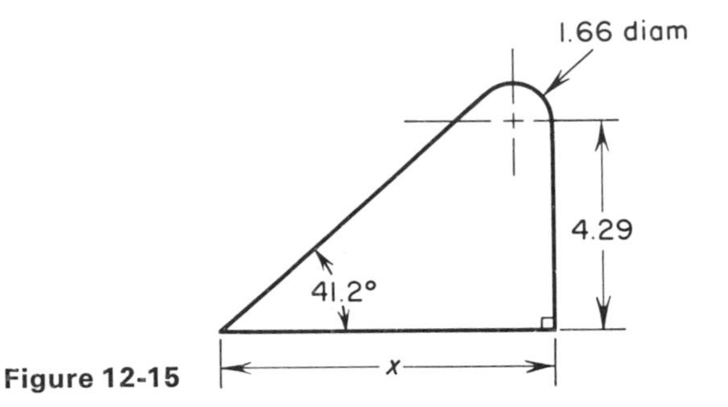

Figure 12-15

31. Solve for x in Fig. 12-16.

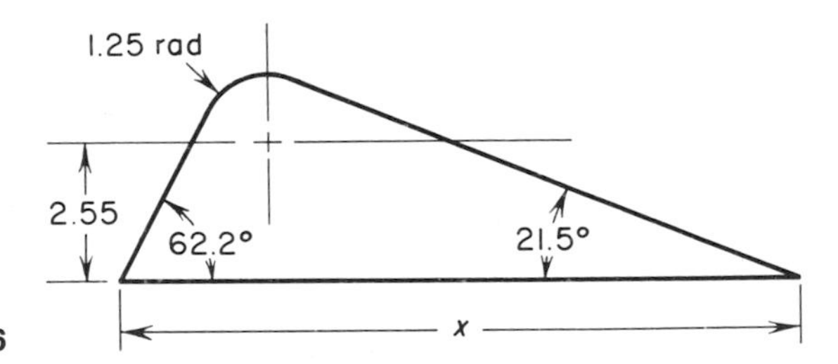

Figure 12-16

12-6 Special topics

One special topic that concerns the solution of right triangles is the inverse trigonometric function in which a function is known and we refer to the angle involved. Another special topic concerns the fact that small angles (angles near 0°) have special values for their functions.

Inverse trigonometric functions In part (c) of Example 9 we were asked to find A if $\csc A = 1.090$. Another way to state this request is "Find the angle whose csc is 1.090." Still another is "Find arccsc 1.090" or "Find $\csc^{-1}$ 1.090." We will favor the arccsc notation over the $\csc^{-1}$ notation here. From Sec. 11-3 we recognize that many angles may have a given trigonometric function. Thus if $\sin A = 1/2$, A may be 30, 150, 390, 510°, and many other values. If asked to "Find arcsin 1/2," we might answer with many angles; however, use of the smallest positive angle as the answer is commonly accepted. Thus arcsin $1/2 = 30°$. Section 16-5 treats inverse trigonometric functions more fully.

Functions of small angles For angles that are small—that is, between 0° and 5°—we can assert that three items are nearly equal in magnitude: *sine* of the angle, *tangent* of the angle, and *size* of the angle as measured in radians. Referring to Fig. 12-17, which shows a small angle θ in the standard position on a cartesian plane, we note a circle of radius r struck through $P(x, y)$. Now note three measures.

$$\tan \theta = \frac{y}{x}, \qquad \sin \theta = \frac{y}{r}, \qquad PM = r\theta \qquad (\theta \text{ in radians}).$$

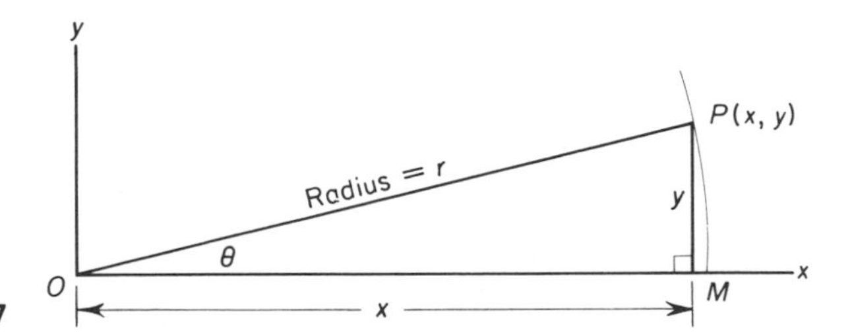

Figure 12-17

Furthermore, if θ is near zero, then the value of $x \approx r$ (x approximately equal to r) and $y \approx \widehat{PM}$. We may consider the circle to be of unit radius ($r = 1$) without loss of generality. This makes $r = 1$ and $x \approx 1$ so that

$$\tan \theta \approx y, \qquad \sin \theta \approx y, \qquad \widehat{PM} \approx y \approx \theta \qquad (\theta \text{ in radians}).$$

Table B indicates (to four figures) that $\tan \theta = \sin \theta = \theta$ (radians) for $0° < \theta < 2°20'$. Even at 6°, $\sin \theta$, $\tan \theta$, and θ (radians) differ by less than 1%.

In many situations, we consider $\sin \theta$, $\tan \theta$, and θ (radians) to be equal if $\theta < 0.1$ radian or if $\theta < 5.73°$.

Exercises 12-6

In Exercises 1–10, find the small angle or function indicated.

1. $\sin \theta = 0.0400$. Find θ in degrees.
2. $\sin \theta = 0.0300$. Find θ in degrees.

3. $\tan\theta = 0.0800$. Find θ in degrees.

4. $\tan\theta = 0.0525$. Find θ in degrees.

5. $\theta = 4.50°$. Find $\sin\theta$, $\tan\theta$.

6. $\theta = 5.25°$. Find $\sin\theta$, $\tan\theta$.

7. $\theta = 1.80°$. Find $\sin\theta$, $\tan\theta$.

8. $\theta = 3.80°$. Find $\sin\theta$, $\tan\theta$.

9. $\theta = 2°20'$. Find $\sin\theta$, $\tan\theta$.

10. $\theta = 5°48'$. Find $\sin\theta$, $\tan\theta$.

In Exercises 11–20, find the angle indicated by the inverse trigonometric statement. Use interpolation where necessary.

11. arcsin 0.3333

12. arccos 0.6929

13. arctan 0.4656

14. arcctn 1.453

15. arccos 0.8283

16. arctan 1.374

17. arcctn 0.2596

18. arctan 0.0594

19. arccos 0.0329

20. arctan 12.25

13

Solution of oblique triangles

In the preceding two chapters we described the solutions of right triangles by trigonometry and discussed computations involving the electronic calculator and the tables of natural trigonometric functions.

In this chapter we discuss the solution of oblique triangles by deriving and employing the sine and cosine laws and the tangent law.

We also enlarge on the applications of trigonometry by discussing additional problems involving vectors, problems concerning mechanics, solid geometry, force, acceleration, and the composition and resolution of forces.

13-1 The sine law

An *oblique triangle* is a triangle that does not contain a right angle. It may be isosceles or equilateral or have no two sides equal to each other. It has many properties that can be expressed mathematically, including those expressed by the sine law. The *sine law* asserts that the sides of a triangle are in the same ratio as the sines of the angles opposite those sides. Let us prove this assertion.

Figure 13-1 shows a general oblique triangle ABC. In this triangle we introduce a convention that is used throughout this chapter. Capital letters are used to designate angles and lowercase letters to designate the sides opposite. Thus side a is opposite angle A, side b is opposite angle B, and side c is opposite angle C.

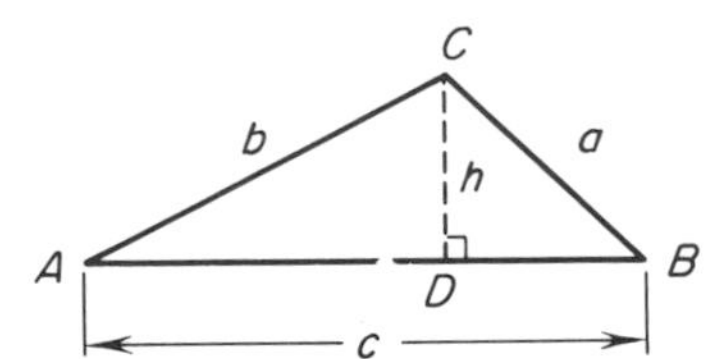

Figure 13-1

If, in Fig. 13-1, we drop a perpendicular to AB from C, the length of the perpendicular is the altitude h. From the right triangle ADC we have

$$\frac{h}{b} = \sin A \quad \text{or} \quad h = b \sin A. \tag{1}$$

Also, from the right triangle BDC

$$\frac{h}{a} = \sin B \quad \text{or} \quad h = a \sin B. \tag{2}$$

Equating the right-hand expressions in (1) and (2), we have

$$b \sin A = a \sin B$$

from which

$$\frac{a}{\sin A} = \frac{b}{\sin B}. \tag{3}$$

To establish expression (3), an altitude was drawn from C to c. Another similar expression can be established by drawing another altitude from a different vertex. In Fig. 13-2 we have chosen the same triangle but have drawn an altitude from A to side a extended; this altitude has length k and its foot is at E. From right triangle ABE we have

$$\frac{k}{c} = \sin B \quad \text{or} \quad k = c \sin B. \tag{4}$$

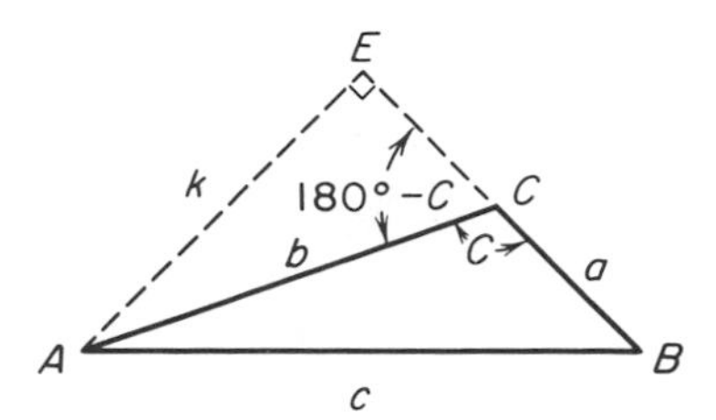

Figure 13-2

Also, from right triangle ACE

$$\frac{k}{b} = \sin(180° - C) \quad \text{or} \quad k = b \sin(180° - C). \tag{5}$$

Then from (4) and (5)

$$c \sin B = b \sin(180° - C)$$

or

$$\frac{b}{\sin B} = \frac{c}{\sin(180° - C)}. \tag{6}$$

If we construct two vectors of magnitude b and reference angles $180° - C$ and C as shown in Fig. 13-3, we see that

$$\sin C = \sin(180° - C)$$

and (6) becomes, by substitution,

$$\frac{b}{\sin B} = \frac{c}{\sin C}. \tag{7}$$

Next, (3) and (7) combine to form the *sine law*

$$\frac{a}{\sin A} = \frac{b}{\sin B} = \frac{c}{\sin C}, \tag{8}$$

where (8) asserts that *the sides of an oblique triangle are in the same ratio as the sines of the angles opposite those sides.*

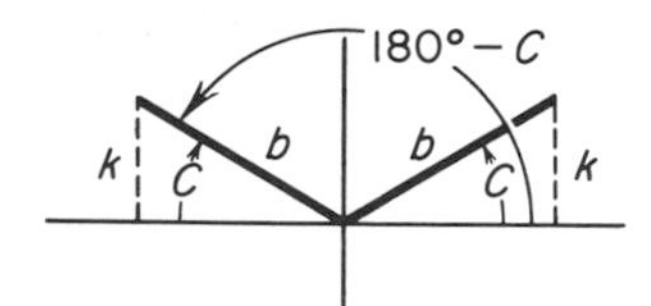

Figure 13-3

Example 1 From the information shown in Fig. 13-4, find the lengths of sides a and b.

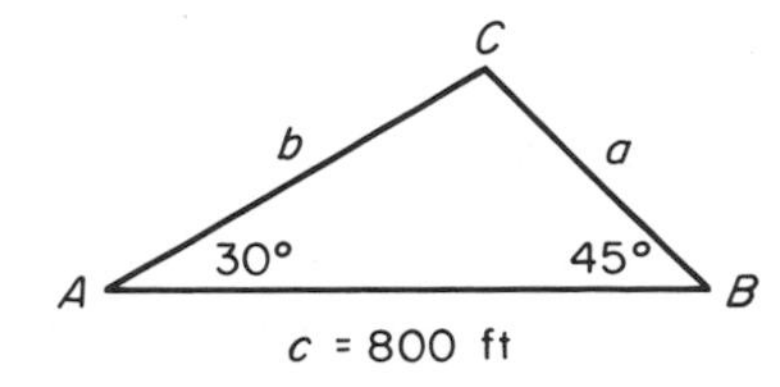

Figure 13-4

Solution From the given triangle, $A = 30°$ and $B = 45°$ and so

$$C = 180° - (30° + 45°) = 105°.$$

Using a part of (8), we have

$$\frac{a}{\sin A} = \frac{c}{\sin C}. \tag{9}$$

Substituting the known information into (9), gives

$$\frac{a}{\sin 30°} = \frac{800}{\sin 105°}$$

from which

$$a = \frac{800 \sin 30°}{\sin 105°} = \frac{800(0.5)}{0.966} = 414.1 \text{ ft.}$$

To find b, we have

$$\frac{b}{\sin B} = \frac{c}{\sin C} \tag{10}$$

but $B = 45°$, $c = 800$, and $C = 105°$ so that we now have from (10)

$$b = \frac{800 \sin 45°}{\sin 105°} = \frac{800(0.7071)}{0.966} = 585.6 \text{ ft.} \tag{11}$$

One question that may occur is "When can the sine law be applied?" Inspecting

$$\frac{a}{\sin A} = \frac{b}{\sin B}, \tag{12}$$

we see that if three parts of (12) are known, the fourth part can be found. Thus the sine law can be used when we know:

1. *Two angles and a side opposite one of them* This can be modified because if we know two angles, we can find the third angle where now the known side will lie opposite one of the three known angles. Thus we can use the sine law when we are given two angles and any side of a triangle.
2. *Two sides and the angle opposite one of them* This introduces the possibility of the *ambiguous case*, which will be discussed in Sec. 13-2.

Exercises 13-1

In Exercises 1–12, find the quantity in parentheses where the given information refers to the general triangle ABC.

1. $A = 45°, B = 60°, b = 25$ (a)

2. $A = 60°, B = 75°, a = 14$ (b)

3. $B = 120°, C = 15°, a = 10$ (b)

4. $B = 150°, A = 10°, c = 6$ (b)

5. $C = 60°, B = 60°, a = 8$ (b)

6. $C = 35°, B = 2C, b = 10$ (c)

7. $A = 110°, C = 40°, a = 20$ (b)

8. $B = 60°, C = 70°, c = 12$ (a)

9. $A = 50°, B = 30°, b = 15$ (c)

10. $A = 45°, C = 105°, b = 10$ (c)

11. $A = 110°, C = 30°, c = 5$ (b)

12. $B = 65°, C = 50°, a = 8$ (c)

In Exercises 13–24, we are given two sides and the angle opposite one of them. Find the angle in parentheses; give only those solutions where the angle requested is less than 90°.

13. $a = 3, c = 6, C = 30°$ (A)

14. $a = 3, c = 6, C = 60°$ (A)

15. $b = 7, a = 10, A = 45°$ (B)

16. $b = 25, a = 37.5, A = 30°$ (B)

17. $c = 18, b = 3, B = 20°$ (C)

18. $c = 3, b = 18, B = 20°$ (C)

19. $a = 12, c = 6, A = 60°$ (C)

20. $a = 10, c = 7, A = 40°$ (C)

21. $a = 8, b = 4, A = 30°$ (B)

22. $b = 15, c = 3, B = 70°$ (C)

23. $a = 18, c = 9, C = 20°$ (A)

24. $a = 10, b = 6, B = 15°$ (A)

25. Use the sine law to find a from the general triangle ABC if $A = 34°12'$, $C = 102°29'$, and $b = 3.65$.

26. Find DC in Fig. 13-5.

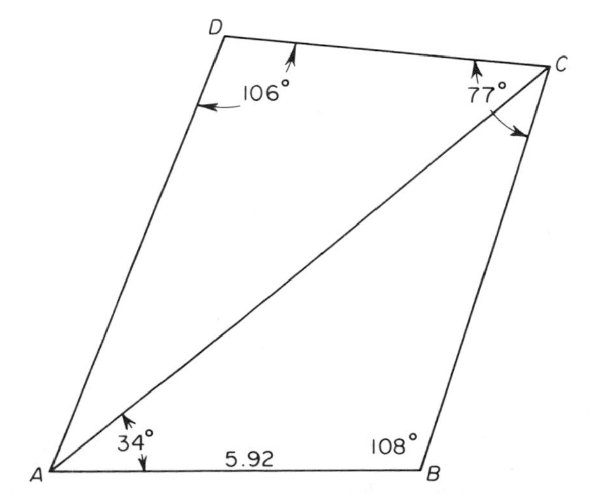

Figure 13-5

27. Find BC in Fig. 13-6.

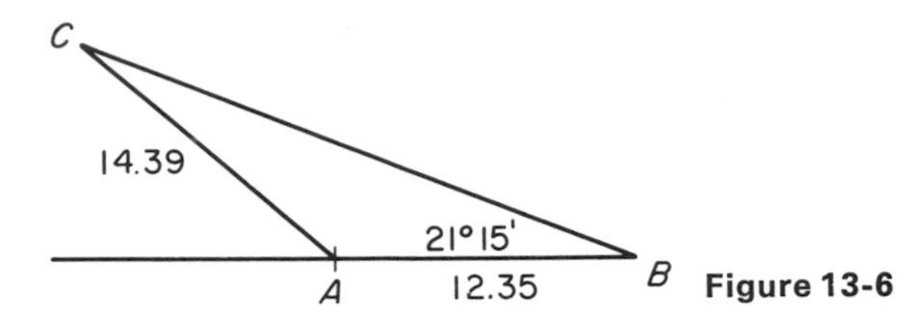

Figure 13-6

28. A traveler proceeds from point O along a path OB as shown in Fig. 13-7. When he arrives at B, 12.5 miles from O, he changes course by 30°25′ to proceed to point C. If C is 18.3 miles from O in the direction shown, find angle α, his original direction.

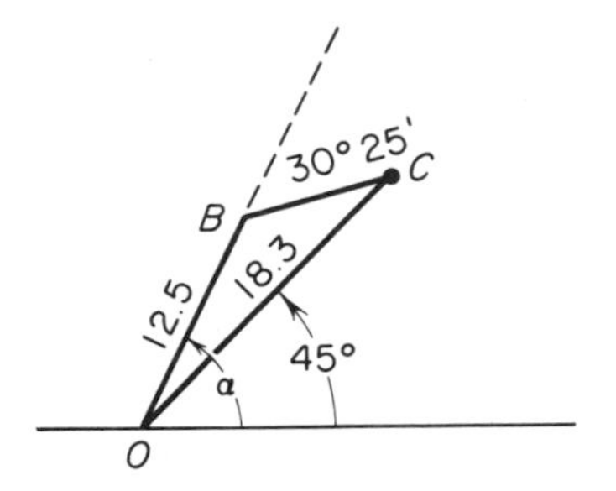

Figure 13-7

29. Two forces act on a body at O as shown in Fig. 13-8. Force F_1 is of magnitude 48.5 lb and direction 62°20′; F_2 is 69.4 lb at 235°15′. It is assumed that resultant R is in a direction 160°. Check the validity of that assumption.

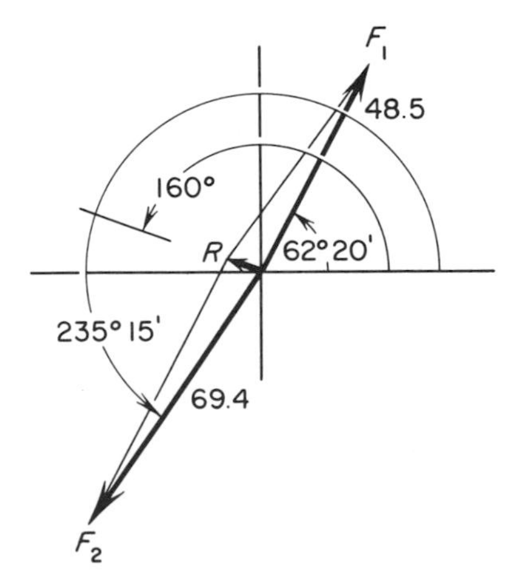

Figure 13-8

30. A vehicle pulls a 675-lb load by exerting 275 lb of force on a tow-chain slanting at 18° from the horizontal. What is the pulling force on the load? What vertical force does the load exert on the ground while being pulled?

31. In Fig. 13-9 the lines Ox, Oy, and Oz are mutually perpendicular at O. A line is drawn from point A in the "floor" to point B in the "wall." Find the length of AB if $BB_z = 8.462$, $BB_y = 5.965$, $AA_x = 1.072$, and $AA_y = 6.337$.

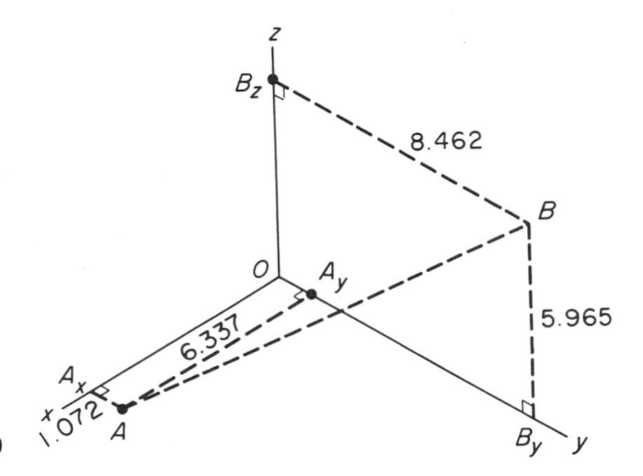

Figure 13-9

32. In the general triangle ABC, angle A may be found from equation

$$\sin \frac{A}{2} = \sqrt{\frac{(s-b)(s-c)}{bc}},$$

where $s = (a + b + c)/2$. Given a triangle ABC with $a = 3.286$, $b = 4.367$, and $c = 4.955$, find A by the given equation.

33. Find the area of the triangle in Exercise 32, using formula

$$\text{Area} = \tfrac{1}{2}bc \sin A.$$

34. The angles of a triangle are in the ratio 2: 3: 4. Find the longest side of the triangle if the shortest side is of length 14.28.

35. Given the pyramid in Fig. 13-10 with equilateral base ABC, where $BC = 12.96$. Given also angle $ADC = 38°12'$, angle $BDA = 33°16'$, angle $BDC = 41°56'$, and $AD = 15.05$. Find angle DBC.

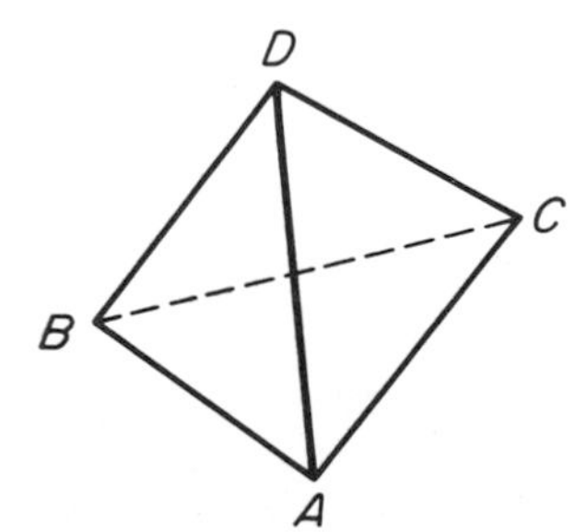

Figure 13-10

13-2 Sine law: The ambiguous case

Earlier we mentioned that the sine law was used to solve triangles when two angles and one side are known or when two sides and an angle opposite one of those two sides are known. It was also mentioned that the latter case presented an ambiguity. Let us inspect this ambiguity with reference to Fig. 13-11.

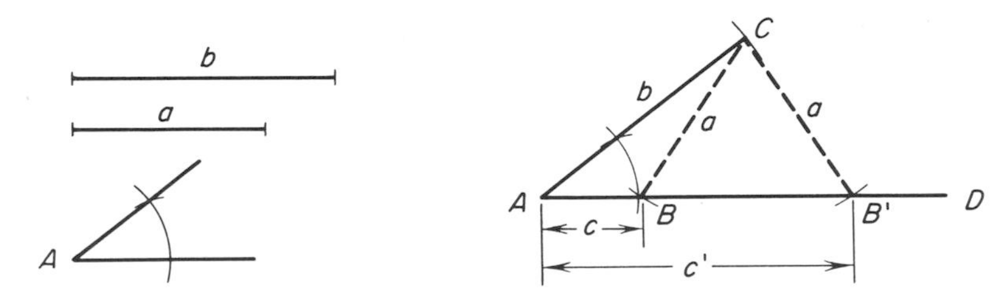

Figure 13-11

Assume that we know b, A, and a of a triangle as shown. Note that these known parts constitute two sides and the angle opposite one of them. If we attempt to construct the proposed triangle from the information provided, we start by laying off line AC, indicating the distance from A to C as the given length b. Next, at A reproduce the given angle A, establishing baseline AD. Next, side a is opposite angle A. We can complete the construction by swinging an arc of length a from C with the intention of intersecting baseline AD. This arc can intersect AD twice, as shown in Fig. 13-11. Note that two triangles $\triangle ABC$ and $\triangle AB'C$ both have the given properties; that is, both have a side of length b, an angle of magnitude A, and a side of length a.

The implication in the preceding paragraph is clear. The given data do not necessarily describe a single triangle. Since two triangles are possible, we call this an *ambiguous case*, and in the absence of further information that would restrict us to only one of the triangles we must accept that either triangle could be the desired triangle and must therefore solve both triangles.

Example 2 Solve the general triangle ABC if $b = 2$, $B = 30°$, and $c = 3$.

Solution First, construct the triangle as in Fig. 13-12. The construction results in $\triangle BCA$ and $\triangle BC'A$, each of which has $b = 2$, $c = 3$, and $B = 30°$. Each triangle is solved separately by using the sine law.

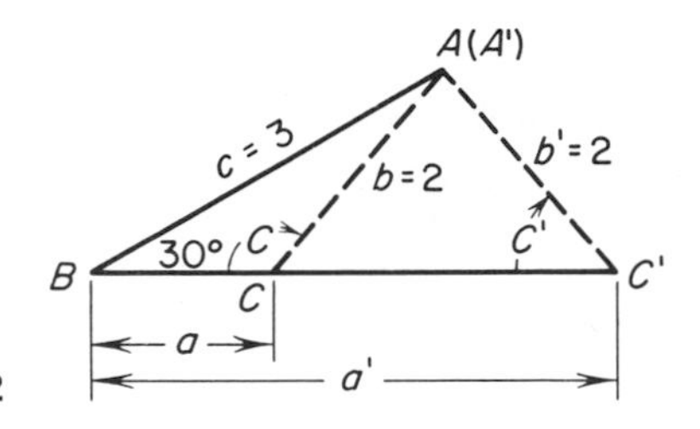

Figure 13-12

For either triangle,

$$\frac{b}{\sin B} = \frac{c}{\sin C}$$

or

$$\frac{b}{\sin B} = \frac{c}{\sin C'} \tag{13}$$

and substituting $b = 2$, $c = 3$, and $B = 30°$ into (13), we have either

$$\sin C = 0.75 \quad \text{or} \quad \sin C' = 0.75. \tag{14}$$

Referring now to Fig. 13-12, since $\triangle ACC'$ is isosceles, angles C and C' are supplements, with C obtuse and C' acute. Therefore from (14)

$$C = 131°25' \quad \text{and} \quad C' = 48°35'.$$

Since the sum of the angles of a triangle equals 180°,

$$A' = \angle BAC' = 180° - (B + C') = 101°25',$$

$$A = \angle BAC = 180° - (B + C) = 18°35'.$$

To find the remaining side, we have

$$\frac{a}{\sin \angle BAC} = \frac{2}{\sin 30°}, \quad \text{from which} \quad a = 1.275,$$

and

$$\frac{a'}{\sin \angle BAC'} = \frac{2}{\sin 30°}, \quad \text{from which} \quad a' = 3.921.$$

The organized solution shows

$$a = 1.275, \qquad C = 131°25', \qquad A = 18°35',$$

$$a' = 3.921, \qquad C' = 48°35', \qquad A' = 101°25'.$$

Exercises 13-2

In Exercises 1–16, the given values give rise to the ambiguous case. Solve for the quantity in parentheses; give both solutions. In each case, refer to the general triangle ABC.

1. $A = 30°, a = 10, b = 15 \quad (C)$
2. $A = 40°, a = 8, b = 10 \quad (C)$
3. $C = 50°, c = 8, b = 10 \quad (a)$

4. $C = 45°, c = 12, b = 15$ (*a*)

5. $B = 30°, b = 6, c = 9$ (*a*)

6. $C = 40°, c = 15, a = 20$ (*B*)

7. $A = 45°, a = 9, c = 12$ (*B*)

8. $B = 45°, b = 12, a = 15$ (*C*)

9. $C = 60°, c = 9, b = 10$ (*a*)

10. $A = 60°, a = 14, b = 15$ (*c*)

11. $A = 12°15', a = 7.5, b = 15$ (*C*)

12. $B = 67°20', b = 8.44, c = 8.52$ (*a*)

13. $C = 10°29', b = 3c$ (*B*)

14. $C = 10°29', b = 2c$ (*C*)

15. $A = 30°, a = 3.83, b = 7.66$ (*C*)

16. $B = 12°25', b = 8.16, c = 9.00$ (*A*)

Referring to Fig. 13-11, assume that we are given a, b, and A. Discuss the solutions of the triangle in Exercises 17–22 if

17. $a = b$ **18.** $a = b \sin A$ **19.** $a < b \sin A$

20. $a > b$ **21.** $A = 90°$ **22.** A is obtuse

Using the material in Sec. 13-2 and the congruency theorems from plane geometry, indicate whether a unique triangle is formed when the parts in Exercises 23–28 are known.

23. Three sides

24. Two sides and the angle opposite one of them

25. Two sides and the included angle

26. Two angles and the included side

27. Two angles and the side opposite one of them

28. Three angles

29. Given the general triangle ABC, where A, a, and b are known, show that

$$c = b \cos A \pm a \cos B.$$

13-3 The cosine law

Another trigonometric law that is useful in solving triangles is the cosine law. We first show its derivation, then discuss when it can be applied, and, finally, show its applications.

Given general triangle ABC pictured in Fig. 13-13, construct a perpendicular from C to AB at D. It can readily be shown that

$$AD = b \cos A$$

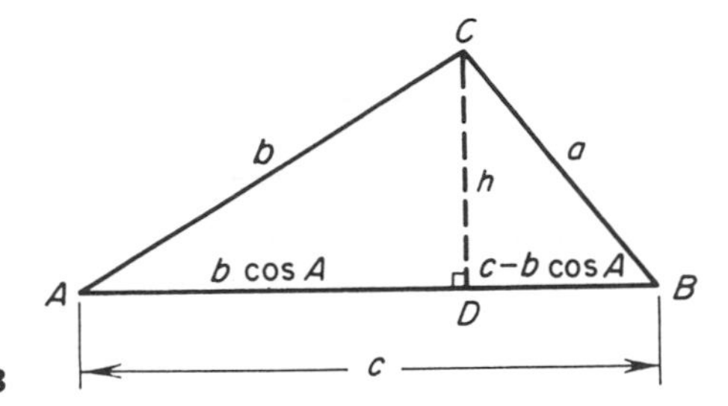

Figure 13-13

and that $DB = c - b \cos A$. From right triangle ADC

$$h = \sqrt{b^2 - (b \cos A)^2}, \tag{15}$$

and from right triangle BDC

$$h = \sqrt{a^2 - (c - b \cos A)^2}. \tag{16}$$

Equating (15) and (16) to eliminate h, we have

$$\sqrt{b^2 - (b \cos A)^2} = \sqrt{a^2 - (c - b \cos A)^2}. \tag{17}$$

Squaring both sides of (17) and solving for a^2, we obtain

$$a^2 = b^2 + c^2 - 2bc \cos A,$$

which is one form of the *cosine law.*

By using different perpendiculars in Fig. 13-13 and thus involving angle B or angle C, we can obtain the three relationships

$$\begin{aligned} a^2 &= b^2 + c^2 - 2bc \cos A, \\ b^2 &= a^2 + c^2 - 2ac \cos B, \\ c^2 &= a^2 + b^2 - 2ab \cos C, \end{aligned} \tag{18}$$

where Eqs. (18) are the three available forms of the cosine law.

Inspecting Eqs. (18), we see that the cosine law can be used to solve a triangle when the given parts are

1. *Three sides* If a, b, and c are given, then angles A, B, or C can be found directly, depending on which of Eqs. (18) is used.
2. *Two sides and the included angle* As an example, if b, c, and A are given, the first of Eqs. (18) is readily solved for a.
3. *Two sides and the angle opposite one of them* To illustrate, if a, b, and A are given, the first of Eqs. (18) becomes a quadratic in c. This quadratic can be difficult to solve, but it will give both values of c in the ambiguous case. This case is usually solved more easily by the sine law.
4. *Two angles and a side* This case, although possible by the cosine law, would produce simultaneous quadratics. It is much more easily solved by the sine law.

The preceding discussion reveals that the cosine law can be used in all cases of solvable triangles. The last two cases listed are better solved by the sine law, however. The sss case and the sas case cannot be solved by the sine law; therefore the cosine law is recommended.

Another trigonometric law—the tangent law—involves functions of half angles and is useful in solving oblique triangles; it is discussed later.

Example 3 Two forces $F_1 = 4$ lb and $F_2 = 6$ lb are applied to a body, producing a resultant force of $R = 8$ lb. What is the angle between F_1 and R?

Solution Sketching the forces as in Fig. 13-14(a), we produce the parallelogram $ABCD$ with sides F_1 and F_2 and diagonal R. The triangle ABC from Fig. 13-14(a) is reproduced in Fig. 13-14(b), where we see that angle A is the desired angle between F_1 and R. Applying $a = 6$, $b = 8$, and $c = 4$ to

$$a^2 = b^2 + c^2 - 2bc \cos A,$$

we have

$$(6)^2 = (8)^2 + (4)^2 - 2(8)(4) \cos A$$

from which

$$\cos A = \frac{44}{64} = 0.6875.$$

and

$$A = 46°34'.$$

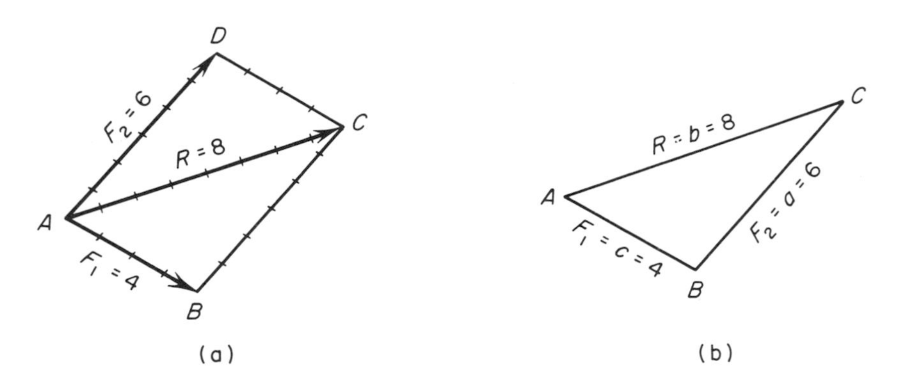

Figure 13-14

Example 4 Two aircraft leave the same point C at the same time, diverging at an angle of 60°, each flying in a straight line. When aircraft A is 90 miles from C and aircraft B is 100 miles from C, how far are the aircraft apart?

Solution See Fig. 13-15. Here $a = 100$, $b = 90$, and $C = 60°$ and we wish to find c. Applying the last of Eqs. (18),

$$c^2 = (100)^2 + (90)^2 - 2(100)(90) \cos 60°$$

from which

$$c = \sqrt{9100} = 95.39 \text{ miles.}$$

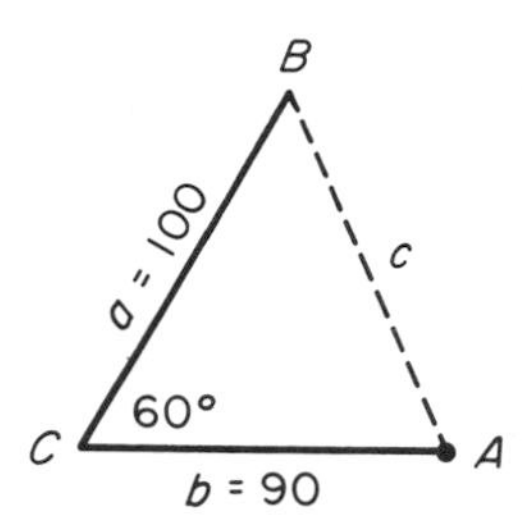

Figure 13-15

Exercises 13-3

In Exercises 1–10, use the given information to solve for the quantity in parentheses. In all cases, reference is made to general triangle ABC.

1. $a = 3, b = 5, c = 6$ (B)
2. $a = 4, b = 5, c = 8$ (C)
3. $a = 8, b = 6, C = 45°$ (c)
4. $c = 5, b = 4, C = 60°$ (a)
5. $a = 4.073, b = 6.325, c = 5.923$ (C)
6. $a = 6.225, b = 7.140, c = 5.885$ (B)
7. $a = b = c$ (any angle)
8. $a = b, c = 2b$ (A)
9. $a = 3.2, c = 4.6, B = 150°$ (C)
10. $b = 5.6, c = 8.2, A = 150°$ (a)
11. Two forces $F_1 = 4.3$ lb and $F_2 = 6.2$ lb are applied to a body. The resultant force is 5.6 lb. What is the angle between F_1 and F_2?
12. Two forces F_1 and F_2 are applied to a body with a resultant force $R = 310$ lb. If the directions of F_1 and F_2 differ by 60° and $F_1 = 250$ lb, find F_2.
13. The cosine law fails for a triangle with sides $a = 2$, $b = 6$, and $c = 9$. Explain why.
14. Show by the cosine law that the angle opposite the largest side of the triangle with sides in a 5: 12: 13 ratio is a right angle.

In Fig. 13-16 the tetrahedron pictured has $AB = 4.6$, $BC = 3.75$, $AD = 6.8$, $\angle ABD = 65.3°$, $\angle ABC = 59.2°$, and $\angle DBC = 68°$. Using this given data, find solutions for Exercises 15–26.

15. $\angle DAB$ **16.** $\angle ADB$ **17.** $\angle BCD$ **18.** $\angle CDB$

19. $\angle DAC$ **20.** $\angle ACD$ **21.** $\angle CDA$ **22.** $\angle BAC$

23. $\angle ACB$ **24.** AC **25.** BD **26.** CD

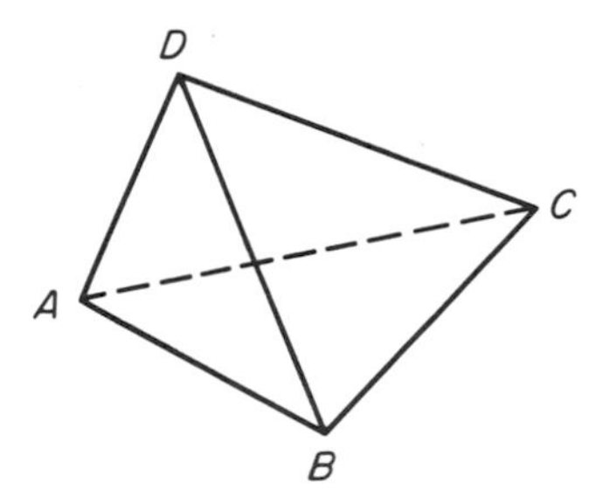

Figure 13-16

27. Using the information given for the tetrahedron in Fig. 13-16, sketch a line from C perpendicular to AB at a point called E. Find angle ECD.

28. Three forces F_1, F_2, and F_3 act on a body to produce a resultant R. If $F_1 =$ 20 lb at 30°, $F_2 =$ 40 lb at 270°, and $R =$ 50 lb at 120°, find the magnitude and direction of F_3. (*Note:* Assume that all forces are acting in one plane.)

29. Simplify expression (17).

30. If n forces $F_1, F_2, \ldots, F_n$ act on a body at angles $\phi_1, \phi_2, \phi_3, \ldots, \phi_n$, respectively, and all forces are in the same cartesian coordinate plane,

$$R_x = F_1 \cos \phi_1 + F_2 \cos \phi_2 + F_3 \cos \phi_3 + \cdots + F_n \cos \phi_n$$

and
$$R_y = F_1 \sin \phi_1 + F_2 \sin \phi_2 + F_3 \sin \phi_3 + \cdots + F_n \sin \phi_n,$$

where R_x and R_y are the horizontal and vertical components of the resultant, respectively.

13-4 Law of tangents

A trigonometric law that is particularly useful in solving triangles when two sides and the included angle are known is the tangent law. We will avoid the derivation of the tangent law, since it requires use of trigonometric identities that are not introduced until Chapter 16, and will therefore present the tangent law as a formula.

Given the general triangle with angles A, B, and C and opposing sides a, b, and c as pictured in Fig. 13-1, we have the following relationships, which form what is called the *tangent law*:

$$\frac{a - b}{a + b} = \frac{\tan \frac{1}{2}(A - B)}{\tan \frac{1}{2}(A + B)}, \tag{19}$$

$$\frac{b-c}{b+c}=\frac{\tan\frac{1}{2}(B-C)}{\tan\frac{1}{2}(B+C)}, \tag{20}$$

$$\frac{c-a}{c+a}=\frac{\tan\frac{1}{2}(C-A)}{\tan\frac{1}{2}(C+A)}. \tag{21}$$

Consider an example of the use of the tangent law as it involves the sas case.

Example 5 Solve the triangle in Fig. 13-17.

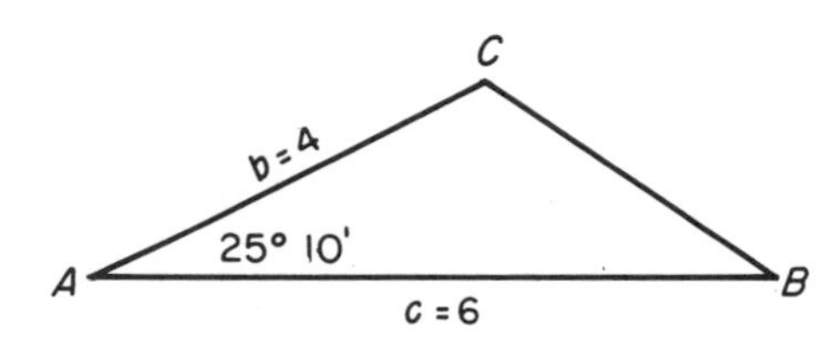

Figure 13-17

Solution With sides b and c known, application of (20) will readily determine angles B and C. With $b = 4$, $c = 6$, and $\frac{1}{2}(B + C) = \frac{1}{2}(180° - 25°10') = \frac{1}{2}(154°50') = 77°25'$, we have from (20)

$$\frac{4-6}{4+6}=\frac{\tan\frac{1}{2}(B-C)}{\tan 77°25'}$$

from which $\tan\frac{1}{2}(B-C) = -\frac{2}{10}(\tan 77°25') = -0.2(4.480) = 0.8960$, or $\frac{1}{2}(B-C) = -(41°52')$ and

$$B - C = -83°44'. \tag{22}$$

Using the fact that the angles of a triangle add to 180° and angle A is given as $25°10'$, then

$$B + C = 154°50'. \tag{23}$$

Solving (22) and (23) simultaneously, we have $B = 35°33'$ and $C = 119°17'$. At this point we know all the angles of Fig. 13-17 and two of the sides. We can use any of the three trigonometric laws of oblique triangles to obtain the third side $a = 2.925$.

Exercises 13-4

In Exercises 1–10, the given data are parts of the general triangle ABC. Solve for the part in parentheses, using the tangent law.

1. $a = 5, b = 4, C = 45°$ (A)
2. $a = 6, b = 4, C = 45°$ (A)
3. $c = 7.4, a = 3.5, B = 32°$ (C)

4. $c = 3.6, a = 7.5, B = 31°$ (C)
5. $b = 18.4, c = 15.2, A = 120.6°$ (B)
6. $b = 20.5, c = 17.4, A = 118.5°$ (B)
7. $a = b, C = 100°$ (A)
8. $b = c, A = 90°$ (B)
9. $a = 2b, C = 80°$ (A)
10. $a = 2b, C = 100°$ (A)
11. Two sides of a triangle are in a ratio of 3 to 4. The included angle is 60°. Find the other angles.
12. Two sides of a triangle are in a ratio of 4 to 5. The included angle is 70°. Find the other angles.
13. Evaluate 2 arctan 1/7.
14. Evaluate 2 arctan 1/3.

14

Graphs of the trigonometric functions

Each of the trigonometric functions can be plotted as a graph. The most commonly used function is the sine function. In this chapter we show the graphs of the six elementary trigonometric functions but present a detailed study only of the sine wave.

Knowledge of the graphs of the trigonometric functions is especially useful in the study of alternating-current electricity, sound, vibrations, and many other phenomena that are periodic in nature. In addition to the graphs of the elementary trigonometric functions, we also discuss Lissajous figures, which are often observed in oscilloscope work and which are useful in determining unknown frequencies.

14-1 Periodicity and definitions

Assume that the pendulum in Fig. 14-1 swings back and forth across a rest position. Assume further that each time it swings to the right it exactly reaches position A and each time it swings to the left it exactly reaches position B. Also, the time required to get from A to O to B to O and back to A is the same for each swing of the pendulum; call this time T. Now if the timing is started when the pendulum is at any other position, say D, and the time required to traverse the D–O–B–O–D–A–D route is still T, then we have a case of *periodicity*.

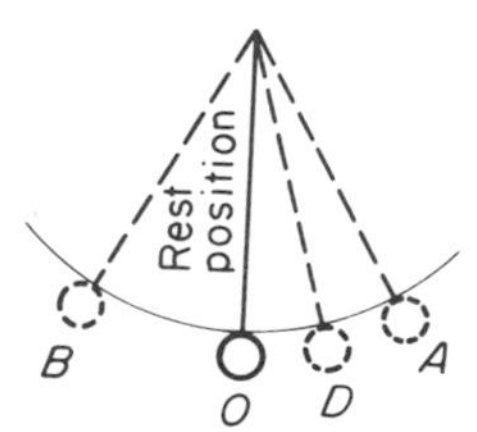

Figure 14-1

A body that is in periodic motion traverses a series of positions according to a fixed pattern. The series of positions that it assumes is called the *cycle*. The time required to complete one cycle—that is, to assume all the positions and velocities—is called the *time period*. The number of cycles completed in a given time period is called the *frequency*.

The earth assumes positions periodically with reference to the sun. The time required to complete one cycle (one complete revolution about the sun) is approximately 365 days. This means that 365 days from today the earth will be in the same position relative to the sun that it is in today. It will be in the same position 2(365) days from today and 3(365) days and $n(365)$, where n is an integer. The frequency of the earth's motion is then one cycle per 365 days, or one cycle per year. The time period is $T = 365$ days.

The position of the minute hand of a clock is periodic with time $T = 60$ minutes as the time period. The position of the hour hand of a clock is periodic with time period $T = 12$ hours. The frequencies are 24 cycles per day and 2 cycles per day for the minute hand and hour hand, respectively.

If the periodic motion is circular, the time period is the time required to make one full revolution. One full revolution is 360° or 2π radians, depending on the system of angula· measurem.nt used. Let us consider the relationship between time period and frequency for circular motion, along with units.

Assume that a body is in circular motion with frequency f. This means (using time in seconds) that f is in cycles per second. Consider the motion to be effective and uniform for t seconds. The angle θ generated in t

seconds is

$$\theta(\text{cycles}) = f\left(\frac{\text{cycles}}{\text{sec}}\right) \times t(\text{sec}) = ft(\text{cycles}).$$

Then since there are 2π radians per cycle,

$$\theta\,(\text{radians}) = f\left(\frac{\text{cycles}}{\text{sec}}\right) \times 2\pi\left(\frac{\text{radians}}{\text{cycle}}\right) \times t\,(\text{sec}) = 2\pi ft\,(\text{radians}).$$

Also, since there are 360° per cycle, then

$$\theta\,(\text{degrees}) = f\left(\frac{\text{cycles}}{\text{sec}}\right) \times 360\left(\frac{\text{degrees}}{\text{cycle}}\right) \times t\,(\text{sec}) = 360ft\,(\text{degrees}).$$

As stated before, the time required to complete one cycle is T sec, or there are T sec/cycle. On the other hand, frequency f is in cycles/sec. Since T and f have units that are reciprocals, then

$$T = \frac{1}{f} \quad \text{or} \quad f = \frac{1}{T}, \tag{1}$$

where (1) relates the time period and the frequency.

Example 1 An airfield beacon sweeps in a uniform circular motion at the rate of 60° per second of time.
(a) What is the time T required to complete one cycle?
(b) What is its frequency in cycles per second?

Solution

(a) To find T, the time required to complete one cycle, we have

$$60\,\frac{\text{degrees}}{\text{sec}} \times T\,\frac{\text{sec}}{\text{cycle}} = 360\,\frac{\text{degrees}}{\text{cycle}} \tag{2}$$

or

$$60T = 360$$

from which

$$T = 6\,\frac{\text{sec}}{\text{cycle}}.$$

To develop (2), we know that one cycle equals 360°. Since the beacon rotates at the rate of 60° for each second of time, we see that it requires 6 sec to complete one cycle.

(b) Since $T = 6$ sec/cycle, then by (1)

$$f = \frac{1}{T} = \frac{1}{6\text{ sec/cycle}} = \frac{1}{6}\,\frac{\text{cycles}}{\text{sec}}.$$

Exercises 14-1

1. A satellite circles the globe exactly 12 times per day. What is its time period in minutes? What is its frequency in cycles per hour? Cycles per minute?

2. The earth spins on its axis one full revolution per 24 hours. What is its time period in days? What is its frequency in cycles per day? Cycles per hour? Cycles per year?
3. A wheel rotates uniformly at the rate of 1000° per second. A point on a spoke of the wheel passes through how many cycles per minute? What is the time period in seconds?
4. An automobile is traveling at the rate of 40 mph. A point on a spoke of a wheel is in periodic motion with reference to the axle. The greatest radius of the tire is 16 in. What is the time period for the point? What is the frequency in cycles per second?
5. Two points are on the same spoke of a wheel that is rotating at 10 cycles/sec. If point A is 12 in. from the axle and point B is 18 in. from the axle, what is the time period for A? For B?
6. Two wheels mounted on the same axis rotate at different angular velocities. Wheel A rotates at 80 cycles/sec and wheel B rotates at 60 cycles/sec. One and only one spoke on each wheel is painted red. Periodically the red spokes align into exactly the same position. What is the length of the time period between consecutive alignments if
   (a) The wheels rotate in the same direction?
   (b) The wheels rotate in opposite directions?

14-2 Rotating vectors; the sine wave

A vector **P** shown in Fig. 14-2 can be considered to rotate. In the chapter on right triangles we considered a vector to be stationary, but certain conventions established there will be maintained here. Reviewing them, we consider the positive direction of rotation to be counterclockwise. Also, the direction of **P** is indicated by θ, which initiates at the positive x axis.

Referring again to Fig. 14-2, if **P** rotates uniformly, it will assume the same position and velocity periodically. This assumption of the same position and velocity can be independent of time, although later on we will often make it dependent on time. Here the position of the endpoint of the

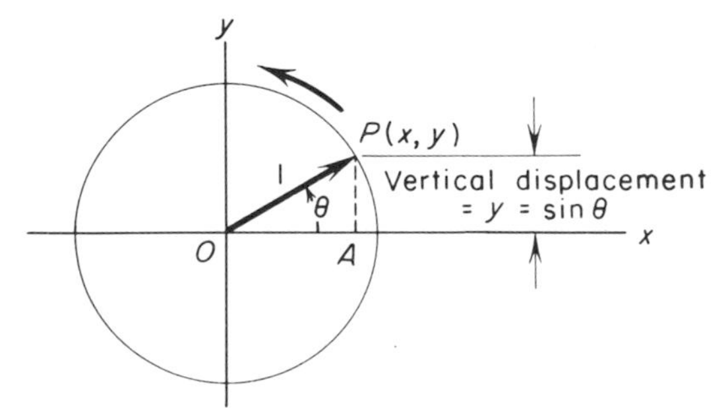

Figure 14-2

vector (point P) depends on the magnitude of θ. If P is at P_0 when θ is of value θ_0, then P is again at P_0 when $\theta = \theta_0 + 2\pi, \theta = \theta_0 + 4\pi, \theta = \theta_0 + 6\pi, \ldots, \theta = \theta_0 + 2n\pi$, where n is an integer. In other words, the position of P is cyclical with respect to θ, and the angular period of θ is 2π.

Continuing to refer to Fig. 14-2, the vertical displacement of P (namely, the ordinate y) is also periodic with respect to θ. We note two facts about this vertical displacement. First, y never exceeds OP in magnitude and the range of y is $-OP \leqq y \leqq OP$. Secondly, by the definition of the sine function,

$$\sin\theta = \frac{AP}{OP} = \frac{\text{vertical displacement}}{\text{magnitude}}. \tag{3}$$

If, in Fig. 14-2, we consider the circle to be a unit circle (that is, the radius is 1), then $OP = 1$. Furthermore,

$$AP = \text{vertical displacement} = y.$$

These facts revise (3) to read

$$y = \sin\theta, \tag{4}$$

where (4) asserts that the vertical displacement of a rotating unit vector equals the sine of the angle passed through.

We earlier asserted that the vertical displacement of the unit vector ranges between -1 and $+1$, depending on the size of θ. Let us portray this graphically as in Fig. 14-3. Here we show the unit vector starting in position OP_0 and subsequently assuming positions OP_1, OP_2, and so on. The vertical displacement of P_i (the general P) is, by (4), the sine of θ_i. In Fig. 14-3 the vectors are sketched 30° apart. These same 30° intervals are plotted on the θ axis at the right. At each abscissa the proper vertical displacement is plotted, producing the curve shown, which is the graph of $y = \sin\theta$.

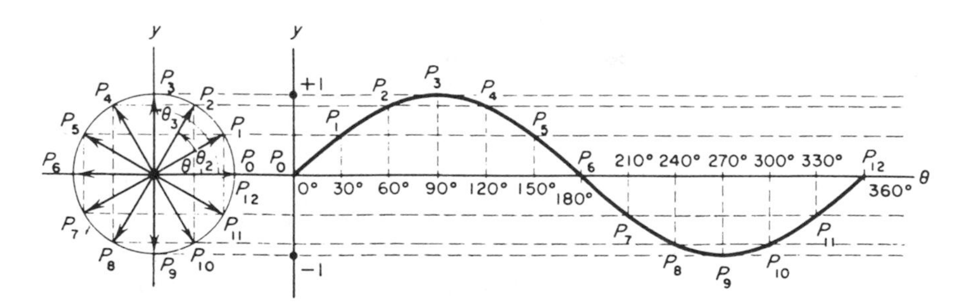

Figure 14-3

We will call the graph of $y = \sin\theta$ the graph of the *fundamental.* Equation $y = \sin\theta$ is not the general sine equation but is highly restrictive. Several modifications can be made, each modifying the graph. For instance, we might be asked to graph

$$y = 3 \sin \theta,$$

$$y = 4 \sin 3\theta,$$

$$y = \tfrac{1}{2} \sin (2\theta - 45^\circ),$$

$$y = a \sin (b\theta + c),$$

where the four listed equations introduce new properties, such as amplitude, angular period, and phase angle, all of which are discussed in subsequent sections.

Example 2 Eliminating care for exactness, show a rapid way of producing a rough graph of $y = \sin x$.

Solution Sketch an x axis and a y axis as shown in Fig. 14-4. Assign a positive and negative unity on the y axis. Assign equally spaced positions along the x axis, showing $x = 0, 90, 180, 270, 360^\circ$. From knowledge of the sine function, we know that $\sin 0^\circ = 0$, $\sin 90^\circ = 1$, $\sin 180^\circ = 0$, $\sin 270^\circ = -1$, and $\sin 360^\circ = 0$, establishing the five points shown in Fig. 14-4. Join these points with a smooth curve, duplicating the general shape of the curve shown in Fig. 14-3.

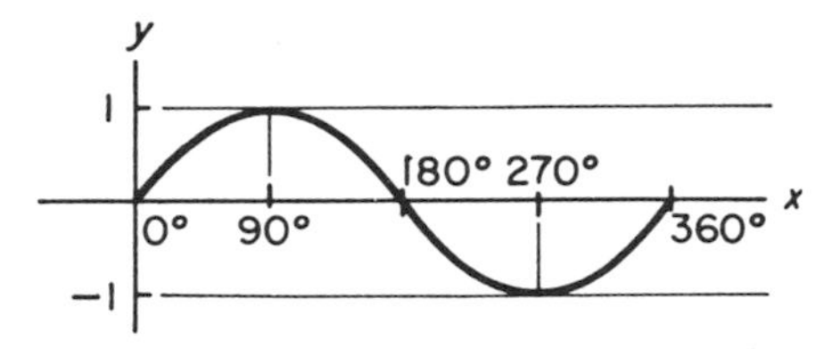

Figure 14-4

14-3 More on the sine wave

In Sec. 14-2 we showed how a rotating vector generates a sine wave. We were highly restrictive, however, showing only the graph of

$$y = a \sin (b\theta + c), \tag{5}$$

where both a and b were unity and c was zero, causing (5) to have the form of (4). We further suggested that we could call the graph of (4) the fundamental wave. The graph of the fundamental is useful for comparison when a or b of (5) is not unity and c is not zero. Let us see how modifications of a, b, and c affect the sketch of the sine waves shown in Figs. 14-3 and 14-4.

Example 3 Show the graph of

$$y = 2 \sin \theta. \tag{6}$$

Solution First, for background and comparison purposes, sketch the fundamental

$$y_1 = \sin \theta \tag{7}$$

as shown in Fig. 14-5. Next, compare the ordinates of (6) and (7). Since $y_1 = \sin \theta$ and $y = 2 \sin \theta$, then by comparison,

$$y = 2y_1, \tag{8}$$

where (8) means that the ordinate of (6) is twice the ordinate of (7) for any given value of θ. This result suggests that all vertical displacements of the dotted curve in Fig. 14-5, when doubled, will be the appropriate vertical displacements of (6).

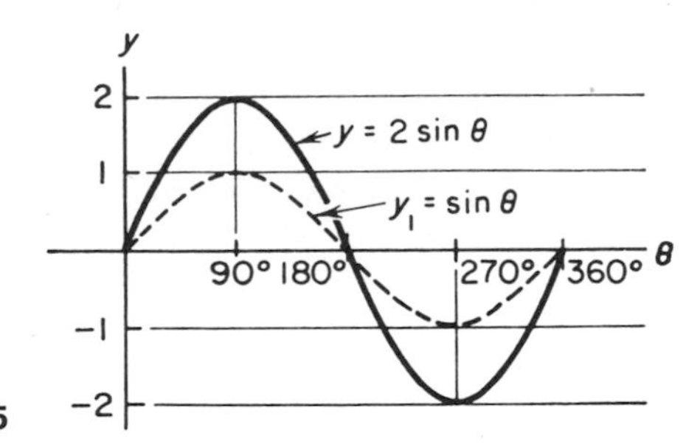

Figure 14-5

The preceding example suggests correctly that the only distinction between the graphs of $y_1 = \sin \theta$ and $y_2 = a \sin \theta$ is the factor a. Here

$$y_2 = ay_1,$$

and a is called the *amplitude factor*. It is often called the *amplitude*, corresponding to the greatest vertical displacement of $y = a \sin \theta$; we will use the latter terminology in subsequent discussions. Note that the introduction of factor a merely elongates or shrinks the vertical displacements of the fundamental. If $a > 1$, we have elongation; if $a < 1$, we have shrinking. The value of a in no way affects the plots of the abscissas; that is, it neither shrinks nor elongates the graph in the horizontal sense.

Consider the introduction of a nonunity b in (5). In the next three examples $a = 1$ and $b \neq 1$.

Example 4 Show the graph of

$$y = \sin 2\theta. \tag{9}$$

Solution As before, we will first graph the fundamental equation

$$y_1 = \sin \theta \tag{10}$$

shown as the dotted curve in Fig. 14-6. Comparing (9) and (10), we see

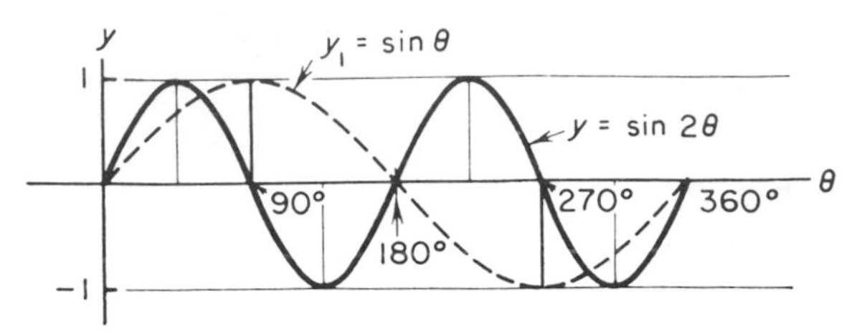

Figure 14-6

that a given assignment for θ will not necessarily produce equal values for y and y_1, since $y = \sin 2\theta$ and $y_1 = \sin \theta$. We know that the critical values of y_1 occur 90° apart with respect to θ. That is, when $\theta = 90^\circ$, $y_1 = 1 =$ maximum; when $\theta = 180^\circ$, $y_1 = 0$; when $\theta = 270^\circ$, $y_1 = -1 =$ minimum; when $\theta = 0^\circ$ or 360°, $y_1 = 0$. The critical values are called here the *zeros*, *maxima*, and *minima* of the ordinate.

Inspecting (9), we assign to θ values that are multiples of 45°; then 2θ will assume values that are multiples of 90°, suggesting that the critical values of $y = \sin 2\theta$ are spaced 45° apart. Consider Table 14-1.

From the data in Table 14-1 and knowledge of the fact that critical values of y occur at intervals of 45°, we have the graph of $y = \sin 2\theta$ as shown by the solid line in Fig. 14-6.

Table 14-1

θ°	0	45	90	135	180	225	270	315	360
$2\theta^\circ$	0	90	180	270	360	45	540	630	720
$y_1 = \sin\theta$	0	0.707	1	0.707	0	−0.707	−1	−0.707	0
$y = \sin 2\theta$	0	1	0	−1	0	1	0	−1	0

Remarks: Inspection shows that the *angular period* of $y = \sin\theta$ is 360°. In other words, one full cycle of the sine wave is completed for each 360° increase in θ. Inspecting $y = \sin 2\theta$, an increase of 360° in θ increases 2θ by 720°, meaning that we pass through *two* full cycles of the wave for each 360° increase in θ. This fact suggests that the angular period of $y = \sin 2\theta$ is $360^\circ/2 = 180^\circ$. In practical applications, $y = \sin 2\theta$ is called the *second harmonic* of $y = \sin\theta$.

Continuing the preceding discussion, let us compare the graphs of

$$y = \sin\theta, \tag{11}$$

$$y = \sin n\theta. \tag{12}$$

The angular period of (11) is 360°. If we increase θ by 360° in (12), $n\theta$ is increased by $n(360°)$, suggesting that a 360° increase in θ causes (12) to pass through n full cycles. Therefore the *angular period* of (12) is $360°/n$. If n is a positive integer, (12) is called the nth harmonic of 11).

Example 5 Show the graph of $y = \sin 6\theta$.

Solution The angular period of $y = \sin 6\theta$ is $360°/6 = 60°$. This means that one full cycle of $y = \sin 6\theta$ is completed for each 60° increase in θ. Moreover, the critical values occur $60°/4 = 15°$ apart. Figure 14-7 shows the graphs of $y = \sin 6\theta$ and the fundamental $y_1 = \sin \theta$. Here $y = \sin 6\theta$ is called the *sixth harmonic* of $y = \sin \theta$.

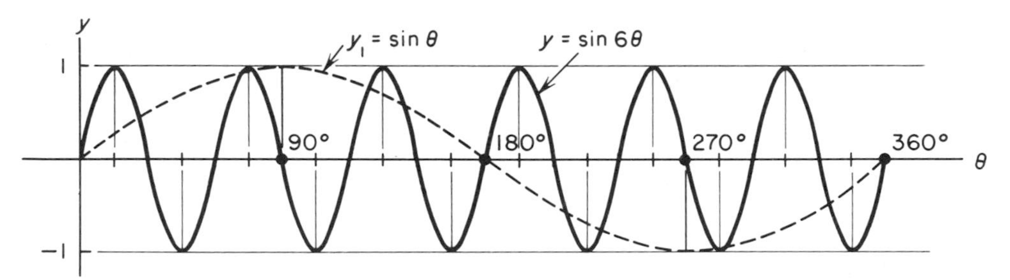

Figure 14-7

Example 6 Show the graphs of

$$y = \sin x, \tag{13}$$

$$y = \sin \frac{\pi}{2} x, \tag{14}$$

laying off the horizontal axis in radians.

Solution Regard (13) as the fundamental wave. An increase of 2π radians in x completes one full cycle of (13). Therefore 2π radians is the angular period of (13), and the critical values are $2\pi/4 = \pi/2$ radians apart. The graph of (13) is shown as the dotted line in Fig. 14-8.

The angular period of (14) is $2\pi \div \pi/2 = 4$ radians. The fussy part of plotting (14) now is locating positions $x = 1, 2, 3, 4$ radians along the horizontal axis. For convenience, values $\pi/2$, π, $3\pi/2$, and 2π are shown as decimal values 1.57, 3.14, 4.71, and 6.28 in Fig. 14-8. Using these decimal values as guides, we plot the critical points for (14) and show the graph as the solid line.

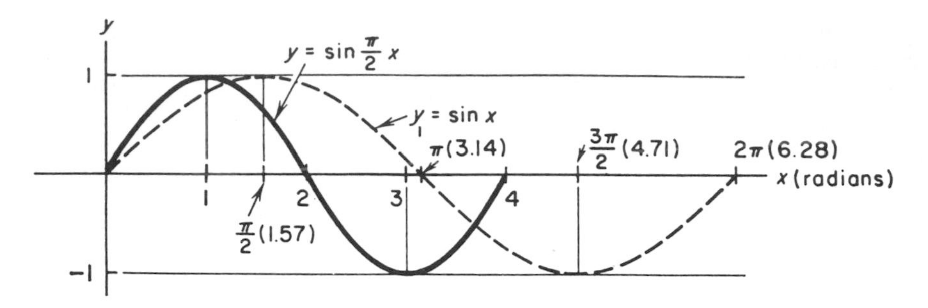

Figure 14-8

In summary, the distinction between the graphs of the fundamental $y = \sin \theta$ and $y = \sin b\theta$ (where $b \neq 1$) is a matter of elongation along the θ axis. The amplitude is not affected by the nonunity value of b; only the angular period of $y = \sin b\theta$ is affected. The angular period of $y = \sin b\theta$ is either $360°/b$ or $2\pi/b$, depending on the system of angular measurement used.

Consider next the introduction of a nonzero value of c in (5).

Example 7 Show the graph of

$$y = \sin \theta, \tag{15}$$

$$y = \sin (\theta + 60°). \tag{16}$$

Solution Again, (15) is called the fundamental. Let us change our approach and show the generation of waves (15) and (16) by rotating vectors. First, observe that a 360° change in θ produces a 360° change in $\theta + 60°$, suggesting that the angular periods of (15) and (16) are both 360°. The dotted graph in Fig. 14-9 is the graph of (15) and is the result of rotating vector **P**, where it is important to note that the initial direction of **P** is 0°. The graph of (16) is shown as the result of rotating vector **R** in Fig. 14-9, where we see that for $\theta = 0°$

$$y = \sin (0° + 60°) = \sin 60°.$$

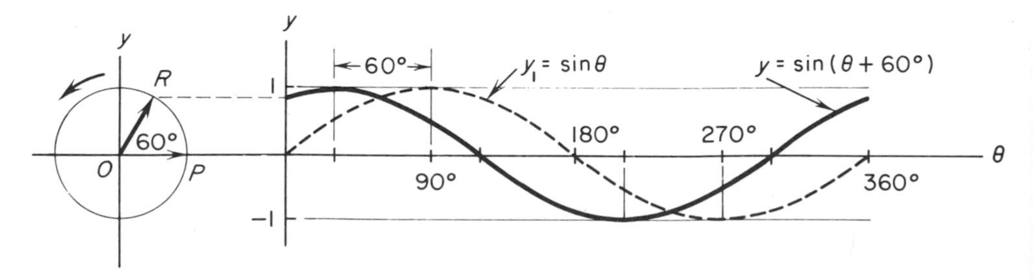

Figure 14-9

which asserts that **R** is 60° ahead of **P** for all assignments of θ. This statement is clear from the fact that $\theta + 60°$ is 60° greater than θ. It further asserts that the critical values of (16) are 60° *ahead* of the corresponding critical values of (15). (We say "ahead" here despite the fact that the graphed critical values of (16) are 60° to the *left* of those of (15). If the θ axis were a time axis with the time increasing to the right, the critical values of (16) would occur 60° *sooner* in time.)

Remarks: The distinction between the graphs of

$$y = \sin \theta, \tag{17}$$

$$y = \sin (\theta + c) \tag{18}$$

is a matter of horizontal shift. Accepting (17) as the reference curve, (18) is shifted left or right by an amount c depending on the sign of c, and the curve is in no way distorted. If $c > 0$, the shift is to the left; if $c < 0$, the shift is to the right. Accordingly, c is called the *phase shift* or *phase angle*. We note that (17) and (18) are both first harmonics. If a phase shift is to be intelligently described, it must be with reference to an unshifted wave of the same harmonic. Consider Example 8.

Example 8 Show the graphs of

$$y = \sin (3\theta + 60°), \tag{19}$$

$$y_1 = \sin 3\theta. \tag{20}$$

Solution Inspecting (20), we have an unshifted third harmonic with angular period 120° and unity amplitude. This is shown as the dotted line in Fig. 14-10. To plot (20), we must be careful, for the phase shift is *not* 60°. Consider values of θ that will cause $3\theta + 60°$ to be the critical values 90, 180, 270, 360°. If we wish the first maximum value of (19),

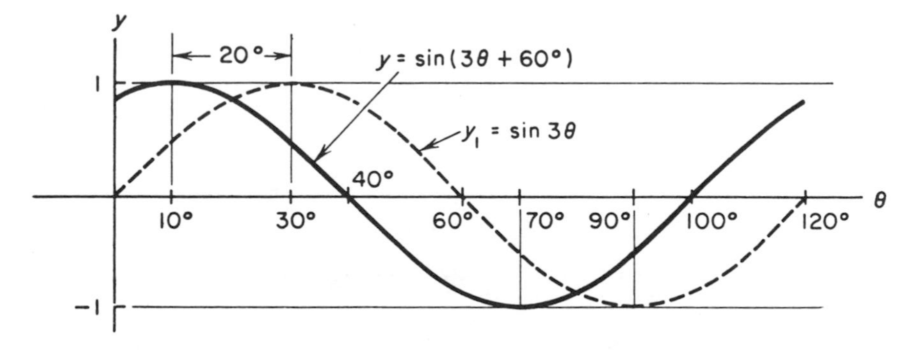

Figure 14-10

then

$$3\theta + 60° = 90°$$

from which $\theta = 10°$ and we plot a maximum at $\theta = 10°$. To obtain the 180° zero point,

$$3\theta + 60° = 180°,$$

from which $\theta = 40°$. Continuing to plot the critical values of (19), we produce the graph shown in Fig. 14-10. Note that (19) leads (20) not by 60°, but by $60°/3 = 20°$.

Remarks: The distinction between the graphs of

$$y = \sin b\theta, \tag{21}$$

$$y = \sin (b\theta + c) \tag{22}$$

is a horizontal shift. The phase shift is of magnitude c/b. When $b > 0$, then (22) is shifted to the left of (21) if $c > 0$ and to the right of (21) if $c < 0$.

Summary: The general equation of the sine wave is given as

$$y = a \sin (bx + c)$$

where a is the *amplitude* or greatest vertical displacement, $360°/b$ or $2\pi/b$ is the *angular period*, and c/b is the *phase shift* with respect to the corresponding harmonic.

Exercises 14-2 and 14-3

Show graphs of the equations in Exercises 1–12. In all but Exercise 1 plot the fundamental wave for comparison purposes on the same set of axes.

1. $y = \sin x$

2. $y = \frac{1}{2} \sin x$

3. $y = 2.3 \sin x$

4. $y = \sin 5\theta$

5. $y = \sin \dfrac{\theta}{3}$

6. $y = \sin 3\pi x$

7. $y = \sin (\theta - 30°)$

8. $y = 1.8 \sin (\theta + 90°)$

9. $y = 3 \sin 4\theta$

10. $y = 6 \sin (2\theta + 90°)$

11. $y = 1.4 \sin (6\theta - 48°)$

12. $y = 5 \sin \left(\dfrac{\pi}{2}x + \dfrac{\pi}{4}\right)$.

13. What are the amplitudes of the sine waves in Exercises 8, 9, and 10?

14. What are the angular periods of the sine waves in Exercises 4, 5, and 6?

15. The sine wave in Exercise 10 is shifted how many degrees in which direction with reference to $y = \sin 2\theta$?

16. The sine waves in Exercises 11 and 12 are shifted how far in what direction with reference to $y = \sin 6\theta$ and $y = \sin (\pi/2)x$, respectively?

17. Referring to Fig. 14-11, determine the equations of curves A, B, and C.

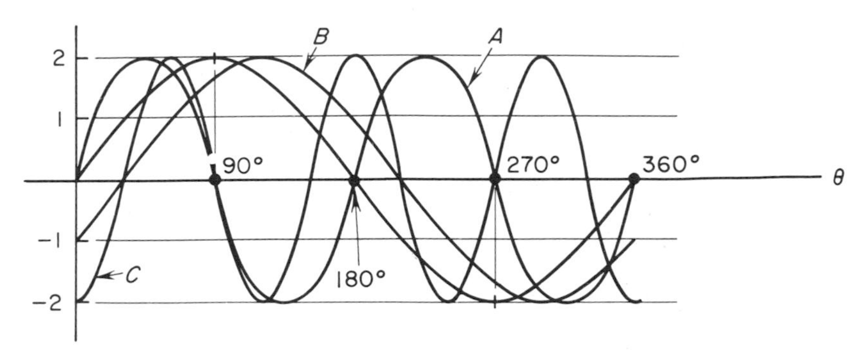

Figure 14-11

18. In Fig. 14-12 the graph of $y_1 = \sin\theta$ is shown as a solid line. Also shown is $y_2 = \sin(-\theta)$ as a dotted line. For any given θ, what relationship exists between y_1 and y_2? What relationship exists between the sine of an angle and the sine of the negative of the same angle?

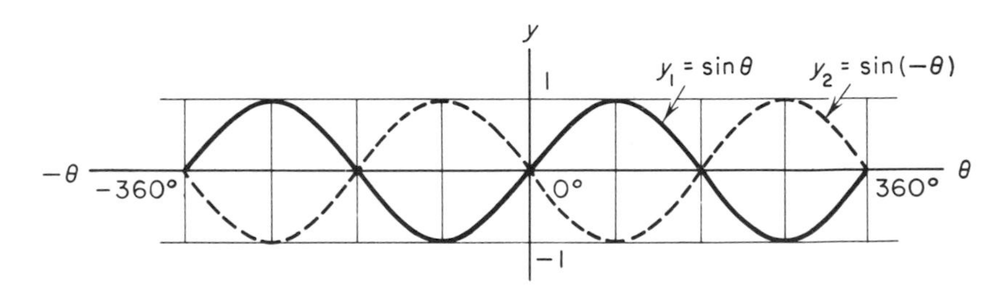

Figure 14-12

19. Given equation $s = 100 \sin 3600\phi$. What is the angular period? What is the smallest positive value of ϕ for which s is a maximum?

20. Given equation $s = 100 \sin(50\phi - 30°)$. What is the smallest positive value of ϕ for which s is a maximum?

14-4 The angle as a function of time

In Sec. 14-1 we briefly discussed an angle as a function of time. If a vector rotates with a uniform velocity ω (omega), where ω is in units such as degrees per second or radians per second, then the angle θ through which the vector passes is a function of time:

$$\theta = \omega t. \tag{23}$$

A typical sine function, usually written

$$y = Y \sin (\theta \pm \psi), \tag{24}$$

can be written in the form

$$y = Y \sin (\omega t \pm \psi). \tag{25}$$

When the angular velocity is given as f cycles/sec, we can convert to radians per second by multiplying the number of cycles by 2π, or the angular velocity can be written $2\pi f$ rad/sec. The angle θ through which the vector passes is

$$\theta = 2\pi f t, \tag{26}$$

where (26) is a revision of (23) and we can apply (23) or (26) to (24), which can be written

$$y = Y \sin (2\pi f t \pm \psi). \tag{27}$$

Expressions (25) and (27) show y as a function of time t. In (27) Y is the amplitude, $2\pi f$ is the angular velocity, and ψ is the phase angle. It is emphasized here that the only variables in (27) are y and t. It is further emphasized that $2\pi f t$, ψ, and $2\pi f t \pm \psi$ are *angles*.

The reader should know how to graph (27) and how to find any one of the five parts y, Y, f, t, ψ when the other four parts are given. This latter demand has some restrictions on it.

In the study of alternating current in electricity, (27) is generally revised to forms

$$i = I_{\max} \sin 2\pi f t, \tag{28}$$

$$e = E_{\max} \sin 2\pi f t, \tag{29}$$

where i and e are instantaneous current and voltage, $I_{\max}$ and $E_{\max}$ are maximum current and voltage, f is frequency, and t is time. Let us consider some applications of (28) and (29).

Example 9 Referring to (28), at what least positive value of time will a 60-cycle/sec generator with a maximum output of 100 A be instantaneously delivering 40 A?

Solution We will show both a mathematical and a graphical solution. First, the mathematical solution. From the given data we have $i = 40$, $I_{\max} = 100$, and $f = 60$ and we wish to find t. Substituting the given data into (28), we have

$$40 = 100 \sin (2)(\pi)(60)(t)$$

or

$$40 = 100 \sin 120\pi t. \tag{30}$$

In (30) it is emphasized that $120\pi t$ is an *angle* that is a function of time. Here 120π is angular velocity in radians per second and $120\pi t$ is in radians. Dividing both sides of (30) by 100, we get

$$\sin 120\pi t = 0.4 \tag{31}$$

Next, $120\pi t$ is an angle whose sine is 0.4; the least such positive angle is approximately 23.5°, and from (31)

$$120\pi t \text{ (in radians)} = 23.5 \text{ (in degrees).} \tag{32}$$

Expression (32) is awkward because the left side is in radians and the right side is in degrees. For demonstration purposes, we will show (32) converted entirely to degrees in (33) and entirely to radians in (34).

Since π radians equal 180°, then 120π radians equal 21,600° and (32) becomes

$$21{,}600t \text{ (degrees)} = 23.5 \text{ (degrees)} \tag{33}$$

from which $t = 0.00109$ sec.

Since there are 180° in π radians, then

$$\frac{180^\circ}{\pi} = \frac{23.5^\circ}{x}$$

and $x = 0.1305\pi$ radians, meaning that there are 0.1305π radians in 23.5°. Substituting into (32), we have

$$120\pi t \text{ (radians)} = 0.1305\pi \text{ (radians)} \tag{34}$$

from which $t = 0.00109$ sec.

Graphically we obtain a fair approximation of the solution. The basic equation is

$$i = 100 \sin 120\pi t, \tag{35}$$

where 100 is the amplitude and the time period required to complete one cycle is 1/60 sec, since the frequency is 60 cycles/sec. These data produce the graph shown in Fig. 14-13. We see that the first 40-A output is reached at a time approximately one-quarter of 1/240 sec or about 1/960 sec. This has a numerical value of slightly more than 0.001

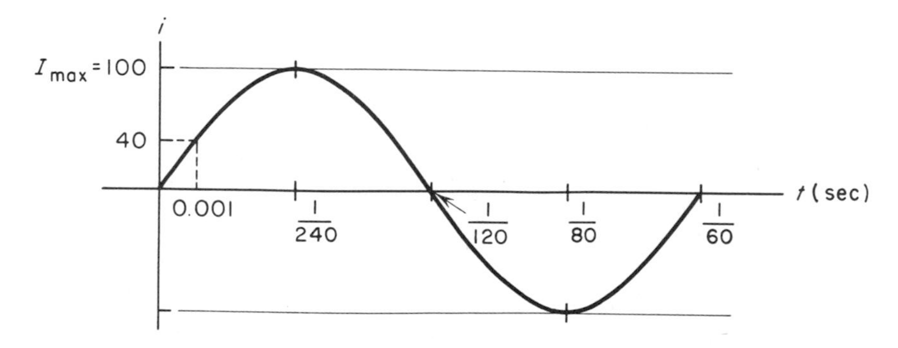

Figure 14-13

sec, which agrees favorably with the more exact mathematical solution $t = 0.00109$ sec. The graphical approach is quick and often useful in serving as a check against gross error.

Example 10 Referring to (29), find E_{max} if $f = 100$ cycles/sec and $e = 20$ volts when $t = 0.001$ sec.

Solution Substituting the given data into (29) gives

$$20 = E_{max} \sin (2\pi)(100)(0.001). \tag{36}$$

We emphasize that the quantity $(2\pi)(100)(0.001)$ in (36) is an *angle* and this angle is in radians. Simplifying (36), we have

$$E_{max} \sin 0.2\pi = 20. \tag{37}$$

Here the angle is 0.2π radian, which is

$$(0.2)(180°) = 36°,$$

and (37) becomes

$$E_{max} \sin 36° = 20,$$

but $\sin 36° = 0.588$ so that we now have

$$0.588E_{max} = 20$$

from which

$$E_{max} = \frac{20}{0.588} = 34 \text{ volts.}$$

Exercises 14-4

In Exercises 1–5, use equation $i = I_{max} \sin 2\pi ft$ and the given information to find the value of the indicated unknown.

1. Given $I_{max} = 110$ A (amperes), $f = 100$ cycles per second, and $t = 0.0025$ sec, find i.
2. Given $I_{max} = 100$ A and $f = 120$ cycles per second, find the smallest positive value of t for which $i = 50$ A. Find the second smallest positive value of t for which $i = 50$ A. Give a general expression that identifies all values of t for which $i = 50$ A.
3. Given $i = I_{max}$ when $f = 60$ cycles per second, find the least positive value of t.
4. In Exercise 1, what is the time interval between consecutive maximum values of i? Between consecutive zero values of i? Show the graph.
5. Given $i = I_{max}$ when $t = 0.005$ sec, give an expression for all frequencies under which this condition can be met.
6. Given equation $e = E_{max} \sin (2\pi ft + c)$, find c if $E_{max} = 12$ volts, $e = 4$ volts, and $f = 600$ cycles per second when $t = 0.0001$ sec.
7. Sketch the graphs of $y = -\sin \theta$ and $y = \sin (\theta - 180°)$. What seems to be the result of shifting the sine wave by 180°?

8. A generator delivers current according to equation

$$i = 50 \sin\left(240\pi t + \frac{\pi}{2}\right).$$

What is the maximum current that it can deliver? What is the frequency? What is the phase shift in terms of seconds?

In Exercises 9–16, use equation $i = I_{max} \sin(2\pi ft + \gamma)$ and the given information to find the value of the indicated unknown.

9. If $f = 100$ cycles/sec, $\gamma = 18°$, and $i = 2$ A when $t = 0.001$ sec, find I_{max}.

10. If $f = 200$ cycles/sec, $\gamma = 24°$, and $i = 12$ A when $t = 0.00075$ sec, find I_{max}.

11. What is the lowest frequency that will allow $i = 8.66$ A when $t = 0.001$ sec if $I_{max} = 10$ A and $\gamma = 36°$?

12. What is the lowest frequency that will allow $i = 10$ A when $t = 0.00015$ sec if $I_{max} = 20$ A and $\gamma = 0.1\pi$ radian?

13. Find the smallest positive value of t for which $i = 20$ A if $I_{max} = 50$ A, $f = 400$ cycles/sec, and $\gamma = 30°$. Refer to Fig. 14-9 and note that both the first and second quadrants contain an angle whose sin $= 0.4$.

14. Find the smallest positive value of t for which $i = 10$ A if $I_{max} = 20$ A, $f = 100$ cycles/sec, and $\gamma = 40°$. See Exercise 13.

15. Find the phase angle γ if $i = 0$ A when $t = 0.0015$ sec, $I_{max} = 30$ A, and $f = 60$ cycles/sec. Is the waveform ahead of or trailing (in time) the unshifted reference wave?

16. Find the phase angle γ if $i = 8.66$ A when $t = 0.002$ sec, $I_{max} = 10$ A and $f = 100$ cycles/sec.

14-5 Graphs of the other trigonometric functions

Earlier we discussed the graph of the general sine function and explored maximum and minimum points, amplitude, phase shift, and angular period as they related to the sine function. The remaining five functions possess interesting graphs, but they are not as popular from the point of view of general usage as the sine function. Our purpose here is to show the graphs of the other five functions and discuss them briefly.

The graph of $y = \cos\theta$

By the definition of the cofunction, the cosine of an angle is the same as the sine of the complementary angle. Therefore

$$y = \cos\theta = \sin(90° - \theta). \tag{38}$$

We already have a firm picture of the graph of the sine function. Equation (38) suggests that the cosine wave is a shifted sine wave, which is precisely the case. Assigning values of θ to the sine function in (38), we note that (38) plots as shown in Fig. 14-14. The graph of $y = \sin\theta$ is shown for comparison as the broken line in Fig. 14-14. We see that the cosine wave is merely a sine wave that has been shifted 90°. More precisely, the graph of $y = \cos\theta$ leads the graph of $y = \sin\theta$ by a 90° phase shift; otherwise the graphs are identical. Discussions concerning angular period and amplitude that are appropriate for the sine wave are also appropriate for the cosine wave.

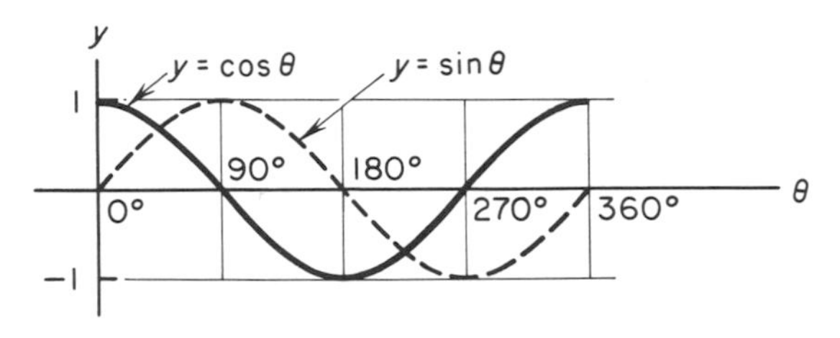

Figure 14-14

The graph of $y = \tan\theta$

To consider the variation in y for $y = \tan\theta$, refer to Fig. 14-15. Here we show the rotating vector **P**, where the vector is of changeable length,

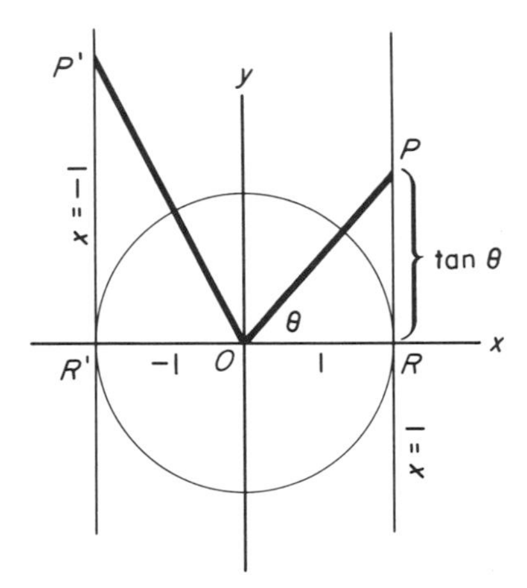

Figure 14-15

always terminating on either $x = 1$ or $x = -1$. From the triangle ORP, with $OR = 1$,

$$\tan\theta = \frac{PR}{OR} = \frac{PR}{1} = PR,$$

which asserts that PR equals the tangent of the reference angle. Note that as θ goes from 0° to 90°, $\tan\theta$ goes from 0 to a very large positive number. If $90° < \theta < 180°$, then

$$\tan\theta = \frac{P'R'}{-1}$$

so that tan θ is negative, varying from a large negative value when θ is slightly greater than 90° to zero when $\theta = 180°$. This variation, extended through 360°, is shown in Fig. 14-16, where we see that the angular period of the function $y = \tan\theta$ is 180°, as opposed to 360° for $y = \sin\theta$ and $y = \cos\theta$. We note, too, that no effort is made to discuss amplitude because the greatest vertical displacement of $y = \tan\theta$ is not defined.

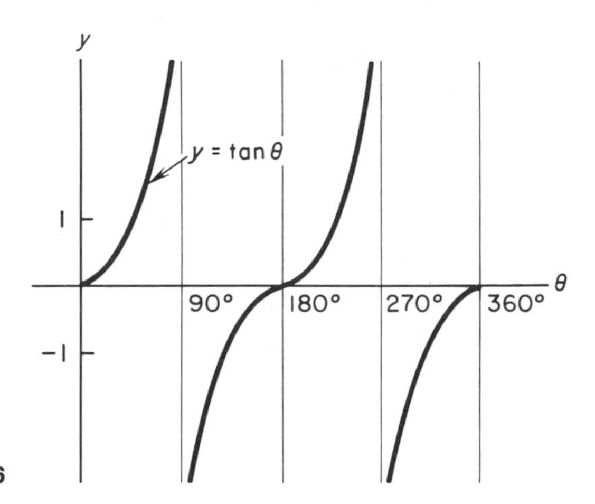

Figure 14-16

Graphs of $y = \sec\theta$, $y = \csc\theta$, and $y = \operatorname{ctn}\theta$ by reciprocation of ordinates

If we have two equations, $y_1 = \sin x$ and $y_2 = \csc x$, we know that the ordinates are related by equation $y_1 = 1/y_2$ because sin x and csc x are reciprocals. This asserts that, for any assigned value of x, the value of y_2 is the reciprocal of the value of y_1, hence the notion of reciprocation of ordinates. To show this reciprocation, refer to the accompanying chart of values. These values plot a curve shown in Fig. 14-17, where $y_1 = \sin x$ is shown as a dotted line and the reciprocal function $y_2 = \csc x$ as the solid line.

$x°$	0	10	20	30	45	60	90	180	210	270
$y_1 = \sin x$	0	0.174	0.342	0.5	$\frac{\sqrt{2}}{2}$	$\frac{\sqrt{3}}{2}$	1	0	$-\frac{1}{2}$	-1
$y_2 = \csc x$	∞	5.75	2.92	2	$\frac{2}{\sqrt{2}}$	$\frac{2}{\sqrt{3}}$	1	∞	-2	-1

Some observations are in order. To graph $y = \csc\theta$, $y = \sec\theta$, and $y = \operatorname{ctn}\theta$, first sketch the graph of the reciprocal and then reciprocate ordinates. Reciprocation does not affect the angular period. Observe for Fig.

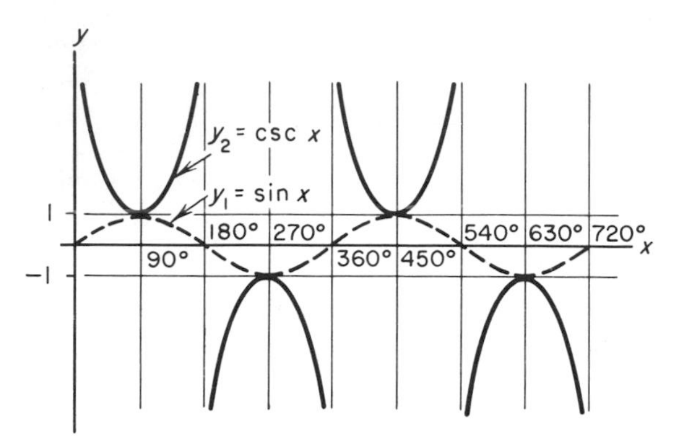

Figure 14-17

14-17 that y_2 has no finite value when $y_1 = 0$. Also, the reciprocal of a positive ordinate is positive and the reciprocal of a negative ordinate is negative.

Exercises 14-5

In Exercises 1–20, show the graphs of the given equations. Graph one full cycle of the function.

1. $y = \sin 2\theta$

2. $y = \cos \frac{\theta}{3}$

3. $z = \tan 3x$

4. $z = \csc 5x$

5. $s = \sec 120\pi t$

6. $y = \text{ctn } 5\phi$

7. $x = 3 \tan \pi\phi$

8. $x = 2 \csc 100\pi t$

9. $y = 4.5 \sec 3\pi\theta$

10. $y = \frac{5}{\csc 12\theta}$

11. $y = \sin (\theta + 40°)$

12. $y = \sin (\theta - 40°)$

13. $x = \sin (2\theta + 40°)$

14. $x = \sin (2\theta - 40°)$

15. $z = \cos (\theta + 30°)$

16. $z = \cos (\theta - 30°)$

17. $y = \cos (2\theta + 30°)$

18. $y = \cos (2\theta - 30°)$

19. $x = \tan (\theta + 30°)$

20. $x = \tan (2\theta - 60°)$

From the graphs of the appropriate trigonometric functions indicate whether the relationships given in Exercises 21–38 are true or false, considering any value of the angle.

21. $-\sin \theta = \sin (-\theta)$

22. $\cos \theta = \cos (-\theta)$

23. $\tan \theta = -\tan (-\theta)$

24. $\csc \theta = \csc (-\theta)$

25. $\sec \theta = -\sec (-\theta)$

26. $\sin (90° + \phi) = \sin (90° - \phi)$

27. $\tan \phi = \tan (90° + \phi)$

28. $\tan (90° + \phi) = -\tan (90° - \phi)$

29. $\tan x = \tan (180° + x)$

30. $\sin \theta = \sin (\theta + 360°)$

31. $\text{ctn } \phi = \text{ctn } (-\phi)$

32. $\tan x = -\text{ctn } (x + 90°)$

33. $\sin \phi = \cos (\phi + 90°)$

34. $\sin \phi = -\cos (\phi + 90°)$

35. $\csc \phi = \sec (\phi - 90°)$

36. $\sin 2\theta = \cos (2\theta - 90°)$

37. $2 \sin \phi = \dfrac{1}{2 \csc \phi}$

38. $\cos (90° - \phi) = \cos (90° + \phi)$

39. For what values of x does $\tan x = \text{ctn}\, x$?

40. For what values of x does $\sec x = \cos x$?

14-6 Graphing by composition of ordinates

In studies of such phenomena as vibration, resonance, and alternating current, equations like

$$y = f_1(x) + f_2(x) + f_3(x) + \cdots \tag{39}$$

occur in which the total ordinate y is composed of contributions made by several functions. This situation is especially true in ac studies where the contributing functions are trigonometric. The graph of an equation like (39) is usually difficult to construct by the elementary method of assigning arbitrary values to x, computing the corresponding y values, and then plotting the resulting (x, y) pairs. An alternate method provides a rapid way of showing what is often an adequate picture of the variation.

From (39) we can say that the composite ordinate y is of the form

$$y = y_1 + y_2 + y_3 + \cdots$$

where the components of the composite y are $y_1, y_2, y_3, \ldots$. We can also say that

$$y_1 = f_1(x), \qquad y_2 = f_2(x), \qquad y_3 = f_3(x), \ldots \tag{40}$$

The separate functions shown in (40) can be plotted and the ordinates $y_1, y_2, y_3, \ldots$ for arbitrarily chosen values of x added algebraically by use of dividers, a ruler, or a graph. The method described here is cal'ed *composition of ordinates*.

Example 11 Show the graph of

$$y = \cos x + \tfrac{1}{2} \sin 2x \tag{41}$$

by the composition of ordinates.

Solution If we express (41) in the form

$$y = y_1 + y_2,$$

we are assuming that

$$y_1 = \cos x \tag{42}$$

and

$$y_2 = \tfrac{1}{2} \sin 2x. \tag{43}$$

We graph (42) and (43) as in Fig. 14-18. At an arbitrarily chosen abscissa, say $x = k$, (42) has the ordinate $y_1 = KN$ shown in Fig. 14-18. Note that KN is positive. Also at $x = k$, (43) has the ordinate $y_2 = KL$, where KL is negative and is measured downward f om the x axis. To find $y_1 + y_2$, we can measure y_2 downward from point N (here $y_2 = KL = NM$). Now $KM = y_1 + y_2$, where M is a point on the desired graph.

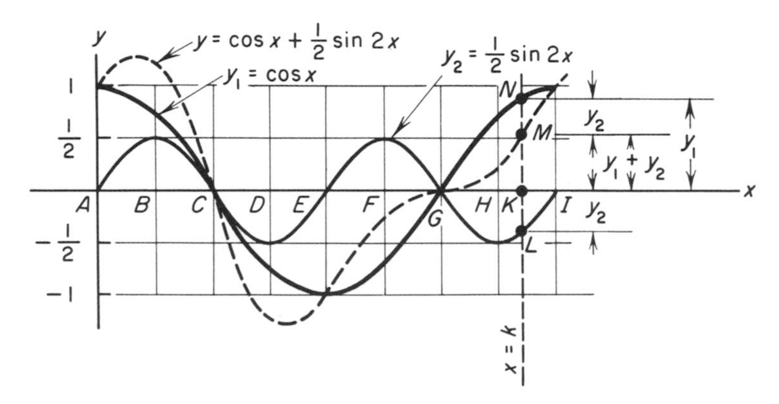

Figure 14-18

In Fig. 14-18 the specific locations A through I show maximum, minimum, and zero values of each y_1 and y_2. Moreover, at A, $y_2 = 0$ and $y_1 = 1$ so that

$$y = y_1 + y_2 = 1.$$

At B, $y_2 = \frac{1}{2}$ and so $y = y_1 + \frac{1}{2}$. At C, $y_1 = y_2 = 0$ and so $y = 0$. Similar considerations of the ordinates provided by components y_1 and y_2 at positions A through I give a reasonably complete description of (41), which is graphed as the dotted curve in Fig. 14-18.

In Example 11 we considered the graph of the sum of a sine and a cosine function in which the amplitudes and the angular periods of the components differed. One particularly common and interesting form is

$$y = a \sin nx + b \cos nx, \tag{44}$$

where we have the sum of a sine and a cosine function and both functions are of the same angular period, with amplitudes that are not necessarily equal. To graph (44), we could use the composition of ordinates; however, we can be more exact by expressing (44) as a shifted sine wave. Consider Example 12.

Example 12 Show that

$$y = Y \sin (nx + \gamma) \tag{45}$$

can be expressed in the form of (44).

Solution From the trigonometric identity

$$\sin(\theta + \phi) = \sin\theta\cos\phi + \cos\theta\sin\phi,$$

(45) becomes

$$\begin{aligned} y &= Y(\sin nx \cos\gamma + \cos nx \sin\gamma) \\ &= (Y\cos\gamma)\sin nx + (Y\sin\gamma)\cos nx. \end{aligned} \tag{46}$$

Here $Y\cos\gamma$ is a constant that we will call a and $Y\sin\gamma$ is a constant that we will call b; so (45) is now in the form of (44), where

$$a = Y\cos\gamma, \tag{47}$$

$$b = Y\sin\gamma. \tag{48}$$

Expressions (47) and (48) can be fitted to the right triangle shown in Fig. 14-19, where we see the additional relationships

$$Y = \sqrt{a^2 + b^2}$$

and $$\tan\gamma = \frac{b}{a} \quad \text{or} \quad \gamma = \arctan\frac{b}{a}.$$

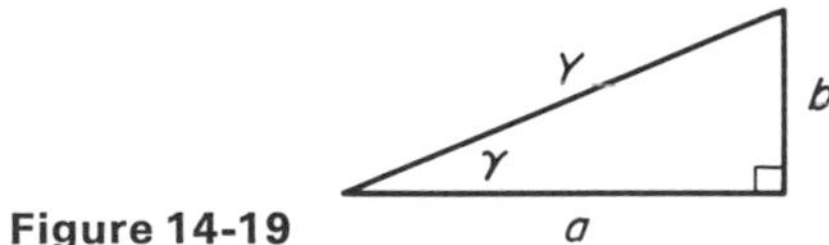

Figure 14-19

Comparing (45) and (44), we now have

$$y = a\sin nx + b\cos nx = \sqrt{a^2 + b^2}\sin(nx + \gamma).$$

The graph of (44) is then a sine wave of angular period $2\pi/n$, of amplitude $\sqrt{a^2 + b^2}$, and of phase shift $\arctan(b/a)$. Refer to Fig. 14-20.

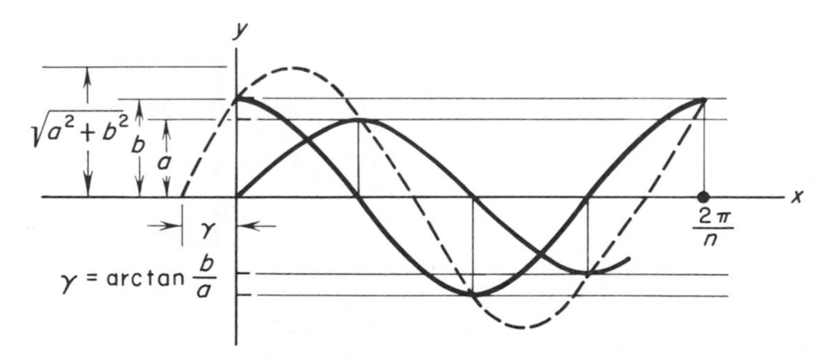

Figure 14-20

Exercises 14-6

In Exercises 1–16, show the graphs of the given equations, using the method of composition of ordinates. Plot the graphs in the interval shown.

1. $y = \sin x + \sin 2x, \quad 0 < x < 2\pi$
2. $y = \sin x + \cos 2x, \quad -\pi < x < \pi$
3. $y = \sin x + \dfrac{x}{2}, \quad 0 < x < 2\pi$
4. $y = \sin x - \dfrac{x}{\pi}, \quad 0 < x < 2\pi$
5. $y = \sin x + \dfrac{1}{3}\sin 3x, \quad 0 < x < 2\pi$
6. $y = \cos x + \cos 2x, \quad -\pi < x < \pi$
7. $y = \sin x + \dfrac{1}{2}\tan x, \quad 0 < x < 2\pi$
8. $y = \cos x - \dfrac{x}{2\pi}, \quad -\pi < x < \pi$
9. $y = \sin x + \dfrac{x^2}{4\pi}, \quad 0 < x < 2\pi$
10. $y = 3\cos x - \dfrac{x^2}{2\pi}, \quad -\pi < x < \pi$
11. $i = \sin 50\pi t + 25t, \quad 0 < t < 0.04$
12. $i = 2\cos 200\pi t - 100t, \quad 0 < t < 0.01$
13. $i = 2\sin 50\pi t + 2500t^2, \quad -0.02 < t < 0.02$
14. $y = \sin 2x - \dfrac{x}{2\pi}, \quad 0 < x < 2\pi$
15. $i = 3\sin 100\pi t + 4\sin 200\pi t, \quad 0 < t < 0.02$
16. $i = 4\sin 100\pi t - 3\cos 200\pi t, \quad 0 < t < 0.02$
17. Given equation $s = 4\sin t + 3\cos t$, express it in the form of (45). Show that it is a shifted sine wave of amplitude 5, phase shift 36.8°, and angular period 2π.
18. If the expression given in Exercise 17 were written as a shifted cosine wave, we would have the same amplitude 5, the same angular period 2π, but a different phase angle. What would that phase angle be?
19. Show that equation $y = a\sin x + b\sin(x + \theta)$ can be expressed as

$$y = \sqrt{a^2 + b^2 + 2ab\cos\theta}\,\sin(x + \phi),$$

where
$$\phi = \arctan\frac{b\sin\theta}{a + b\cos\theta}.$$

14-7 Lissajous figures; parametric equations

In most of the preceding work we have been able to express relationships in the form

$$y = f(x), \tag{49}$$

where the values assigned to the independent variable x determined values of the dependent variable y. Many variations are difficult to express in the explicit form (49) and can be more simply expressed as

$$y = g(t), \tag{50}$$

$$x = h(t), \tag{51}$$

where t is the independent variable. Assignment of values to t imposes values on the dependent variables y and x. Here t is called a *parameter* and (50) and (51) *parametric equations.*

A simple example of parametric equations can be shown with a circle. In Fig. 14-21 we show a circle of radius R with center on the origin. For any point (x, y), we have

$$x = R \cos \theta, \tag{52}$$

$$y = R \sin \theta, \tag{53}$$

where x and y are both determined when a value is assigned to parameter θ.

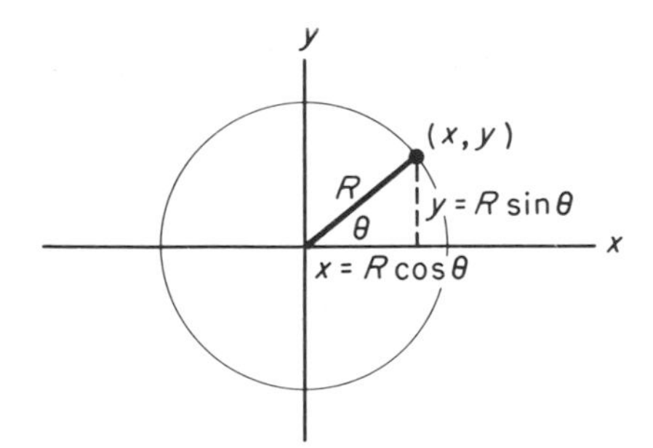

Figure 14-21

The parameter in (52) and (53) is readily eliminated by adding the square. Thus

$$x^2 = R^2 \cos^2 \theta,$$

$$y^2 = R^2 \sin^2 \theta,$$

and
$$x^2 + y^2 = R^2(\sin^2 \theta + \cos^2 \theta) = R^2,$$

where $x^2 + y^2 = R^2$ is easily seen to be the equation of the origin-centered circle shown in Fig. 14-21. Elimination of the parameter is generally more difficult.

Inspection of Eqs. (50) and (51) allows several observations. If we plot $y = g(t)$ on the (x, y) plane independent of $x = h(t)$, we see that modifications in t modify the *vertical position* of the point plotted. Introducing $x = h(t)$ and assigning values to t, we move the point *horizontally*. We may accept t as a time parameter, showing that modifications in time can cause both a horizontal and a vertical change in the position of the plotted point. If Eqs. (50) and (51) are periodic functions, the point moves horizontally according to some period and moves vertically according to some period. As a result, the point will periodically traverse the same path. Let us use this last observation as a basis of discussion in Example 13.

Example 13 Discuss the graph of the function represented by parametric equations

$$x = 2 \sin 2\theta, \tag{54}$$

$$y = \cos 3\theta. \tag{55}$$

Solution First, we observe that both (54) and (55) are periodic. The angular period of (54) is 180°; the angular period of (55) is 120°. This asserts that a 360° change in θ will send x through two full cycles and will send y through three full cycles. A graphical interpretation of the preceding statement is: as θ passes from 0° to 360°, the (x, y) point plotted passes through two full cycles in the horizontal sense and three full cycles in the vertical sense.

By inspection of (54) and (55), we see that the value of x ranges from -2 to $+2$ and the value of y ranges from -1 to $+1$. Combining two of our observations, we see that x assumes a maximum value $(+2)$ *two* times and y assumes its maximum value $(+1)$ *three* times as θ ranges from 0 to 360°; the same is true for the number of minima involved. The graph will be inclosed in the rectangle of length 4 and height 2 shown in Fig. 14-22. It will contact both the right and left sides of the

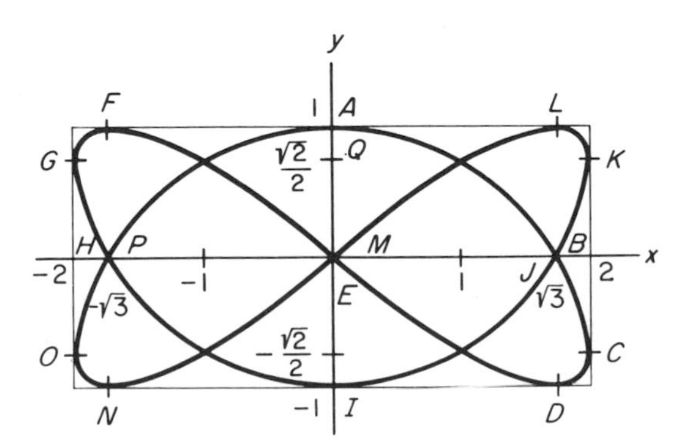

Figure 14-22

rectangle twice and the top and bottom of the rectangle three times as θ passes from 0 to 360°.

In plotting points, we choose values of θ that will provide zeros or maxima or minima for either x or y or both x and y. These points are shown in Table 14-2. The plot of the points results in the graph shown in Fig. 14-22, where the direction of movement along the curve is in the alphabetic sequence of points A through Q.

Note in Fig. 14-22 that the plotted curve tangents the right edge of the rectangle twice and the upper edge three times. In electricity this situation may have been shown on an oscilloscope where a known horizontal input signal of, say, 240 cycles is used to determine the frequency of an unknown vertical input. If the combined signals produce a trace that shows three vertical maxima for each two horizontal maxima, then the ratio of the frequencies is 3 to 2 and the unknown vertical input is $(\frac{3}{2})(240) = 360$ cycles.

Table 14-2

$\theta°$	0	30	45	60	90	120	135	150	180	210	225	240	270	300	315	330	360
x	0	$\sqrt{3}$	2	$\sqrt{3}$	0	$-\sqrt{3}$	-2	$-\sqrt{3}$	0	$\sqrt{3}$	2	$\sqrt{3}$	0	$-\sqrt{3}$	-2	$-\sqrt{3}$	0
y	1	0	$\frac{-\sqrt{2}}{2}$	-1	0	1	$\frac{\sqrt{2}}{2}$	0	-1	0	$\frac{\sqrt{2}}{2}$	1	0	-1	$\frac{-\sqrt{2}}{2}$	0	1
Point	A	B	C	D	E	F	G	H	I	J	K	L	M	N	O	P	Q

The group of figures characterized by properties shown in Example 13 is known as the group of *Lissajous figures*. They are limited usually to two sine waves that vary about axes at right angles. The figures are used for phase and frequency measurements.

Exercises 14-7

In Exercises 1–20, prepare the graphs of the given sets of equations.

1. $x = \sin\theta$
$y = \cos\theta$

2. $x = \sin\theta$
$y = \sin\theta$

3. $x = \sin\theta$
$y = \sin 2\theta$

4. $x = \sin\theta$
$y = \sin(\theta + 45°)$

5. $x = 2\sin\theta$
$y = 3\sin\theta$

6. $x = 4\sin 2\theta$
$y = 3\sin\theta$

7. $x = 2\sin\theta$
$y = 3\sin(\theta + 90°)$

8. $x = 2\sin\theta$
$y = 3\sin 2\theta$

9. $x = 2\cos 3\theta$
$y = 3\sin 4\theta$

10. $x = 3\sin 120\pi t$
$y = 4\cos 240\pi t$

11. $x = \sin \theta$
$y = \sin (3\theta + 90°)$

12. $x = 4 \sin \theta$
$y = 4 \sin 4\theta$

13. $x = 2 + \sin t$
$y = 2 \cos t$

14. $x = \sin 5\theta$
$y = \cos 4\theta$

15. $x = 5 \sin 120\pi t$
$y = 5 \sin (240\pi t + 30°)$

16. $x = 5 \sin \theta$
$y = 5 \sin (4\theta + 90°)$

17. $x = \sin (\theta + 45°)$
$y = \cos 2\theta$

18. $x = \tan \theta$
$y = \sin \theta$

19. $x = 2 \sin 3\theta$
$y = \sin \left(3\theta + \frac{\pi}{3}\right)$

20. $x = 2 \sin 3\theta$
$y = \sin \left(3\theta - \frac{\pi}{3}\right)$

21. Compare the graph of Exercise 5 with the graph of

$$x = 2 \sin (\theta + 45°), \qquad y = 3 \cos \theta.$$

22. Determine the dimensions of the rectangle enclosing the Lissajous figure represented by equations

$$x = 3 \sin 2t, \qquad y = 4 \cos 5t.$$

23. A certain Lissajous figure is represented by equations

$$x = a \sin kt, \qquad y = b \cos 20t.$$

If the graph of the figure contacts the vertical side of the containing rectangle four times for each five times that it contacts the horizontal side, determine k.

24. Eliminate the parameter in Exercises 1, 2, 5, 13.

14-8 Definitions involving polar coordinates

Polar coordinates probably have greater usage than rectangular coordinates. An example lies in the location of Boston with respect to New York City. The most probable description is that Boston is approximately 200 miles northeast of New York City, using distance (200 miles) and direction (northeast) instead of rectangular coordinates. In mathematics distance ρ and direction θ are used, and a point in a plane is assigned coordinates (ρ, θ) in a system of *polar coordinates*. The distance is measured from a reference point and the direction is measured as an angle from a reference line. The reference point is called a *pole* and the reference line a *polar axis*—hence the name *polar coordinates*. In standard usage, the reference point is the origin, labeled O, and the reference line is drawn through O to the right with the positive angle θ measured in a counterclockwise direction.

From Fig. 14-23(b) we locate P with respect to O by drawing OM to the right, indicating angle θ and distance ρ and assigning coordinates (ρ, θ) to P. Here ρ is given the name *distance* or *magnitude* and OP is called the *radius vector*. Angle θ is called the *reference angle*. If P were to be located in

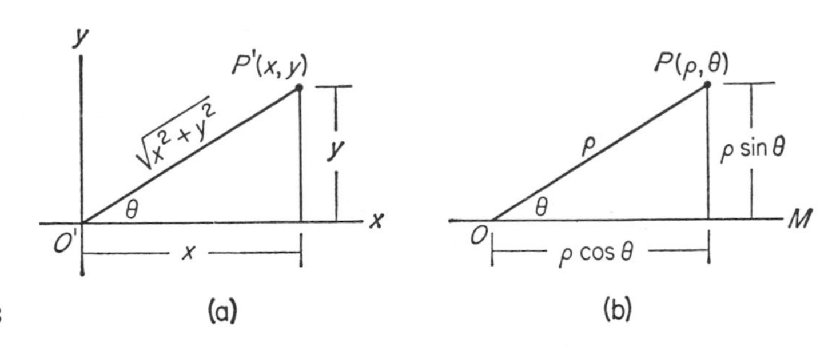

Figure 14-23

rectangular coordinates as in Fig. 14-23(b), we see that by superimposing (a) on (b) so that $O'x$ is imposed on OM with O' and P' coinciding, respectively, with O and P, we have the following relationships between the rectangular and polar locations of P:

$$\begin{gathered} \rho = \sqrt{x^2 + y^2}, \qquad x = \rho \cos \theta, \qquad y = \rho \sin \theta, \\ \cos \theta = \frac{x}{\sqrt{x^2 + y^2}}, \qquad \sin \theta = \frac{y}{\sqrt{x^2 + y^2}}, \qquad \tan \theta = \frac{y}{x} \end{gathered} \tag{56}$$

Consider two applications using the conversions in (56).

Example 14 Convert $x^2 + y^2 = 25$ to polar form.

Solution From (56)

$$\rho = \sqrt{x^2 + y^2} \quad \text{and} \quad \rho^2 = x^2 + y^2 = 25$$

from which we have solution $\rho = 5$.

Discussion: Equation $x^2 + y^2 = 25$ is recognized as an origin-centered circle of radius 5. Similarly, equation $\rho = 5$ is a circle of radius 5, which can be seen from the fact that $\rho = 5$ is an equation independent of θ. Therefore distance OP in Fig. 14-23(b) is five units, regardless of the magnitude of θ.

Example 15 Convert

$$\rho = \frac{5}{\sqrt{3 - 2 \cos 2\theta}}$$

into rectangular coordinates.

Solution Given

$$\rho = \frac{5}{\sqrt{3 - 2 \cos 2\theta}}, \tag{57}$$

squaring both sides of (57) and clearing fractions, we have

$$\rho^2(3 - 2\cos 2\theta) = 25. \tag{58}$$

Substituting identity $\cos 2\theta = 1 - 2\sin^2\theta$ into (58) and collecting like terms, we obtain

$$\rho^2(1 + 4\sin^2\theta) = 25.$$

Then applying appropriate substitutions from (56), we have the solution

$$x^2 + 5y^2 = 25,$$

which is recognized as an origin-centered ellipse. A more complete discussion of forms like (57) will be made in a later section where the graphing of conics in the polar form is treated.

Exercises 14-8

Convert the equations in Exercises 1–10 to polar form.

1. $x^2 + y^2 = 4$

2. $x = 6$

3. $y = 5$

4. $x^2 - y^2 = 9$

5. $(x - 2)^2 + y^2 = 25$

6. $xy = 8$

7. $y^2 = 2x$

8. $y = 2x$

9. $x^2 + y^2 = \arctan \dfrac{y}{x}$

10. $\sqrt{x^2 + y^2} = e^{\arctan y/x}$

Convert the equations in Exercises 11–20 to rectangular form.

11. $\rho = 5\sin\theta$

12. $\rho = 3$

13. $\rho \sin\theta = 5$

14. $\tan\theta = 2$

15. $\theta = 135°$

16. $\rho = 3 + 2\cos\theta$

17. $\rho = e^\theta$

18. $\rho = \sin 2\theta$

19. $\rho = \dfrac{\theta}{4}$

20. $\rho^2 = 9\sin 2\theta$

14-9 Graphs of polar equations

One approach to the problem of plotting curves in polar coordinates consists of expressing the equation in the explicit form $\rho = f(\theta)$, choosing arbitrary values of the independent variable θ, and determining the corresponding values of the dependent variable. This process creates a set of (ρ, θ) coordinates that can be plotted and joined with a smooth curve. The method, however, is often tedious and does not necessarily reveal certain characteristics of the curve at critical portions.

The approach applied here is to express the equation in a mathematical form. We will type those equations that plot into straight lines, conics, and spirals; these three groups of curves are the curves most commonly studied in polar equations. First, let us discuss the plotting of a few selected points in polar. In Fig. 14-24 concentric circles with center O are laid off to indicate lengths of the radius vector; we have also laid off angles in 15° increments. Further refinements for distance and angle could be established, depending on the accuracy required. The point P_1 is given four different designations, the most common being $P_1(\rho, \theta) = (5, 45°)$ or $(5, \pi/4)$, depending on the system of angular measurement requested. Alternate expressions for $P_1(\rho, \theta)$ are $(-5, 225°)$, $(-5, -135°)$ and $(5, -315°)$. We note that if the angle is removed 180° from the proper smallest positive angle, the sense of ρ is changed.

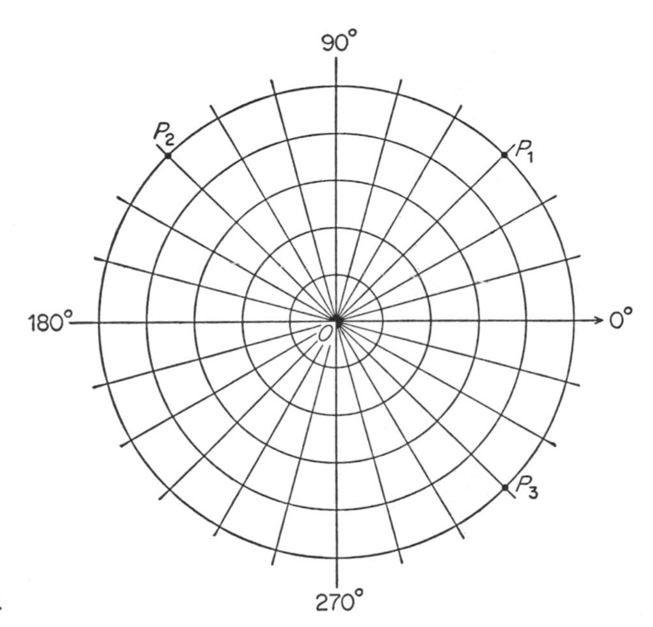

Figure 14-24

Symmetry

In Fig. 14-24 point $P_1(5, 135°)$ is located symmetrically with P_1 across the terminal side of the line $\theta = 90°$. Similarly, $P_3(5, -45°)$ is symmetrical with P_1 across the line $\theta = 0°$. Expanding this discussion of symmetry to portions of curves rather than isolated points, we see that equation $\rho = 5 + 2 \sin \theta$ is symmetrical about the line $\theta = 90°$, since $\sin (90° + \phi) = \sin (90° - \phi)$, where ϕ is any angle. In general, to check $\rho = f(\theta)$ for symmetry about an axis $\theta = k$, determine whether the expression $f(k + \phi) = f(k - \phi)$ is satisfied.

Equations that graph as straight lines

The general equation of a straight line in rectangular form is given as

$$Ax + By = C. \tag{59}$$

Substituting $x = \rho \cos \theta$ and $y = \rho \sin \theta$ into (59) and solving for ρ, we have

$$\rho = \frac{C}{A \cos \theta + B \sin \theta} \tag{60}$$

as the equation of a straight line in polar coordinates. Using (59), if $A = 0$, then $y = C/B$ and the line parallels the x axis; if $A = 0$ is imposed on (60), then $\rho = (C \csc \theta)/B$ is the equation of the line in polar form, symmetrical with the axis $\theta = 90°$ and C/B length units from the origin. Similar modifications regarding B in (59) and (60) produce $\rho = (C \sec \theta)/A$ as the equation of a line parallel with $\theta = 90°$.

From (59) again, if the line passes through the origin, then $C = 0$ and $Ax + By = 0$, or

$$\frac{y}{x} = -\frac{A}{B}.$$

But, from polars, $\tan \theta = y/x$; therefore $\tan \theta = -A/B$ is the equation of a line passing through the origin in polar coordinates. We note that equation $\tan \theta = -A/B$ is independent of ρ and that θ is a constant; also, the equation fails if $B = 0$.

Equations that graph as conics

For our purposes, we start with the assertion that in polar coordinates the equation of the general conic is

$$\rho = \frac{k}{a + b \cos \theta}, \tag{61}$$

where a, b, and k are nonzero constants.

If we use some of the substitutions (56) in (61) and simplify, we have

$$(a^2 - b^2)x^2 + b^2y^2 + 2kbx - k^2 = 0, \tag{62}$$

which is of the form

$$Ax^2 + By^2 + Cx + D = 0. \tag{63}$$

Note that (63) is a conic that has been shifted off the origin along the x axis (if $C \neq 0$). If we replaced $\cos \theta$ with $\sin \theta$ in (61), then (63) would have a first degree y term in it and no first degree x term, indicating a vertical shift off the origin. Some observations of (62) reveal the particular conic. Equation

(62) is a

parabola if

$$a^2 - b^2 = 0,$$

ellipse if

$$a^2 - b^2 > 0 \quad \text{and} \quad b \neq 0, \tag{64}$$

hyperbola if

$$a^2 - b^2 < 0 \quad \text{and} \quad b \neq 0.$$

Equation (61) reveals that if $b = 0$, then $\rho = k/a$ and (61) is an origin-centered circle.

Example 16 Graph equation

$$\rho = \frac{2}{1 + \cos\theta}.$$

Solution Comparing the given equation to (61), we have $a = 1$ and $b = 1$. From observations (64) we see that the given equation is a parabola. Moreover, since $\cos\theta$ is symmetrical about $\theta = 0°$, we have symmetry about the line $\theta = 0°$, indicating that the parabola opens either to the right or to the left. Referring now to the given equation, ρ is least when the denominator $|1 + \cos\theta|$ is greatest. This situation occurs when $\cos\theta = +1$ or when $\theta = 0°$. Similarly, ρ is greatest when $|1 + \cos\theta|$ is least or when $\cos\theta = -1$. It shows that the parabola opens to the left. The graph, with a few extra points plotted, is shown in Fig. 14-25.

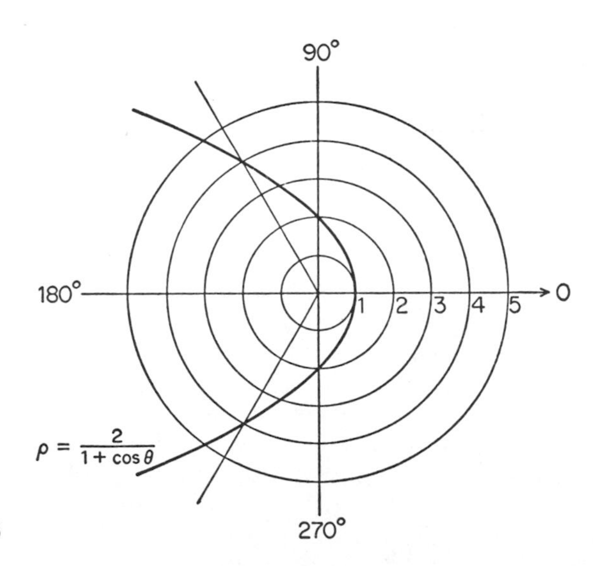

Figure 14-25

Spirals

Three equations of spirals are discussed here.

The *spiral of Archimedes* has equation $\rho = a\theta$, $a > 0$. We know from polar coordinates that two reference angles that differ by 2π radians produce the same direction for the radius vector; therefore for

$$\theta = k, \quad \theta = k + 2\pi, \quad \theta = k + 4\pi, \quad \text{etc.,}$$

ρ is similarly directed. Referring to $\rho = a\theta$, when

$$\theta = k, \quad k + 2\pi, \quad k + 4\pi, \quad \ldots,$$

$$\rho = ak, \quad ak + 2a\pi, \quad ak + 4a\pi, \quad \ldots,$$

respectively, where it is seen that each 2π change in θ produces a $2a\pi$ change in ρ. Phrased differently, it asserts that a fixed change in ρ accompanies a fixed change in θ.

The *reciprocal spiral* has equation $\rho\theta = a$. Here it is apparent that each increase in θ is accompanied by a corresponding decrease in ρ and vice versa, since the product of ρ and θ is constant. As θ increases to large values, ρ diminishes toward zero; as $\theta \longrightarrow 0$, $\rho \longrightarrow \infty$. The reciprocal spiral, unlike the spiral of Archimedes, does not pass through the origin.

The *logarithmic spiral* has equation $\rho = e^{a\theta}$, $a \neq 0$. Here ρ is exponentially related to θ. We note, too, that for two values of θ that are 2π radians apart, say $\theta_1 = k$ and $\theta_2 = k + 2\pi$, then

$$\rho_1 = e^{ak} \quad \text{and} \quad \rho_2 = e^{ak+2a\pi}$$

and the relationship between ρ_1 and ρ_2 is such that

$$\rho_2 = \rho_1 e^{2a\pi}. \tag{65}$$

From (65) it is seen that for a 2π increase in the reference angle, the radius vector is multiplied by a constant factor $e^{2a\pi}$.

The logarithmic spiral convincingly shows the advantage of polar coordinates over rectangular coordinates in certain cases. If $\rho = e^{a\theta}$ were converted to rectangular coordinates, the result would be the very awkward expression

$$\sqrt{x^2 + y^2} = e^{a \arctan y/x},$$

illustrating the comparative simplicity of the polar expression.

Comparing two of the spirals mentioned above, note that for a 2π change in the reference angle, the magnitude of the radius vector of the Archimedes spiral is changed by an additive constant, whereas the magnitude of the radius vector of the logarithmic spiral is changed by a constant ratio.

Exercises 14-9

Plot Exercises 1–20 in polar coordinates.

1. $\theta = 45°$
2. $\rho = \dfrac{\sin\theta}{3}$
3. $\rho \cos\theta = 5$
4. $\rho(2\sin\theta + \cos\theta) = 3$
5. $\rho(1 - \sin\theta) = 1$
6. $\rho(1 - 2\sin\theta) = 1$
7. $\rho(2 - \sin\theta) = 1$
8. $\rho = 5$
9. $\rho = \dfrac{\theta}{2}$
10. $\rho = 1 + \dfrac{\theta}{\pi}$
11. $\rho\theta = \dfrac{1}{2}$
12. $\rho\theta = -\pi$
13. $2\rho = e^{\theta}$
14. $3\rho = e^{-\theta}$
15. $2 + \cos\theta = \rho$
16. $2 + 3\cos\theta = \rho$
17. $3 + 2\cos\theta = \rho$
18. $\rho = 5\cos 2\theta$
19. $\rho = 4\sin 3\theta$
20. $\rho = 3\cos 5\theta$
21. Plot the graphs of equation $\rho = a + b\cos\theta$ (a and b positive) if (a) $a > b$, (b) $a = b$, (c) $a < b$. If $a < b$, the figure is a limaçon. If $a > b$, the figure is a cardioid.
22. How many loops are on the graph of equation $\rho = a\cos 2\theta$? Of $\rho = a\sin 2\theta$? Why does the number of loops differ?
23. Show that $\rho = 2c\cos(\theta - \alpha)$ is the equation of a circle passing through $(0, 0)$ with center (c, α).
24. Plot the curve $\rho^2 = \cos 2\theta$. What is the domain of θ for which ρ is defined?
25. Convert $(x^2 + y^2)^2 = 4(x^2 - y^2)$ to polar coordinates.

15

The j operator

It is frequently useful in engineering and science to express vectors in the form of a complex number. Manipulations with vectors, especially the fundamental operations (addition, subtraction, multiplication, and division), are often simplified when expressed in complex notation. In this chapter we describe the complex number, show its representation in the complex plane, and discuss the fundamental operations of the numbers and the vectors they represent. The number j that we use is called i in pure mathematics. In technical mathematics the symbol i is reserved for electrical current; thus j is used here to avoid confusion.

The number j is often called the j operator because its use facilitates the fundamental operations of vector quantities.

15-1 The number j

In Sec. 6-6 we defined the number j as

$$j = \sqrt{-1}, \tag{1}$$

where $\sqrt{-1}$ is one of the two equal factors of -1. The quantity j might be pictured geometrically as in Fig. 15-1, where we have a semicircle of radius unity with the center of the circle at the intersection of two perpendicular axes. Along one axis real numbers are laid out according to accepted convention, positive numbers to the right and negative numbers to the left.

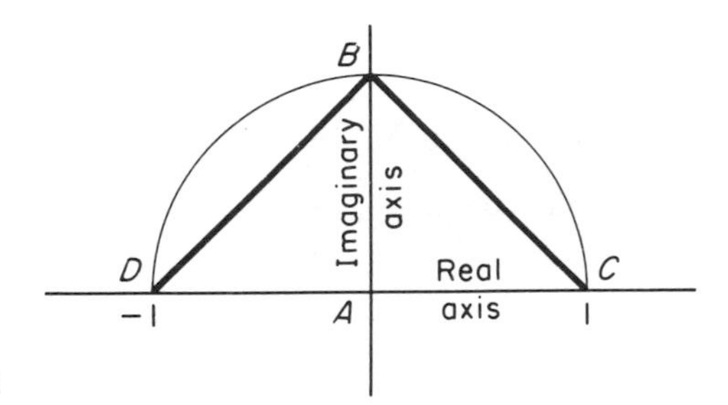

Figure 15-1

From plane geometry angle DBC is a right angle because it is inscribed in a semicircle. Also, AB is the geometric mean between DA and AC; that is,

$$(AB)^2 = (DA)(AC) = (-1)(1) = -1. \tag{2}$$

From (2)

$$AB = \sqrt{-1} = j,$$

and we see that AB can be interpreted in Fig. 15-1 as being of magnitude j, where j is often called the *imaginary unit* or the *j operator.*

Powers of j are cyclical in nature, as can be seen from the following:

$$\begin{aligned}
j &= \sqrt{-1}, \\
j^2 &= -1, \\
j^3 &= j \cdot j^2 = j(-1) = -j, \\
j^4 &= (j^2)^2 = (-1)^2 = +1, \\
j^5 &= j \cdot j^4 = j(1) = j, \\
j^6 &= j \cdot j^5 = j \cdot j = j^2 = -1.
\end{aligned}$$

Inspection shows that powers of j are cyclical in a four-element cycle, where the elements are j, -1, $-j$, and $+1$. In general, if n is any integer,

$$\begin{aligned}
j^{4n+1} &= j, \\
j^{4n+2} &= -1, \\
j^{4n+3} &= -j, \\
j^{4n+4} &= +1.
\end{aligned}$$

Negative powers of j can also be expressed as one of the four elements previously mentioned:

$$j^{-1} = \frac{1}{j} = \frac{1}{j} \cdot \frac{j}{j} = \frac{j}{j^2} = \frac{j}{-1} = -j,$$

$$j^{-2} = \frac{1}{j^2} = \frac{1}{-1} = -1,$$

$$j^{-3} = \frac{1}{j^3} = \frac{1}{j^3} \cdot \frac{j}{j} = \frac{j}{j^4} = \frac{j}{1} = j,$$

$$j^{-4} = \frac{1}{j^4} = 1.$$

Example 1 Simplify expressions (a) j^{27}, (b) j^{41}, and (c) j^{-19}.

Solution We can solve them simply by removing the multiple of four that is nearest the value of the exponent. Thus for j^{27}, we may call this $j^{24} \cdot j^3$, where $j^{24} = 1$, leaving the answer as $j^3 = -j$. We could also have called $j^{27} = j^{28} \cdot j^{-1}$, where $j^{28} = 1$ and $j^{-1} = -j$, so that once again $j^{27} = -j$. We have, by using similar procedures,

(a) $$j^{27} = j^{24} \cdot j^3 = -j,$$

(b) $$j^{41} = j^{40} \cdot j^1 = j,$$

(c) $$j^{-19} = j^{-20} \cdot j^1 = j.$$

Note that any power of j where we have j^{4n} (n an integer) has the value $+1$.

15-2 The Argand diagram; various forms of the complex number

Any number like $j5$, $-j5$, $j1.5$, $-j\frac{9}{2}$ or jb, where b is a real number, is called a *pure imaginary number* and can be represented as in Fig. 15-1. In the chapter on quadratics we encountered numbers like $3 + j5$, $\frac{3}{2} - j\frac{3}{2}$, or $a + jb$, where a and b are real numbers and $a + jb$ is called a *complex number*. Examining $a + jb$ carefully, we see that the complex number has two components; one is a real number a that could be measured along a horizontal axis (real axis), and the other is a pure imaginary number jb that is measured along the vertical axis (imaginary axis); they are shown in Fig. 15-2. Continuing to refer to Fig. 15-2, vector **P** can be represented by the complex number $a + jb$. The point P has coordinates (a, b). The magnitude of vector **P** is, by the Pythagorean Theorem,

$$\rho = \sqrt{a^2 + b^2}, \tag{3}$$

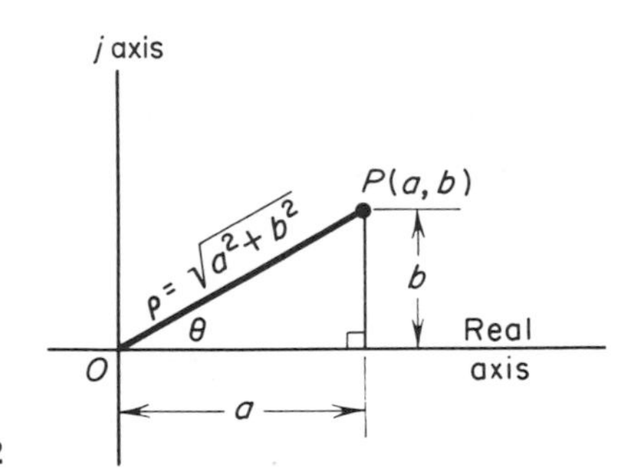

Figure 15-2

and the reference angle (the angle between the right-hand horizontal and vector **P** in the counterclockwise direction) is such that

$$\tan \theta = \frac{b}{a},$$

or

$$\theta = \arctan \frac{b}{a}. \tag{4}$$

The representation shown in Fig. 15-2 is a mathematical standard. The plane on which the axes are drawn is called the *complex plane* or *gaussian plane* and the diagrammatic representation of the complex number is called the *Argand diagram.* Vector **P** can be described by two different forms that are already established. In the *rectangular form* we have

$$\mathbf{P} = a + jb$$

and in the polar form

$$\mathbf{P} = \rho\underline{/\theta} = \sqrt{a^2 + b^2}\,\underline{/\arctan (b/a)}. \tag{5}$$

Example 2 Represent vector **P** $= 3 + j5$ in polar form.

Solution First, we sketch **P** as in Fig. 15-3. Note that point P is three units to the right (three real units) and five units upward (five imaginary

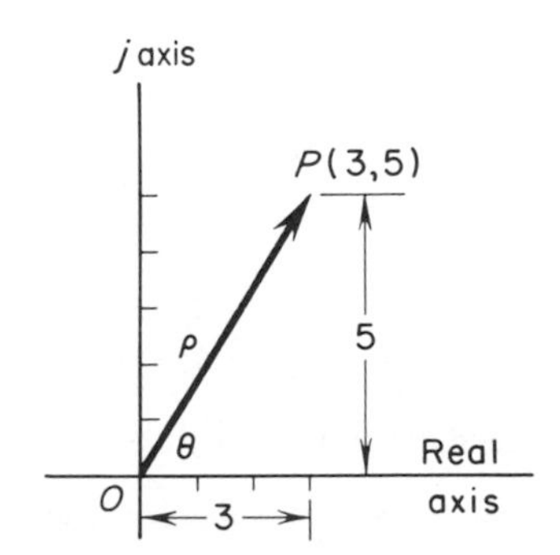

Figure 15-3

units). By the Pythagorean Theorem, we have

$$\rho = \sqrt{3^2 + 5^2} = \sqrt{9 + 25} = \sqrt{34}.$$

Using the definition of the tangent function, we have

$$\tan \theta = \frac{5}{3} \quad \text{or} \quad \arctan \frac{5}{3} = \theta = 59.1^\circ$$

and

$$3 + j5 = \rho\underline{/\theta} = \sqrt{34}\,\underline{/59.1^\circ},$$

where $3 + j5$ is the rectangular form of **P** and $\sqrt{34}\,\underline{/59.1^\circ}$ is the polar form.

In the preceding example the rectangular form of a vector was given and we were asked to find the polar form. Frequently conversions from polar to rectangular are required. Referring to Fig. 15-4, we are given vector **P** in

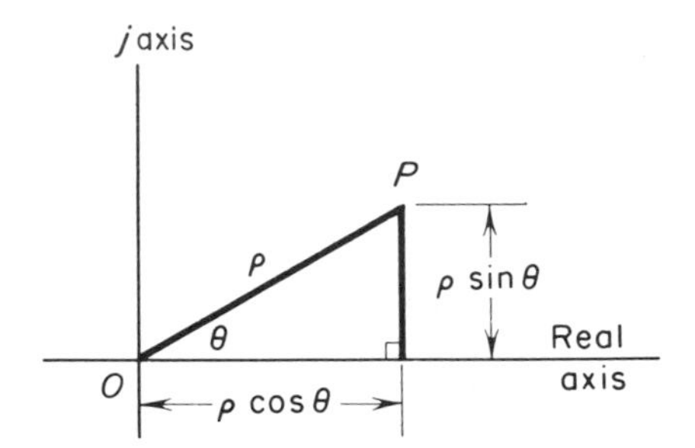

Figure 15-4

its polar form $\rho\underline{/\theta}$, where ρ and θ are known. The horizontal (real) component of **P** is $\rho \cos \theta$ and the vertical component is $\rho \sin \theta$, where we can now express **P** by the notations

$$\mathbf{P} = \rho\underline{/\theta} = \rho \cos \theta + j\rho \sin \theta = \rho(\cos \theta + j \sin \theta). \tag{6}$$

Here the factored form $\rho(\cos \theta + j \sin \theta)$ is called the *trigonometric form* of **P**.

Example 3 Express $6\underline{/210^\circ}$ in rectangular and trigonometric forms.

Solution Referring to Fig. 15-5 and (6), **P** is in Quadrant III with $a = 6 \cos 210^\circ$ and $b = 6 \sin 210^\circ$ so that

$$\begin{aligned} a + jb &= 6 \cos 210^\circ + j6 \sin 210^\circ = -6 \cos 30^\circ - j6 \sin 30^\circ \\ &= -6(0.866) - j6(0.500) = -5.196 - j3.000, \end{aligned} \tag{7}$$

where (7) is the rectangular form of **P** and the trigonometric form is

$$6(\cos 210^\circ + j \sin 210^\circ).$$

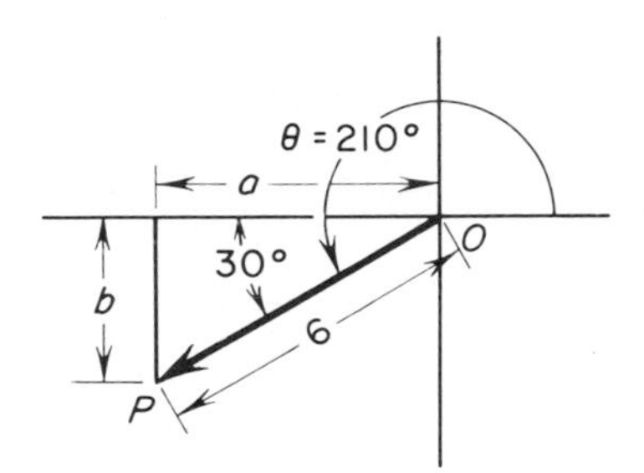

Figure 15-5

An advantage of the trigonometric over the rectangular form is that both the magnitude and direction of **P** can be immediately seen by inspection of the trigonometric form. Also, the magnitude of the reference angle is exactly described so that there is no doubt as to which revolution the vector is in; that is, a vector $\mathbf{P} = 1 + j$ may have a reference angle of 45°, 45° + 360°, 45° + 2(360°), 45° + 3(360°), . . . , 45° + n(360°), where n is an integer. The trigonometric form is also useful in discussion of DeMoivre's theorem, which is described later.

Conversion between the trigonometric and polar forms is done by inspection, since

$$\rho\underline{/\theta} = \rho(\cos\theta + j\sin\theta).$$

As examples, we have

$$3\underline{/330^\circ} = 3(\cos 330^\circ + j\sin 330^\circ)$$

and

$$1.2(\cos 96^\circ + j\sin 96^\circ) = 1.2\underline{/96^\circ}.$$

Exercises 15-2

In Exercises 1–10, express the given quantity as one of the elements in the four-element cycle of powers of j—j, -1, $-j$, or $+1$.

1. j^{13}
2. j^{11}
3. j^{88}
4. j^{67}
5. j^{-9}
6. j^{-14}
7. $(j^2)^2(j^3)$
8. $(j^2)^3(j^3)^2$
9. $(j^{-3})^{-2}$
10. $\dfrac{(j^6)(j^5)}{j^{13}}$

In Exercises 11–20, convert the given expressions to rectangular form.

11. $3\underline{/30^\circ}$
12. $4\underline{/45^\circ}$
13. $1.8\underline{/110^\circ}$
14. $2.7\underline{/152^\circ}$
15. $0.39\underline{/245^\circ}$
16. $260\underline{/188^\circ}$
17. $36\underline{/-30^\circ}$
18. $120\underline{/-45^\circ}$
19. $480\underline{/-135^\circ}$
20. $210\underline{/-225^\circ}$

In Exercises 21–28, convert the given rectangular expressions to polar form.

21. $2 + j2$
22. $3 - j5$
23. $4 - j4\sqrt{3}$
24. $3 + j4$
25. $-0.635 + j0.727$
26. $-14 - j12$
27. $-\sqrt{3} - j$
28. $3.62 + j8.27$

15-3 Addition and subtraction of vectors

Two vectors $\mathbf{P}_1 = a + jb$ and $\mathbf{P}_2 = c + jd$ can be added in the rectangular form. To add

$$\mathbf{P}_1 + \mathbf{P}_2 = (a + jb) + (c + jd) = \mathbf{P}_3,$$

we refer to Fig. 15-6. We see that $\mathbf{P}_1$ has a horizontal component a, $\mathbf{P}_2$ has a horizontal component c, and the composite vector $\mathbf{P}_3$ has a horizontal component $a + c$. Similarly, the vertical components add algebraically to $b + d$. The sum can be shown as

$$\mathbf{P}_1 + \mathbf{P}_2 = (a + jb) + (c + jd) = (a + c) + j(b + d) = \mathbf{P}_3, \qquad (8)$$

where (8) asserts that the sum of two complex numbers has as its real part the algebraic sum of the real parts of the components and as its imaginary part the sum of the imaginary parts of its components.

We note here from Fig. 15-6 that the quadrilateral $OP_1P_3P_2$ is a parallelogram; OP_2 is parallel to P_1P_3 and OP_1 is parallel to P_2P_3. In adding vectors, then, we are using the parallelogram method common to physics, mechanics, surveying, and so on.

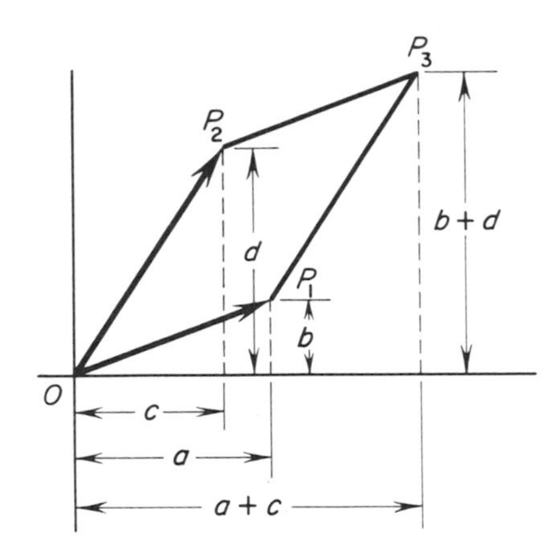

Figure 15-6

Example 4 Add vectors $3 - j5$ and $4 + j2$.

Solution By (8), we have

$$(3 - j5) + (4 + j2) = (3 + 4) + j(-5 + 2) = 7 - j3.$$

Discussion: In effect, we merely columnized the vertical and horizontal components as

$$\begin{array}{r} 3 - j5 \\ 4 + j2 \\ \hline 7 - j3 \end{array}$$

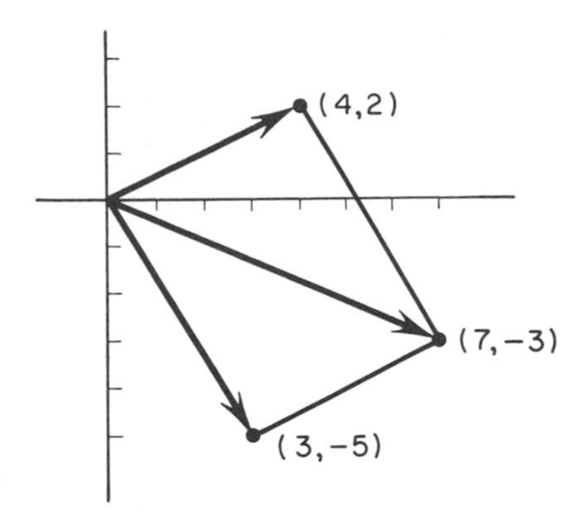

Figure 15-7

and proceeded as in the addition of algebraic polynomials. This process is pictured in Fig. 15-7.

To subtract vectors in the complex plane, we proceed in a way similar to that for addition except that we apply the rule of algebra: change the sign of the subtrahend and proceed as in addition. Thus if $\mathbf{P}_1 = a + jb$ and $\mathbf{P}_2 = c + jd$, then

$$\mathbf{P}_1 - \mathbf{P}_2 = (a + jb) - (c + jd) = (a - c) + j(b - d).$$

The addition and subtraction of vectors in polar form are somewhat more difficult in that the horizontal and vertical components must be determined before addition or subtraction is performed. So if $\mathbf{P}_1 = \rho_1\underline{/\theta}$ and $\mathbf{P}_2 = \rho_2\underline{/\phi}$, then

$$\begin{array}{rl}\mathbf{P}_1 &= \rho_1 \cos\theta + j\rho_1 \sin\theta, \\ \mathbf{P}_2 &= \rho_2 \cos\phi + j\rho_2 \sin\phi, \\ \hline \mathbf{P}_1 + \mathbf{P}_2 &= (\rho_1 \cos\theta + \rho_2 \cos\phi) + j(\rho_1 \sin\theta + \rho_2 \sin\phi).\end{array}$$

Example 5 Add vectors $\mathbf{P}_1 = 3\underline{/35^\circ}$ and $\mathbf{P}_2 = 5\underline{/308^\circ}$, expressing the sum in polar form.

Solution Converting to rectangular form,

$$\mathbf{P}_1 = 3\cos 35^\circ + j3\sin 35^\circ = 2.45 + j1.72,$$

$$\mathbf{P}_2 = 5\cos 308^\circ + j5\sin 308^\circ = 3.08 - j3.94.$$

The sum $\mathbf{P}_3$ is

$$\begin{array}{r} 2.45 + j1.72 \\ \underline{3.08 - j3.94} \\ 5.53 - j2.22 \end{array}$$

where $\mathbf{P}_3$ is in rectangular form. Since we were asked to express the sum in polar form, we must convert to polar, with

$$\mathbf{P}_3 = 5.53 - j2.22 = 5.97\underline{/338.1^\circ}.$$

A rapid check for gross error in the addition or subtraction of vectors can be made by sketching the vectors on a complex plane and adding (or subtracting) them by the parallelogram method shown in Fig. 15-6.

Exercises 15-3

Add the expressions in Exercises 1–12, leaving the results in polar form.

1. $3 + j2, 4 - j3$

2. $4 + j5, 3 - j6$

3. $4 + j2, 4 - j2$

4. $6 + j, -6 + j$

5. $3\angle 30^\circ, 4\angle 120^\circ$

6. $1.5\angle 45^\circ, 3\angle -30^\circ$

7. $4\angle 130^\circ, 4\angle 310^\circ$

8. $6\angle 72^\circ, 6\angle -108^\circ$

9. $12.6\angle 308^\circ, 14.2\angle 75^\circ$

10. $3.5\angle 12^\circ, 4.2\angle 102^\circ$

11. $3\angle 30^\circ, 4\angle 120^\circ, 3\angle 210^\circ$

12. $3 + j7, 4.6\angle 138^\circ, 6.2\angle 235^\circ$

In Exercises 13–18, subtract the latter vector from the former. Express results in polar form.

13. $3 + j6, 5 - j2$

14. $2 - j5, 6 + j8$

15. $3\angle 26^\circ, 3\angle 154^\circ$

16. $4\angle 85^\circ, 4\angle 95^\circ$

17. $3 + j4, 4\angle 53.1^\circ$

18. $120 + j8, 45 + j8$

19. If two vectors are equal in magnitude and have reference angles exactly 180° apart, what is their sum?

20. Given two vectors $\rho\angle\phi$ and $\rho\angle -\phi$, where the magnitudes are equal and one reference angle is the negative of the other, show that the sum of the vectors is $2\rho \cos \phi$.

21. Given vector $3 + j6$, show that vector $-6 + j3$ has the same magnitude but a reference angle 90° larger. In general, show that vectors $a + jb$ and $-b + ja$ are of equal magnitudes but 90° apart.

22. What are the rectangular forms of the vectors represented in polar form by $1\angle 0^\circ, 1\angle 90^\circ, 1\angle 180^\circ, 1\angle 270^\circ$?

15-4 Multiplication and division of vectors

We discuss the multiplication and division of vectors first in the rectangular form and then in the polar form.

Given two vectors $\mathbf{P}_1 = a + jb$ and $\mathbf{P}_2 = c + jd$, their product is found by the ordinary laws of algebraic multiplication of two binomials. Thus

$$(a + jb)(c + jd) = ac + jad + jbc + j^2bd.$$

Then, with $j^2 = -1$, we have

$$(a + jb)(c + jd) = (ac - bd) + j(ad + bc). \tag{9}$$

From (9) we see that the product of two complex numbers is also a complex number.

(b) $$110\underline{/90^\circ} \div 4\underline{/200^\circ} = \frac{110}{4}\underline{/90^\circ - 200^\circ} = 27.5\underline{/-110^\circ}$$

(c) $$4.5\underline{/-125^\circ} \div 3.2\underline{/100^\circ} = \frac{4.5}{3.2}\underline{/-125^\circ - 100^\circ}$$
$$= 1.41\underline{/-225^\circ} = 1.41\underline{/135^\circ}$$

Exercises 15-4

For Exercises 1–24, we are given vectors **A** through **H**.

$$\mathbf{A} = 3 + j2,\quad \mathbf{B} = 3 - j2,\quad \mathbf{C} = -4 + j,\quad \mathbf{D} = -4 - j,$$
$$\mathbf{E} = 6\underline{/30^\circ},\quad \mathbf{F} = 12\underline{/120^\circ},\quad \mathbf{G} = 3\underline{/225^\circ},\quad \mathbf{H} = 4\underline{/315^\circ}.$$

Perform the indicated operations. Express answers in the most convenient form.

1. **AB**
2. **BC**
3. **AD**
4. **EF**
5. **EG**
6. **FH**
7. **AG**
8. **CF**
9. **A/B**
10. **B/C**
11. **A/D**
12. **F/E**
13. **G/H**
14. **H/E**
15. **F/C**
16. **D/E**
17. **(A + B)/AB**
18. **(C + B)/CB**
19. **(E + G)/EG**
20. **(H + E)/HE**
21. $\mathbf{A}^2$
22. $\mathbf{E}^2$
23. $\mathbf{G}^3$
24. $\mathbf{D}^3$
25. Show the derivation of Eq. (26).
26. If a vector in the polar form is squared, how is the reference angle changed?
27. If a vector in the polar form is raised to the nth power, how is the reference angle changed?
28. What is the polar form of j? Multiplication by j advances the reference angle by how many degrees?
29. Show that vector $1 + j$ can be represented in polar form as
$$\sqrt{2}\,[\cos(45^\circ \pm n360^\circ) + j\sin(45^\circ \pm n360^\circ)].$$

15-5 Powers and roots of complex numbers; DeMoivre's theorem

According to (25), the product of two vectors is

$$\rho_1\underline{/\phi_1} \cdot \rho_2\underline{/\phi_2} = \rho_1\rho_2\underline{/\phi_1 + \phi_2}. \tag{27}$$

If $\rho_1\underline{/\phi_1} = \rho_2\underline{/\phi_2}$ here, then we are squaring a vector and

$$(\rho_1\underline{/\phi_1})^2 = \rho_1^2\underline{/2\phi_1}, \tag{28}$$

The division of vector $\mathbf{P}_1 = a + jb$ by $\mathbf{P}_2 = c + jd$ is done by methods established in the chapter on radicals and exponents for the division of a quantity by a binomial where either or both terms of the divisor contain a square root. Thus

$$\frac{\mathbf{P}_1}{\mathbf{P}_2} = \frac{a + jb}{c + jd} \cdot \frac{c - jd}{c - jd} = \frac{ac + jbc - jad - j^2bd}{c^2 - j^2d^2}$$
$$= \frac{(ac + bd) + j(bc - ad)}{c^2 + d^2} = \frac{ac + bd}{c^2 + d^2} + j\frac{bc - ad}{c^2 + d^2}, \tag{10}$$

where we rationalized the denominator by multiplying by its conjugate. From (10) we observe that the quotient of two complex numbers is also a complex number.

Example 6 (a) Multiply $3 - j5$ by $4 + j2$. (b) Divide $6 - j3$ by $4 + j$.

Solution

(a) To multiply $3 - j5$ by $4 + j2$, we have

$$(3 - j5)(4 + j2) = 12 - j20 + j6 - j^210$$
$$= 12 + 10 - j20 + j6 = 22 - j14.$$

(b) To divide $6 - j3$ by $4 + j$, we have

$$\frac{6 - j3}{4 + j} \cdot \frac{4 - j}{4 - j} = \frac{24 - j12 - j6 + j^23}{16 - j^2} = \frac{21 - j18}{17}.$$

To multiply two vectors in polar form, we could choose the route of converting to the rectangular form and then multiplying as in (9). However, this method is usually tedious. Another method, somewhat cumbersome to prove but easy to apply, is available. Let us derive the method.

Given two vectors $\rho_1\underline{/\phi_1}$ and $\rho_2\underline{/\phi_2}$, let us find their product $\rho_3\underline{/\phi_3}$. We first express the given vectors in rectangular form, letting

$$\rho_1\underline{/\phi_1} = a + jb, \tag{11}$$

and

$$\rho_2\underline{/\phi_2} = c + jd. \tag{12}$$

By (9), the product of the two vectors is

$$\rho_3\underline{/\phi_3} = (\rho_1\underline{/\phi_1})(\rho_2\underline{/\phi_2}) = (ac - bd) + j(ad + bc). \tag{13}$$

Vectors (11), (12), and (13) are shown in Fig. 15-8 in parts (a), (b), and (c), respectively. Let us first show the relationship between ρ_1, ρ_2, and ρ_3. From Fig. 15-8(a)

$$\rho_1 = \sqrt{a^2 + b^2}; \tag{14}$$

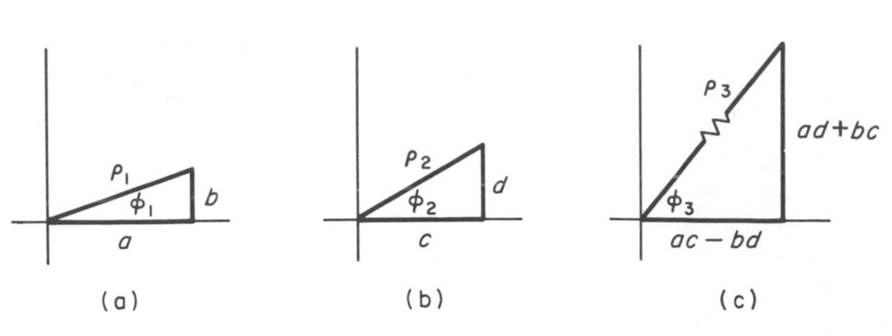

Figure 15-8

from Fig. 15-8(b)

$$\rho_2 = \sqrt{c^2 + d^2}; \tag{15}$$

and from Fig. 15-8(c)

$$\begin{aligned}\rho_3 &= \sqrt{(ad + bc)^2 + (ac - bd)^2} \\ &= \sqrt{a^2d^2 + 2abcd + b^2c^2 + a^2c^2 - 2abcd + b^2d^2} \\ &= \sqrt{a^2d^2 + b^2c^2 + a^2c^2 + b^2d^2} \\ &= \sqrt{a^2(c^2 + d^2) + b^2(c^2 + d^2)} \\ &= \sqrt{(a^2 + b^2)(c^2 + d^2)} = \sqrt{a^2 + b^2} \cdot \sqrt{c^2 + d^2}.\end{aligned} \tag{16}$$

So using (14), (15), and (16), we conclude that

$$\rho_3 = \rho_1 \cdot \rho_2. \tag{17}$$

The conclusion in (17) is interpreted as follows. *The magnitude of the product of two vectors is the product of the magnitudes of the two vectors.*

Next, we will show the relationship between ϕ_1, ϕ_2, and ϕ_3. From Fig. 15-8(a)

$$\tan \phi_1 = \frac{b}{a}; \tag{18}$$

from Fig. 15-8(b)

$$\tan \phi_2 = \frac{d}{c}; \tag{19}$$

and from Fig. 15-8(c)

$$\tan \phi_3 = \frac{ad + bc}{ac - bd}. \tag{20}$$

If we divide the numerator and denominator of (20) by ac, we have

$$\tan \phi_3 = \frac{ad/ac + bc/ac}{ac/ac - bd/ac} = \frac{d/c + b/a}{1 - (b/a)(d/c)}. \tag{21}$$

Substituting (18) and (19) into (21) gives

$$\tan \phi_3 = \frac{\tan \phi_2 + \tan \phi_1}{1 - \tan \phi_2 \tan \phi_1}. \tag{22}$$

Next, we refer to the trigonometric formula for the tangent of the sum of two angles:

$$\tan (A + B) = \frac{\tan A + \tan B}{1 - \tan A \tan B}. \tag{23}$$

Applying (23) to (22), we obtain

$$\tan \phi_3 = \tan (\phi_1 + \phi_2) = \frac{\tan \phi_2 + \tan \phi_1}{1 - \tan \phi_2 \tan \phi_1}$$

from which we reach the conclusion that

$$\phi_3 = \phi_1 + \phi_2. \tag{24}$$

The conclusion in (24) is interpreted as follows. *The reference angle of the product of two vectors is the sum of the reference angles of the two vectors.*

Finally, from (13), (17), and (24), we conclude that

$$\rho_1\underline{/\phi_1} \cdot \rho_2\underline{/\phi_2} = \rho_1\rho_2\underline{/\phi_1 + \phi_2}. \tag{25}$$

Example 7 Here we show three examples of products of vectors in the polar form. In each case the product is obtained by multiplying the magnitudes and adding the reference angles according to (25).

(a) $3\underline{/30^\circ} \cdot 4\underline{/45^\circ} = (3)(4)\underline{/30^\circ + 45^\circ} = 12\underline{/75^\circ}$

(b) $6\underline{/-60^\circ} \cdot 5\underline{/20^\circ} = (6)(5)\underline{/-60^\circ + 20^\circ} = 30\underline{/-40^\circ}$

(c) $2.5\underline{/330^\circ} \cdot 10\underline{/285^\circ} = (2.5)(10)\underline{/330^\circ + 285^\circ} = 25\underline{/615^\circ} = 25\underline{/255^\circ}$

We will not show the derivation of the appropriate equation for the division of vectors in the polar form, primarily because it duplicates to a considerable extent the derivation of the product equation. We will, however, show the equation in mathematical symbolism and give a verbal statement of it. The reader is asked to show the derivation.

Given vectors $\rho_1\underline{/\phi_1}$ and $\rho_2\underline{/\phi_2}$, their quotient is

$$\frac{\rho_1\underline{/\phi_1}}{\rho_2\underline{/\phi_2}} = \frac{\rho_1}{\rho_2}\underline{/\phi_1 - \phi_2}. \tag{26}$$

The verbal statement of (26) is as follows. *The magnitude of the quotient of two vectors is the quotient of the magnitudes of the two vectors. The reference angle of the quotient of two vectors is the difference of the reference angles of the two vectors.* The order of division and subtraction is as shown in (26).

Example 8 Here we show three examples of the division of vectors in polar form. Each division is in accordance with Eq. (26).

(a) $3\underline{/120^\circ} \div 6\underline{/50^\circ} = \frac{3}{6}\underline{/120^\circ - 50^\circ} = 0.5\underline{/70^\circ}$

where (28) asserts "to square a complex number, square the magnitude and double the reference angle." Repeated application of this notion shows that

$$(\rho \underline{/\phi})^n = \rho^n \underline{/n\phi}. \tag{29}$$

Converting to the trigonometric form, (29) can be written

$$[\rho(\cos \phi + j \sin \phi)]^n = \rho^n(\cos n\phi + j \sin n\phi). \tag{30}$$

Expression (30) is a part of *DeMoivre's theorem* and is useful in finding powers of vectors in the complex plane.

Example 9 Expand $(-2 + j2\sqrt{3})^6$.

Solution First, let us convert the given expression $-2 + j2\sqrt{3}$ to trigonometric form. Refer to Fig. 15-9, which shows a sketch of the vector, and we see that $\rho = 4$ and the least positive value of ϕ is $\phi = 120°$. Thus

$$-2 + j2\sqrt{3} = 4\underline{/120°} = 4(\cos 120° + j \sin 120°).$$

Then according to DeMoivre's theorem (30),

$$\begin{aligned} 4^6(\cos 120° + j \sin 120°)^6 &= 4^6(\cos 720° + j \sin 720°) \\ &= 4^6(\cos 0° + j \sin 0°) \\ &= 4^6 = 4096. \end{aligned}$$

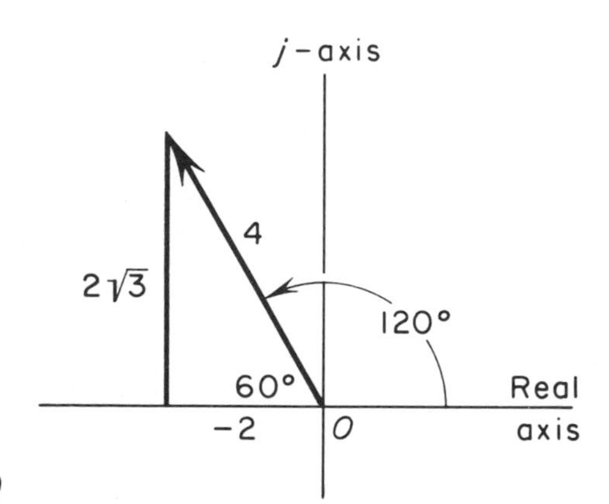

Figure 15-9

DeMoivre's theorem is also useful in finding roots of complex numbers. Without providing the proof, we assert that (30) is also applicable for fractional values of n, meaning that it is useful in finding roots as well as powers. Let us consider an example and then establish a general expression for roots.

Example 10 Evaluate the expression

$$\left(-\frac{1}{2} - j\frac{1}{2}\sqrt{3}\right)^{1/3}.$$

Solution First, we sketch the vector as in Fig. 15-10. Inspecting the figure, we might have a tendency to describe reference angle ϕ as 240° only. Actually, ϕ has many values, such as 240°, 240° + 360°, 240° + 2(360°), 240° + 3(360°), . . . , where it is apparent that we need not restrict ourselves to the first-revolution value of ϕ. The general value of ϕ is given as

$$\phi = 240^\circ + k(360^\circ)$$

where k is an integer.

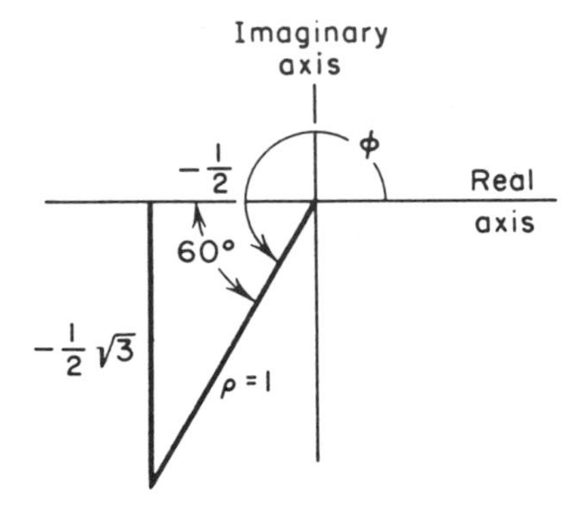

Figure 15-10

The magnitude of the desired vector, as seen in Fig. 15-10, is

$$\sqrt[3]{\rho} = \sqrt[3]{1} = 1.$$

Since we are taking the cube root, the reference angle(s) of the given vector will be divided by 3. Both facts can be seen to follow from the power form (30); if the three cube roots are each raised to the third power by (30), we should return to the given vector. Application of (30) necessitates cubing the magnitudes of the solution vectors and tripling their reference angles. The desired roots, then, are

$$r_1 = (1)^{1/3}\left(\cos\frac{240^\circ}{3} + j\sin\frac{240^\circ}{3}\right) = 1(\cos 80^\circ + j\sin 80^\circ)$$

$$= 0.174 + j0.985 = 1\underline{/80^\circ},$$

$$r_2 = (1)^{1/3}\left(\cos\frac{600^\circ}{3} + j\sin\frac{600^\circ}{3}\right) = 1(\cos 200^\circ + j\sin 200^\circ)$$

$$= -0.940 - j0.342 = 1\underline{/200^\circ},$$

The division of vector $\mathbf{P}_1 = a + jb$ by $\mathbf{P}_2 = c + jd$ is done by methods established in the chapter on radicals and exponents for the division of a quantity by a binomial where either or both terms of the divisor contain a square root. Thus

$$\frac{\mathbf{P}_1}{\mathbf{P}_2} = \frac{a + jb}{c + jd} \cdot \frac{c - jd}{c - jd} = \frac{ac + jbc - jad - j^2bd}{c^2 - j^2d^2}$$

$$= \frac{(ac + bd) + j(bc - ad)}{c^2 + d^2} = \frac{ac + bd}{c^2 + d^2} + j\frac{bc - ad}{c^2 + d^2}, \tag{10}$$

where we rationalized the denominator by multiplying by its conjugate. From (10) we observe that the quotient of two complex numbers is also a complex number.

Example 6 (a) Multiply $3 - j5$ by $4 + j2$. (b) Divide $6 - j3$ by $4 + j$.

Solution

(a) To multiply $3 - j5$ by $4 + j2$, we have

$$(3 - j5)(4 + j2) = 12 - j20 + j6 - j^2 10$$
$$= 12 + 10 - j20 + j6 = 22 - j14.$$

(b) To divide $6 - j3$ by $4 + j$, we have

$$\frac{6 - j3}{4 + j} \cdot \frac{4 - j}{4 - j} = \frac{24 - j12 - j6 + j^2 3}{16 - j^2} = \frac{21 - j18}{17}.$$

To multiply two vectors in polar form, we could choose the route of converting to the rectangular form and then multiplying as in (9). However, this method is usually tedious. Another method, somewhat cumbersome to prove but easy to apply, is available. Let us derive the method.

Given two vectors $\rho_1 \underline{/\phi_1}$ and $\rho_2 \underline{/\phi_2}$, let us find their product $\rho_3 \underline{/\phi_3}$. We first express the given vectors in rectangular form, letting

$$\rho_1 \underline{/\phi_1} = a + jb, \tag{11}$$

and

$$\rho_2 \underline{/\phi_2} = c + jd. \tag{12}$$

By (9), the product of the two vectors is

$$\rho_3 \underline{/\phi_3} = (\rho_1 \underline{/\phi_1})(\rho_2 \underline{/\phi_2}) = (ac - bd) + j(ad + bc). \tag{13}$$

Vectors (11), (12), and (13) are shown in Fig. 15-8 in parts (a), (b), and (c), respectively. Let us first show the relationship between ρ_1, ρ_2, and ρ_3. From Fig. 15-8(a)

$$\rho_1 = \sqrt{a^2 + b^2}; \tag{14}$$

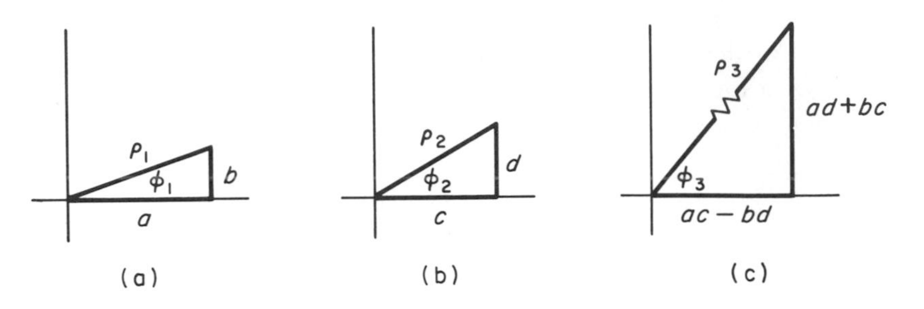

Figure 15-8

from Fig. 15-8(b)

$$\rho_2 = \sqrt{c^2 + d^2}; \tag{15}$$

and from Fig. 15-8(c)

$$\begin{aligned} \rho_3 &= \sqrt{(ad + bc)^2 + (ac - bd)^2} \\ &= \sqrt{a^2d^2 + 2abcd + b^2c^2 + a^2c^2 - 2abcd + b^2d^2} \\ &= \sqrt{a^2d^2 + b^2c^2 + a^2c^2 + b^2d^2} \\ &= \sqrt{a^2(c^2 + d^2) + b^2(c^2 + d^2)} \\ &= \sqrt{(a^2 + b^2)(c^2 + d^2)} = \sqrt{a^2 + b^2} \cdot \sqrt{c^2 + d^2}. \end{aligned} \tag{16}$$

So using (14), (15), and (16), we conclude that

$$\rho_3 = \rho_1 \cdot \rho_2. \tag{17}$$

The conclusion in (17) is interpreted as follows. *The magnitude of the product of two vectors is the product of the magnitudes of the two vectors.*

Next, we will show the relationship between ϕ_1, ϕ_2, and ϕ_3. From Fig. 15-8(a)

$$\tan \phi_1 = \frac{b}{a}; \tag{18}$$

from Fig. 15-8(b)

$$\tan \phi_2 = \frac{d}{c}; \tag{19}$$

and from Fig. 15-8(c)

$$\tan \phi_3 = \frac{ad + bc}{ac - bd}. \tag{20}$$

If we divide the numerator and denominator of (20) by ac, we have

$$\tan \phi_3 = \frac{ad/ac + bc/ac}{ac/ac - bd/ac} = \frac{d/c + b/a}{1 - (b/a)(d/c)}. \tag{21}$$

Substituting (18) and (19) into (21) gives

$$\tan \phi_3 = \frac{\tan \phi_2 + \tan \phi_1}{1 - \tan \phi_2 \tan \phi_1}. \tag{22}$$

Next, we refer to the trigonometric formula for the tangent of the sum of two angles:

$$\tan(A + B) = \frac{\tan A + \tan B}{1 - \tan A \tan B}. \tag{23}$$

Applying (23) to (22), we obtain

$$\tan \phi_3 = \tan(\phi_1 + \phi_2) = \frac{\tan \phi_2 + \tan \phi_1}{1 - \tan \phi_2 \tan \phi_1}$$

from which we reach the conclusion that

$$\phi_3 = \phi_1 + \phi_2. \tag{24}$$

The conclusion in (24) is interpreted as follows. *The reference angle of the product of two vectors is the sum of the reference angles of the two vectors.*

Finally, from (13), (17), and (24), we conclude that

$$\rho_1\underline{/\phi_1} \cdot \rho_2\underline{/\phi_2} = \rho_1\rho_2\underline{/\phi_1 + \phi_2}. \tag{25}$$

Example 7 Here we show three examples of products of vectors in the polar form. In each case the product is obtained by multiplying the magnitudes and adding the reference angles according to (25).

(a) $3\underline{/30^\circ} \cdot 4\underline{/45^\circ} = (3)(4)\underline{/30^\circ + 45^\circ} = 12\underline{/75^\circ}$

(b) $6\underline{/-60^\circ} \cdot 5\underline{/20^\circ} = (6)(5)\underline{/-60^\circ + 20^\circ} = 30\underline{/-40^\circ}$

(c) $2.5\underline{/330^\circ} \cdot 10\underline{/285^\circ} = (2.5)(10)\underline{/330^\circ + 285^\circ} = 25\underline{/615^\circ} = 25\underline{/255^\circ}$

We will not show the derivation of the appropriate equation for the division of vectors in the polar form, primarily because it duplicates to a considerable extent the derivation of the product equation. We will, however, show the equation in mathematical symbolism and give a verbal statement of it. The reader is asked to show the derivation.

Given vectors $\rho_1\underline{/\phi_1}$ and $\rho_2\underline{/\phi_2}$, their quotient is

$$\frac{\rho_1\underline{/\phi_1}}{\rho_2\underline{/\phi_2}} = \frac{\rho_1}{\rho_2}\underline{/\phi_1 - \phi_2}. \tag{26}$$

The verbal statement of (26) is as follows. *The magnitude of the quotient of two vectors is the quotient of the magnitudes of the two vectors. The reference angle of the quotient of two vectors is the difference of the reference angles of the two vectors.* The order of division and subtraction is as shown in (26).

Example 8 Here we show three examples of the division of vectors in polar form. Each division is in accordance with Eq. (26).

(a) $3\underline{/120^\circ} \div 6\underline{/50^\circ} = \frac{3}{6}\underline{/120^\circ - 50^\circ} = 0.5\underline{/70^\circ}$

(b) $$110\underline{/90^\circ} \div 4\underline{/200^\circ} = \frac{110}{4}\underline{/90^\circ - 200^\circ} = 27.5\underline{/-110^\circ}$$

(c) $$4.5\underline{/-125^\circ} \div 3.2\underline{/100^\circ} = \frac{4.5}{3.2}\underline{/-125^\circ - 100^\circ}$$
$$= 1.41\underline{/-225^\circ} = 1.41\underline{/135^\circ}$$

Exercises 15-4

For Exercises 1–24, we are given vectors **A** through **H**.

$$\mathbf{A} = 3 + j2,\quad \mathbf{B} = 3 - j2,\quad \mathbf{C} = -4 + j,\quad \mathbf{D} = -4 - j,$$
$$\mathbf{E} = 6\underline{/30^\circ},\quad \mathbf{F} = 12\underline{/120^\circ},\quad \mathbf{G} = 3\underline{/225^\circ},\quad \mathbf{H} = 4\underline{/315^\circ}.$$

Perform the indicated operations. Express answers in the most convenient form.

1. **AB** **2.** **BC** **3.** **AD**

4. **EF** **5.** **EG** **6.** **FH**

7. **AG** **8.** **CF** **9.** **A/B**

10. **B/C** **11.** **A/D** **12.** **F/E**

13. **G/H** **14.** **H/E** **15.** **F/C**

16. **D/E** **17.** **(A + B)/AB** **18.** **(C + B)/CB**

19. **(E + G)/EG** **20.** **(H + E)/HE** **21.** $\mathbf{A}^2$

22. $\mathbf{E}^2$ **23.** $\mathbf{G}^3$ **24.** $\mathbf{D}^3$

25. Show the derivation of Eq. (26).

26. If a vector in the polar form is squared, how is the reference angle changed?

27. If a vector in the polar form is raised to the nth power, how is the reference angle changed?

28. What is the polar form of j? Multiplication by j advances the reference angle by how many degrees?

29. Show that vector $1 + j$ can be represented in polar form as

$$\sqrt{2}\,[\cos(45^\circ \pm n360^\circ) + j\sin(45^\circ \pm n360^\circ)].$$

15-5 Powers and roots of complex numbers; DeMoivre's theorem

According to (25), the product of two vectors is

$$\rho_1\underline{/\phi_1} \cdot \rho_2\underline{/\phi_2} = \rho_1\rho_2\underline{/\phi_1 + \phi_2}. \tag{27}$$

If $\rho_1\underline{/\phi_1} = \rho_2\underline{/\phi_2}$ here, then we are squaring a vector and

$$(\rho_1\underline{/\phi_1})^2 = \rho_1^2\underline{/2\phi_1}, \tag{28}$$

where (28) asserts "to square a complex number, square the magnitude and double the reference angle." Repeated application of this notion shows that

$$(\rho\underline{/\phi})^n = \rho^n\underline{/n\phi}. \tag{29}$$

Converting to the trigonometric form, (29) can be written

$$[\rho(\cos\phi + j\sin\phi)]^n = \rho^n(\cos n\phi + j\sin n\phi). \tag{30}$$

Expression (30) is a part of *DeMoivre's theorem* and is useful in finding powers of vectors in the complex plane.

Example 9 Expand $(-2 + j2\sqrt{3})^6$.

Solution First, let us convert the given expression $-2 + j2\sqrt{3}$ to trigonometric form. Refer to Fig. 15-9, which shows a sketch of the vector, and we see that $\rho = 4$ and the least positive value of ϕ is $\phi = 120°$. Thus

$$-2 + j2\sqrt{3} = 4\underline{/120°} = 4(\cos 120° + j\sin 120°).$$

Then according to DeMoivre's theorem (30),

$$\begin{aligned} 4^6(\cos 120° + j\sin 120°)^6 &= 4^6(\cos 720° + j\sin 720°) \\ &= 4^6(\cos 0° + j\sin 0°) \\ &= 4^6 = 4096. \end{aligned}$$

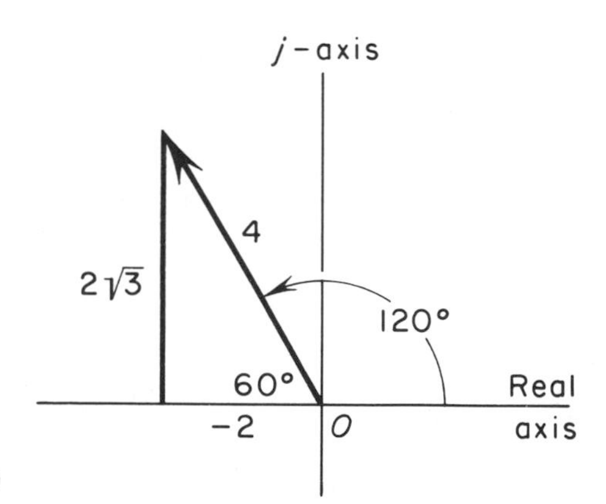

Figure 15-9

DeMoivre's theorem is also useful in finding roots of complex numbers. Without providing the proof, we assert that (30) is also applicable for fractional values of n, meaning that it is useful in finding roots as well as powers. Let us consider an example and then establish a general expression for roots.

Example 10 Evaluate the expression

$$\left(-\frac{1}{2} - j\frac{1}{2}\sqrt{3}\right)^{1/3}.$$

Solution First, we sketch the vector as in Fig. 15-10. Inspecting the figure, we might have a tendency to describe reference angle ϕ as 240° only. Actually, ϕ has many values, such as 240°, 240° + 360°, 240° + 2(360°), 240° + 3(360°), . . . , where it is apparent that we need not restrict ourselves to the first-revolution value of ϕ. The general value of ϕ is given as

$$\phi = 240° + k(360°)$$

where k is an integer.

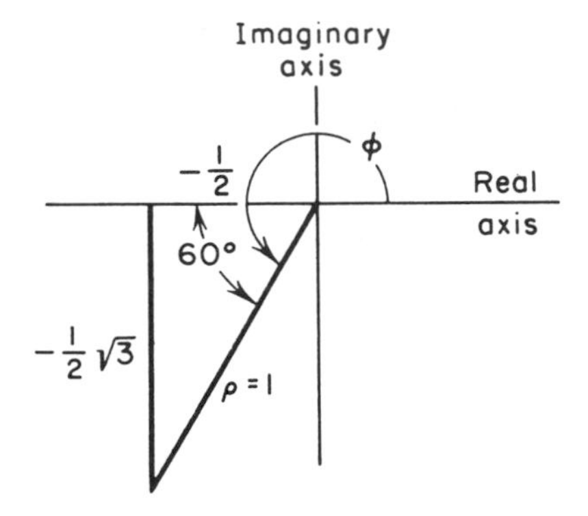

Figure 15-10

The magnitude of the desired vector, as seen in Fig. 15-10, is

$$\sqrt[3]{\rho} = \sqrt[3]{1} = 1.$$

Since we are taking the cube root, the reference angle(s) of the given vector will be divided by 3. Both facts can be seen to follow from the power form (30); if the three cube roots are each raised to the third power by (30), we should return to the given vector. Application of (30) necessitates cubing the magnitudes of the solution vectors and tripling their reference angles. The desired roots, then, are

$$r_1 = (1)^{1/3}\left(\cos\frac{240°}{3} + j\sin\frac{240°}{3}\right) = 1(\cos 80° + j\sin 80°)$$

$$= 0.174 + j0.985 = 1\underline{/80°},$$

$$r_2 = (1)^{1/3}\left(\cos\frac{600°}{3} + j\sin\frac{600°}{3}\right) = 1(\cos 200° + j\sin 200°)$$

$$= -0.940 - j0.342 = 1\underline{/200°},$$

$$r_3 = (1)^{1/3}\left(\cos\frac{960^\circ}{3} + j\sin\frac{960^\circ}{3}\right) = 1(\cos 320^\circ + j\sin 320^\circ)$$
$$= 0.766 - j0.643 = 1\underline{/320^\circ}.$$

We note that the first solution has a reference angle 80° = 240°/3 and that each successive solution is advanced 360°/3 = 120° from the previous solution. Any attempt to go beyond three solutions (ours was a *cube* root) would simply duplicate a previous solution. The solutions are shown in Fig. 15-11. In any root problem the solutions will be equally spaced; the angle between consecutive solutions will be $360^\circ/n$, where we are interested in the nth roots.

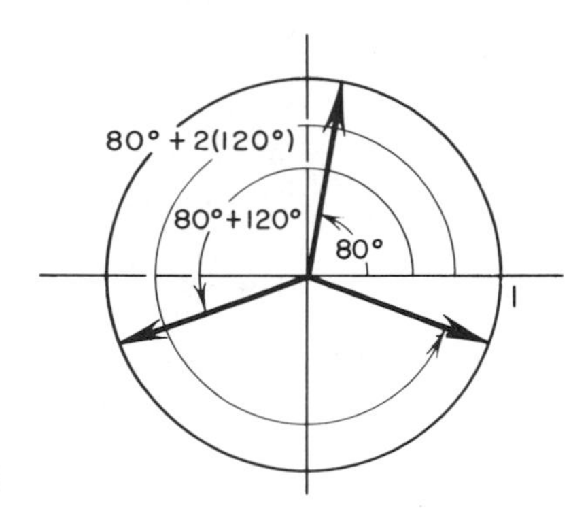

Figure 15-11

Using the preceding example, we present the general expression for the nth root of a vector in the complex plane:

$$(a + jb)^{1/n} = (\rho\underline{/\phi})^{1/n} = \rho^{1/n}\left[\cos\left(\frac{\phi + k\cdot 360^\circ}{n}\right) + j\sin\left(\frac{\phi + k\cdot 360^\circ}{n}\right)\right], \tag{31}$$

where n is the order of the root and k is an integer. Expression (31) is also a part of DeMoivre's theorem.

Example 11 Find the three cube roots of -1.

Solution Since -1 can be written in polar form as $1\underline{/180^\circ}$, then by (31) and Fig. 15-12, we have $\rho = 1$, $n = 3$, $\phi = 180^\circ$, and $k = 0, 1, 2$. Now we have solutions

$$r_1 = (1)^{1/3}\left(\cos\frac{180}{3} + j\sin\frac{180^\circ}{3}\right)$$
$$= 1(\cos 60^\circ + j\sin 60^\circ) = \frac{1}{2} + j\frac{\sqrt{3}}{2},$$

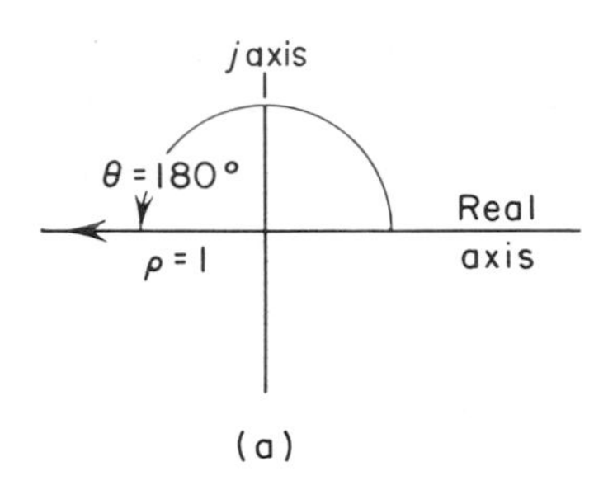

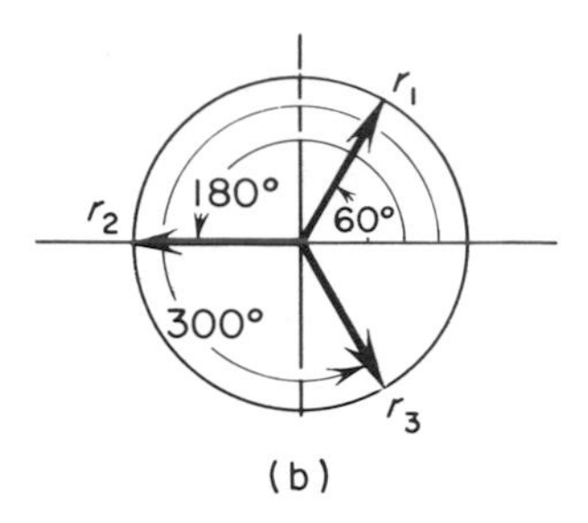

Figure 15-12

$$r_2 = (1)^{1/3}\left(\cos\frac{540^\circ}{3} + j\sin\frac{540^\circ}{3}\right)$$

$$= 1(\cos 180^\circ + j\sin 180^\circ) = -1,$$

$$r_3 = (1)^{1/3}\left(\cos\frac{900^\circ}{3} + j\sin\frac{900^\circ}{3}\right)$$

$$= 1(\cos 300^\circ + j\sin 300^\circ) = \frac{1}{2} - j\frac{\sqrt{3}}{2}.$$

Figure 15-12(a) shows the given quantity, -1. Figure 15-12(b) shows the three solutions, spaced $120^\circ = 360^\circ/3$ apart, with the first solution having reference angle $180^\circ/3 = 60^\circ$.

We can establish the correctness of the three solutions in Example 11 by cubing each and obtaining as the value of the cube the original number, -1.

It is worth noting that through DeMoivre's theorem we see that every real number has n nth roots. At most, two of the roots are real. For instance, the number $+1$ has four fourth roots: $1, j, -1, -j$.

Exercises 15-5

Find the indicated roots and powers by using DeMoivre's theorem. Express results in the most convenient form. Wherever practical, sketch the given vectors and solutions.

1. $\sqrt{-1}$

2. $\sqrt[4]{-1}$

3. $\sqrt{j}$

4. $\sqrt{1+j}$

5. $(1+j)^3$

6. $(1-j)^3$

7. $\sqrt[4]{1}$

8. $(1-j\sqrt{3})^4$

9. $(1+j\sqrt{3})^5$

10. $\left(-\frac{1}{2}+j\frac{\sqrt{3}}{2}\right)^4$

11. $(-4+j4)^3$

12. $(1+j2)^4$

13. $(2+j)^4$

14. $(3\underline{/0^\circ})^3$

15. $\sqrt{5\underline{/360^\circ}}$ **16.** $\sqrt[3]{-8}$

17. $[2(\cos 72^\circ + j\sin 72^\circ)]^5$ **18.** $[3(\cos 30^\circ + j\sin 30^\circ)]^3$

19. $\sqrt[3]{-j}$ **20.** $\sqrt[3]{1-j}$

15-6 Special properties of vectors in the complex plane

Reflections

Vectors can be reflected across either axis or through the origin by simple techniques. Such reflections find extensive application in ac electricity. In Fig. 15-13 a given vector $a + jb$ is shown as a solid line and its reflection is shown as a dotted line.

From Fig. 15-13(a) the reflection of $a + jb$ across the real axis is $a - jb$, accomplished by changing the sign of b. In polar form, $\rho\underline{/\phi}$ and $\rho\underline{/-\phi}$ are reflections across the real axis; here we see that reflection across the real axis is accomplished by reversing the sign of the reference angle.

From Fig. 15-13(b) the reflection of $a + jb$ across the imaginary axis is $-a + jb$, accomplished by changing the sign of a. In polar form, $\rho\underline{/\phi}$ and $\rho\underline{/180^\circ - \phi}$ are reflections of each other across the imaginary axis; here we see that reflection is accomplished by choosing the supplement of the reference angle.

From Fig. 15-13(c) the reflection of $a + jb$ through the origin is $-a - jb$, accomplished by changing the sign of both a and b. In polar form, $\rho\underline{/\phi}$ and $\rho\underline{/180^\circ + \phi}$ are reflections of each other through the origin; this reflection is accomplished by increasing the reference angle 180°.

Advancing the reference angle

It is often useful to advance the reference angle 90°, 180°, or 270° (−90°). Referring to Fig. 15-14, let us see how these advances can be made.

90° advance

In Fig. 15-14 vector $\mathbf{P} = \rho\underline{/\phi} = a + jb$ is given. $\mathbf{P}_1$ has reference angle $\phi + 90^\circ$, therefore $\mathbf{P}$ is perpendicular to $\mathbf{P}_1$ and the slope of $\mathbf{P}_1$ is the negative reciprocal of the slope of $\mathbf{P}$, suggesting the choice of the vertical component a and horizontal component $-b$ for $\mathbf{P}_1$ (as opposed to b and a, respectively for $\mathbf{P}$). So if vector $a + jb$ is advanced 90°, the resulting vector is $-b + ja$. We note that

$$j(a + jb) = -b + ja,$$

which suggests that an advance of 90° can be accomplished by multiplication by j. In polar form, $j = 1\underline{/90^\circ}$. If the original vector $a + jb = \rho\underline{/\phi}$ is mul-

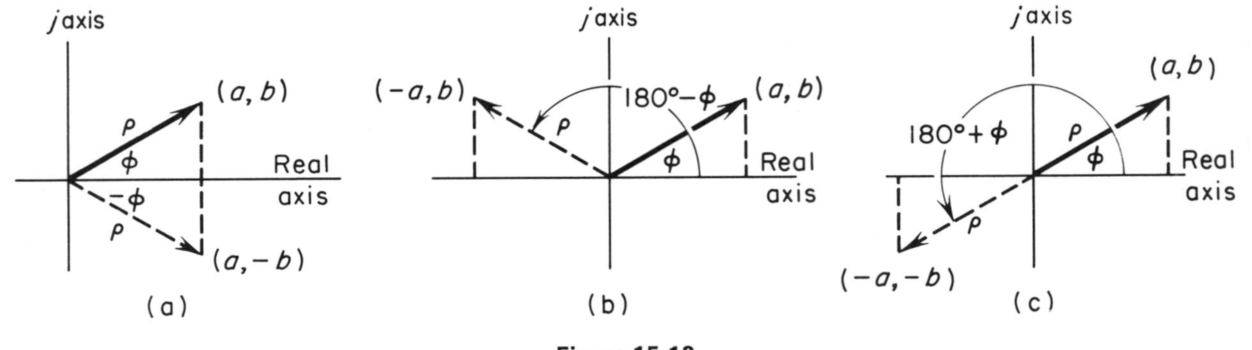
j axis
(a, b)
ρ
φ
−φ
ρ
Real axis
(a, −b)
(a)
j axis
(−a, b)
ρ
180°−φ
(a, b)
φ
Real axis
(b)
j axis
(a, b)
180°+φ
ρ
φ
Real axis
ρ
(−a, −b)
(c)

Figure 15-13

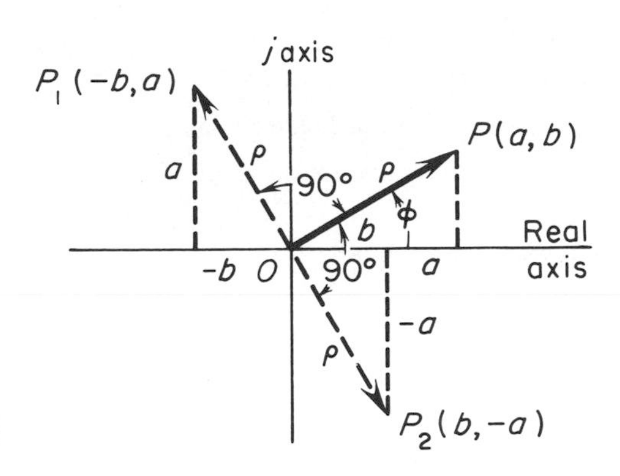

Figure 15-14

tiplied by j, we have

$$\rho\underline{/\phi} \cdot 1\underline{/90^\circ} = \rho\underline{/\phi + 90^\circ},$$

where clearly the magnitude of the vector is undisturbed but the reference angle is advanced by 90°.

270° advance

In Fig. 15-14 $\mathbf{P}_2$ is advanced 270° (or moved backward 90°) from **P**. This step is done by choosing new components b and $-a$ so that the new vector is $b - ja$. It is also done by multiplying the given vector by $-j$ or $1\underline{/-90^\circ}$.

180° advance

Referring to Fig. 15-13(c), we see that an advance of 180° is accomplished by changing the signs of both a and b so that $-a - jb$ is 180° removed from $a + jb$. We note that -1 is $1\underline{/180^\circ}$ and so

$$\rho\underline{/\phi} \cdot 1\underline{/180^\circ} = \rho\underline{/\phi + 180^\circ}. \tag{32}$$

We also note that -1 is $-1\underline{/0^\circ}$ so that

$$\rho\underline{/\phi} \cdot -1\underline{/0^\circ} = -\rho\underline{/\phi}. \tag{33}$$

Comparing (32) and (33), we note that an advance of 180° is the equivalent of a change in sense of the vector.

Exercises 15-6

What is the polar form of the vectors in Exercises 1–4?

1. 1 **2.** j **3.** -1 **4.** $-j$

What is the rectangular form of the vectors in Exercises 5–8?

5. $-2\underline{/180^\circ}$

6. $3\underline{/\phi + 180^\circ}$

7. $(-2\underline{/180^\circ})^2$

8. $3 + j4$ advanced by 90°.

In Exercises 9–11, show that multiplication of a vector by

9. j advances the reference angle by 90°.

10. -1 reverses the sense of the vector.

11. $-j$ diminishes the reference angle by 90°.

Given vector $\mathbf{P} = -3 + j4$, give the rectangular expression in Exercises 12–14 for the vector that is the reflection of **P**

12. across the real axis.

13. across the imaginary axis.

14. through the origin.

In Exercises 15–17, given vector $\mathbf{P} = \rho\underline{/\phi}$, give the polar expression for the vector that is the reflection of **P**

15. across the real axis.

16. across the imaginary axis.

17. through the origin.

Show the expression for the vector advanced 90° beyond the given vector in Exercises 18–22.

18. $3 - j5$ 19. $-4 + j2$ 20. $3\underline{/120^\circ}$ 21. 3

22. $-j5$

23. Show that division of a complex number by its reflection across the real axis doubles the reference angle.

24. Show that the product of a complex number by its reflection across the real axis results in a positive real number that is the square of the magnitude of the original number.

25. Show that the product of a complex number by its reflection across the imaginary axis results in a negative real number that is the negative of the square of the magnitude of the original vector.

26. Given two vectors with equal magnitudes and with reference angles 90° apart, show that their product is a pure imaginary number if either of the original vectors was a pure imaginary number.

16

Miscellaneous topics in trigonometry

In preceding chapters we defined the trigonometric functions, applied trigonometry to right triangles and oblique triangles, graphed the trigonometric functions on the cartesian coordinate plane, and discussed vectors in the complex plane.

Here we turn to certain miscellaneous topics in trigonometry, including identities, trigonometric equations, inverse trigonometric functions, and approximate solutions, among others. These subjects are included to broaden our coverage of trigonometry and to show certain additional applications.

16-1 Square relationships

An equation is called an *identity* if the two members of the equation are equal for all values assigned to the variables for which both members are defined. Thus

$$x^2 - 1 = (x + 1)(x - 1)$$

is an identity because the left member and the right member are equal, regardless of the value assigned to x. The expression

$$x^2 - 2x = 3$$

is not an identity because the equality is true only for values $x = 3$ and $x = -1$.

Identities are generally used in trigonometry for two major reasons: (a) to reduce an expression that may contain two or more different functions of an angle to an expression containing only one trigonometric function and (b) to interchange functions of multiple angles with powers of functions of single angles.

We consider four groups of identities in this chapter—the square relationships, functions of sums and differences of angles, multiple-angle formulas, and half-angle formulas.

Turning to square relationships, in Fig. 16-1 we are given a point $P(x, y)$ on a circle of radius r. From the Pythagorean Theorem we have

$$x^2 + y^2 = r^2. \tag{1}$$

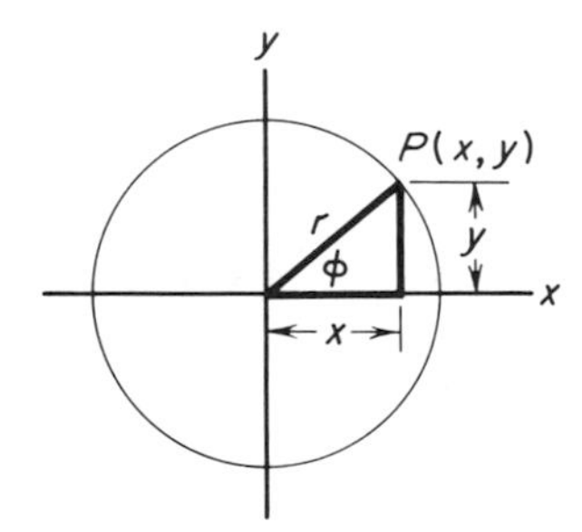

Figure 16-1

Dividing (1) by r^2 gives

$$\frac{x^2}{r^2} + \frac{y^2}{r^2} = \frac{r^2}{r^2}$$

from which

$$\left(\frac{y}{r}\right)^2 + \left(\frac{x}{r}\right)^2 = 1. \tag{2}$$

Next, from Fig. 16-1

$$\frac{y}{r} = \sin\phi \quad \text{and} \quad \frac{x}{r} = \cos\phi$$

so that (2) can now be written

$$\sin^2 \phi + \cos^2 \phi = 1. \tag{3}$$

Equation (3) is an identity because it holds for any value assigned to ϕ. Two other square relationships are common. If we divide (1) by x^2, we have

$$\frac{x^2}{x^2} + \frac{y^2}{x^2} = \frac{r^2}{x^2}$$

or

$$1 + \left(\frac{y}{x}\right)^2 = \left(\frac{r}{x}\right)^2, \tag{4}$$

but from Fig. 16-1

$$\frac{y}{x} = \tan \phi \quad \text{and} \quad \frac{r}{x} = \sec \phi$$

so that (4) becomes

$$1 + \tan^2 \phi = \sec^2 \phi. \tag{5}$$

If we divide (1) by y^2 and proceed as in the derivations of (3) and (5), we have a third square relationship:

$$1 + \text{ctn}^2 \phi = \csc^2 \phi. \tag{6}$$

We call expressions (3), (5), and (6) the *square relationships*. Together with the reciprocal relationships, they are useful in certain manipulations. Consider two examples.

Example 1 Convert the expression

$$\cos^2 \phi - \text{ctn}^2 \phi = 5 \tag{7}$$

to one containing no trigonometric functions other than powers of $\sin \phi$.

Solution From (3)

$$\cos^2 \phi = 1 - \sin^2 \phi, \tag{8}$$

and from (6)

$$\text{ctn}^2 \phi = \csc^2 \phi - 1 = \frac{1}{\sin^2 \phi} - 1. \tag{9}$$

Substituting (8) and (9) into (7), we obtain

$$1 - \sin^2 \phi - \left(\frac{1}{\sin^2 \phi} - 1\right) = 5$$

from which

$$\sin^4 \phi + 3 \sin^2 \phi + 1 = 0, \tag{10}$$

where we note that (10) is a desired expression containing only powers of $\sin \phi$.

Example 2 Referring to Fig. 16-1, show that

$$\tan \phi = \frac{\sin \phi}{\cos \phi}.$$

Solution From the figure

$$\tan \phi = \frac{y}{x}; \tag{11}$$

dividing both the numerator and denominator of (11) by r, we have

$$\tan \phi = \frac{y/r}{x/r}.$$

But from Fig. 16-1

$$\frac{y}{r} = \sin \phi \quad \text{and} \quad \frac{x}{r} = \cos \phi$$

so that

$$\tan \phi = \frac{\sin \phi}{\cos \phi}.$$

A summary of certain square relationships and quotients is given in Table 16-1.

Table 16-1 Identities Involving Squares and Quotients

1.	$\sin^2 \phi + \cos^2 \phi = 1$
2.	$1 + \tan^2 \phi = \sec^2 \phi$
3.	$1 + \text{ctn}^2 \phi = \csc^2 \phi$
4.	$\tan \phi = \frac{\sin \phi}{\cos \phi}$
5.	$\text{ctn}\, \phi = \frac{\cos \phi}{\sin \phi}$

Exercises 16-1

Using the identities given in Table 16-1, in addition to the definitions and reciprocals of the trigonometric functions, do Exercises 1–5.

1. Modify the expression

$$4 + \text{ctn}^2 x = \sec^2 x$$

to an expression containing only powers of $\cos x$.

2. Modify the expression $\cos^2 x - \text{ctn}^2 x = 3$ to an expression containing only powers of $\sin x$.

3. Show that the expression

$$\frac{\sec \phi}{\tan \phi + \text{ctn } \phi}$$

is equal to $\sin \phi$.

4. Modify the expression

$$\frac{\sec^2 \phi - \sin \phi \sec \phi - 1}{\tan \phi - 1}$$

to one containing the tangent function only.

5. Show that the expression

$$\frac{2 \tan x + \sec^2 x}{1 + \tan x}$$

is equal to $1 + \tan x$.

6. Show that the expression

$$\left(\tan \phi + \frac{\cos \phi}{\sin \phi}\right)(\cos \phi)$$

is equal to $\csc \phi$.

7. Show that the expression

$$\sec \phi = \cos \phi + \tan \phi$$

can be reduced to the expression $\sin \phi = 1$.

8. Using Fig. 16-2 and the definitions of the trigonometric functions, express

$$\sec \phi = \cos \phi + \tan \phi$$

in terms of x, y, and r only and simplify the resulting expression to $y/r = 1$. Compare this with the result in Exercise 7.

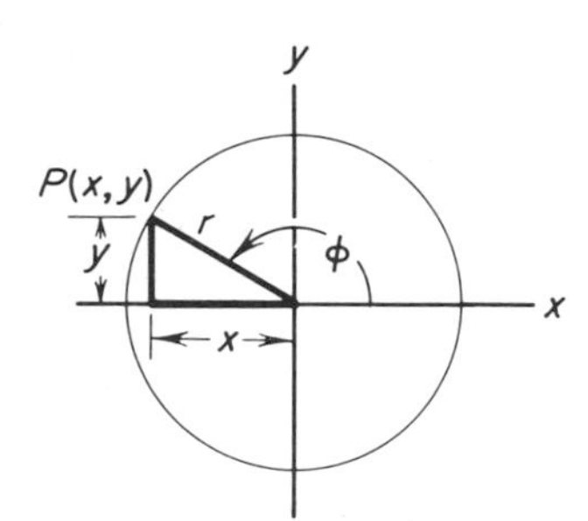

Figure 16-2

9. Given expression

$$\frac{x^2 + y^2}{x^2} - \frac{\sqrt{x^2 + y^2}}{y} = \frac{x^2}{r^2},$$

where x, y, and r are obtained from Fig. 16-2, modify the given expression to another form containing only powers of $\sin \phi$.

10. Modify the expression

$$\sin^3 \phi - 3 \sin \phi + 1 = 0$$

to read

$$(1 + \tan^2 \phi)^{3/2} = 3 \tan \phi + 2 \tan^3 \phi.$$

11. Given the isosceles trapezoid *ABCD* in Fig. 16-3 with a rectangle *CDEF* mounted on one of the equal sides. If $FC = h$, where h is the altitude of the trapezoid, show that $FD = h\sqrt{1 + \csc^2 \phi}$.

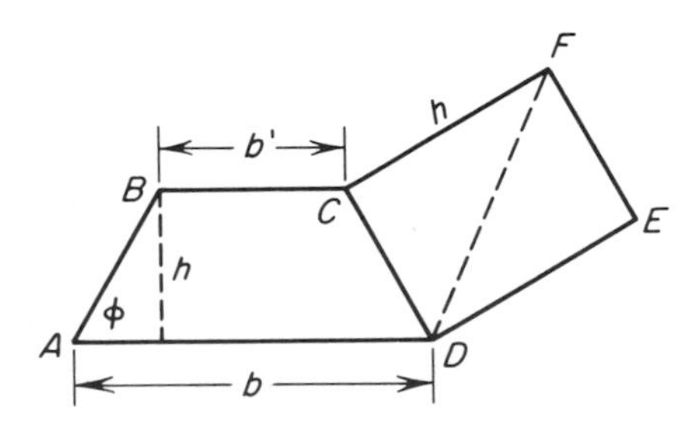

Figure 16-3

12. Show that the expression

$$\sin^4 \phi = \frac{1 - \sin^2 \phi}{-\csc^2 \phi} + \frac{1}{1 + \operatorname{ctn}^2 \phi}$$

is an identity by reducing it to one of the forms in Table 16-1.

13. Show that

$$9 \tan^2 \phi = 25(\sec^2 \phi - 1 - \sin^2 \phi)$$

is reducible to the expression $\csc \phi = 5/3$.

16-2 Functions of sums and differences of angles

It is often useful to express an angle as the sum or difference of two angles and to obtain a trigonometric function of this sum or difference in terms of functions of the separate angles. Figure 16-4 shows a unit circle with $OC = OG = OF = 1$, where $\angle COF = \theta + \phi$. Let us now obtain expressions for $\sin(\theta + \phi)$ and $\cos(\theta + \phi)$.

From F construct $FA \perp OC$ and $FD \perp OG$. From D construct $DE \perp FA$. Then $\angle EFD = \angle COG = \theta$ because the sides of the angles are perpendicular left side to left side and right side to right side. We also see that $AE = BD$. Next,

$$\sin(\theta + \phi) = AF$$

but

$$AF = BD + EF$$

so that

$$\sin(\theta + \phi) = BD + EF. \tag{12}$$

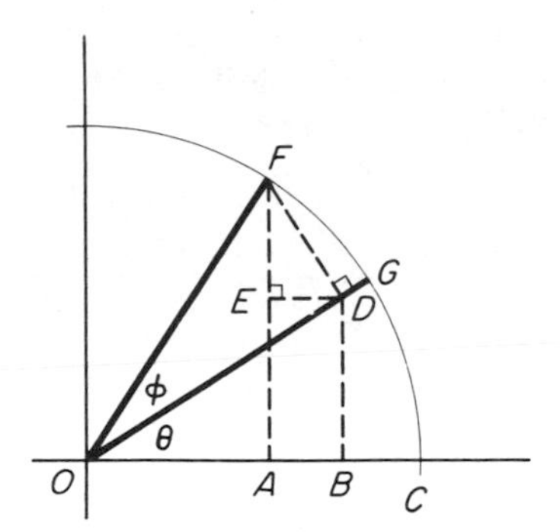

Figure 16-4

Then from triangle EFD

$$EF = FD \cos \theta,$$

and from triangle OBD

$$DB = OD \sin \theta$$

so that (12) becomes

$$\sin (\theta + \phi) = OD \sin \theta + FD \cos \theta. \tag{13}$$

From triangle ODF

$$FD = OF \sin \phi = \sin \phi$$

and

$$OD = OF \cos \phi = \cos \phi$$

so that (13) becomes

$$\sin (\theta + \phi) = \sin \theta \cos \phi + \cos \theta \sin \phi, \tag{14}$$

where (14) shows the sine of the sum of two angles in terms of the sine and cosine of the separate angles.

Using Fig. 16-4, we can develop the equation

$$\cos (\theta + \phi) = \cos \theta \cos \phi - \sin \theta \sin \phi. \tag{15}$$

Development of (15) is left as an exercise for the reader.

If we substitute $-\phi$ for ϕ in (14), we obtain

$$\sin (\theta - \phi) = \sin \theta \cos (-\phi) + \cos \theta \sin (-\phi),$$

but

$$\sin (-\phi) = -\sin \phi \quad \text{and} \quad \cos (-\phi) = \cos \phi$$

so that

$$\sin (\theta - \phi) = \sin \theta \cos \phi - \cos \theta \sin \phi. \tag{16}$$

By substituting $-\phi$ for ϕ in (15), we get

$$\cos (\theta - \phi) = \cos \theta \cos \phi + \sin \theta \sin \phi. \tag{17}$$

Equations (14), (15), (16), and (17) give expressions for the sine and cosine functions of the sums and differences of two angles in terms of the sines and cosines of the separate angles.

Optional forms involving sines and cosines of the sums and differences of two angles are made available by combining pairs of Eqs. (14) to (17). For example, by adding (15) and (17),

$$\begin{aligned} \cos(\theta + \phi) &= \cos\theta\cos\phi - \sin\theta\sin\phi \\ \cos(\theta - \phi) &= \cos\theta\cos\phi + \sin\theta\sin\phi \\ \hline \cos(\theta + \phi) + \cos(\theta - \phi) &= 2\cos\theta\cos\phi \end{aligned}$$

from which

$$\cos\theta\cos\phi = \tfrac{1}{2}[\cos(\theta + \phi) + \cos(\theta - \phi)]. \tag{18}$$

Other optional forms comparable to (18) are given in Table 16-2.

Table 16-2 Trigonometric Identities

1. $\sin(\theta + \phi) = \sin\theta\cos\phi + \cos\theta\sin\phi$
2. $\sin(\theta - \phi) = \sin\theta\cos\phi - \cos\theta\sin\phi$
3. $\cos(\theta + \phi) = \cos\theta\cos\phi - \sin\theta\sin\phi$
4. $\cos(\theta - \phi) = \cos\theta\cos\phi + \sin\theta\sin\phi$
5. $\cos\theta\cos\phi = \tfrac{1}{2}[\cos(\theta + \phi) + \cos(\theta - \phi)]$
6. $\sin\theta\sin\phi = \tfrac{1}{2}[\cos(\theta - \phi) - \cos(\theta + \phi)]$
7. $\sin\theta\cos\phi = \tfrac{1}{2}[\sin(\theta + \phi) + \sin(\theta - \phi)]$
8. $\cos\theta\sin\phi = \tfrac{1}{2}[\sin(\theta + \phi) - \sin(\theta - \phi)]$
9. $\sin 2\theta = 2\sin\theta\cos\theta$
10. $\cos 2\theta = \cos^2\theta - \sin^2\theta = 1 - 2\sin^2\theta = 2\cos^2\theta - 1$
11. $\cos^2\theta = \tfrac{1}{2}(1 + \cos 2\theta)$
12. $\sin^2\theta = \tfrac{1}{2}(1 - \cos 2\theta)$
13. $\cos\dfrac{\theta}{2} = \sqrt{\dfrac{1 + \cos\theta}{2}}$
14. $\sin\dfrac{\theta}{2} = \sqrt{\dfrac{1 - \cos\theta}{2}}$

Exercises 16-2

1. Referring to Fig. 16-4, derive form 3 of Table 16-2.
2. By substituting $-\phi$ for ϕ into form 3, derive form 4 of Table 16-2.
3. By adding or subtracting the proper pairs of forms 1 to 4 in Table 16-2, derive forms 6, 7, and 8.

In Exercises 4–11, assume knowledge of functions

$$\sin 30^\circ = \cos 60^\circ = \frac{1}{2}, \qquad \cos 30^\circ = \sin 60^\circ = \frac{\sqrt{3}}{2},$$

$$\sin 45^\circ = \cos 45^\circ = \frac{\sqrt{2}}{2}, \qquad \sin 90^\circ = 1, \qquad \cos 90^\circ = 0.$$

Using this known information and appropriate forms from Table 16-2, obtain radical values of the given functions.

4. $\sin 15°$ **5.** $\cos 15°$ **6.** $\cos 75°$ **7.** $\sin 75°$

8. $\cos 105°$ **9.** $\cos 150°$ **10.** $\sin 120°$ **11.** $\cos 135°$

12. A tapestry hung on a wall is viewed by an observer as shown in Fig. 16-5. The observer's eye is at A, 12 ft from the wall DB. The tapestry DC is 7 ft high and is hung so that its bottom is 9 ft above the observer's eye. Find the magnitude of the angle ϕ without first finding θ.

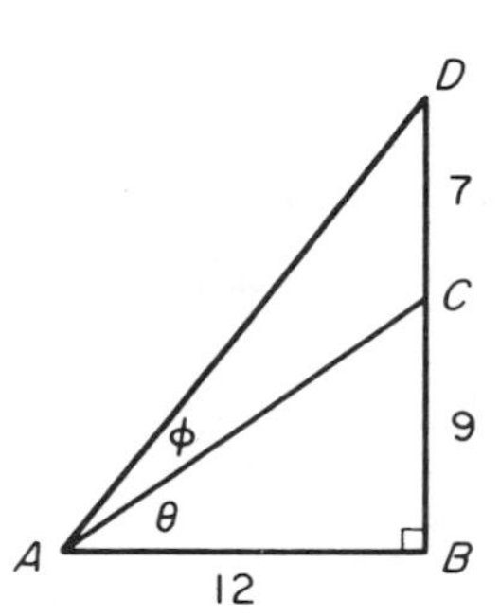

Figure 16-5

13. By using form 2 in Table 16-2, show that

$$\sin(180° - \phi) = \sin \phi.$$

14. By using appropriate forms from Table 16-2, show that

$$\sin(90° \pm \phi) = \cos \phi.$$

15. Show that $\sin(\alpha + \beta + \gamma)$ can be written

$$\sin\alpha\cos\beta\cos\gamma - \sin\alpha\sin\beta\sin\gamma + \cos\alpha\sin\beta\cos\gamma + \cos\alpha\cos\beta\sin\gamma.$$

16-3 Functions of multiple and partial angles

It is often useful to refer to trigonometric functions of multiple angles such as 2θ or 3θ. The angle 2θ is called a *double angle*; $\sin 2\theta$ and $\cos 2\theta$ can be expressed in terms of $\sin\theta$ and $\cos\theta$ rather easily, provided that we have expressions for $\sin(\theta + \phi)$ and $\cos(\theta + \phi)$. Referring to Table 16-2, let us develop an expression (form 9) for $\sin 2\theta$ in terms of functions of the single angle.

In form 1, where

$$\sin(\theta + \phi) = \sin\theta\cos\phi + \cos\theta\sin\phi,$$

let $\phi = \theta$ so that

$$\sin(\theta + \theta) = \sin\theta\cos\theta + \sin\theta\cos\theta$$

from which

$$\sin 2\theta = 2 \sin \theta \cos \theta, \tag{19}$$

where (19) is form 9 of Table 16-2 and is an expression for the sine of the double angle in terms of functions of the single angle.

Example 3 Express $\cos 3\theta$ in terms of $\cos \theta$ only.

Solution Let us start with form 3 of Table 16-2, where

$$\cos (\theta + \phi) = \cos \theta \cos \phi - \sin \theta \sin \phi. \tag{20}$$

In (20), substitute 2θ for ϕ, obtaining

$$\cos 3\theta = \cos \theta \cos 2\theta - \sin \theta \sin 2\theta. \tag{21}$$

Then substituting

$$\cos 2\theta = \cos^2 \theta - \sin^2 \theta$$

and

$$\sin 2\theta = 2 \sin \theta \cos \theta$$

into (21), we have

$$\cos 3\theta = \cos \theta(\cos^2 \theta - \sin^2 \theta) - \sin \theta(2 \sin \theta \cos \theta). \tag{22}$$

Simplifying (22) gives

$$\cos 3\theta = \cos^3 \theta - 3 \sin^2 \theta \cos \theta. \tag{23}$$

Substituting the identity $\sin^2 \theta = 1 - \cos^2 \theta$ into (23), we obtain

$$\cos 3\theta = 4 \cos^3 \theta - 3 \cos \theta.$$

Half-angle formulas follow from double-angle formulas, which, in turn, follow from forms 5 and 6 of Table 16-2. Thus from form 5

$$\cos \theta \cos \phi = \frac{1}{2} [\cos (\theta + \phi) + \cos (\theta - \phi)].$$

Let $\phi = \theta$ and

$$\cos \theta \cos \theta = \frac{1}{2} [\cos (\theta + \theta) + \cos 0]$$

from which

$$\cos^2 \theta = \frac{1}{2}(1 + \cos 2\theta). \tag{24}$$

If we let $\phi = 2\theta$, or $\phi/2 = \theta$, in (24), then

$$\cos^2 \frac{\phi}{2} = \frac{1}{2}(1 + \cos \phi),$$

which, when solved for $\cos \phi/2$ by taking the square root of both sides, becomes

$$\cos \frac{\phi}{2} = \sqrt{\frac{1 + \cos \phi}{2}},$$

which is comparable to form 13 of Table 16-2.

Example 4 Express $\sin^4 \theta$ in terms of first powers of functions of multiples of θ.

Solution From form 12 of Table 16-2

$$\sin^4 \theta = (\sin^2 \theta)^2 = \left[\frac{1}{2}(1 - \cos 2\theta)\right]^2$$

$$= \frac{1}{4} - \frac{1}{2} \cos 2\theta + \frac{1}{4} \cos^2 2\theta. \tag{25}$$

Since we wish to express $\sin^4 \theta$ in terms of first powers of functions of multiple angles, we must convert $\cos^2 2\theta$. Using form 13 and replacing $\theta/2$ by 2θ, we have

$$\cos^2 2\theta = \frac{1 + \cos 4\theta}{2}. \tag{26}$$

Substituting (26) into (25) and simplifying, we obtain

$$\sin^4 \theta = \frac{1}{4} - \frac{1}{2} \cos 2\theta + \frac{\frac{1}{4}(1 + \cos 4\theta)}{2}$$

$$= \frac{3}{8} - \frac{1}{2} \cos 2\theta + \frac{1}{8} \cos 4\theta.$$

Exercises 16-3

1. Using form 3 of Table 16-2, derive form 10.
2. Using form 6 of Table 16-2, derive form 12.
3. Using form 6 of Table 16-2, derive form 14.

In Exercises 4–8, convert the forms $\cos n\theta$ and $\sin n\theta$ to forms involving only powers of functions of the single angle θ.

4. $\sin 3\theta$
5. $\sin 5\theta$
6. $\cos 4\theta$
7. $\sin 4\theta$
8. $\cos 5\theta$
9. In Fig. 16-6 we are given the graphs of

$$y_1 = a \sin \theta, \qquad y_2 = b \cos \theta,$$
$$y = y_1 + y_2 = a \sin \theta + b \cos \theta.$$

If y is written in the form

$$y = \sqrt{a^2 + b^2} \cos \phi \sin \theta + \sqrt{a^2 + b^2} \sin \phi \cos \theta,$$

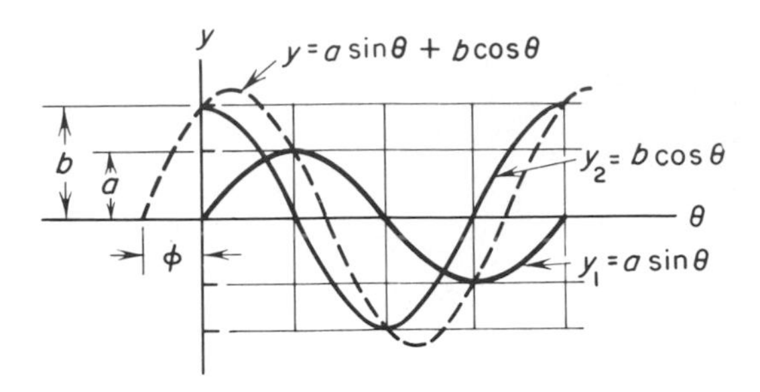

Figure 16-6

where $a = \sqrt{a^2 + b^2} \cos \phi$ and $b = \sqrt{a^2 + b^2} \sin \phi$ (see Fig. 16-7), show that y can be written

$$y = \sqrt{a^2 + b^2} \sin (\theta + \phi),$$

where $\phi = \arctan b/a$. Also, interpret $\sqrt{a^2 + b^2}$ and ϕ as properties of the graph of y.

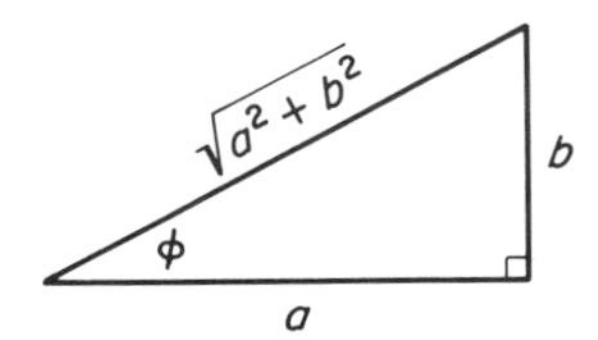

Figure 16-7

10. Referring to Exercise 9, a certain composite wave consists of components $y_1 = 3 \sin \omega t$, $y_2 = 4 \cos \omega t$. Determine the amplitude and the phase shift of the composite. What is the equation of the composite?

11. Using form 13 of Table 16-2 and the fact that $\cos 30° = \sqrt{3}/2$, show that $\cos 15° = 0.965$.

12. Repeating the method used in Exercise 11, evaluate $\cos 3.75°$.

13. Referring to Exercise 9, we are given that a certain composite wave is of form $y = 3 \sin (x + 30°)$. Show that this wave is made up of components $y_1 = a \sin x$, $y_2 = b \cos x$, where $a = 3\sqrt{3}/2$ and $b = 3/2$.

14. Using appropriate forms from Table 16-2, show that

$$\sin \phi = \cos (90° - \phi).$$

16-4 Trigonometric equations

In this section we describe two types of equations involving trigonometric functions. The first type contains only trigonometric expressions and constants; the second type contains trigonometric expressions and others that may be algebraic or logarithmic or exponential in nature.

For equations of the first type, we suggest this process of solution. Convert all trigonometric functions to the same function by way of identities and then solve the resulting equation by suitable means; that is, apply the quadratic formula for equations that are quadratic in the trigonometric function or use approximate methods where no formula is available.

Example 5 Solve for ϕ if

$$3 \tan^2 \phi(1 - \sin \phi) + 1 - 3 \sin \phi = 0. \tag{27}$$

Solution First, note that (27) contains two trigonometric functions—$\tan \phi$ and $\sin \phi$. Let us rewrite (27) in terms of one trigonometric function only. It is often expedient to obtain the expression in terms of $\sin \phi$. Since

$$\tan \phi = \frac{\sin \phi}{\cos \phi},$$

then (27) can be written

$$3\left(\frac{\sin^2 \phi}{\cos^2 \phi}\right)(1 - \sin \phi) + 1 - 3 \sin \phi = 0. \tag{28}$$

Clearing fractions in (28), we have

$$3 \sin^2 \phi - 3 \sin^3 \phi + \cos^2 \phi - 3 \sin \phi \cos^2 \phi = 0. \tag{29}$$

Then substituting

$$\cos^2 \phi = 1 - \sin^2 \phi$$

into (29) and simplifying, we obtain

$$2 \sin^2 \phi - 3 \sin \phi + 1 = 0. \tag{30}$$

Inspecting (30), we see that only one trigonometric function is present and that the equation is a quadratic in $\sin \phi$. We can factor (30) as

$$(2 \sin \phi - 1)(\sin \phi - 1) = 0$$

from which

$$2 \sin \phi - 1 = 0 \quad \text{and} \quad \sin \phi = \frac{1}{2} \tag{31}$$

or

$$\sin \phi - 1 = 0 \quad \text{and} \quad \sin \phi = 1. \tag{32}$$

Inspecting (31) and (32), we see that many solutions for ϕ are possible; we will choose to accept only solutions for $0° \leqq \phi < 360°$ so that from (31)

$$\phi = 30°, 150°,$$

and from (32)

$$\phi = 90°.$$

A word of caution: Not all answers found in Example 5 need be solutions of (27). We can check the validity of the answers by substituting $\phi = 30°, 90°, 150°$ into (27) to determine whether they satisfy (27). Upon substituting, we see that only $\phi = 30°$, $150°$ are satisfactory and that $\phi = 90°$ fails and thus is not acceptable. This is a case of an extraneous root.

In Eq. (30) we might choose to substitute for $\sin\phi$ by making some choice like $x = \sin\phi$ so that (30) can be written

$$2x^2 - 3x + 1 = 0$$

from which

$$x = \sin\phi = \frac{1}{2}, 1.$$

This substitution is useful if working with trigonometric functions is confusing or complicated.

Example 6 Given right triangle ABC in Fig. 16-8, find ϕ if

$$\sin\phi + \cos\phi = \frac{\sqrt{5}}{2}. \tag{33}$$

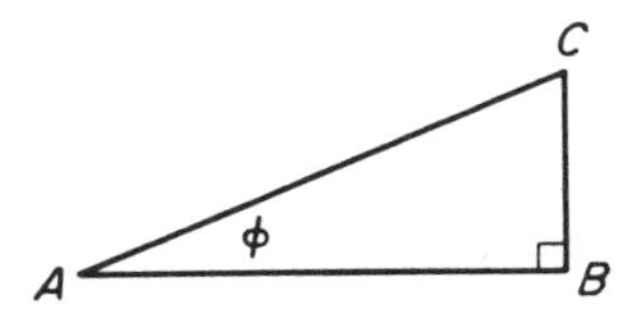

Figure 16-8

Solution 1 We cannot choose a single identity from our tables of identities that will enable us to convert directly to an expression in only one trigonometric function. If we square both sides of (33), however, we have

$$\sin^2\phi + 2\sin\phi\cos\phi + \cos^2\phi = \frac{5}{4},$$

but

$$\sin^2\phi + \cos^2\phi = 1$$

so that now

$$2\sin\phi\cos\phi + 1 = \frac{5}{4}$$

or

$$2\sin\phi\cos\phi = \frac{1}{4}. \tag{34}$$

From form 9, Table 16-2,

$$2\sin\phi\cos\phi = \sin 2\phi,$$

and (34) becomes

$$\sin 2\phi = \frac{1}{4}$$

from which

$$2\phi = 14.5°, 165.5°$$

and

$$\phi = 7.25°, 82.75°.$$

Solution 2 From Exercise 9 in Exercises 16-3 we have

$$\sin \phi + \cos \phi = \sqrt{2} \sin (\phi + 45°). \tag{35}$$

Comparing (35) and (33),

$$\sqrt{2} \sin (\phi + 45°) = \frac{\sqrt{5}}{2}$$

from which

$$\sin (\phi + 45°) = \frac{\sqrt{5}}{2\sqrt{2}} = \frac{\sqrt{10}}{4} = 0.790$$

and

$$\phi + 45° = 52.25°, 127.75°$$

or

$$\phi = 7.25°, 82.75°.$$

Occasionally certain equations present a mixture of trigonometric expressions with algebraic or logarithmic or exponential expressions. In these cases, the solution usually depends on graphical or approximate methods. Graphical methods provide accuracy to a degree that depends on the care and scale used. Approximate methods that can be used are similar to those applied to polynomial equations in Chapter 8.

Example 7 Solve for x if

$$x = \tan x \tag{36}$$

where x is given in radians.

Graphical Solution We choose to graph on the same plane the equations

$$y_1 = x \qquad y_2 = \tan x,$$

where the graphs are shown in Fig. 16-9. Inspection of the graphs shows that where $y_1 = y_2$, we also have $x = \tan x$ and we have a solution. The graph suggests that there is an infinite number of solutions, where each intersection of the straight line and tangent curve constitutes a solution. Let us refine one of the solutions to the nearest hundredth; we will choose the solution that is nearest $x = 3\pi/2 = 4.71$ radians.

Solution by the Method of Linear Interpolation Restating the problem, we wish to find the solution, to the nearest hundredth, of

$$x = \tan x,$$

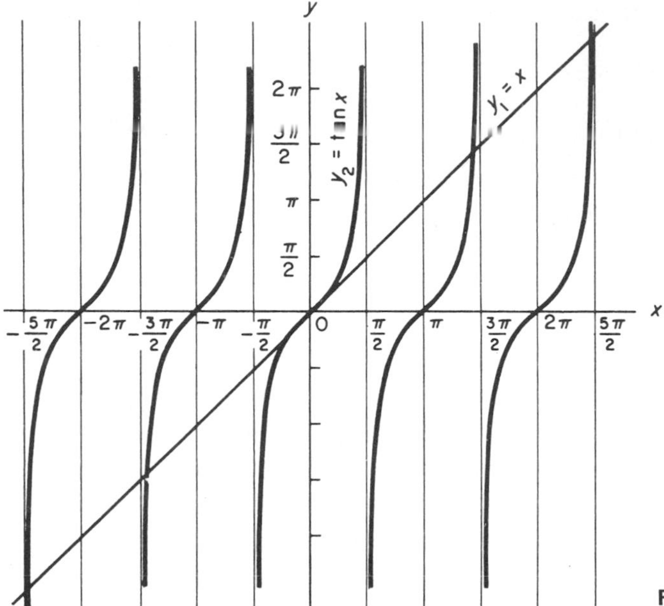

Figure 16-9

where the particular solution chosen is at approximately $x = 4.6$. First, we let

$$y = -x + \tan x$$

from which $x = \tan x$ when $y = 0$. Consider a chart of values of y for values of x chosen near $x = 4.6$ radians; from it we see that there seems

x (radians)	4.7	4.6	4.5	4.4
$\tan x$	80.6	8.87	4.71	3.10
y	+75.9	+4.27	+0.21	−1.3

to be a root for $4.4 < x < 4.5$ because of the change in the sign of y as x increases from 4.4 to 4.5. Moreover, it seems that the root in question appears to favor the 4.5 end of the interval. Refer now to Fig. 16-10, which plots points (4.5, 0.21) and (4.4, −1.3) and provides a basis for interpolation.

By equating ratios of certain corresponding parts of the similar triangles in Fig. 16-10, we have $h \approx 0.08$. We create a new chart of values, with the new estimate of the root being

$$4.4 + h = 4.4 + 0.08 = 4.48.$$

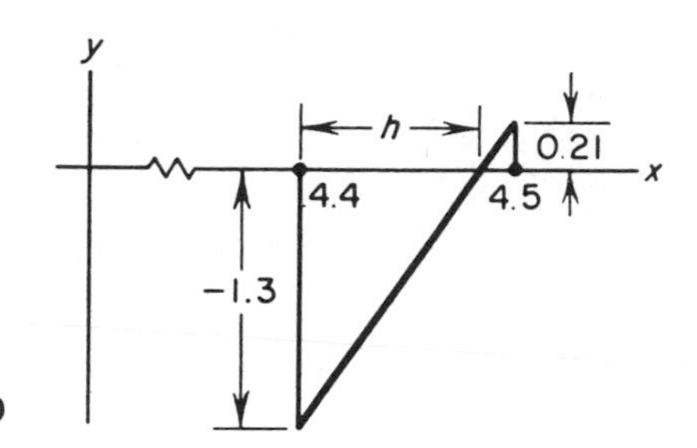

Figure 16-10

From the chart we see a sign change in y as x changes from $x = 4.49$ to

x (radians)	4.48	4.49	4.50
$\tan x$	4.225	4.425	4.637
y	-0.255	-0.065	$+0.137$

$x = 4.50$. Careful inspection shows that the root probably favors the 4.49 end of the interval, with the desired solution now $x = 4.49$.

In summary, we suggest that approximate solutions be obtained by first graphing the equation to get an approximation, then applying the method of linear interpolation.

Exercises 16-4

In Exercises 1–10, solve for the angle. In all cases, an infinite number of solutions is possible; list only those for which the angle is between 0 and 360°.

1. $3 \cos^2 \phi - 2 \cos \phi - 1 = 0$
2. $2 \cos^2 \phi - 3 \cos \phi + 1 = 0$
3. $2 \sin^2 \alpha + 5 \sin \alpha + 2 = 0$
4. $5 \sin^2 \alpha + 4 \sin \alpha - 1 = 0$
5. $4 \tan^2 \tau + 4 \tan \tau + 1 = 0$
6. $\tan^2 \tau - 3 \tan \tau - 4 = 0$
7. $\sin^3 \beta + 2 \sin^2 \beta - \sin \beta = 2$
8. $\cos^3 \beta + 3 \cos^2 \beta - \cos \beta - 3 = 0$
9. $4 \sin^2 3\theta - 1 = 0$
10. $\tan^2 5\theta - 1 = 0$

11. Explain why the expressions given have no valid or real solutions.
(a) $\sin^2 \phi + 2 \sin \phi - 8 = 0$
(b) $\sin^2 \phi + 2 \sin \phi + 8 = 0$
(c) $\sin^3 \phi - 8 = 0$

12. For what two values of α ($0° \leqq \alpha \leqq 180°$) does $\sin \alpha$ exceed $\cos \alpha$ by unity?

13. For what range of values of k in

$$\sin \alpha = k + \cos \alpha$$

is $\sin \alpha$ or $\cos \alpha$ imaginary?

In Exercises 14–19, obtain the solutions of the given equations graphically. Use the method of linear interpolation to obtain the root shown as an answer; where no answer is shown, find the least positive answer.

14. $\tan x = -x$ **15.** $\sin x = 2x$ **16.** $\cos x = -\frac{1}{3}x$

17. $\sin x = e^{-x}$ **18.** $\tan x = e^x$ **19.** $\tan x = \dfrac{1}{x}$

20. In Fig. 16-11 we are given a circle of radius 3 with a chord AB. What central angle ϕ will intercept an arc ACB such that the sum of the lengths of chord AB and arc ACB equals one radius?

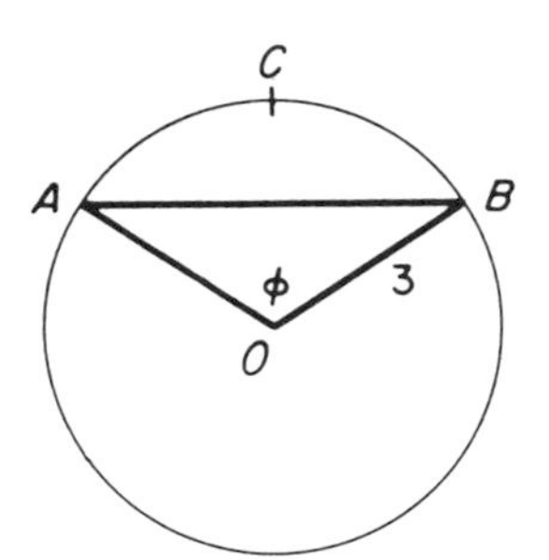

Figure 16-11

21. A cable that sags, owing to its own weight only, is in the shape of a catenary whose equation can be written

$$y = \frac{e^x - e^{-x}}{2}.$$

For what value of x does $y = 2.1$?

22. The circular segment shown in Fig. 16-12(a) has a moment of inertia I_y about the y axis expressed by equation

$$I_y = \frac{r^4}{4}(\alpha + \sin \alpha \cos \alpha).$$

Find α for a sector that is such that $r = 4$ and $I_y = 101$.

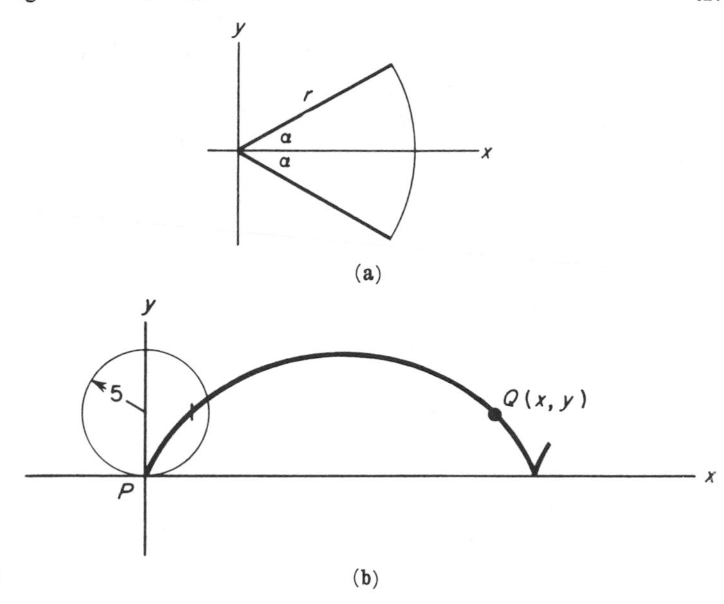

Figure 16-12

23. A curve traced by a point on the circumference of a circle that rolls along a fixed straight line without slipping is called a *cycloid.* The equation of the cycloid shown in Fig. 16-12(b) is given as

$$x = 5(\phi - \sin\phi) \qquad y = 5(1 - \cos\phi),$$

where x is the x coordinate of the general point Q, y is the y coordinate of Q, and ϕ is the angle through which the circle has turned. Find y and ϕ when $x = 2$.

24. An equation describing an oscillating wave of decreasing amplitude may be given as

$$y = e^{-x}\sin x.$$

When $y = 0.274$, x has a value such that $0.25 < x < 0.50$. Find x to the nearest hundredth.

16-5 Inverse trigonometric functions

In earlier chapters we referred to such expressions as $y = x^2$, $y = e^x$, and $y = \sin x$. Each is explicit, with the dependent variable being y; each can be solved for x, giving $x = \sqrt{y}$, $x = \ln y$, and $x = \arcsin y$, respectively. Our interest here is on the last expression; if $y = \sin x$, then $x = \arcsin y$, where the latter is read "x is an angle whose sine is y."

Suppose that we are asked: Given $\sin x = \frac{1}{2}$, what is the value of x? Our first impulse is to respond that $x = 30°$. It is granted that $x = 30°$ is *one* angle whose sine is $\frac{1}{2}$, but there are many others, such as 150°, 30° ± 360°, 150° ± 360°, 30° ± (2)(360°), 150° ± (2)(360°), and any angle that differs from 30° or 150° by an integral multiple of 360°. So in more compact form, we can say that when $\sin x = \frac{1}{2}$, then

$$x = 30^\circ \pm n(360^\circ) \quad \text{and} \quad x = 150^\circ \pm n(360^\circ),$$

where n is an integer. This fact is shown graphically in Fig. 16-13, where $y = \sin x$ and $y = \frac{1}{2}$ are graphed; the intersections of $y = \frac{1}{2}$ and $y = \sin x$ show some of the values of x for which $\sin x = \frac{1}{2}$.

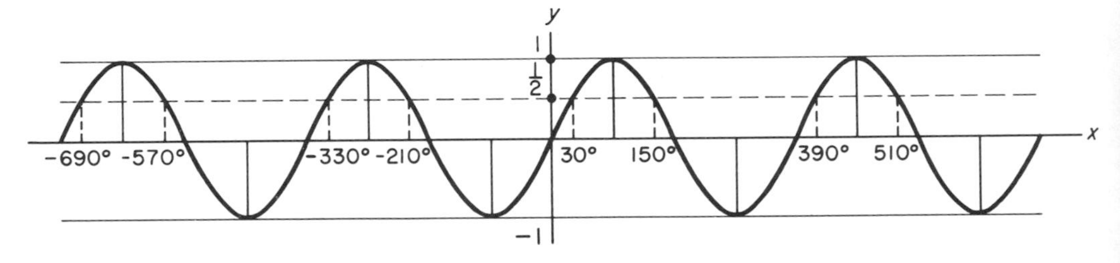

Figure 16-13

We call $x = \arcsin y$ the *inverse function* of $y = \sin x$ within the framework of certain restrictions. These restrictions concern the fact that $y = \sin x$ is a single-valued function for a given value of x, but $x = \arcsin y$ is not single valued for a given value of y. We will set aside these restrictions and accept as solutions of $x = \arcsin y$ those values of x such that $0^\circ \leqq x \leqq 360^\circ$.

Example 8 Find ϕ if $\phi = \arctan 1$.

Solution The given statement asks us to "find the angles (for $0^\circ \leqq \phi \leqq 360^\circ$) that are such that $\tan \phi = 1$." From our knowledge of the tangent variation we have solutions $\phi = 45^\circ, 225^\circ$.

Example 9 Find the value of cos (arcsin 1).

Solution Starting inside the parentheses,

$$\arcsin 1 = 90^\circ$$

so that

$$\cos(\arcsin 1) = \cos(90^\circ) = 0.$$

Example 10 If $\theta = \operatorname{arcsec} a/b$, find $\tan \theta$ in terms of a and b.

Solution By definition,

$$\sec \theta = \frac{a}{b} = \frac{\text{hypotenuse}}{\text{horizontal component}}.$$

Then from Fig. 16-14 we place a on the hypotenuse and b on the horizontal leg, requiring that the remaining leg be $\sqrt{a^2 - b^2}$. Resorting to the definition of the tangent function,

$$\tan \theta = \frac{\sqrt{a^2 - b^2}}{b}.$$

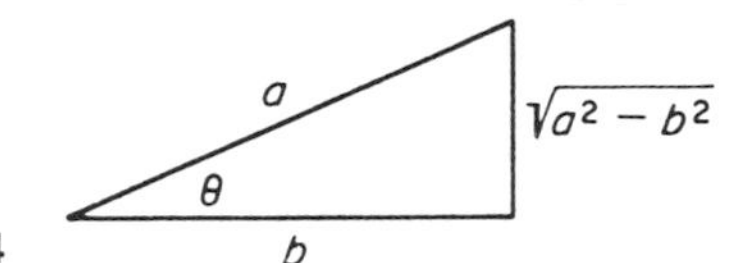

Figure 16-14

Exercises 16-5

In Exercises 1–10, find the values of the angle ϕ, where ϕ is between 0 and 360°, that are solutions of the given inverse functions.

1. $\phi = \arcsin\left(\frac{\sqrt{3}}{2}\right)$

2. $\phi = \arccos \frac{1}{2}$

3. $\phi = \arctan(-1)$

4. $\phi = \text{arccsc} \frac{2}{\sqrt{2}}$

5. $\phi = \text{arcsec}\left(\frac{-2}{\sqrt{3}}\right)$

6. $\phi = \text{arcctn } 1$

7. $3\phi = \arccos 0.93358$

8. $5\phi = \text{arcctn}(-3)$

9. $\phi = 30° + 3 \arctan 1$

10. $\phi = 50° + \frac{1}{2} \arcsin(-1)$

11. Find the value of x that satisfies equation

$$\arctan x = \arccos 2x.$$

12. Find the value of
(a) sin (arcsin 1).
(b) sec (arccos 0.5).
(c) tan (arcsin 0.8660).
(d) ctn (arcsec 2).

13. Given that $y_1 = \arcsin k$, $y_2 = \arccos 2k$, show that y_1 and y_2 are related by equation

$$2 \sin y_1 = \cos y_2.$$

Additional exercises 16-5

1. A vertical pole PE, shown in Fig. 16-15, is supported by four guy wires. The pole is sunk into terrain that slopes at 15°. Find the lengths of the wires PA, PB, PC, and PD if A, B, C, and D are located 20, 40, 20, and 30 ft, respectively, from E and the pole PE is 50 ft high.

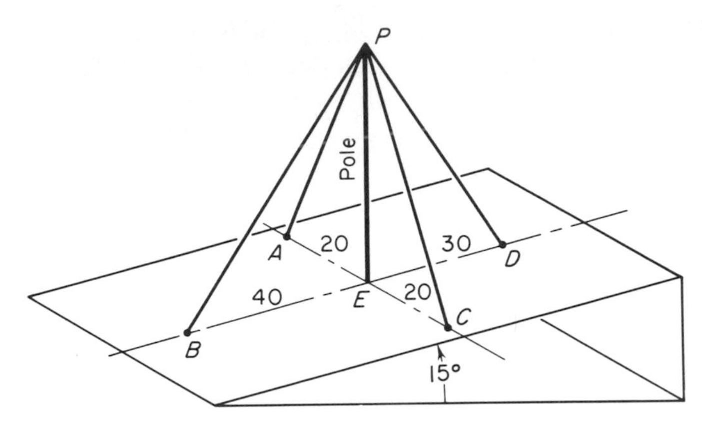

Figure 16-15

2. Figure 16-16 shows a rough sketch of a speedometer dial. Lay out calibrations along a 12-in. scale, AB, showing markings for 0, 10, 20, 30, . . . , 120 mph. Assume that the speedometer needle will sweep through an angle AOB, where $\angle AOB = 120°$.

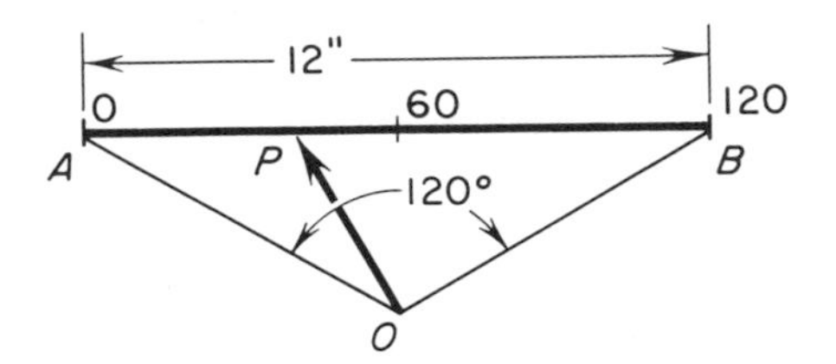

Figure 16-16

3. A piston rod AB is connected to the rim of a wheel of radius one foot as shown in Fig. 16-17. The wheel rotates at 60 rps and the rod is 4 ft long. Show that the displacement x of the piston (point B) from the center of the wheel at any

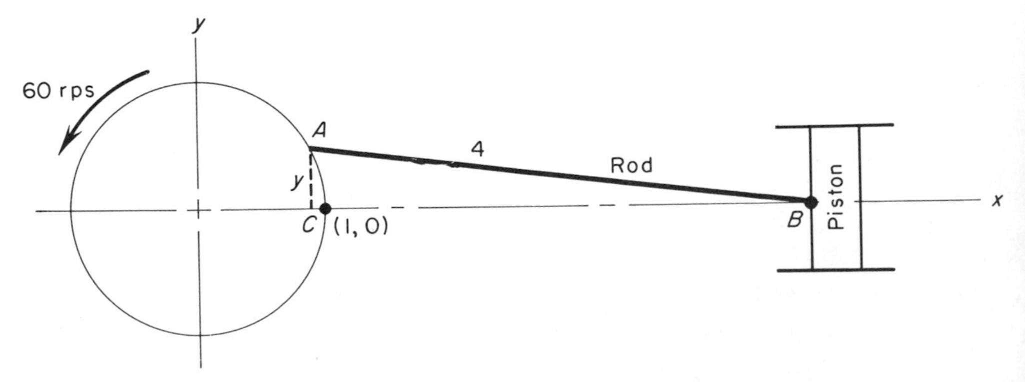

Figure 16-17

time t is

$$x = \sqrt{16 - \sin^2 120\pi t} + \cos 120\pi t,$$

assuming that $t = 0$ when A is at (1, 0). Find x when $t = 1/720$ sec.

4. A one-inch ball is inserted into a V-shaped slot as shown in Fig. 16-18. What is the radius x of another ball that will touch the smaller ball and the sides of the slot if the slot angle is 30°?

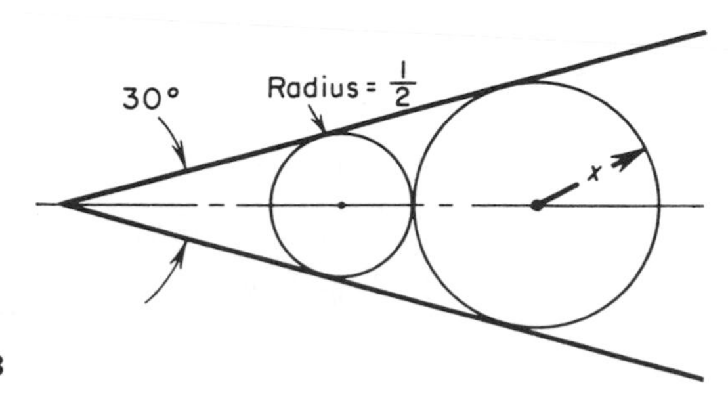

Figure 16-18

5. A gunner at A in Fig. 16-19 fires at an aircraft moving at 500 ft/sec that is 500 yd away at the closest point B. If the gunner has a weapon that fires at 2700 ft/sec and he fires when the aircraft is at the closest point, by how far will he miss the point on the plane at which he aims?

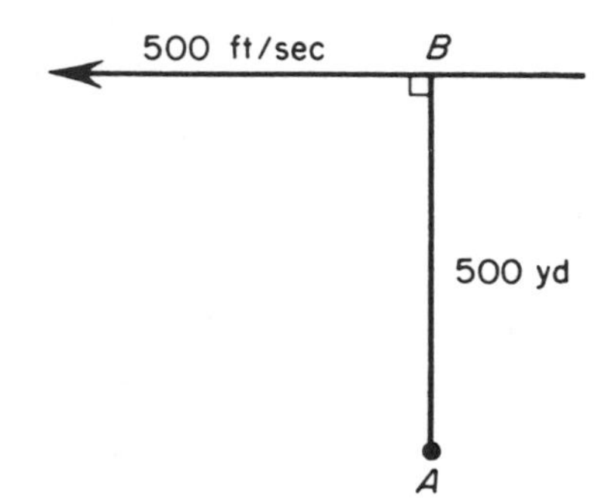

Figure 16-19

6. Given $i = I_{max} \sin(2\pi ft + \phi)$, find t if $I_{max} = 120$, $i = 100$, $f = 60$, and $\phi = 30°$.

7. A ball thrown from the ground with initial velocity v_0 at an angle ϕ with the ground (see Fig. 16-20) travels a path according to equation

$$y = x \tan\phi - \frac{16x^2}{v_0^2} \sec^2\phi,$$

where y and x are the vertical and horizontal displacements, respectively, of the ball from the launch point. At what two angles can the ball be thrown if it is released at initial velocity 100 ft/sec and strikes the earth 200 ft from the point of launch?

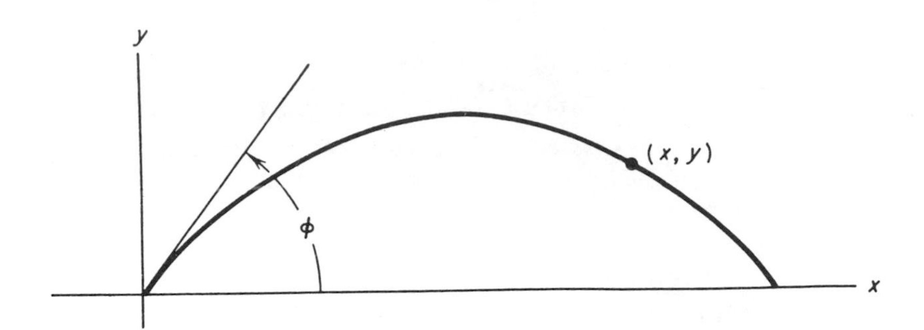

Figure 16-20

8. An automobile travels along a level surface at 60 mph. A point P is located on a wheel of the vehicle 10 in. from the center of the wheel. If P is directly below the axle when $t = 0$ sec, how fast is the point moving upward when $t = 3$ sec?

9. Given equation

$$45 \sin \phi = 4 + 180\left(\phi - \frac{\pi}{6}\right),$$

find, by approximate methods, the value of ϕ that satisfies it.

10. From the expression $y = \sin \phi$ we obtain the differential

$$dy = \sin \phi \, d\phi,$$

where $d\phi$ is a change in ϕ and dy is an approximation of the corresponding change in y when $d\phi$ is small. Find dy when $\phi = 30°$ and $d\phi = 1°$. Compare the value of $y + dy$ to the value of sin 31° obtained in a table of values of trigonometric functions.

11. Give all the real roots of equation

$$\tan^3 \phi - 4 \tan^2 \phi - 3 \tan \phi + 1 = 0,$$

where $0° < \phi < 360°$.

17

Binomial expansion and progressions

In this chapter the binomial expansion and certain progressions are described. The binomial expansion is useful in such areas as statistics, probability, finance, and calculus. Progressions have numerous applications in finance, mechanics, number studies, and other areas. Considerations of progressions are readily expanded into series studies in which series are important in approximate solutions, differential equations, wave analysis, and similar areas.

17-1 Binomial expansion; the general term

Given binomial $a + b$, we can choose to raise the binomial to a variety of powers. These powers may have integral, fractional, positive, or negative exponents in the more common applications of the binomial expansion. Let us first consider positive integral (and zero) values of the exponents. A few expansions are listed.

$$
\begin{aligned}
(a+b)^0 &= 1, \\
(a+b)^1 &= a+b, \\
(a+b)^2 &= a^2+2ab+b^2, \\
(a+b)^3 &= a^3+3a^2b+3ab^2+b^3, \\
(a+b)^4 &= a^4+4a^3b+6a^2b^2+4ab^3+b^4.
\end{aligned} \tag{1}
$$

The coefficients of the expansions in (1) can be arranged in *Pascal's triangle*, where any number in the triangle can be obtained by adding the nearest two numbers above it.

$$
\begin{array}{ccccccccccc}
 & & & & & 1 & & & & & \\
 & & & & 1 & & 1 & & & & \\
 & & & 1 & & 2 & & 1 & & & \\
 & & 1 & & 3 & & 3 & & 1 & & \\
 & 1 & & 4 & & 6 & & 4 & & 1 & \\
1 & & 5 & & 10 & & 10 & & 5 & & 1
\end{array}
$$

Pascal's triangle can be used to determine the coefficients of the expansion of a binomial. However, a superior method is through the general expansion of the binomial, which is given here without proof.

$$
(a+b)^n = a^n + na^{n-1}b + \frac{n(n-1)a^{n-2}b^2}{2!} + \frac{n(n-1)(n-2)a^{n-3}b^3}{3!} + \cdots \tag{2}
$$

Here n can be a variety of number types, including negative and positive integers, fractions, and zero. The symbol (!) in (2) may be unfamiliar. We define $k!$ (k factorial) as the product of all positive integers from 1 through k, or

$$
k! = 1 \cdot 2 \cdot 3 \cdot 4 \cdot \cdots \cdot k.
$$

We also make the definition $0! = 1$.

Example 1 Using (2), expand the expression

$$
\left(x^2 - \frac{2}{x}\right)^4. \tag{3}
$$

Solution Comparing (3) to the left member of (2), we have $a = x^2$, $b = -2/x$, and $n = 4$. Substituting them into the right member of (2) gives

$$\left(x^2 - \frac{2}{x}\right)^4 = (x^2)^4 + 4(x^2)^3\left(-\frac{2}{x}\right) + \frac{(4)(3)(x^2)^2}{2!}\left(-\frac{2}{x}\right)^2 + \frac{(4)(3)(2)(x^2)^1}{3!}\left(-\frac{2}{x}\right)^3 + \frac{(4)(3)(2)(1)(x^2)^0}{4!}\left(-\frac{2}{x}\right)^4,$$

which simplifies to the expression

$$x^8 - 8x^5 + 24x^2 - \frac{32}{x} + \frac{16}{x^4}.$$

Certain useful observations of the properties of (2) are in order. Some of them are seen more readily through expressions (1) and Pascal's triangle; all are made assuming that n is a positive integer.

1. The expression contains $n + 1$ terms.
2. The sum of the exponents of a and b in any term is n.
3. The coefficients are symmetrical; that is, the coefficients of the first and last terms are the same, the coefficients of the terms numbered 2 from the ends are the same, and so on.
4. The exponent of b agrees with the factorial indicated in the same term.

The general term of (2) can be derived intuitively by numbering the terms of (2) and comparing the term numbers with certain portions of the expression for that term. The observations of the properties of the expansion listed above are also useful in the derivation. Numbering the terms from left to right in (2) as term 1, term 2, term 3, and so on, we note that

1. Term $k + 1$ contains $k!$ in its denominator.
2. Term $k + 1$ contains an exponent for b equal to k.
3. Since the sum of the exponents in any term is n, then term $k + 1$ contains an exponent $n - k$, in view of the exponent of b shown in (2).
4. The factors of the numerator of the coefficient are such that the least factor is one greater than the exponent of a.

From these observations we see that *the* $(k + 1)$*st term of the expansion* (2) *is*

$$\frac{n(n-1)(n-2)\cdots(n-k+1)}{k!}a^{n-k}b^k. \tag{4}$$

Example 2 Find the fourth term of the expansion

$$\left(\frac{y}{3} - \frac{6}{y^2}\right)^{13}.$$

Solution Comparing the given expression to (2) and then referring to (4), we have $n = 13$, $k + 1 = 4$ or $k = 3$, $a = y/3$, and $b = -6/y^2$. Substituting into (4), we have the fourth term of the given expansion as

$$\frac{13 \cdot 12 \cdot 11}{3!}\left(\frac{y}{3}\right)^{10}\left(-\frac{6}{y^2}\right)^3 = -\frac{2288}{2187}y^4.$$

Exercises 17-1

In Exercises 1–5, show in simplified form all terms of the given expansion.

1. $(x^2 - y)^3$ **2.** $\left(m - \frac{1}{n}\right)^5$ **3.** $\left(a^2 - \frac{2}{b}\right)^4$

4. $(x + \Delta x)^3$ **5.** $\left(1 + \frac{1}{x}\right)^5$

In Exercises 6–10, give the term requested.

6. Fifth term of $\left(\frac{3}{x} - y^2\right)^7$

7. Fourth term of $\left(\frac{2}{y} + \frac{y}{4}\right)^8$

8. Seventh term of $\left(3 + \frac{m}{2}\right)^6$

9. Third term of $(x + \Delta x)^5$

10. Third term of $\left(m^2 - \frac{1}{m^4}\right)^6$

11. Which term of $[x^2 - (1/x)]^{12}$ contains no x?

12. Which term of $[x^3 + (3/x^2)]^{15}$ contains no x?

13. Which term of $[m - (2/m^3)]^{11}$ contains m^{-9}?

14. How many terms are in $(a^2 - b)^{23}$?

15. How many terms are in $(a^2 - b)^{14}$?

16. Evaluate $(1 - \frac{1}{2})^4$ by the binomial expansion.

17. Expand $[(a - b) - (c - d)]^3$ by the binomial expansion.

18. How many terms of $(x + \Delta x)^n$ will contain a power of Δx?

19. Evaluate the expression $(1 + 1/x)^x$, where $x = 2, 5, 10$. Use the binomial expansion.

17-2 More on the binomial expansion; roots

In Sec. 17-1 discussion of the binomial expansion was restricted to those expansions involving positive integral values of the exponent. Here we expand our discussion to include other values of the exponent.

Given expression (2), if n is either a negative integer or a fraction, we assert that the following properties are true.

1. The number of terms in the expansion is infinite.
2. The array of coefficients is not symmetrical.
3. Expression (4) is still valid for the $(k + 1)$st term.

Let us show an expansion of (2) in which n is not a positive integer.

Example 3 Show the first four terms of the expansion of

$$\sqrt[3]{x^2 - 9}. \tag{5}$$

Solution We can write (5) in the power form $(x^2 - 9)^{1/3}$ and compare it to the left member of (2), from which we have $a = x^2$, $b = -9$, and $n = \frac{1}{3}$. Substituting the quantities into the right member of (2), we have

$$(x^2 - 9)^{1/3} = (x^2)^{1/3} + \left(\frac{1}{3}\right)(x^2)^{-2/3}(-9) + \frac{(\frac{1}{3})(-\frac{2}{3})(x^2)^{-5/3}}{2!}(-9)^2$$
$$+ \frac{(\frac{1}{3})(-\frac{2}{3})(-\frac{5}{3})}{3!}(x^2)^{-8/3}(-9)^3 + \cdots,$$

where the right-hand member simplifies to

$$x^{2/3} - 3x^{-4/3} - 9x^{-10/3} - 45x^{-16/3} - \cdots. \tag{6}$$

The binomial expansion can be used to compute roots. In Example 3, with x assigned a numerical value, (6) can be used to obtain the cube root of $x^2 - 9$. Other roots can be obtained with any degree of accuracy imposed.

Example 4 Evaluate $\sqrt{24}$ correct to thousandths, using the binomial expansion.

Solution Note that the number 24 can be expressed as the sum or difference of many number pairs; this sum or difference would establish the number 24 as a binomial. Some such binomials include $13 + 11$, $30 - 6$, $25 - 1$, and $16 + 8$. We will select (for reasons that will become apparent when the expansion is displayed) a binomial consisting in part of the perfect square closest to 24—namely, 25. (If the problem had requested the *cube* root of 24, we would select as a part of the binomial the perfect *cube* closest to 24.) Now we have

$$\sqrt{24} = (25 - 1)^{1/2} = (25)^{1/2} + \frac{1}{2}(25)^{-1/2}(-1)$$
$$+ \frac{\frac{1}{2}(-\frac{1}{2})}{2!}(25)^{-3/2}(-1)^2 + \frac{\frac{1}{2}(-\frac{1}{2})(-\frac{3}{2})}{3!}(25)^{-5/2}(-1)^3 + \cdots$$
$$= 5 - \frac{1}{10} - \frac{1}{1000} - \frac{1}{50,000} - \cdots$$
$$= 5.00000 - 0.10000 - 0.00100 - 0.00002 - \cdots$$
$$= 4.89898. \tag{7}$$

Since the problem requested accuracy to thousandths, we have the answer $\sqrt{24} = 4.899$. Certain factors regarding the selection of the number of terms to which the binomial is expanded should be mentioned here; the number of terms depends on the rate at which the successive terms are converging on a zero value. In (7) the fourth term is already too small to affect the answer as given to thousandths. The third term might affect the thousandths place through carrying operations or borrowing operations. Convergence, then, is the key to the choice; rate of convergence is determined by the choice of the binomial representing the number 24—namely, $25 - 1$. Some other choice, such as $49 - 25$ or $-11 + 36$, may converge painfully slowly or may even diverge.

Exercises 17-2

In Exercises 1–8, show the first four terms of the expansion of the given binomial.

1. $(x + 1)^{-1}$ **2.** $(1 + x)^{-1}$ **3.** $(4m - n)^{-1/2}$

4. $\left(1 + \frac{i}{n}\right)^{nt}$ **5.** $(x + \Delta x)^n$ **6.** $(x^2 - \frac{1}{2})^{1/2}$

7. $(4 + 3x)^{1/2}$ **8.** $(y^2 - 2)^{-1/2}$

9. The number e is defined as

$$e = \lim_{x\to\infty} \left(1 + \frac{1}{x}\right)^x = 2.718 \ldots .$$

Evaluate the binomial involved with $x = 100$ and $x = 1000$. Use only the first six terms of the expansions.

10. Evaluate $(1.05)^{-4}$ for the first five terms of the expansion. Next, evaluate $(1.05)^4$. Reciprocate the latter result and compare it to the former.

In Exercises 11–22, evaluate the given expression by the binomial expansion, correct to the nearest thousandth.

11. $\sqrt{37}$ **12.** $\sqrt{15}$ **13.** $\sqrt{47}$ **14.** $\sqrt{80}$

15. $\sqrt{120}$ **16.** $\sqrt[3]{7}$ **17.** $\sqrt{\frac{1}{10}}$ **18.** $\sqrt[3]{120}$

19. $(10)^{3/2}$ **20.** $(27)^{-3/2}$ **21.** $(50)^{-3/2}$ **22.** $(31)^{0.2}$

17-3 Arithmetic progressions

A *sequence* is a set of elements that can be arranged according to some order or property. An *arithmetic progression* is a sequence whose consecutive elements differ by a common quantity called the *common difference*. Accordingly, the sequences in (8), (9), and (10) are arithmetic progressions. In (8), (9), and (10) the common differences are

$$4, 6, 8, 10, 12, \ldots \tag{8}$$

$$121, 110, 99, 88, 77, \ldots \tag{9}$$

$$4a + 8b, 5a + 6b, 6a + 4b, 7a + 2b, \ldots \tag{10}$$

$+2$, -11, and $a - 2b$, respectively. The common difference can be obtained by subtracting any term (save the last) in the arithmetic progression from the succeeding term.

Table 17-1

Term number	*Value of term from expression* (9)	*Value of term in general arithmetic progression*
1	121	a
2	$121 + 1(-11)$	$a + d$
3	$121 + 2(-11)$	$a + 2d$
4	$121 + 3(-11)$	$a + 3d$
.	.	.
.	.	.
.	.	.
n	$121 + (n - 1)(-11)$	$a + (n - 1)d$

Arithmetic progressions provide excellent opportunities for formulation. In Table 17-1 we predict the value of the nth (general) term of expression (9). We see that the first term is 121, the second term is 121 plus *one* common difference, the third term is 121 plus *two* common differences, and so forth to the nth term, which is 121 plus $n - 1$ common differences. In the right column of Table 17-1 we see the intuitive development of the formula for the nth term of the general arithmetic progression,

$$l = a + (n - 1)d, \tag{11}$$

where l is the value of the nth term, a is the value of the first term, and d is the common difference.

Example 5 Give the value of the twenty-first term of the A.P.

$$91, 85, 79, 73, 67, \ldots .$$

Solution By inspection, the common difference is -6 and the first term is 91. Referring to (11), we have $n = 21$, $a = 91$, and $d = -6$. Substituting,

$$l = a + (n - 1)d = 91 + (21 - 1)(-6) = 91 - 120 = -29.$$

If we indicate the sum of the separate terms of a sequence, we have a *series*. It is often useful to obtain the sum of n terms of an arithmetic progression. We can obtain the sum by inspection and, by the same inspection method, discover a means of formulating a general expression for the sum. Let us consider the sum of the first 24 terms of the A.P.

$$64, 59, 54, 49, 44, \ldots . \tag{12}$$

First, we obtain the twenty-fourth term of (12) by applying (11):

$$l = 64 + (23)(-5) = 64 - 115 = -51.$$

Listing the first three and the last three terms of (12), we have

+13

$$64, 59, 54, \ldots, -41, -46, -51. \tag{13}$$

+13

+13

We note from (13) that the sum of the first and last terms is $+13$. The sum of the terms numbered two from each end is $+13$, the sum of the terms numbered three from each end is $+13$, and so on. Since there are 24 terms, there are 12 such pairs, each of value $+13$, or the sum of the first 24 terms of (12) is

$$S = 12(+13) = 156.$$

At this point we can generalize. The sum S of an arithmetic series is written here in two ways; in these two ways, the terms are simply reversed. Adding expressions (14) and (15) produces the sum (16).

$$S = a + a + d + a + 2d + \cdots + l - 2d + l - d + l \tag{14}$$

$$S = l + l - d + l - 2d + \cdots + a + 2d + a + d + a \tag{15}$$

$$2S = (a + l) + (a + l) + (a + l) + \cdots + (a + l) + (a + l) + (a + l) \tag{16}$$

Assuming that there are n terms in each (14) and (15), then there are n terms in (16), each of value $a + l$, so that (16) can be written

$$2S = n(a + l)$$

or

$$S = \frac{n}{2}(a + l), \tag{17}$$

where S is the sum of n terms of an A.P. whose first term is a and whose

nth term is l. Here $n/2$ is the number of pairs of terms, and $a + l$ is the value of the sum of the first and nth terms.

Example 6 Find the sum of the first 33 terms of the A.P.

$$-105, -98, -91, -84, \ldots.$$

Solution First, we find the value of the thirty-third term by (11):

$$l = -105 + (33 - 1)(+7) = -105 + 224 = +119.$$

Then apply (17) with $n = 33$, $a = -105$, and $l = 119$ so that

$$S = \frac{33}{2}(-105 + 119) = \frac{33}{2}(14) = 231.$$

In Example 6 some work could have been eliminated if expressions (11) and (17) had previously been combined in such a way that l was deleted. If we substitute (11) into (17), we have another expression for the sum of n terms of an A.P.—namely,

$$S = \frac{n}{2}[a + a + (n - 1)d] = \frac{n}{2}[2a + (n - 1)d]. \tag{18}$$

Example 7 A man takes a job at a quarterly salary of \$4800. At the end of each quarter he is given a raise of \$240. What will be his total earnings for his first 10 quarters on the job?

Solution This will be an A.P. with terms

$$4800, 5040, 5280, 5520, \ldots,$$

where $a = 4800$, $n = 10$, and $d = 240$. Applying (18) to find the total earnings, we have

$$S = \frac{10}{2}[2(4800) + (10 - 1)(240)] = \$58{,}800.$$

Exercises 17-3

In Exercises 1–12, find the value of the term indicated.

1. Twentieth term of 3, 5, 7, 9, . . .
2. Fifty-fifth term of 4, 6, 8, 10, . . .
3. Seventeenth term of 68, 56, 44, 32, . . .
4. Twenty-fourth term of 82, 66, 50, 34, . . .
5. Fourteenth term of $5m + 8n, 6m + 6n, 7m + 4n, \ldots$
6. Fifteenth term of $12a - 21b, 9a - 18b, 6a - 15b, \ldots$

7. Thirty-third term of $-73, -66, -59, -52$
8. Fifteenth term of $-17, -25, -33, -41$
9. Twenty-first term of $-15a + 24, -17a + 19, -19a + 14$
10. Thirteenth term of $-36 - 7x, -30 - 11x, -24 - 15x$
11. Thirty-first term of $-12x^2 + 40, -9x^2 + 32, -6x^2 + 24$
12. Twenty-sixth term of $(24/a) + 13a, (19/a) + 19a, (14/a) + 25a$

In Exercises 13–18, provide values for the missing items. In each case, assume that the given information is appropriate for an arithmetic progression.

	S	n	a	l	d
13.		6	−12	13	
14.		13	27		−3
15.	3612		50		32
16.	6750	15		−900	
17.	0			360	9
18.		46		91	5

19. Show that the sum of the first n terms of an A.P. is zero if the first and last terms are equal but opposite in sign.
20. The initial taxable value of a machine is \$8000. Each year the taxable value depreciates by \$1200. A tax of 5% of the current value of the machine is charged at the end of each year. What total taxes are paid in the first 5 years of the life of the machine?
21. An automobile steadily uses more gasoline per mile because of wear, dirt, and so on. During each 500 miles driven, the automobile uses one more pint of gasoline than it did during the previous 500 miles. The automobile averaged 20 miles per gallon during its first 500 miles. What is its gasoline consumption (in miles per gallon) during the period when its distance traveled is between 8500 and 9000 miles?
22. A salaried employee earns \$10,400 during his first year of employment. At the end of each year he receives a raise of \$500. What is his salary during his tenth year of employment? What are his total earnings during his first eight years of employment? During which year of his employment does he earn \$14,900 as his year's salary?
23. A spherical balloon is inflated. During each second its radius expands by an amount equal to 0.01 in. less than the radius expanded during the previous second. If radius R expands at the rate of 0.35 in./sec when $R = 18$ in., what is R exactly 30 sec later?
24. Given the sequence

$$-24, -19, -21, -16, -18, -13, -15, -10, -12, \ldots,$$

which combines two arithmetic progressions. What is the fortieth term of the combined progression? What is the sum of the first 40 terms of the combined progression?

25. It is known that the terms a, ab, b form an A.P. Show that the fourth term may be $3ab - 2a$.

26. A free-falling body falls is such a way that its velocity increases by 32 ft/sec each second. Assume that a free-falling body has an initial velocity of 120 ft/sec. What is its velocity at the end of 9 sec?

27. Show that the terms $a, a + b, b$ are in an A.P. only if $a = -b$.

17-4 Geometric progressions

A *geometric progression* is a sequence in which successive terms are related by a common ratio. The common ratio is determined by dividing any term (after the first) by the preceding term. Examples of a G.P. are shown in (19), (20), and (21), where the ratios are $\frac{1}{2}$, -1, and $3m$, respectively.

$$16, 8, 4, 2, 1, \frac{1}{2}, \frac{1}{4}, \ldots \tag{19}$$

$$1, -1, 1, -1, 1, -1, \ldots \tag{20}$$

$$2m, 6m^2, 18m^3, 54m^4, 162m^5, \ldots \tag{21}$$

Table 17-2 shows an intuitive development of the value of the nth term of (19) and the general geometric progression. From the table we see that

$$l = ar^{n-1}, \tag{22}$$

where l is the value of the nth term of the G.P., a is the first term, and r is the common ratio.

Table 17-2

Term number	*Value of term from expression* (19)	*Value of term in general geometric progression*
1	16	a
2	$16(\frac{1}{2})$	ar
3	$16(\frac{1}{2})^2$	ar^2
4	$16(\frac{1}{2})^3$	ar^3
.	.	.
.	.	.
.	.	.
n	$16(\frac{1}{2})^{n-1}$	ar^{n-1}

Example 8 Give the value of the ninth term of the G.P.

$$243, -81, 27, -9, \ldots. \tag{23}$$

Solution Note that (23) is a G.P. with first term $a = 243$ and common ratio $r = -\frac{1}{3}$. Since we wish to find the value of the ninth term, then $n = 9$. Substituting these values into (22) gives

$$l = 243\left(-\frac{1}{3}\right)^{9-1} - 243\left(-\frac{1}{3}\right)^{8} = \frac{1}{27}.$$

To find the sum of the first n terms of a G.P., we have

$$S = a + ar + ar^2 + ar^3 + \cdots + ar^{n-1}. \qquad (24)$$

The right member of (24) is expressible as the quotient of two quantities:

$$\frac{a(1 - r^n)}{1 - r} = a + ar + ar^2 + ar^3 + \cdots + ar^{n-1}. \qquad (25)$$

Then from (24) and (25) we have

$$S = \frac{a(1 - r^n)}{1 - r} \qquad (r \neq 1) \qquad (26)$$

where S is the sum of the first n terms of a G.P. with common ratio r and first term a.

Example 9 Find the sum of the first ten terms of the G.P.

$$20, 10, 5, \frac{5}{2}, \ldots. \qquad (27)$$

Solution By inspection, we have first term $a = 20$, common ratio $r = \frac{1}{2}$, and $n = 10$. Substituting into (24), we obtain

$$S = \frac{20[1 - (1/2)^{10}]}{1 - 1/2} = \frac{20[1 - 1/1024]}{1/2} = 40\left(\frac{1023}{1024}\right).$$

A graphical representation of (27) is interesting. It is shown in Fig. 17-1, where $S_1, S_2, S_3, \ldots$ are the sums of 1, 2, 3, ... terms of (27), respec-

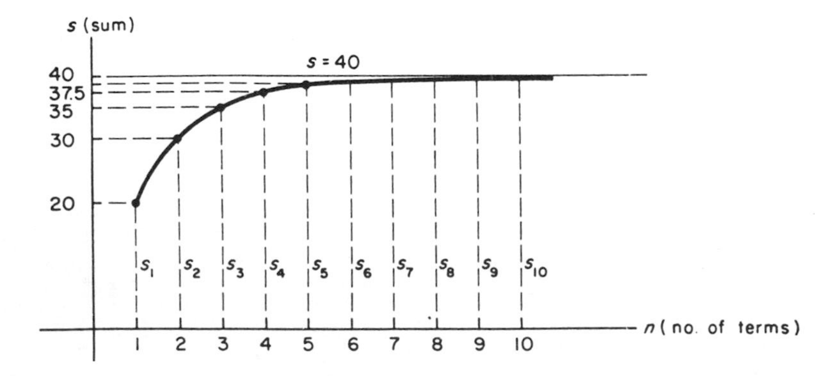

Figure 17-1

tively. The terms S_8, S_9, S_{10} appear to contribute very little to the value of S. It seems, too, that as n increases without bound, S approaches the value $S = 40$.

Example 9 might illustrate the case of a man leaping toward a wall that is 40 ft away. With each leap he covers one-half ($r = \frac{1}{2}$) the distance remaining between himself and the wall. His limit is 40 ft; even with n becoming a very large number, he will never cover the entire 40 ft but can cover as close to 40 ft as he pleases.

The sum of all the terms of a G.P. can be discussed with reference to (28). From

$$S = \frac{a(1 - r^n)}{1 - r} \tag{28}$$

let n become very large ($n \longrightarrow \infty$). Three cases occur.

Case 1 If $|r| > 1$, then $|r^n| \longrightarrow \infty$ and the numerator of (28) goes to negative infinity. Thus we can assert that if $|r| > 1$, the sum of a G.P. cannot be determined.

Case 2 If $r = 1$, then $1 - r = 1 - r^n = 0$ and S is the ratio $0/0$, which cannot be determined. If $r = -1$, then $1 - r = 2$ and $1 - r^n$ has either the value $+2$ or 0 with n odd or even, respectively; therefore S can have the value $+a$ or zero.

Case 3 If $|r| < 1$, then $r^n \longrightarrow 0$ as $n \longrightarrow \infty$. Thus (28) becomes

$$S = \frac{a}{1 - r} \qquad (|r| < 1) \tag{29}$$

where (29) gives the sum of *all* terms of a G.P.

Example 10 Give the sum of all the terms of

$$64, -16, 4, -1, +\frac{1}{4}, \ldots.$$

Solution Here $r = -\frac{1}{4}$ and (29) is appropriate because $|r| < 1$. Also, $a = 64$. Substituting into (29),

$$S = \frac{64}{1 - (-1/4)} = \frac{64}{5/4} = \frac{256}{5}.$$

Exercises 17-4

In Exercises 1–12, find the value of the term indicated for the given geometric progression.

1. Eighth term of 1/243, 1/81, 1/27, . . .
2. Tenth term of 128, −64, 32, . . .
3. Fifty-fifth term of 1, −1, 1, −1, . . .
4. Thirty-sixth term of $m^{44}, m^{42}, m^{40}, \ldots$

5. Twenty-sixth term of $a^{12}/b^{15}, a^{11}/b^{16}, a^{10}/b^{17}, \ldots$
6. Fifteenth term of $(1 + i)^6, (1 + i)^7, (1 + i)^8, \ldots$
7. Eleventh term of $64x^5, 32x^4, 16x^3, \ldots$
8. Twenty-first term of $x^{10}/128, -x^9/64, x^8/32, \ldots$
9. Twelfth term of $(1 + x)^{-8}/128, (1 + x)^{-7}/64, (1 + x)^{-6}/32, \ldots$
10. Tenth term of $y^{-4}/243, -y^{-7/2}/81, y^{-3}/27, \ldots$
11. Fifteenth term of $y^7/(2)^8, y^{13/2}/(2)^7, y^6/(2)^6, \ldots$
12. Twelfth term of $(1 + y)^8/729, -(1 + y)^7/243, (1 + y)^6/81, \ldots$

In Exercises 13–18, provide values for the missing terms, assuming that the given information is appropriate for a geometric progression.

	S	n	a	r	l
13.		8	100	$\frac{1}{10}$	
14.	$\frac{181}{144}$			$-\frac{3}{4}$	$\frac{9}{16}$
15.	-1	15	-1		
16.	2.222			0.1	0.002
17.		5	m^{16}		m^8
18.	$P(1 + i)^n$		P	$1 + i$	

19. A man earns \$1000 during a certain year. In each successive year he doubles his income. What are his earnings during his seventh year? What is his total income during his first seven years?

20. A ball bounces in such a way that it always gains a height that is three-fourths the height gained on the previous bounce. Assume that the ball is dropped from a height of 40 ft. What distance will the ball travel before coming to rest?

21. The assessed value of a machine in a factory declines, for tax purposes, at the rate of 20% per year. A tax of 4% is charged on the assessed value each year. If the machine is initially worth \$10,000, how much taxes are paid during the first five years of the life of the machine?

22. A hot metallic body at a temperature of 350°F is placed outdoors to cool, where the temperature outdoors is 70°F. During each 5-min interval, the temperature of the body will fall 60% of the difference between the temperature of the body and the outdoor temperature. Show that the body will cool to approximately 73°F after 25 min outdoors.

23. Two vehicles A and B are in motion. At a certain instant $t = 0$, A is traveling at 100 mph and B at 20 mph. Thereafter A decreases in speed by an amount of 10% per minute of the speed at the beginning of the minute and B gains 15% per minute. When will A and B be traveling at the same speed?

24. Show a graphical solution of Exercise 23. Graph the progress of A and B on the same set of axes. The point where the graphs intersect is the desired solution.

25. A substance goes into solution at the rate of 25% of the undissolved portion each minute. If the initial amount was 64/27 lb, what amount is undissolved after 6 min?

26. A man builds an annuity by depositing $1000 at the beginning of each year into an account. The account collects interest at 3% annually (compounded). Show that the amount of the annuity after 10 years is

$$A = 1000\left[\frac{(1.03)^{11} - 1}{0.03}\right] = \$11{,}464.$$

27. Show that the series

$$\frac{1}{10} + \left(\frac{1}{10}\right)^2 + \left(\frac{1}{10}\right)^3 + \left(\frac{1}{10}\right)^4 + \cdots$$

has the same value as the fraction 1/9.

17-5 Arithmetic and geometric means

Assume that we are given two nonconsecutive terms of a progression. The terms that lie between the given terms are called *means*. Means are reasonably simple to find without the aid of a formula; on the other hand, a formula is easily developed. In Example 11 we solve a specific case of arithmetic means and develop a formula for the general case.

Example 11 Find three arithmetic means between 46 and 18.

Solution We can list five terms of the A.P. with 46 as the first term, 18 as the fifth term, and blanks provided for the three missing terms:

$$46, \quad , \quad , \quad , 18.$$

By inspection, we see that, in order to proceed from the first term to the fifth term, we require four common differences, or

$$46 + 4d = 18$$

from which the common difference is $d = -7$. Then the missing means are

$$\begin{aligned} 46 + d &= 46 - 7 = 39, \\ 46 + 2d &= 46 - 14 = 32, \\ 46 + 3d &= 46 - 21 = 25. \end{aligned}$$

Example 11 can be used to generalize. Assume that we wish to find k arithmetic means between a and b. There will be $k + 1$ common differences accumulated in proceeding from a to b, or

$$a + (k + 1)d = b$$

from which

$$d = \frac{b - a}{k + 1}, \tag{30}$$

where d is the common difference, a is the first term, b is the last term, and k is the number of means to be provided. Expression (30) is useful for arithmetic progressions.

Example 12 Find six arithmetic means between -12 and $+23$.

Solution Employing (30), we have $k = 6$, $a = -12$, $b = 23$, and

$$d = \frac{23 - (-12)}{6 + 1} = \frac{35}{7} = 5.$$

The common difference is then 5 and the missing means are -7, -2, $+3$, $+8$, $+13$, and $+18$.

Geometric means can be found by a method similar to that used in the solution of Example 11. Consider Example 13.

Example 13 Find two geometric means between -2 and $+54$.

Solution We can list four terms of a G.P. with -2 as the first term, $+54$ as the fourth term, and blanks for the second and third terms:

$$-2, \quad , \quad , +54.$$

By inspection, we see that -2 must be multiplied by the common ratio three times to proceed to $+54$, or

$$-2r^3 = 54$$

from which the common ratio is

$$r = \sqrt[3]{-27} = -3.$$

Then the desired geometric means are

$$-2(-3) = +6 \qquad -2(-3)^2 = -18.$$

Assume that we wish to find k geometric means between a and b. Doing so requires $k + 1$ common ratios, or

$$ar^{k+1} = b$$

from which

$$r = \sqrt[k+1]{\frac{b}{a}}. \tag{31}$$

Note from (31) that if $k + 1$ is an even number, r may be positive or negative, with a and b of the same sign, and r is imaginary if a and b are of different signs. If $k + 1$ is odd, r will be of the same sign as the ratio of b/a.

Example 14 Find three geometric means between 1/16 and 16.

Solution Applying (31), we have $k = 3$, $a = 1/16$, and $b = 16$ with

$$r = \sqrt[4]{\frac{16}{1/16}} = \sqrt[4]{256} = \pm 4.$$

The means are then either 1/4, 1, 4 or $-1/4$, 1, -4.

Exercises 17-5

Find the means indicated in Exercises 1–9.

1. One arithmetic mean between 4 and -4
2. One arithmetic mean between 7 and 15
3. Three arithmetic means between -6 and 14
4. Five arithmetic means between 7 and 19
5. One arithmetic mean between a^2 and a^3
6. Two geometric means between 2 and 16
7. Three geometric means between 6 and 3/8
8. Four geometric means between 1 and -1
9. Four geometric means between 10 and 0.0001
10. Show that the imaginary unit j can be defined as the geometric mean between 1 and -1.
11. A machine depreciates in value at the same amount each year. The initial value of the machine is V_0 and the value after n years is V_n. What is its value during the third year? fifth year? kth year?
12. A bouncing ball rebounds on each bounce to a height that is a fixed fractional part of the height achieved on the previous bounce. On the first bounce it reaches a height of 24 ft; on the fourth bounce it reaches 9 ft. What heights does it reach on the second and third bounces?
13. A simple interest account grows in the form of an A.P. The account is worth \$1300 after 5 years and \$1780 after 13 years. What was its original value? What is its value after 9 years?

14. An altitude drawn from the vertex of a right triangle to the hypotenuse is the geometric mean of the segments of the hypotenuse. How long are the segments of the hypotenuse of a right triangle if the altitude is 3 in. and the hypotenuse is 12 in.?

15. Show that vectors $1\underline{/90^\circ}$, $1\underline{/180^\circ}$, and $1\underline{/270^\circ}$ can be regarded as the three geometric means between $1\underline{/0^\circ}$ and $1\underline{/360^\circ}$.

16. Show that the value of the one arithmetic mean between two numbers is the arithmetic average of the two numbers.

17. The geometric mean of a group of n numbers is the nth root of the product of the n numbers. What is the geometric mean of the group of numbers 1, 2, 1, 4, -1, 8, -2? Of 3, 6, 9, 0, 4?

18. Give a general formula for the geometric mean G in Exercise 17 if the n numbers are $a_1, a_2, a_3, \ldots, a_n$.

19. Using the definition of the geometric mean given in Exercise 17, show that $G = ar^2$ for numbers a, ar, ar^2, ar^3, ar^4.

20. The arithmetic mean $\bar{x}$ of numbers $x_1, x_2, x_3, \ldots, x_n$ is defined as

$$\bar{x} = \frac{x_1 + x_2 + x_3 + \cdots + x_n}{n} = \frac{1}{n}\sum_{i=1}^{n} x_1.$$

Find the arithmetic mean of numbers 6, 12, 3, 4, -12, -9, -15, 6, 2, 6.

Tables

Tables B, C, and D are taken from Lawrence M. Clar and James A. Hart, *Mathematics for the Technologies*. © 1978 by Prentice-Hall, Inc., Englewood Cliffs, N.J. 07632. Reprinted by permission of Prentice-Hall, Inc.

Table A: Properties of Geometric Figures

For the figures shown
A = plane area
V = volume
C = circumference
S = surface area
p = perimeter

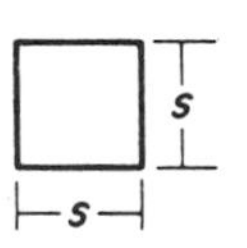

square
$A = s^2$

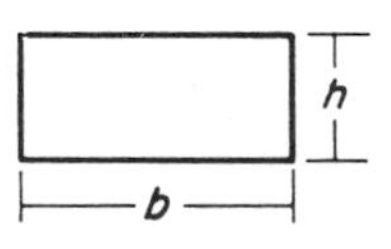

rectangle
$A = bh$

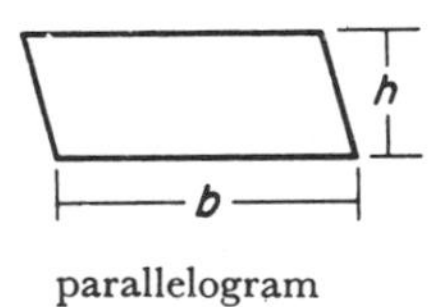

parallelogram
$A = bh$

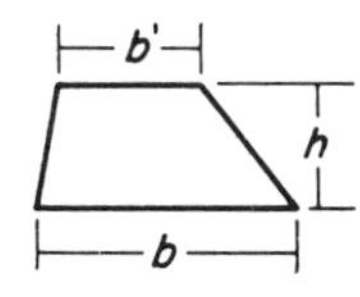

trapezoid
$A = \frac{1}{2}(b + b')h$

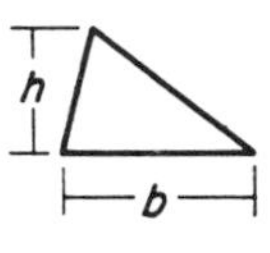

triangle
$A = \frac{1}{2}bh$

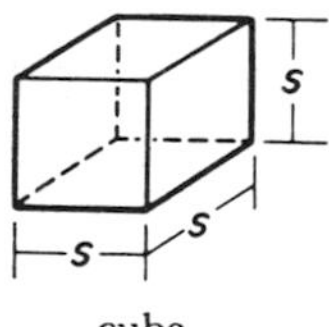

cube
$V = s^3$
$S = 6s^2$

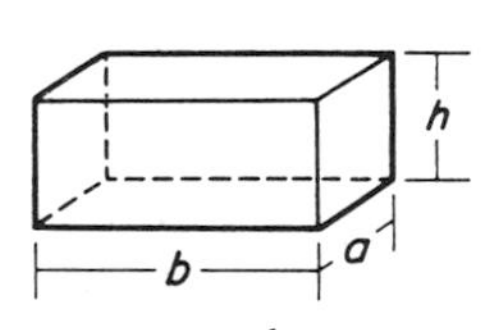

rectangular parallelepiped
$V = abh$
$S = 2ab + 2ah + 2bh$

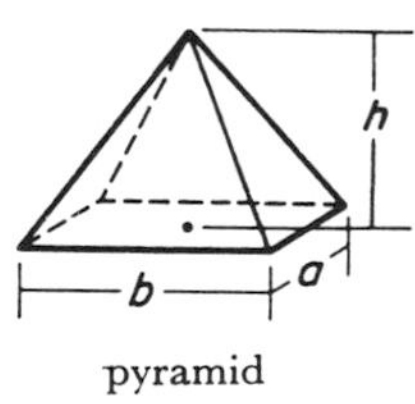

pyramid
$V = \frac{1}{3}abh$

Table A: Properties of Geometric Figures (cont.)

right
circular
cylinder

$V = \pi r^2 h$

$S = 2\pi r^2 + 2\pi r h$

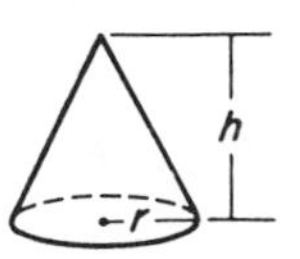

right
circular
cone

$V = \frac{1}{3}\pi r^2 h$

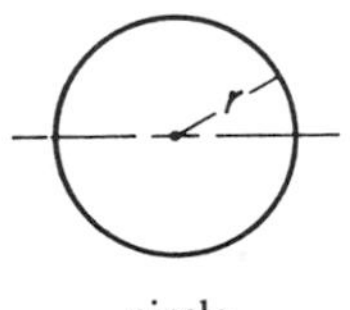

circle

$C = 2\pi r$

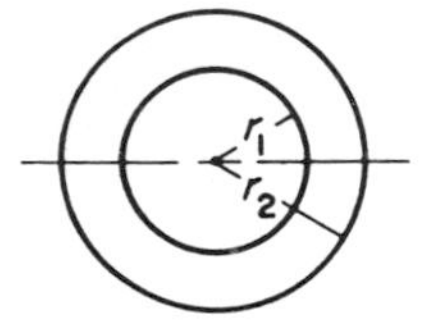

annulus

$A = \pi(r_2^2 - r_1^2)$

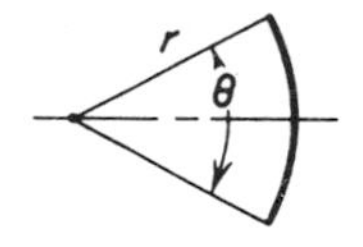

circular
sector

$A = \frac{1}{2} r^2 \theta$

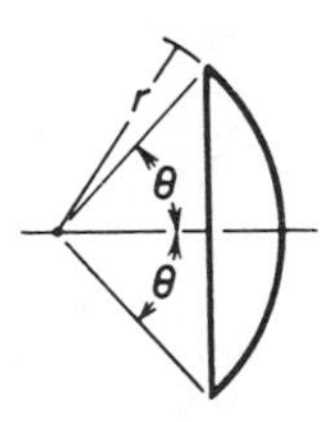

circular
segment

$$A = r^2\left(\theta - \frac{\sin 2\theta}{2}\right)$$

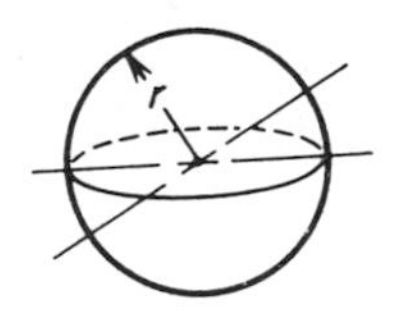

sphere

$V = \frac{4}{3}\pi r^3$

$S = 4\pi r^2$

Table B: Natural Trigonometric Functions

Degrees	Radians	Sin	Cos	Tan	Cot	Sec	Csc		
0°00′	.0000	.0000	1.0000	.0000	——	1.000	——	1.5708	90°00′
10	029	029	000	029	343.8	000	343.8	679	50
20	058	058	000	058	171.9	000	171.9	650	40
30	.0087	.0087	1.0000	.0087	114.6	1.000	114.6	1.5621	30
40	116	116	.999	116	85.94	000	85.95	592	20
50	145	145	999	145	68.75	000	68.76	563	10
1°00′	.0175	.0175	.9998	.0175	57.29	1.000	57.30	1.5533	89°00′
10	204	204	998	204	49.10	000	49.11	504	50
20	233	233	997	233	42.96	000	42.98	475	40
30	.0262	.0262	.9997	.0262	38.19	1.000	38.20	1.5446	30
40	291	291	996	291	34.37	000	34.38	417	20
50	320	320	995	320	31.24	001	31.26	388	10
2°00′	.0349	.0349	.9994	.0349	28.64	1.001	28.65	1.5359	88°00′
10	378	378	993	378	26.43	001	26.45	330	50
20	407	407	992	407	24.54	001	24.56	301	40
30	.0436	.0436	.9990	.0437	22.90	1.001	22.93	1.5272	30
40	465	465	989	466	21.47	001	21.49	243	20
50	495	494	988	495	20.21	001	20.23	213	10
3°00′	.0524	.0523	.9986	.0524	19.08	1.001	19.11	1.5184	87°00′
10	553	552	985	553	18.07	002	18.10	155	50
20	582	581	983	582	17.17	002	17.20	126	40
30	.0611	.0610	.9981	.0612	16.35	1.002	16.38	1.5097	30
40	640	640	980	641	15.60	002	15.64	068	20
50	669	669	978	670	14.92	002	14.96	039	10
4°00′	.0698	.0698	.9976	.0699	14.30	1.002	14.34	1.5010	86°00′
10	727	727	974	729	13.73	003	13.76	981	50
20	756	756	971	758	13.20	003	13.23	952	40
30	.0785	.0785	.9969	.0787	12.71	1.003	12.75	1.4923	30
40	814	814	967	816	12.25	003	12.29	893	20
50	844	843	964	846	11.83	004	11.87	864	10
5°00′	.0873	.0872	.9962	.0875	11.43	1.004	11.47	1.4835	85°00′
10	902	901	959	904	11.06	004	11.10	806	50
20	931	929	957	934	10.71	004	10.76	777	40
30	.0960	.0958	.9954	.0963	10.39	1.005	10.43	1.4748	30
40	989	987	951	992	10.08	005	10.13	719	20
50	.1018	.1016	948	.1022	9.788	005	9.839	690	10
6°00′	.1047	.1045	.9945	.1051	9.514	1.006	9.567	1.4661	84°00′
10	076	074	942	080	9.255	006	9.309	632	50
20	105	103	939	110	9.010	006	9.065	603	40
30	.1134	.1132	.9936	.1139	8.777	1.006	8.834	1.4573	30
40	164	161	932	169	8.556	007	8.614	544	20
50	193	190	929	198	8.345	007	8.405	515	10
7°00′	.1222	.1219	.9925	.1228	8.144	1.008	8.206	1.4486	83°00′
10	251	248	922	257	7.953	008	8.016	457	50
20	280	276	918	287	7.770	008	7.834	428	40
30	.1309	.1305	.9914	.1317	7.596	1.009	7.661	1.4399	30
40	338	334	911	346	7.429	009	7.496	370	20
50	367	363	907	376	7.269	009	7.337	341	10
8°00′	.1396	.1392	.9903	.1405	7.115	1.010	7.185	1.4312	82°00′
10	425	421	899	435	6.968	010	7.040	283	50
20	454	449	894	465	6.827	011	6.900	254	40
30	.1484	.1478	.9890	.1495	6.691	1.011	6.765	1.4224	30
40	513	507	886	524	6.561	012	6.636	195	20
50	542	536	881	554	6.435	012	6.512	166	10
9°00′	.1571	.1564	.9877	.1584	6.314	1.012	6.392	1.4137	81°00′
		Cos	Sin	Cot	Tan	Csc	Sec	Radians	Degrees

Table B: Natural Trigonometric Functions (cont.)

Degrees	Radians	Sin	Cos	Tan	Cot	Sec	Csc		
9°00′	.1571	.1564	.9877	.1584	6.314	1.012	6.392	1.4137	81°00′
10	600	593	872	614	197	013	277	108	50
20	629	622	868	644	084	013	166	079	40
30	.1658	.1650	.9863	.1673	5.976	1.014	6.059	1.4050	30
40	687	679	858	703	871	014	955	021	20
50	716	708	853	733	769	015	855	992	10
10°00′	.1745	.1736	.9848	.1763	5.671	1.015	5.759	1.3963	80°00′
10	774	765	843	793	576	016	665	934	50
20	804	794	838	823	485	016	575	904	40
30	.1833	.1822	.9833	.1853	5.396	1.017	5.487	1.3875	30
40	862	851	827	883	309	018	403	846	20
50	891	880	822	914	226	018	320	817	10
11°00′	.1920	.1908	.9816	.1944	5.145	1.019	5.241	1.3788	79°00′
10	949	937	811	974	066	019	164	759	50
20	978	965	805	004	989	020	089	730	40
30	.2007	.1994	.9799	.2035	4.915	1.020	5.016	1.3701	30
40	036	.022	793	065	843	021	945	672	20
50	065	051	787	095	773	022	876	643	10
12°00′	.2094	.2079	.9781	.2126	4.705	1.022	4.810	1.3614	78°00′
10	123	108	775	156	638	023	745	584	50
20	153	136	769	186	574	024	682	555	40
30	.2182	.2164	.9763	.2217	4.511	1.024	4.620	1.3526	30
40	211	193	757	247	449	025	560	497	20
50	240	221	750	278	390	026	502	468	10
13°00′	.2269	.2250	.9744	.2309	4.331	1.026	4.445	1.3439	77°00′
10	298	278	737	339	275	027	390	410	50
20	327	306	730	370	219	028	336	381	40
30	.2356	.2334	.9724	.2401	4.165	1.028	4.284	1.3352	30
40	385	363	717	432	113	029	232	323	20
50	414	391	710	462	061	030	182	294	10
14°00′	.2443	.2419	.9703	.2493	4.011	1.031	4.134	1.3265	76°00′
10	473	447	696	524	962	031	088	235	50
20	502	476	689	555	914	032	039	206	40
30	.2531	.2504	.9681	.2586	3.867	1.033	3.994	1.3177	30
40	560	532	674	617	821	034	950	148	20
50	589	560	667	648	776	034	906	119	10
15°00′	.2618	.2588	.9659	.2679	3.732	1.035	3.864	1.3090	75°00′
10	647	616	652	711	689	036	822	061	50
20	676	644	644	742	647	037	782	032	40
30	.2705	.2672	.9636	.2773	3.606	1.038	3.742	1.3003	30
40	734	700	628	805	566	039	703	974	20
50	763	728	621	836	526	039	665	945	10
16°00′	.2793	.2756	.9613	.2867	3.487	1.040	3.628	1.2915	74°00′
10	822	784	605	899	450	041	592	886	50
20	851	812	596	931	412	042	556	857	40
30	.2880	.2840	.9588	.2962	3.376	1.043	3.521	1.2828	30
40	909	868	580	994	340	044	487	799	20
50	938	896	572	.3026	305	045	453	770	10
17°00′	.2967	.2924	.9563	.3057	3.271	1.046	3.420	1.2741	73°00′
10	996	952	555	089	237	047	388	712	50
20	.3025	979	546	121	204	048	356	683	40
30	054	.3007	.9537	.3153	3.172	1.049	3.326	1.2654	30
40	083	035	528	185	140	049	295	625	20
50	113	062	520	217	108	050	265	595	10
18°00′	.3142	.3090	.9511	.3249	3.078	1.051	3.236	1.2566	72°00′
		Cos	Sin	Cot	Tan	Csc	Sec	Radians	Degrees

Table B: Natural Trigonometric Functions (cont.)

Degrees	Radians	Sin	Cos	Tan	Cot	Sec	Csc		
18°00′	.3142	.3090	.9511	.3249	3.078	1.051	3.236	1.2566	72°00′
10	171	118	502	281	047	052	207	537	50
20	200	145	492	314	018	053	179	508	40
30	.3229	.3173	.9483	.3346	2.989	1.054	3.152	1.2479	30
40	258	201	474	378	960	056	124	450	20
50	287	228	465	411	932	057	098	421	10
19°00′	.3316	.3256	.9455	.3443	2.904	1.058	3.072	1.2392	71°00′
10	345	283	446	476	877	059	046	363	50
20	374	311	436	508	850	060	021	334	40
30	.3403	.3338	.9426	.3541	2.824	1.061	2.996	1.2305	30
40	432	365	417	574	798	062	971	275	20
50	462	393	407	607	773	063	947	246	10
20°00′	.3491	.3420	.9397	.3640	2.747	1.064	2.924	1.2217	70°00′
10	520	448	387	673	723	065	901	188	50
20	549	475	377	706	699	066	878	159	40
30	.3578	.3502	.9367	.3739	2.675	1.068	2.855	1.2130	30
40	607	529	356	772	651	069	833	101	20
50	636	557	346	805	628	070	812	072	10
21°00′	.3665	.3584	.9336	.3839	2.605	1.071	2.790	1.2043	69°00′
10	694	611	325	872	583	072	769	014	50
20	723	638	315	906	560	074	749	985	40
30	.3752	.3665	.9304	.3939	2.539	1.075	2.729	1.1956	30
40	782	692	293	973	517	076	709	926	20
50	811	719	283	006	496	077	689	897	10
22°00′	.3840	.3746	.9272	.4040	2.475	1.079	2.669	1.1868	68°00′
10	869	773	261	074	455	080	650	839	50
20	898	800	250	108	434	081	632	810	40
30	.3927	.3827	.9239	.4142	2.414	1.082	2.613	1.1781	30
40	956	854	228	176	394	084	595	752	20
50	985	881	216	210	375	085	577	723	10
23°00′	.4014	.3907	.9205	.4245	2.356	1.086	2.559	1.1694	67°00′
10	043	934	194	279	337	088	542	665	50
20	072	961	182	314	318	089	525	636	40
30	.4102	.3987	.9171	.4348	2.300	1.090	2.508	1.1606	30
40	131	014	159	383	282	092	491	577	20
50	160	041	147	417	264	093	475	548	10
24°00′	.4189	.4067	.9135	.4452	2.246	1.095	2.459	1.1519	66°00′
10	218	094	124	487	229	096	443	490	50
20	247	120	112	522	211	097	427	461	40
30	.4276	.4147	.9100	.4557	2.194	1.099	2.411	1.1432	30
40	305	173	088	592	177	100	396	403	20
50	334	200	075	628	161	102	381	374	10
25°00′	.4363	.4226	.9063	.4663	2.145	1.103	2.366	1.1345	65°00′
10	392	253	051	699	128	105	352	316	50
20	422	279	038	734	112	106	337	286	40
30	.4451	.4305	.9026	.4770	2.097	1.108	2.323	1.1257	30
40	480	331	013	806	081	109	309	228	20
50	509	358	001	841	066	111	295	199	10
26°00′	.4538	.4384	.8988	.4877	2.050	1.113	2.281	1.1170	64°00′
10	567	410	975	913	035	114	268	141	50
20	596	436	962	950	020	116	254	112	40
30	.4625	.4462	.8949	.4986	2.006	1.117	2.241	1.1083	30
40	654	488	936	022	991	119	228	054	20
50	683	514	923	059	977	121	215	025	10
27°00′	.4712	.4540	.8910	.5095	1.963	1.122	2.203	1.0996	63°00′
		Cos	Sin	Cot	Tan	Csc	Sec	Radians	Degrees

Table B: Natural Trigonometric Functions (cont.)

Degrees	Radians	Sin	Cos	Tan	Cot	Sec	Csc		
27°00′	.4712	.4540	.8910	.5095	1.963	1.122	2.203	1.0996	63°00′
10	741	566	897	132	949	124	190	966	50
20	771	592	884	169	935	126	178	937	40
30	.4800	.4617	.8870	.5206	1.921	1.127	2.166	1.0908	30
40	829	643	857	243	907	129	154	879	20
50	858	669	843	280	894	131	142	850	10
28°00′	.4887	.4695	.8829	.5317	1.881	1.133	2.130	1.0821	62°00′
10	916	720	816	354	868	134	118	792	50
20	945	746	802	392	855	136	107	763	40
30	.4974	.4772	.8788	.5430	1.842	1.138	2.096	1.0734	30
40	003	797	774	467	829	140	085	705	20
50	032	823	760	505	816	142	074	676	10
29°00′	.5061	.4848	.8746	.5543	1.804	1.143	2.063	1.0647	61°00′
10	091	874	732	581	792	145	052	617	50
20	120	899	718	619	780	147	041	588	40
30	.5149	.4924	.8704	.5658	1.767	1.149	2.031	1.0559	30
40	178	950	689	696	756	151	020	530	20
50	207	975	675	735	744	153	010	501	10
30°00′	.5236	.5000	.8660	.5774	1.732	1.155	2.000	1.0472	60°00′
10	265	025	646	812	720	157	990	443	50
20	294	050	631	851	709	159	980	414	40
30	.5323	.5075	.8616	.5890	1.698	1.161	1.970	1.0385	30
40	352	100	601	930	686	163	961	356	20
50	381	125	587	969	675	165	951	327	10
31°00′	.5411	.5150	.8572	.6009	1.664	1.167	1.942	1.0297	59°00′
10	440	175	557	048	653	169	932	268	50
20	469	200	542	088	643	171	923	239	40
30	.5498	.5225	.8526	.6128	1.632	1.173	1.914	1.0210	30
40	527	250	511	168	621	175	905	181	20
50	556	275	496	208	611	177	896	152	10
32°00′	.5585	.5299	.8480	.6249	1.600	1.179	1.887	1.0123	58°00′
10	614	324	465	289	590	181	878	094	50
20	643	348	450	330	580	184	870	065	40
30	.5672	.5373	.8434	.6371	1.570	1.186	1.861	1.0036	30
40	701	398	418	412	560	188	853	0007	20
50	730	422	403	453	550	190	844	977	10
33°00′	.5760	.5446	.8387	.6494	1.540	1.192	1.836	.9948	57°00′
10	789	471	371	536	530	195	828	919	50
20	818	495	355	577	520	197	820	890	40
30	.5847	.5519	.8339	.6619	1.511	1.199	1.812	.9861	30
40	876	544	323	661	501	202	804	832	20
50	905	568	307	703	492	204	796	803	10
34°00′	.5934	.5592	.8290	.6745	1.483	1.206	1.788	.9774	56°00′
10	963	616	274	787	473	209	781	745	50
20	992	640	258	830	464	211	773	715	40
30	.6021	.5664	.8241	.6873	1.455	1.213	1.766	.9687	30
40	050	688	225	916	446	216	758	657	20
50	080	712	208	959	437	218	751	628	10
35°00′	.6109	.5736	.8192	.7002	1.428	1.221	1.743	.9599	55°00′
10	138	760	175	046	419	223	736	570	50
20	167	783	158	089	411	226	729	541	40
30	.6196	.5807	.8141	.7133	1.402	1.228	1.722	.9512	30
40	225	831	124	177	393	231	715	483	20
50	254	854	107	221	385	233	708	454	10
36°00′	.6283	.5878	.8090	.7265	1.376	1.236	1.701	.9425	54°00′
		Cos	Sin	Cot	Tan	Csc	Sec	Radians	Degrees

Table B: Natural Trigonometric Functions (cont.)

Degrees	Radians	Sin	Cos	Tan	Cot	Sec	Csc		
36°00′	.6283	.5878	.8090	.7265	1.376	1.236	1.701	.9425	54°00′
10	312	901	073	310	368	239	695	396	50
20	341	925	056	355	360	241	688	367	40
30	.6370	.5948	.8039	.7400	1.351	1.244	1.681	.9338	30
40	400	972	021	445	343	247	675	308	20
50	429	995	004	490	335	249	668	279	10
37°00′	.6458	.6018	.7986	.7536	1.327	1.252	1.662	.9250	53°00′
10	487	041	969	581	319	255	655	221	50
20	516	065	951	627	311	258	649	192	40
30	.6545	.6088	.7934	.7673	.1303	1.260	1.643	.9163	30
40	574	111	916	720	295	263	636	134	20
50	603	134	898	766	288	266	630	105	10
38°00′	.6632	.6157	.7880	.7813	1.280	1.269	1.624	.9076	52°00′
10	661	180	862	860	272	272	618	047	50
20	690	202	844	907	265	275	612	018	40
30	.6720	.6225	.7826	.7954	1.257	1.278	1.606	.8988	30
40	749	248	808	002	250	281	601	959	20
50	778	271	790	050	242	284	595	930	10
39°00′	.6807	.6293	.7771	.8098	1.235	1.287	1.589	.8901	51°00′
10	836	316	753	146	228	290	583	872	50
20	865	338	735	195	220	293	578	843	40
30	.6894	.6361	.7716	.8243	1.213	1.296	1.572	.8814	30
40	923	383	698	292	206	299	567	785	20
50	952	406	679	342	199	302	561	756	10
40°00′	.6981	.6428	.7660	.8391	1.192	1.305	1.556	.8727	50°00′
10	010	450	642	441	185	309	550	698	50
20	039	472	623	491	178	312	545	668	40
30	.7069	.6494	.7604	.8541	1.171	1.315	1.540	.8639	30
40	098	517	585	591	164	318	535	610	20
50	127	539	566	642	157	322	529	581	10
41°00′	.7156	.6561	.7547	.8693	1.150	1.325	1.524	.8552	49°00′
10	185	583	528	744	144	328	519	523	50
20	214	604	509	796	137	332	514	494	40
30	.7243	.6626	.7490	.8847	1.130	1.335	1.509	.8465	30
40	272	648	470	899	124	339	504	436	20
50	301	670	451	952	117	342	499	407	10
42°00′	.7330	.6691	.7431	.9004	1.111	1.346	1.494	.8378	48°00′
10	359	713	412	057	104	349	490	348	50
20	389	734	392	110	098	353	485	319	40
30	.7418	.6756	.7373	.9163	1.091	1.356	1.480	.8290	30
40	447	777	353	217	085	360	476	261	20
50	476	799	333	271	079	364	471	232	10
43°00′	.7505	.6820	.7314	.9325	1.072	1.367	1.466	.8203	47°00′
10	534	841	294	380	066	371	462	174	50
20	563	862	274	435	060	375	457	145	40
30	.7592	.6884	.7254	.9490	1.054	1.379	1.453	.8116	30
40	621	905	234	545	048	382	448	087	20
50	650	926	214	601	042	386	444	058	10
44°00′	.7679	.6947	.7193	.9657	1.036	1.390	1.440	.8029	46°00′
10	709	967	173	713	030	394	435	999	50
20	738	988	153	770	024	398	431	970	40
30	.7767	.7009	.7133	.9827	1.018	1.402	1.427	.7941	30
40	796	030	112	884	012	406	423	912	20
50	825	050	092	942	006	410	418	883	10
45°00′	.7854	.7071	.7071	1.000	1.000	1.414	1.414	.7854	45°00′
		Cos	Sin	Cot	Tan	Csc	Sec	Radians	Degrees

Table C: Common Logarithms of Numbers

N	0	1	2	3	4	5	6	7	8	9
10	0000	0043	0086	0128	0170	0212	0253	0294	0334	0374
11	0414	0453	0492	0531	0569	0607	0645	0682	0719	0755
12	0792	0828	0864	0899	0934	0969	1004	1038	1072	1106
13	1139	1173	1206	1239	1271	1303	1335	1367	1399	1430
14	1461	1492	1523	1553	1584	1614	1644	1673	1703	1732
15	1761	1790	1818	1847	1875	1903	1931	1959	1987	2014
16	2041	2068	2095	2122	2148	2175	2201	2227	2253	2279
17	2304	2330	2355	2380	2405	2430	2455	2480	2504	2529
18	2553	2577	2601	2625	2648	2672	2695	2718	2742	2765
19	2788	2810	2833	2856	2878	2900	2923	2945	2967	2989
20	3010	3032	3054	3075	3096	3118	3139	3160	3181	3201
21	3222	3243	3263	3284	3304	3324	3345	3365	3385	3404
22	3424	3444	3464	3483	3502	3522	3541	3560	3579	3598
23	3617	3636	3655	3674	3692	3711	3729	3747	3766	3784
24	3802	3820	3838	3856	3874	3892	3909	3927	3945	3962
25	3979	3997	4014	4031	4048	4065	4082	4099	4116	4133
26	4150	4166	4183	4200	4216	4232	4249	4265	4281	4298
27	4314	4330	4346	4362	4378	4393	4409	4425	4440	4456
28	4472	4487	4502	4518	4533	4548	4564	4579	4594	4609
29	4624	4639	4654	4669	4683	4698	4713	4728	4742	4757
30	4771	4786	4800	4814	4829	4843	4857	4871	4886	4900
31	4914	4928	4942	4955	4969	4983	4997	5011	5024	5038
32	5051	5065	5079	5092	5105	5119	5132	5145	5159	5172
33	5185	5198	5211	5224	5237	5250	5263	5276	5289	5302
34	5315	5328	5340	5353	5366	5378	5391	5403	5416	5428
35	5441	5453	5465	5478	5490	5502	5514	5527	5539	5551
36	5563	5575	5587	5599	5611	5623	5635	5647	5658	5670
37	5682	5694	5705	5717	5729	5740	5752	5763	5775	5786
38	5798	5809	5821	5832	5843	5855	5866	5877	5888	5899
39	5911	5922	5933	5944	5955	5966	5977	5988	5999	6010
40	6021	6031	6042	6053	6064	6075	6085	6096	6107	6117
41	6128	6138	6149	6160	6170	6180	6191	6201	6212	6222
42	6232	6243	6253	6263	6274	6284	6294	6304	6314	6325
43	6335	6345	6355	6365	6375	6385	6395	6405	6415	6425
44	6435	6444	6454	6464	6474	6484	6493	6503	6513	6522
45	6532	6542	6551	6561	6571	6580	6590	6599	6609	6618
46	6628	6637	6646	6656	6665	6675	6684	6693	6702	6712
47	6721	6730	6739	6749	6758	6767	6776	6785	6794	6803
48	6812	6821	6830	6839	6848	6857	6866	6875	6884	6893
49	6902	6911	6920	6928	6937	6946	6955	6964	6972	6981
50	6990	6998	7007	7016	7024	7033	7042	7050	7059	7067
51	7076	7084	7093	7101	7110	7118	7126	7135	7143	7152
52	7160	7168	7177	7185	7193	7202	7210	7218	7226	7235
53	7243	7251	7259	7267	7275	7284	7292	7300	7308	7316
54	7324	7332	7340	7348	7356	7364	7372	7380	7388	7396

Table C: Common Logarithms of Numbers (cont.)

N	0	1	2	3	4	5	6	7	8	9
55	7404	7412	7419	7427	7435	7443	7451	7459	7466	7474
56	7482	7490	7497	7505	7513	7520	7528	7536	7543	7551
57	7559	7566	7574	7582	7589	7597	7604	7612	7619	7627
58	7634	7642	7649	7657	7664	7672	7679	7686	7694	7701
59	7709	7716	7723	7731	7738	7745	7752	7760	7767	7774
60	7782	7789	7796	7803	7810	7818	7825	7832	7839	7846
61	7853	7860	7868	7875	7882	7889	7896	7903	7910	7917
62	7924	7931	7938	7945	7952	7959	7966	7973	7980	7987
63	7993	8000	8007	8014	8021	8028	8035	8041	8048	8055
64	8062	8069	8075	8082	8089	8096	8102	8109	8116	8122
65	8129	8136	8142	8149	8156	8162	8169	8176	8182	8189
66	8195	8202	8209	8215	8222	8228	8235	8241	8248	8254
67	8261	8267	8274	8280	8287	8293	8299	8306	8312	8319
68	8325	8331	8338	8344	8351	8357	8363	8370	8376	8382
69	8388	8395	8401	8407	8414	8420	8426	8432	8439	8445
70	8451	8457	8463	8470	8476	8482	8488	8494	8500	8506
71	8513	8519	8525	8531	8537	8543	8549	8555	8561	8567
72	8573	8579	8585	8591	8597	8603	8609	8615	8621	8627
73	8633	8639	8645	8651	8657	8663	8669	8675	8681	8686
74	8692	8698	8704	8710	8716	8722	8727	8733	8739	8745
75	8751	8756	8762	8768	8774	8779	8785	8791	8797	8802
76	8808	8814	8820	8825	8831	8837	8842	8848	8854	8859
77	8865	8871	8876	8882	8887	8893	8899	8904	8910	8915
78	8921	8927	8932	8938	8943	8949	8954	8960	8965	8971
79	8976	8982	8987	8993	8998	9004	9009	9015	9020	9025
80	9031	9036	9042	9047	9053	9058	9063	9069	9074	9079
81	9085	9090	9096	9101	9106	9112	9117	9122	9128	9133
82	9138	9143	9149	9154	9159	9165	9170	9175	9180	9186
83	9191	9196	9201	9206	9212	9217	9222	9227	9232	9238
84	9243	9248	9253	9258	9263	9269	9274	9279	9284	9289
85	9294	9299	9304	9309	9315	9320	9325	9330	9335	9340
86	9345	9350	9355	9360	9365	9370	9375	9380	9385	9390
87	9395	9400	9405	9410	9415	9420	9425	9430	9435	9440
88	9445	9450	9455	9460	9465	9469	9474	9479	9484	9489
89	9494	9499	9504	9509	9513	9518	9523	9528	9533	9538
90	9542	9547	9552	9557	9562	9566	9571	9576	9581	9586
91	9590	9595	9600	9605	9609	9614	9619	9624	9628	9633
92	9638	9643	9647	9652	9657	9661	9666	9671	9675	9680
93	9685	9689	9694	9699	9703	9708	9713	9717	9722	9727
94	9731	9736	9741	9745	9750	9754	9759	9763	9768	9773
95	9777	9782	9786	9791	9795	9800	9805	9809	9814	9818
96	9823	9827	9832	9836	9841	9845	9850	9854	9859	9863
97	9868	9872	9877	9881	9886	9890	9894	9899	9903	9908
98	9912	9917	9921	9926	9930	9934	9939	9943	9948	9952
99	9956	9961	9965	9969	9974	9978	9983	9987	9991	9996

Table D: Natural Logarithms of Numbers

n	$\log_e n$	n	$\log_e n$	n	$\log_e n$
0.0		4.5	1.5041	9.0	2.1972
0.1	7.6974*	4.6	1.5261	9.1	2.2083
0.2	8.3906*	4.7	1.5476	9.2	2.2192
0.3	8.7960*	4.8	1.5686	9.3	2.2300
0.4	9.0837*	4.9	1.5892	9.4	2.2407
0.5	9.3069*	5.0	1.6094	9.5	2.2513
0.6	9.4892*	5.1	1.6292	9.6	2.2618
0.7	9.6433*	5.2	1.6487	9.7	2.2721
0.8	9.7769*	5.3	1.6677	9.8	2.2824
0.9	9.8946*	5.4	1.6864	9.9	2.2925
1.0	0.0000	5.5	1.7047	10	2.3026
1.1	0.0953	5.6	1.7228	11	2.3979
1.2	0.1823	5.7	1.7405	12	2.4849
1.3	0.2624	5.8	1.7579	13	2.5649
1.4	0.3365	5.9	1.7750	14	2.6391
1.5	0.4055	6.0	1.7918	15	2.7081
1.6	0.4700	6.1	1.8083	16	2.7726
1.7	0.5306	6.2	1.8245	17	2.8332
1.8	0.5878	6.3	1.8405	18	2.8904
1.9	0.6419	6.4	1.8563	19	2.9444
2.0	0.6931	6.5	1.8718	20	2.9957
2.1	0.7419	6.6	1.8871	25	3.2189
2.2	0.7885	6.7	1.9021	30	3.4012
2.3	0.8329	6.8	1.9169	35	3.5553
2.4	0.8755	6.9	1.9315	40	3.6889
2.5	0.9163	7.0	1.9459	45	3.8067
2.6	0.9555	7.1	1.9601	50	3.9120
2.7	0.9933	7.2	1.9741	55	4.0073
2.8	1.0296	7.3	1.9879	60	4.0943
2.9	1.0647	7.4	2.0015	65	4.1744
3.0	1.0986	7.5	2.0149	70	4.2485
3.1	1.1314	7.6	2.0281	75	4.3175
3.2	1.1632	7.7	2.0412	80	4.3820
3.3	1.1939	7.8	2.0541	85	4.4427
3.4	1.2238	7.9	2.0669	90	4.4998
3.5	1.2528	8.0	2.0794	95	4.5539
3.6	1.2809	8.1	2.0919	100	4.6052
3.7	1.3083	8.2	2.1041		
3.8	1.3350	8.3	2.1163		
3.9	1.3610	8.4	2.1282		
4.0	1.3863	8.5	2.1401		
4.1	1.4110	8.6	2.1518		
4.2	1.4351	8.7	2.1633		
4.3	1.4586	8.8	2.1748		
4.4	1.4816	8.9	2.1861		

*Attach −10 to these logarithms.

Answers to selected exercises

Exercises 1-2

1. (b) -0.75 (d) 2.75 **2.** (b) $3\frac{1}{5}$ (d) $-6\frac{1}{4}$
3. (b) $-2 < (4 - 5.7) < -\frac{3}{4} < -0.4 < -\frac{1}{3}$ **5.** (a) $x > 2$ (c) $|x| = 3$
(e) $x < -1$ (g) $x < 0$ (i) $0.01 < x < 0.03$

Exercises 1-3

1. Associative **3.** Substitution **5.** Multiplicative inverse
7. Equivalent fractions **9.** Distributive **11.** Commutative
13. Multiplicative inverse **15.** Additive inverse, multiplicative inverse

Exercises 1-4

1. Subtraction **3.** Division **5.** Division **7.** Division
9. Subtraction **18.** 3:5 **19.** 2:1 **20.** 1:1.09 approx.
21. 1.1:1 approx. **23.** 125% **25.** 5% **27.** 250% **29.** .25%
31. Triple **33.** $26\frac{2}{3}$ ft **35.** 45 g

Exercises 1-5

1. 9.3×10^7 **3.** 3.15×10^{-5} **5.** 10^{-8} **7.** 3×10^9 **9.** 1000
11. 0.0000625 **13.** 0.000005 **15.** 1,000,000,000,000 **17.** 6×10^2
19. 6×10^{-7} **21.** 9×10^6 **23.** 2×10^{11} **25.** 4.807×10^3

Exercises 2-1

1. 11 **3.** 1 **5.** -3 **7.** -11 **9.** -26 **11.** -12 **13.** 11
15. -3 **17.** -16 **19.** 2 **21.** 35 ft/sec, -40 ft/sec **23.** 12 sec
25. 5 lb left **28.** \$400 **29.** 6:00, 3:00, 10:00

Exercises 2-2

14. vt **16.** $x^2 + y^2$ **18.** $\dfrac{r_1 r_2}{(r_1 + r_2)}$ **20.** $\dfrac{(x - y)}{(x + y)}$ **22.** $\sqrt{\dfrac{p}{r}}$
24. $(x - y)^3$ **26.** $\dfrac{\pi D^4}{32}$ **28.** $2r + \pi r$ **30.** $6(x + a)^2$
32. $\frac{1}{2}(x^2 + y^2)$ **34.** (a) $6s + 6r$, (c) $6r + 10s$; $9rs$ **36.** $H + W + L$
38. $L_0 + L_0 c\, \Delta t$ **40.** $\dfrac{m}{h - c}; \dfrac{m}{h + c}$ **41.** (a) $abcd$, (b) $\pi r^2 hd$
43. $rbc + 2sac + 2sab + tbc$

Exercises 2-3 and 2-4

1. 2 **3.** -6 **5.** 19 **7.** 2 **9.** 14 **11.** 8 **13.** $-b$
15. b **17.** $84y - 8x - 48$ **19.** 43 **21.** $8x - 5$
23. $3a + 5b + 2c$ **25.** $a^2 + b^2 + c^2 - 3ab - 3ac - 3bc$ **27.** a
29. $\dfrac{3a + b}{3}$

Exercises 2-5

1. $6x - 15y$ **3.** $2x^2 - x - 3$ **5.** $16t^2 - 24ts + 9s^2$ **7.** $a^2 - a - 6$
9. $6a^2 - 5a - 6$ **11.** $a^2bc^2 + ab^2c^3 - 2a^2bc^3$ **13.** $24a^4b^4c^3$
15. $4a^2 - 11ab + 6b^2 + 2a - 4b$
17. $2 - 15x - 28x^2 + 17x^3 + 12x^4 - 4x^5$ **19.** $27s^3 + 8t^6$
21. $x^3 + 3x^2 + 2x$ **23.** $2w^3 + 6w^2$ **25.** $28a^3 - 10a^2$
27. $\pi(R^3 - R^2 - 3R + 6)$ **29.** $40a^2 + 12\pi a^2$

Exercises 2-6

1. $4xy$ **3.** $x + y - z$ **5.** $x - 2y$ **7.** $x - 2y$ **9.** $x^2 + x + 1$
11. $x^4 - x^3 + x^2 - x + 1$ **13.** $ab^2 - 4bc$ **15.** $x^2 - x - 6$
17. $T_2^3 + T_2^2 T_1 + T_2 T_1^2 + T_1^3$ **19.** $125s^3 - 75s^2 + 45s - 27$
21. $x + 2y - 3z$ **23.** $2x^3 - 3x^2 + 3x - 4$ **27.** $a + b + c - d$
29. $x(a - 2b) + 2$

Exercises 2-7

1. $\pi r(r + 2)$ **3.** $a^2(1 - x)(1 + x)$ **5.** $x(n + 1 - fn)$
7. $(a + b)(x - y)$ **9.** $(a - 4b)(a + 4b)$ **11.** $y(2x + y)$
13. $(x + 3)(x + 2)$ **15.** $(x - 3)(x + 2)$ **17.** $(x - 6)(x + 1)$
19. $(m - 5)(m - 5)$ **21.** $(a - b)(a^2 + ab + b^2)$ **23.** $2b(3a^2 + b^2)$
25. $m^3n^3(m - n)(m^2 + mn + n^2)$ **27.** $(4x - 5)(3x + 4)$
29. $2(6x - 5)(x + 2)$ **31.** $5(s - 2)(s - 2)$ **33.** $x(x + 1)(x - 1)$

35. $4(m+5)(m-5)$ **37.** $\frac{2}{3}$ **39.** $12a^2(b-3)(b+1)$
41. $(a+b)(r+s-t)$ **43.** $n(n^2-2n+2)$
45. $(sz-3s-3z)(sz-3s-3z)$ **47.** $(3m^2-n^2)(m^2-3n^2)$
49. $-4(a+1)(4a+1)$ **51.** 1 **53.** $14st$ **55.** $12hk$ **57.** $4y(s+t)$
59. $\frac{49}{16}$

Exercises 2-8

2. $\frac{1}{2(x-y)}$ **4.** $\frac{a-2b}{a+2b}$ **6.** $-\frac{r+1}{r+4}$ **8.** $\frac{m+3n}{n}$ **10.** $\frac{a-2b}{b}$
12. $\frac{m+n-3}{m+n+5}$ **13.** $\frac{3b}{ax}$ **15.** m **17.** $6R$ **19.** 1 **21.** $(a-x)^2$
23. $\frac{x+5}{x}$ **25.** $\frac{x+3}{x-3}$ **27.** $\frac{x^2+2x+1}{x^2+2x}$ **29.** $\frac{k(h-k)}{h}$
31. $\frac{1-m}{n^2}$ **33.** $\frac{2z}{y}$ **35.** $\frac{16m^3p}{R^2}$ **37.** $\frac{3+5h-2h^2}{9h^2+6h+1}$
39. $\frac{2a^3(5a-2)}{-4a^3+12a^2+a-3}$

Exercises 2-9 and 2-10

1. a^3b^2 **3.** $(x+y)^2(x-y)$ **5.** $-2ab(a^2-b^2)$ **7.** $\frac{7}{10a}$
9. $\frac{\pi r^2(3-16r)}{12}$ **11.** $\frac{-x-29y}{15}$ **13.** $\frac{3a^2-4}{a^2}$ **15.** $\frac{4xy}{x^2-y^2}$
17. $\frac{2(bx+ay)}{x^2-y^2}$ **19.** $\frac{(r-1)^2}{r}$ **21.** $\frac{3x^3-10x^2+9x-6}{x(x-1)^2(x-3)}$
23. $\frac{2(b-a)}{(a+b)^2}$ **25.** $\frac{ct_1-ct_2-t_3+t_2}{t_3-t_2}$ **27.** $\frac{2m^2-3mn-n^2}{m^2-n^2}$
28. $\frac{1}{b}$ **30.** ab **32.** $-b$ **34.** $y-x$ **36.** $xy(y-x)$
38. $\frac{m^2-mn}{n^2+mn}$ **40.** $\frac{x-4y}{y}$ **42.** r_2+1 **44.** $\frac{x+1}{x^3}$ **46.** -1

Exercises 3-2

1. 0.56 ton/hr **3.** $2(10)^{19}$ electrons/sec **5.** $1.1(10)^6$ ft-lb/min
7. 8.23 kg/km **9.** 72 in./sec

Exercises 3-3

1. 16 kJ **3.** 15 kW **5.** 5000 Pa (5 kPa) **7.** 2 kΩ
9. $5(10)^{-3}$ m³ or 5 m(m)³

Exercises 3-4

2. $F=\frac{kQ_1Q_2}{R^2}$ **5.** $\text{hp}=\frac{FS}{TK}$ **6.** $P=\frac{FS}{T}$

Exercises 3-5

2. 8.32(10)³ joules/(kg-mole K°) **3.** 0.22%/F°, 0.4%/K° **5.** 60¢
7. 2.68(10)¹⁹ molecules/cm³ **9.** 8(10)⁴ dynes/cm, 0.46 lb/in.
11. 128 ft/day

Exercises 3-6

1. 12,300 miles/(hr-min) **3.** 4.9(10)⁶ hp/mile² **5.** 16 cm³/ft²
7. 6.1(10)¹⁶ atoms/cm³ **10.** W lb/in. **14.** 3.6 in. **16.** 89.4 in.²
18. 40 W/cm³

Exercises 3 7 and 3-8

3. fissions/sec **7.** lb/sec **9.** dyne sec/cm² **11.** g/cm sec

Exercises 3-9 and 3-10

1. 1.3 days **3.** 3.1(10)⁷ electrons/cm **5.** 17.4 lb **7.** 600 miles/hr
9. 8.4(10)²² atoms/cm³

Exercises 4-2

1. $a = 4$ **3.** $m = 3$ **5.** $x = 2$ **7.** $m = -\frac{11}{2}$ **9.** $m = -6$
11. $k = 1$ **13.** $m = 6$ **15.** $x = -6$ **17.** $k = 10.44$
19. $R = -16.75$ **21.** $T = 3.195$ **23.** $y = -\frac{10.2}{7.32}$ **25.** 37, 39
27. 1.51 in. **29.** 24.5283 in. **31.** 1.975 ft **33.** $91\frac{2}{3}°$ **37.** 40°

Exercises 4-3

1. $a = 6$ **3.** $w = 156$ **5.** $y = 19.8$ **7.** $k = 5$ **9.** $m = 27$
11. $b = 14$ **13.** $x = \frac{14}{5}$ **15.** $x = -\frac{4}{11}$ **17.** $s = \frac{11}{17}$
19. $m = -7$ **21.** $r_1 = 1$ **22.** -0.501 **23.** $r = \frac{2}{3}$
25. $AB = \frac{120}{11}$

Exercises 4-4

1. $I = \frac{E}{R}$ **3.** $r = \frac{E}{I} - \frac{R}{2}$ **5.** $f = \frac{w}{2\pi}$ **7.** $V_1 = \frac{P_2V_2}{P_1}$
9. $t = \frac{W}{RI^2}$ **11.** $r = \frac{V_s}{2\pi h} - \frac{h}{2}$ **13.** $A_2 = \frac{A_1}{1 - 2.5k}$
15. $R = \frac{r_1r_2r_3}{r_1r_2 + r_2r_3 + r_1r_3}$ **17.** $b_1 = \frac{2A}{h} - b_2$ **19.** $z = \frac{A - 2xy}{2(x + y)}$
21. $r = \frac{3V + \pi h^3}{3\pi h^2}$ **23.** $y'' = \frac{y'[1 + (y')^2]}{x - h}$ **25.** $l = \frac{48EI\delta + 4Px^3}{3Px^2}$

27. $t_2 = t_1 + \dfrac{L(i_2 - i_1)}{\bar{e}}$ **29.** $n = \dfrac{l - a + d}{d}$ **31.** $r = \dfrac{A - \pi h^2}{2\pi h}$

33. $c = \dfrac{h(a + b) - 2A}{(a - b)}$ **38.** $x = \dfrac{Ch - Bk}{A + B}$

Exercises 5-1

1. IV **3.** II **5.** IV **16.** $(-10, -4)$ **18.** (a) 2, (b) 1, (c) 0, (d) 2

Exercises 5-2

1. $d = 5, m = \dfrac{4}{3}$ **3.** $d = 13, m = -\dfrac{12}{5}$ **5.** $d = 9\sqrt{5}, m = -2$

7. $d = 5\sqrt{10}, m = -\dfrac{13}{9}$ **9.** $d = 3, m = 0$ **11.** $d = 10$, m is not defined

13. $d = \sqrt{a^2 + b^2}, m = -\dfrac{a}{b}$ **15.** Cost per ticket **17.** Electric current

19. Interest rate **20.** Feet **25.** -3 **26.** 38 units above

27. Where $y = -\dfrac{10}{3}$

Exercises 5-3

1. (6.5, 3.5) **3.** $(8, 4\frac{1}{3})$ **5.** (0, 1) **7.** (7, 1)
9. $(2, -\frac{5}{2}), (5, 2), (0, \frac{7}{2})$ **11.** \$1410, \$1830, \$1060
12. log 15.036 = 1.17713 **13.** ctn 16°45′28″ = 3.3210

Exercises 5-4

1. $y = \dfrac{2x}{3}$ **3.** $7y + 11x + 2 = 0$ **5.** $x - 5y = 5$ **7.** $y = -\dfrac{x}{2} + 4$

9. $y = -\dfrac{5x}{3} + \dfrac{50}{9}$ **11.** $8x - y = 4$ **13.** $-\dfrac{A}{B}$ **15.** $\dfrac{C}{A}$

21. $J = 2S/5 + 50; F = 18S/25 + 32; J = C + 50$

Exercises 5-6

2. $\dfrac{a}{d} = \dfrac{b}{e}$ **3.** $-\dfrac{d}{e} = \dfrac{b}{a}$ **4.** $\dfrac{c}{b} = \dfrac{f}{e}$ **7.** $2x - y = 2$

8. $x + 2y = 6$ **11.** $y + 3x = 10$ **12.** $2y + 7x = 8$ **13.** $k = \dfrac{3}{5}$

15. $+\dfrac{4}{3}$

Exercises 5-7

1. (5, 2) **3.** (3, 5) **5.** (4, −1) **7.** (4, 2) **9.** $\left(\dfrac{1}{2}, \dfrac{3}{2}\right)$

11. (8, 8) **13.** (−40, −40) **15.** (15, −8) **17.** $r = -\frac{57}{5}, t = -\frac{243}{5}$
19. (10.19, −4.56) **25.** 7 miles **26.** 4:00 A.M. **27.** 56 miles
29. $v_a = v_b$ **30.** $s_a > s_b; v_b > v_a$ **31.** $s_a > s_b; v_a \geqq v_b$
32. $s_a > s_b; v_a > v_b$ **33.** v_a and v_b are of opposite signs.
35. clocks, $11; radios, $28 **37.** Boat, 6 mph; current, 3 mph
38. $\frac{465}{66}$ lb zinc; $\frac{405}{66}$ lb tin **40.** $7Q + 4D$

Exercises 5-8

1. $x = 10$, $y = -1$, $z = -3$ **3.** $m = -4$, $n = 2$, $p = -5$ **5.** $r = 1.3$, $s = -0.8$, $t = 2.2$ **7.** $r = -1$, $s = 2$, $t = 5$
9. $C = 70, F = 90, W = 80$ **11.** $x^2 + y^2 - 4x - 21 = 0$
12. $y = \frac{x^2}{4} - \frac{3x}{2} - \frac{3}{4}$

Exercises 5-9

1. (3, 2) **3.** (3, 2) **5.** (1, −2) **7.** (4, 3) **9.** (−0.653, −0.580)

Exercises 5-10

1. 9 **3.** 96 **5.** 63 **7.** −13 **9.** 0 **11.** (1, 2, 0)
13. (1, 2, 2, 1) **15.** ($\frac{3}{2}$, 5, −1, 2)

Exercises 6-1

1. m^6 **3.** $9m^4$ **5.** $3m^4$ **7.** 3 **9.** 1 **11.** z^{10} **13.** z^{10}
15. $-\frac{64}{F^3}$ **17.** 1 **19.** $\frac{ac}{2}$ **21.** b **23.** R^{3x} **25.** $(ab)^{1-k}$
27. $\frac{1}{5x^{13}y^{8n}}$ **29.** m^{m^2} **31.** $(3)^6$ **33.** $(5)^{5m-n+2}$ **35.** 6
37. b^{3k+12} **39.** $\frac{1}{m^{6p}}$

Exercises 6-2

1. $\sqrt[2]{b^3}$ **3.** $\sqrt[5]{(c^2d)^2}$ **5.** $\sqrt[3]{\left(\frac{1+x^2}{x}\right)^2}$ **7.** $a^{3/4}b^{1/4}$
9. $b^{1/4}(c + d)^{-3/4}$ **11.** $2\sqrt{3}$ **13.** $2\sqrt{6}$ **15.** $5\sqrt{3}$ **17.** $3\sqrt{3}$
19. $5\sqrt{5}$ **21.** $3\sqrt[3]{2}$ **23.** $2\sqrt[3]{3}$ **25.** $2\sqrt[4]{2}$ **27.** $3\sqrt[4]{2}$
29. $2\sqrt[6]{9}$ **31.** $abc\sqrt[3]{a^2b}$ **33.** $2mp^2\sqrt{3mp}$ **35.** $3a^3b^3\sqrt[4]{3ab^2}$
37. $2x^4\sqrt[n]{2x^5}$ **39.** $ab^2\sqrt[2n-1]{ab}$ **41.** $\frac{\sqrt{3}}{3}$ **43.** $\frac{\sqrt{2}}{2}$ **45.** $\frac{\sqrt{6}}{2}$

47. $\frac{\sqrt{5}}{2}$ **49.** $\frac{\sqrt{10}}{8}$ **51.** $\frac{\sqrt{14}}{4}$ **53.** $\frac{a\sqrt{bc}}{bc}$ **55.** $\frac{3\sqrt{2abc}}{14c^2}$

57. $\frac{b^n(\sqrt[4]{c^n})}{c^n}$ **59.** $\frac{\sqrt[n+1]{x}}{x}$ **61.** $\frac{\sqrt[3]{9}}{3}$ **63.** $\frac{\sqrt[4]{12}}{2}$ **65.** $\frac{\sqrt{m^2-n^2}}{m+n}$

67. $\frac{\sqrt{10}}{12}$ **69.** $\frac{(a+b)\sqrt{a(a+b)}}{a}$

Exercises 6-3

1. $3\sqrt{5}$ **3.** $2\sqrt[3]{6}$ **5.** $3x^2y^3\sqrt{2}$ **7.** $2x$ **9.** 2 **11.** $2\sqrt[6]{486}$

13. $\frac{\sqrt[4]{6}}{3}$ **15.** 1 **17.** $3x\sqrt[3]{3x}$ **19.** $5+2\sqrt{6}$ **21.** 7

23. $-4\sqrt{3}$ **25.** $\sqrt{6}$ **27.** $2x^2-2\sqrt{x^4-y^4}$ **29.** $a\sqrt[6]{ab}$

31. $x^2y^2\sqrt[n]{xy^2}$ **33.** $\sqrt[n^2+n]{a^{-3n-1}b}$

Exercises 6-4

1. $2\sqrt{3}$ **3.** $(a-b+c)\sqrt[n]{x}$ **5.** $-\sqrt{2}$ **7.** $\sqrt{3}$ **9.** $\frac{10\sqrt{3}}{3}$

11. $11\sqrt{2}$ **13.** $7\sqrt{2}$ **15.** $7\sqrt{3}$ **17.** $\frac{38\sqrt{7}}{21}$ **19.** $7\sqrt[3]{4}$

21. $\frac{5}{6}\sqrt[3]{2}+\frac{\sqrt[3]{4}}{3}$ **23.** $(x-y+1)\sqrt{xy}$ **25.** $\sqrt[3]{ab^2}(ab-3a^2b+2a)$

27. $\frac{(a+b-2)\sqrt{a+b}}{a+b}$ **29.** $\frac{4\sqrt{a^2-b^2}}{3}$ **31.** $(5a+4x^2)\sqrt{3x-2a^2}$

33. $\frac{\sqrt{m}(m-2)(m-1)}{m^3}$ **37.** $\frac{(m+n-mn-1-n^2)\sqrt{1-n^2}}{1-n}$

39. $\frac{y\sqrt{2xy}}{2x}$

Exercises 6-5

1. $\frac{1}{5}\sqrt{10}$ **3.** $\frac{1}{4}\sqrt{6}$ **5.** $\frac{1}{4}\sqrt{10}$ **7.** $\frac{1}{6}\sqrt{15}$ **9.** $\frac{1}{2}\sqrt[3]{5}$

11. $\frac{\sqrt{2}}{2}$ **13.** $\frac{\sqrt{2}}{2}$ **15.** $3\frac{\sqrt[3]{4}}{2}$ **17.** $2\sqrt[4]{2}$ **19.** $\sqrt{5}-2$

21. $\frac{8+2\sqrt{2}}{7}$ **23.** $3\sqrt{3}-3\sqrt{2}$ **25.** $\frac{5\sqrt{3}+2\sqrt{5}}{11}$

27. $\frac{3\sqrt{15}-11}{7}$ **29.** $\sqrt{3}$ **31.** $\sqrt{2}$ **33.** $bc\sqrt{ac}$ **34.** $b\sqrt{2ab}$

35. $\sqrt[n]{ab^2}$

Exercises 7-1

1. $x=\pm 4$ **3.** $x=0,\frac{2}{5}$ **5.** $x=\pm 7$ **7.** $x=\pm\sqrt{2}$

9. $x=\pm 3$ **12.** $y=\pm\sqrt{a^2-b^2}$ **14.** $r=\pm\frac{\sqrt{(b-a)(b+a)}}{(b+a)}$

16. $\sin\phi = \pm\frac{1}{2}$ **18.** $\tan\phi = 0, 4$ **20.** $x = \pm\sqrt{2}$ **24.** $r = 0.06$

25. $x = 0, \frac{1}{6}; y = 4, \frac{863}{216}$ **26.** 6.83 in. **27.** $i = \sqrt{\frac{p}{r}}$

Exercises 7-2 and 7-3

1. $x = 7, -3$ **3.** $t = -\frac{2}{3}, \frac{3}{2}$ **5.** $k = -\frac{4}{3}, \frac{11}{8}$ **7.** $m = -3 \pm \sqrt{3}$

9. $b = -1, -\frac{1}{3}$ **11.** $x = 8, -4$ **13.** $m = 7, 2$ **15.** $t = 1, -3$

17. $k = 4, -\frac{3}{2}$ **19.** $t = -2 \pm \sqrt{2}$ **21.** $a = \frac{4}{b}, -\frac{1}{b}$

23. $a = -6 - b, 1 - b$ **25.** $\sin\phi = -\frac{1}{3}, \frac{1}{2}$

27. $(x + 2)^2 + (y - 3)^2 = 25, h = -2, k = 3, r = 5$

29. $m^2 + m - 12 = 0$

Exercises 7-4

1. $m = \frac{5}{2}, -5$ **3.** $x = \frac{-3 \pm \sqrt{9 + 16b}}{2b}$ **5.** $x = -1, -8$

7. $t = \frac{-v_0 \pm \sqrt{v_0^2 + 2gs}}{g}$ **9.** $x = -a - b, -a + b$ **11.** 6, 7

13. 12 in., 16 in. **15.** (a) $k = 16$ (b) $k = 0$ (c) $k = -48$

16. $r = 5$ in. **18.** $c = 1.58$

19. It is impossible to construct such a triangle.

21. (a) 12 (b) 1 (c) 4.20 **22.** 3180°C **25.** $I = 20, 12$

26. $f = \frac{Cx \pm \sqrt{C^2x^2 + 4LC}}{4\pi LC}$

27. $y = 0, x = 3, -1; y = -3, x = 0, 2; y = -4, x = 1, 1$

Exercises 7-5

1. $x = \pm\sqrt{3}$ **3.** $y = \pm\sqrt{6}$ **5.** $t = 1, 9$ **7.** $m = -1, \frac{1}{8}$

9. $x = \pm\frac{\sqrt{3}}{3}$ **11.** $m = -1, -1$ **13.** $x = \frac{3}{2}, -\frac{1}{6}$ **15.** $x = \pm 2a$

17. $a - b = \frac{1}{y}, \frac{3}{2y}$ **19.** $x = \frac{(a + b)^3}{64y^3}, -\frac{(a + b)^3}{27y^3}$

21. $\tan\phi = -1, -2$

Exercises 7-6

11. $t = -\frac{v_0}{g}, s = s_0 - \frac{v_0^2}{2g}$ **13.** $t = 0, 9$ sec; $t = 4.5$ sec; $s = 324$ ft

15. 550 items at $1.10 each

Exercises 7-7

1. Real, unequal, rational **3.** Real, unequal, irrational
5. Real, equal, rational **7.** Imaginary **9.** Real, unequal, rational
11. $c = 9$ **13.** $c = \frac{50}{9}$ **15.** Four roots, all real
16. Two real, two imaginary roots **17.** Four imaginary roots
18. Four equal roots

Exercises 7-8

1. 39 **3.** 4 **5.** 3 **7.** $8, \frac{1}{2}$ **9.** 7, −1 **11.** 49, 25
12. 4.15 **13.** 4.69 **14.** 2.07 **15.** $\alpha = \frac{2\pi v^2}{g}$ **17.** $v = \frac{c\sqrt{m^2 - m_0^2}}{m}$
19. $R = \sqrt{\frac{(A + \pi r^2)}{\pi}}$

Additional Exercises 7-8

1. (a) $b = \pm 4\sqrt{6}$ (b) $a = \pm 2$ **2.** (a) $k = \frac{5}{8}$ **3.** 36.09 ft^2
4. $6 + \sqrt{6}, 6 - \sqrt{6}$ **6.** (a) $BD = 4.40''$; $DC = 3.76''$
7. (b) $AB = \frac{1 + \sqrt{15}}{2}$ **8.** (b) 0.936 **10.** $R_1 = 2.6$ ohms
12. $D = 0.94$ in. **13.** (a) $S_y = S_x\left(\frac{\sin\theta}{\cos\theta}\right) - \frac{gS_x^2}{2(V\cos\theta)^2}$
(b) $S_x = 606$ ft or 3730 ft (c) $S_x = 0$ ft and 4330 ft (d) $S_{y(\max)} = 625$ ft
14. 14.2 and 33.8 ohms **15.** $N = 12$ **16.** $I = 10$ amperes, $P = 500$ watts

Exercises 7-9A

1. $(0, 0), r = 9$ **3.** $(-4, 0), r = 4$ **5.** $(-3, -2), r = \sqrt{13}$
7. $\left(-\frac{3}{2}, \frac{4}{3}\right), r = \sqrt{\frac{146}{3}}$ **9.** $x^2 + y^2 = 100$ **11.** $(x - 10)^2 + y^2 = 50$

Exercises 7-9B

1. $C(0, 0), V(\pm 3, 0)$ **3.** $C(5, 0), V(5, \pm 5)$
5. $C(-3, 5), V(-3 \pm \sqrt{6}, 5)$ **7.** $C(6, -4), V\left(6, -4 \pm \frac{2}{3}\sqrt{33}\right)$
9. $\frac{(x - \frac{1}{2})^2}{(\frac{3}{2})^2} + \frac{(y - 1)^2}{9} = 1$ **11.** $x^2 + 4(y - 2)^2 = 9$

Exercises 7-9C

2. $C(0, 0), V(0, \pm\sqrt{24})$ **4.** $C(3, -6), V(3 \pm \sqrt{6}, -6)$ **6.** $C(0, 0)$
9. $4x^2 - y^2 = 9$

Exercises 7-9D

1. $(-2, 0), (0, 3)$ **3.** $(0, \pm 3)$ **5.** $(-4, 8), (-4, 2)$ **7.** $(0, 0), (4, 0)$
9. $(\pm 1, \pm 2)$ **11.** $(\pm 6, \pm 2), (\pm 26, \pm 18)$ **13.** $\left(-4, \frac{4}{21}\right)$

Exercises 8-2

1. $q(x) = x - 3, R = 0$ **3.** $q(x) = 3x^2 + 2x - 1, R = 1$
5. $q(m) = 4m^2 + 16m + 127, R = 502$
7. $q(k) = k^3 - 2k^2 + 4k - 8, R = 15$
9. $q(x) = 4x^3 - 13x^2 + 15x - 15, R = +10$ **16.** $x = 1, -\frac{1}{2} \pm \frac{1}{2}\sqrt{5}$
18. $x = 3, -1 \pm j$ **20.** $x = 3, -1, -\frac{1}{2} \pm \frac{1}{2}j\sqrt{3}$
22. $x = 1, -\frac{1}{2} \pm \frac{1}{2}j\sqrt{3}$ **24.** $x = 1, -1, j, -j$ **26.** No integral roots
28. $x = 0, -5, 1$ **29.** $2, \frac{5 \pm \sqrt{17}}{2}$ **31.** $5 \times 5 \times 5$ **33.** $h = 6$

Exercises 8-3 and 8-4

1. 5.503 **3.** 4.903 **5.** −0.858 **7.** 1.485 **9.** −2.078
11. 2.148 **13.** $x = 3.557$ **15.** −11.350 **17.** 2.844 **19.** ±2.659
21. Approx. 3.846 ft **22.** $r = 11.3$ in.; $h = 14.3$ in. **23.** $l = 42.02$ in.
24. 95.4 and 5.2 in. approx. **25.** $x = -1.9$ approx. **26.** $h = 1.32$ ft

Exercises 8-5

1. Two real, equal **3.** Two imaginary **5.** One real, two imaginary
7. Two real, two imaginary **9.** One real, two imaginary
11. Three real, equal **13.** Two real **15.** Three real
17. Two imaginary **19.** Two real, equal **21.** Two real
22. $Z = s^2 - 4$ **24.** $Z = s^4$ **26.** $Z = s^4 - 9s^3 + 19s^2 - 9s + 18$
28. $Z = s^4 - 6s^3 + 23s^2 - 34s + 26$

Exercises 9-1

1. 9, 27, 81 **3.** $\frac{1}{3}, \frac{1}{9}, \frac{1}{27}$ **5.** 64 **7.** 0.25, 0.125, 0.0625 **9.** 1.331
11. $A = 150(1.035)^6$ **12.** 31.2 years **13.** 640,000 **14.** 19.08 in.
17. $\frac{10}{32}$ ft; $d = 10\left(\frac{1}{2}\right)^n$

Exercises 9-2

16. $y = 4.4(\frac{1}{2})^x$ **18.** $y = 4.4(\frac{1}{2})^{-x/2}$ **20.** $y = 4.4(\frac{1}{2})^{(x+3)/2}$
22. A approaches (i) (2.718 . . .) **23.** $A = P + Pni$; no

Exercises 9-3

1. 3^{4x} **3.** 16^{-2x} **5.** $8^{-10y/3}$ **7.** $(0.09)^{3.6y}$ **9.** $16^{(6t-9)/4}$
11. $10^{0.7981}$ **13.** $10^{1.6990}$ **15.** $10^{0.9542}$ **17.** $10^{1.5562}$
19. $10^{4.1761}$ **21.** $10^{-0.4771}$ **23.** $10^{-0.2219}$ **25.** $e^{4.6052}$
27. $e^{0.7675}$ **29.** $10^{1.0628}$
31. Using base 10, the exponents can be called the logarithms and

$$\log_{10} ab = \log_{10} a + \log_{10} b.$$

This asserts that the logarithm of a product of two factors is the sum of the logarithms of the two factors.

32. $\log_{10} \frac{a}{b} = \log_{10} a - \log_{10} b$ **33.** $\log_{10} a^n = n \log_{10} a$

34. $\log_{10} \sqrt[n]{a} = \frac{1}{n} \log_{10} a$

35. $\log_{10} N = 2 \log_{10} a + 3 \log_{10} b - (\log_{10} 5 + \log_{10} c)$

Exercises 9-4

1. 10.4771 − 10 **3.** 13.4771 − 10 **5.** 5.4771 − 10
7. 9.6812 − 10 **9.** 11.6857 − 10 **11.** 10.6857 − 10
13. 7.9671 − 10 **15.** 3.7938 − 10 **17.** 10.8603 − 10
19. 1.0000 − 10 **21.** 7 **23.** 7×10^3 **25.** 7×10^{-4} **27.** 7×10^6
29. 7×10^{-27} **31.** 7.10×10^4 **33.** 7.17×10^{-5} **35.** 7.18×10^{-1}
37. 7.25×10^9 **39.** 7.42

Exercises 9-5

1. $\log N = 0.3986$ **3.** $\log N = 1.4777$ **5.** $\log N = 8.4901 - 10$
7. $\log N = 4.2276$ **9.** $\log N = 6.8603 - 10$ **11.** $N = 2.312$
13. $N = 181.8$ **15.** $N = 0.06459$ **17.** $N = 83600$
19. $N = 1.451 \times 10^{16}$

Exercises 9-6

1. 15,300 **3.** 2.959 **5.** 5.850×10^{-4} **7.** 123.9 **9.** 0.05483
11. 0.001573 **13.** 68.53 **15.** 33.34 **17.** 0.2511 **19.** 371.6
21. 102.4 **23.** 0.3146 **25.** 45,900 **27.** 9.5028 − 10
29. $A = 12.455$ **31.** 605.7 in./sec **33.** pH = 8

Exercises 9-7

1. 96.55 **3.** 4.323×10^{-3} **5.** 24.14 **7.** 8.829×10^{-2} **9.** 34.58
11. 28.63 **13.** 1.226 **15.** 0.1347 **17.** 14.31 **19.** 0.5844
21. 46.32 **23.** 0.03794 **25.** 1.632×10^5 **27.** (c) 1195
29. 13.76 **31.** $V = 8.261$ in.^{3} **33.** 6.783 **35.** 2.393
36. $K = \$1327$ **37.** \$8317

Exercises 10-1, 10-2, and 10-3

1. 0.5436 **3.** 3.4498 **5.** 8.8520 − 10 **7.** 0.4939 **9.** −4.211
11. 1314 **13.** −1741 **15.** 1.183×10^4 **17.** 2.585 **19.** 9.309
21. 4.950 **23.** 36.13 **25.** 12.97 **27.** 21.09 **29.** 0.2853

Exercises 10-4

1. (a) 3.16×10^{-9} (c) 6.31×10^{-4} **2.** (a) 11.8 (c) 8.2
3. (a) 1.8 (c) −3.2 **4.** (a) 6 (c) −5.2 **5.** 2.665×10^{-15} mg

Exercises 10-5 and 10-6

1. $x = 3$ **3.** $x = -4$ **5.** $x = -\frac{3}{2}$ **7.** $x = -1$ **9.** $x = 0.7$

11. $Z = -2.32$ **13.** $Z = -1.1$ **15.** $K = \frac{4}{3}$ **17.** $m = 1.21$

19. $n = -1.21$ **21.** 3.3% **23.** \$6139 **25.** 10.37 years
27. 5.96% **28.** $t = 0.00895$ **29.** $V = 443$ **30.** $L = 0.19$

31. $x = \pm 100$ **33.** $A = 2$ **35.** $k = \frac{5}{3}$ **37.** $A = x^n$

Exercises 10-7

1. $y = 4.192$ **3.** $y = 5.756$ **5.** $y = 2.334$ **7.** $x = 1.737$
9. $y = 6.131$ **11.** $y = -3.262$ **13.** $y = 3.322$ **15.** $x = 2.377$

Exercises 10-8

1. $N = 66.7$ **3.** $N = 1.048$ **5.** $N = 0.9085$ **7.** $N = 1.203$
9. $N = 1.397$ **11.** $x = 1.504$ **13.** $x = -1.024$ **15.** $b = 4.201$
17. $N = 0.3925$ **19.** $N = 2.379$ **21.** $N = 3.1 \times 10^{-6}$
23. $N = 5.19 \times 10^{10}$ **25.** $N = 6.629$

Additional Exercises 10-8

1. $i = 0.9$ ampere **3.** $t = 0.115$ second **11.** 34 **13.** $T = 325$°K
15. $W = -15{,}000$ ft-lb **17.** Doubles in 17.7 years **18.** $t = 0.445, 1.555$
19. 28,500 **20.** 38.74; 0.515 **22.** 132.3 minutes

Exercises 11-1

1. (a) All positive (c) $\frac{x}{y}, \frac{y}{x}$ positive, rest negative **3.** 40° **5.** 5°
7. 195°

Exercises 11-2

3. $\tan\theta = \frac{7}{5}$ **4.** $\sec\phi = \sqrt{5}$; $\cos\theta = \frac{1}{3}\sqrt{5}$
7. Horizontal = 27.89 lb; vertical = 18.99 lb **8.** $\cos\theta = 0.833$

Exercises 11-3

2. True **4.** True **6.** False **8.** True **10.** False **12.** False
14. True **17.** 60°, 240° **19.** 225°, 315° **21.** 120°, 240°
23. 225° **25.** 210°, 330° **27.** $-\frac{1}{2}$ **29.** $+1$ **31.** $\frac{2}{\sqrt{3}}$

Exercises 11-4

2. (a) $\sin\theta$ assumes all values between -1 and $+1$, including -1 and $+1$. Symbolically, $-1 \leqq \sin\theta \leqq +1$ (c) $-\infty < \tan\theta < +\infty$

Exercises 11-5

1. $\theta = 43°$ **3.** $\theta = 76°$ **5.** $\theta = 12°$ **7.** $\theta = 58°$
9. $\theta =$ any value **11.** $\theta = 0°$

Exercises 11-6

1. 38.95; 22.5 **3.** $AC = 21.52$; $\theta = 267.4°$
4. (b) 168 lb at 130° (c) System is balanced. **5.** 8590 lb at 304.1°

Exercises 12-1

1. 43°17′13″ **3.** 12°00′32″ **5.** 108°58′19″ **7.** 13°36′00″
9. 12.446° **11.** 76.700° **13.** 327.011° **15.** 112.052°
17. 69.09 miles **18.** Approx. 37.6 miles **19.** $x = 3.6''$
20. $1'' = 0.000278°$ **21.** 57.3°

Exercises 12-2

1. 60° **3.** 75° **5.** 370° **7.** 2922° **9.** 1.146° **11.** $\frac{\pi}{2}$
13. $\frac{5\pi}{6}$ **15.** 5.969 **17.** 14.398 **19.** 3.652
21. (a) $(108/\pi) \times 10^3$ deg/min (b) $5/\pi$ rps **23.** 330/7 rad/sec
24. (a) 30π in./sec (b) 15π in./sec

Exercises 12-3

1. 149°11′41″ **3.** 41°18′43″ **5.** 114°53′19″ **7.** 161°49′27″
9. 11°01′07″ **11.** 47°37′36″ **13.** 28°16′33.75″ **15.** 18°03′58″
17. −(106°02′44″) **19.** 296°26′46″, 338°56′17″

Exercises 12-4

1. 0.2150 **3.** 0.5366 **5.** 0.9131 **7.** 0.2478 **9.** −0.6053
11. −0.5018 **13.** 20°15′ **15.** 38°40′ **17.** 52°20′ **19.** 68°09′
21. 83°06′ **23.** 84°42′

Exercises 12-5

1. 36°52′ **3.** 35°40′ **5.** 53°56′ **7.** 14.67 **9.** 13.19 **11.** 4.370
13. (a) 314.8 ft (b) 398.0 ft **15.** 63°30′ **16.** $A = 32.7°, b = 22.5$
18. $A = 64.1°, a = 34.7$ **20.** $A = 49.3°$ **22.** $b = 1.565$
23. $b = 2.62$ **25.** $c = 24.3$ **26.** 13.5° **28.** 47.2° **30.** 6.99

Exercises 12-6

1. $\theta = 2.29°$ **3.** $\theta = 4.58°$ **5.** 0.0785 **7.** 0.0314 **9.** 0.0407
11. 19°28′ **13.** 24°58′ **15.** 34°05′ **17.** 75°27′ **19.** 88°07′

Exercises 13-1

1. $a = 20.41$ **3.** $b = 12.25$ **5.** $b = 8$ **7.** $b = 10.64$
9. $c = 29.54$ **11.** $b = 6.428$ **13.** 14°30′ **15.** 29°40′
17. No answer **19.** 25°40′ **21.** 14°30′ **23.** 43°10′
26. $DC = 5.46$ **28.** $\alpha = 55°10'$ **30.** 261.5 lb, 590.0 lb **31.** 11.42
32. $A = 40°40'$ **34.** 21.88

Exercises 13-2

1. 18°35′, 101°25′ **3.** 4.122, 8.734 **5.** 3.83, 11.8 **7.** 25°32′, 64°28′
9. 2.55, 17.45 **11.** 12°51′, 142°39′ **13.** 33°05′, 146°55′ **15.** 60°00′00″

Exercises 13-3

1. 56°15′ **3.** 5.667 **5.** 65°19′ **7.** 60°00′ **9.** 17°45′
15. 76.8° **17.** 81.2° **19.** 72.4° **21.** 35.9° **23.** 70.6°
25. 7.29 **27.** 74.4° **28.** 84.6 lb at 120.0°

Exercises 13-4

1. 82.5° **3.** 125.3° **5.** 32.8° **7.** 40° **9.** 71.7°
11. 73.9° and 46.1° **13.** 16.3°

Exercises 14-1

1. 120 min; $\frac{1}{2}$ cycle/hr; $\frac{1}{120}$ cycle/min **3.** $\frac{500}{3}$ cycles/min; $\frac{9}{25}$ sec
5. For both A and B, $T = \frac{1}{10}$ sec **6.** (a) $\frac{1}{20}$ sec (b) $\frac{1}{140}$ sec

Exercises 14-2 and 14-3

13. 1.8, 3, 6 **14.** 72°, 1080°, $\frac{2}{3}$ radian **15.** 45° left
17. $(A)\ y = 2 \sin 2\theta$, $(B)\ y = 2 \sin (\theta - 30°)$
18. $y_1 = -y_2$; $\sin \theta = -\sin(-\theta)$ **19.** 0.1°; 0.025° **20.** 2.4°

ercises 15-4

13 **3.** $-10 - j11$ **5.** $18\underline{/255^\circ}$ **7.** $10.8\underline{/258.7^\circ}$ **9.** $\frac{5 + j12}{13}$

1. $\frac{-14 - j5}{17}$ **13.** $0.75\underline{/-90^\circ}$ **15.** $2.91\underline{/-45.9^\circ}$ **17.** $\frac{6}{13}$

19. $0.178\underline{/120.9^\circ}$ **21.** $5 + j12$ **23.** $27\underline{/-45^\circ}$ **27.** Multiplied by n

Exercises 15-5

1. $\pm j$ **3.** $\pm\frac{\sqrt{2}(1 + j)}{2}$ **5.** $-2 + j2$ **7.** $\pm 1, \pm j$ **9.** $32\underline{/-60^\circ}$

11. $128\sqrt{2}\underline{/45^\circ}$ **13.** $-7 + j24$ **15.** $\pm\sqrt{5}$ **17.** 32

19. $j, \frac{\pm\sqrt{3} - j}{2}$

Exercises 15-6

1. $1\underline{/0^\circ}$ **3.** $1\underline{/180^\circ}$ or $-1\underline{/0^\circ}$ **6.** $-3\cos\phi - j3\sin\phi$ **8.** $-4 + j3$

13. $3 + j4$ **15.** $\rho\underline{/-\phi}$ **17.** $\underline{/\phi + 180^\circ}$ **18.** $5 + j3$ **20.** $3\underline{/210^\circ}$

22. 5

Exercises 16-1

1. $3\cos^4 x - 5\cos^2 x + 1 = 0$

Exercises 16-2

5. $\frac{\sqrt{2}}{4}(\sqrt{3} + 1)$ **7.** $\frac{\sqrt{2}}{4}(\sqrt{3} + 1)$ **8.** $\frac{\sqrt{2}}{4}(1 - \sqrt{3})$ **10.** $\frac{\sqrt{3}}{2}$

Exercises 16-3

5. $16\sin^5\theta - 20\sin^3\theta + 5\sin\theta$ **7.** $4\sin\theta\cos\theta - 8\sin^3\theta\cos\theta$

10. Amplitude $= 5$; phase shift $= 53.1^\circ$; $y = 5\sin(\omega t + 53.1^\circ)$

Exercises 16-4

1. $\phi = 0^\circ, 109.5^\circ, 250.5^\circ$ **3.** $\alpha = 210^\circ, 330^\circ$ **5.** $\tau = 153.4^\circ, 333.4^\circ$

7. $\beta = 90^\circ, 270^\circ$ **9.** $\theta = 10^\circ, 50^\circ, 70^\circ, 110^\circ$, and eight more

11. (a) $|\sin\phi| \leqq 1$ for a valid solution **13.** $k^2 > 2$ **15.** $x = 0$

17. $x = 0.55$ **19.** $x = 0.86$ **21.** $x = 1.49$ **22.** $\alpha = 1.52$

23. $\phi = 1.38$; $y = 4.05$ **24.** $x = 0.43$

Exercises 16-5

1. $60^\circ, 120^\circ$ **3.** $135^\circ, 315^\circ$ **5.** $150^\circ, 210^\circ$ **7.** 7° 113° **9.** 165°

11. $x = 0.455$ **12.** (a) 1 (c) ± 1.732

Exercises 14-4

1. $i = 110$ amperes
2. $\frac{1}{1440}$ sec; $\frac{5}{1440}$ sec; $\frac{12n+1}{1440}$ sec; $\frac{12n+5}{1440}$ sec, where n is an in
5. $f = (2n + 1)50$ cycles/sec, where n is an integer 6. $c = -2$
9. $I_{max} = 2.47$ amperes 11. $f = 66.5$ cycles/sec 13. $t = 0.0$
15. $\gamma = -32.4°$, trailing

Exercises 14-5

21. True 23. True 25. False 27. False 29. True
31. False 33. False 35. True 37. False
39. $x = (2n + 1)(45°)$ 40. $x = n(180°)$

Exercises 14-6

18. $-53.2°$

Exercises 14-7

22. 6 units long, 8 units high 23. $k = 16$
24. Exercise 13: $4(x - 2)^2 + y^2 = 4$

Exercises 14-8

1. $\rho = 4$ 3. $\rho \sin \theta = 5$ 5. $\rho^2 - 4\rho \cos \theta - 21 = 0$
7. $\rho = 2 \operatorname{ctn} \theta \csc \theta$ 9. $\rho^2 = \theta$ 11. $x^2 + y^2 - 5y = 0$ 13. $y = 5$
15. $y = -x$ 17. $x^2 + y^2 = e^{\arctan y/x}$ 19. $\sqrt{x^2 + y^2} = \frac{1}{4} \arctan \frac{y}{x}$

Exercises 14-9

25. $\rho^2 = 4 \cos 2\theta$

Exercises 15-2

1. j 3. $+1$ 5. $-j$ 7. $-j$ 9. -1 11. $2.6 + j1.5$
13. $-0.616 + j1.69$ 15. $-0.165 - j0.354$ 17. $31.2 - j18$
19. $-339.4 - j339.4$ 21. $2\sqrt{2}\underline{/45°}$ 23. $8\underline{/300°}$ 25. $0.965\underline{/131.2°}$
27. $2\underline{/210°}$

Exercises 15-3

1. $7.07\underline{/-8.1°}$ 3. $8\underline{/0°}$ 5. $5\underline{/83.1°}$ 7. 0 9. $12.1\underline{/18.2°}$
11. $4\underline{/120°}$ 13. $8.25\underline{/104.0°}$ 15. $5.39\underline{/0°}$ 17. $1\underline{/53.1°}$ 19. Zero

Additional Exercises 16-5

3. $x = \frac{\sqrt{63} + \sqrt{3}}{2}$ **4.** $x = 0.84$ in. **5.** 284 ft

7. Approx. 20° and 70° **9.** $\phi = 37.4°$

10. $dy = 0.00872$; $y + dy = 0.50872$ **11.** $\phi = 14°13'; 77°45'; 139°23'$

Exercises 17-1

1. $x^6 - 3x^4y + 3x^2y^2 - y^3$ **3.** $a^8 - \frac{8a^6}{b} + \frac{24a^4}{b^2} - \frac{32a^2}{b^3} + \frac{16}{b^4}$

5. $1 + \frac{5}{x} + \frac{10}{x^2} + \frac{10}{x^3} + \frac{5}{x^4} + \frac{1}{x^5}$ **7.** $\frac{28}{y^2}$ **9.** $10x^3(\Delta x)^2$ **11.** Ninth

13. Sixth **15.** 15 **19.** 2.25, 2.488, 2.594

Exercises 17-2

1. $\frac{1}{x} - \frac{1}{x^2} + \frac{1}{x^3} - \frac{1}{x^4} + \cdots$

3. $\frac{m^{-1/2}}{2} + \frac{m^{-3/2}n}{16} + \frac{3m^{-5/2}n^2}{256} + \frac{5m^{-7/2}n^2}{2048} + \cdots$

5. $x^n + nx^{n-1}(\Delta x) + \frac{n(n-1)x^{n-2}}{2!}(\Delta x)^2 + \frac{n(n-1)(n-2)}{3!}x^{n-3}(\Delta x)^3 + \cdots$

7. $2 + \frac{3x}{4} - \frac{9x^2}{64} + \frac{27x^3}{512}$ **11.** 6.083 **13.** 6.856 **15.** 10.955

17. 0.316 **19.** 31.623 **21.** 0.003

Exercises 17-3

1. 41 **3.** -124 **5.** $18m - 18n$ **7.** 151 **9.** $-55a - 76$

11. $78x^2 - 200$ **13.** $S = 3, d = 5$ **15.** $n = 14, l = 466$

17. $n = 81, a = -360$ **21.** 18.43 miles per gallon **23.** 24.15 in.

Exercises 17-4

1. 9 **3.** 1 **5.** $\frac{a^{-13}}{b^{40}}$ **7.** $\frac{x^{-5}}{16}$ **9.** $16(1 + x)^3$ **11.** 64

13. $S = 111.11111, l = 10^{-5}$ **15.** $r = -1, l = -1$

17. $S = \frac{m^8 - m^{18}}{1 - m^2}, r = \frac{1}{m^2}$ **19.** \$64,000; \$127,000

23. After about 6.5 minutes **25.** $\frac{27}{64}$ lb

Exercises 17-5

1. 0 **3.** $-1, 4, 9$ **5.** $\frac{a^3 + a^2}{2}$ **7.** $\pm 3, +\frac{3}{2}, \pm\frac{3}{4}$

9. 1, 0.1, 0.01, 0.001 **11.** $V_k = \frac{k}{n}(V_0 - V_n)$ **13.** \$1000; \$1540

17. 2; 0

Index

Page numbers in italics refer to definitions or special descriptions.